Effect of Mineral-Organic-Microorganism Interactions on Soil and Freshwater Environments

Edited by

J. Berthelin

Centre de Pédologie Biologique du CNRS
Vandœuvre-lès-Nancy, France

P. M. Huang

University of Saskatchewan
Saskatoon, Canada

J.-M. Bollag

Pennsylvania State University
University Park, Pennsylvania

and

F. Andreux

University of Bourgogne
Dijon, France

Springer Science+Business Media, LLC

Library of Congress Cataloging-in-Publication Data

Effect of mineral-organic-microorganism interactions on soil and freshwater environments / edited by J. Berthelin ... [et al.].
p. cm.
The Second International Symposium Mineral-Organic-Microorganism International (ISMOM) held in Nancy, France, September 1996.
Includes bibliographical references and index.
ISBN 978-1-4613-7120-5 ISBN 978-1-4615-4683-2 (eBook)
DOI 10.1007/978-1-4615-4683-2
1. Soil microbiology--Congresses. 2. Soil ecology--Congresses. 3. Soil minerology--Congresses. 4. Microbial metabolism--Congresses. 5. Microbial ecology--Congresses. I. Berthelin, J. II. International Symposium Mineral-Organic-Microorganism (2nd : 1996 : Nancy, France)

QR111 .E256 1999
579'.1757--dc21

99-042966

Proceedings of an International Symposium on the Effect of Mineral-Organic-Microorganism Interactions on Soil and Freshwater Environments, held September 3–6, 1996, in Nancy, France.

ISBN 978-1-4613-7120-5

Originally published by Kluwer Academic / Plenum Publishers, New York in 1999
Softcover reprint of the hardcover 1st edition 1999

PREFACE

The Working Group M.O. (Interactions of soil minerals with organic components and microorganisms) (WGMO) of the International Soil Science Society (ISSS) was founded in 1990 at the 14th World Congress of Soil Science (Kyoto, Japan), with Professor P.M. Huang being the Chairman. Since then, the Working Group M.O. has served as a forum to bring together soil chemists, soil mineralogists, soil microbiologists, soil biochemists, soil physicists and environmental, ecological, and health scientists.

The objective of the Working Group M.O. is to promote research, teaching, and also the exchange of technology concerning the knowledge and the impact of the interactions between minerals-organics and microorganisms on environmental quality, agricultural sustainability, and ecosystem "health".

This group is first a scientific group as defined just previously, but it also intends to develop exchange and transfer between scientists and engineers.

The first International Meeting organized by Professor P. M. Huang, was held in Edmonton, Canada, in August 1992, where 87 papers were presented by scientists from 20 countries. Following this meeting, a two volume book was edited by P. M. Huang, J. Berthelin, J.-M. Bollag, W. B. McGill, and A. L. Page, entitled "Environmental impact of soil component interaction" : Volume I "Natural and anthropogenic organic—volume II "Metals, other inorganic and microbial activities", and published by C.R.C. Lewis Publishers (1995).

Specialized and shorter co-sponsored meetings were held in Mexico (in 1994) at the 15th World Congress of Soil Science (Symposium entitled "Interactions of soil components, agricultural ecosystems and health issues") and in St. Louis (MO, USA), in the 1995 SSSA National Meeting (workshop entitled "Soil chemistry and Ecosystems Health", SSSA special publication number 52, 1998).

The second International Meeting "ISMOM96" was held in Nancy, France, September 1996. It was devoted to the knowledge and application of the effect of Mineral-Organic-Microorganism Interactions on Soil and Freshwater Environments. During this symposium, 6 lectures, 48 oral communications and 60 posters (116 presentations) were presented in five topics from participants of 25 countries.

After this second International Symposium "ISMOM96", manuscripts were submitted for review and forty were accepted to produce this volume presenting the main topics of the conference and providing also a common framework for the Working Group M.O.

The symposium and this volume have been organized with the help of different institutions, organizations, companies and members of the Centre de Pédologie Biologique. We are grateful to the authors who have contributed to this publication and to the external referees who have provided critical inputs to maintain the quality of this book. Gratitude is extended to the sponsors and co-sponsors, Centre National de la Recherche Scientifique, Université Henri-Poincaré (UHP Nancy I), Institut National Polytechnique de Lorraine

(INPL), Ministère de l'Aménagement du Territoire et de l'Environnement, Association Française de l'Etude du Sol, Institut National de la Recherche Agronomique (INRA), International Soil Science Society, International Humic Substances Society, Agence de l'Eau Rhin-Meuse, Communauté Urbaine du Grand Nancy, TREDI, Banque Populaire de Lorraine, HORIBA, and GASPARD.

I personally wish to thank Mrs. Evelyne Jeanroy, Chantal Ginsburger, and Mr. Emmanuel Jeanroy for their great help in the Symposium organization and, particularly Mrs. Evelyne Jeanroy for her participation and help in the editing work.

Jacques Berthelin

ACKNOWLEDGMENT TO SPONSORS AND CO-SPONSORS

The International Symposium Mineral-Organic-Microorganism Interactions (ISMOM96), Nancy, France, was financially supported by:

- Centre National de la Recherche Scientifique
- Ministère de l'Aménagement du Territoire et de l'Environnement
- Université Henri-Poincaré (UHP Nancy I)
- Institut National Polytechnique de Lorraine (INPL Nancy)
- Agence de l'Eau Rhin-Meuse
- Communauté Urbaine du Grand Nancy
- Association Française pour l'Etude du Sol
- TREDI Company
- HORIBA Company
- GASPARD Company
- Banque Populaire de Lorraine

and co-sponsored by:

- Institut National de la Recherche Agronomique (INRA)
- International Soil Science Society (ISSS)
- International Humic Substances Society (IHSS)

EXECUTIVE COMMITTEE OF ISMOM96

P. M. HUANG, Chairman, Working Group MO, University of Saskatchewan, Saskatoon, Canada

J.-M. BOLLAG, Pennsylvania State University, University Park, Pennsylvania, USA

J. BERTHELIN, Centre de Pédologie Biologique du C.N.R.S., Vandœuvre-lès-Nancy, France, Chairman of ISMOM96

INTERNATIONAL SCIENTIFIC COMMITTEE OF ISMOM96

J. BERTHELIN (France)
F. ANDREUX (France)
Ph. BAVEYE (USA)
Ph. BEHRA (France)
W. E. H. BLUM (Austria)
J.-M. BOLLAG (USA)
J. CASES (France)
C. CHENU (France)
G. DEFAGO (Switzerland)
A. M. DEXTER (UK)
T. HATTORI (Japan)
A. HERBILLON (France)
P. M. HUANG (Canada)
M. JAMAGNE (France)
C. de KIMPE (Canada)
S. McGRATH (UK)
J-C. MUNCH (Germany)
O. RICHTER (Germany)
M. ROBERT (France)
P. ROUXHET (Belgium)
N. SENESI (Italy)
A. VIOLANTE (Italy)
D. van ELSAS (NL)
W. van RIEMSDJIK (NL)

LOCAL SCIENTIFIC COMMITTEE OF ISMOM96

M. BABUT (Agence de l'Eau Rhin-Meuse)
J. BERTHELIN (CPB-CNRS-UHP Nancy)
J. C. BLOCK (Pharmacie-UHP Nancy)
J. CASES (LEM-CNRS-ENSG-INPL Nancy)
J. J. EHRHARDT (LCPE-CNRS-UHP Nancy)
P. FAIVRE NANC.I.E. (Nancy)
A. HERBILLON (CPB-CNRS-UHP Nancy)
M. JAUZEIN (IRH-Environnement Nancy)
J. L. MOREL (ENSAIA-INPL Nancy)
M. SARDIN (LSGC-CNRS-INPL Nancy)

INTRODUCTION AND SCIENTIFIC PRESENTATION OF THE VOLUME

Major and trace elements, organo-mineral, and organic compounds of different origins, undergo continual transformation and cycling within the terrestrial environments, particularly in the soil, freshwater and ocean compartments. In the soils, at the interfaces between atmosphere, hydrosphere, lithosphere, and biosphere, these transformations, evolutions and cycling depend on the dependence of interactions between minerals-organics and microorganisms that control and influence directly or indirectly their chemical, biological, and physical properties and also their functioning and quality.

In fact, soils can be considered as complex multiphasic interactive biogeochemical reactors, reservoirs of microorganisms and major compartment of terrestrial ecosystems under the influence of anthropic activities.

To improve our knowledge and management of the different soil functions, e.g. cycling of elements, cycling and quality of water, air quality, support of plant growth and plant production, support of human activities, waste deposits, etc., it appears necessary to enhance the information and understanding of the interactions of soil minerals with organic components and microorganisms.

More often, such information is dispersed because studies are mainly performed under one's own field of interest and/or discipline. This is the reason why the Working Group M.O., that develops studies of the interactions between minerals, organic components and microorganisms in soils," was founded to promote research, teaching and technology in these interdisciplinary fields.

The present volume arose from the Nancy Meeting "ISMOM96" and is devoted to the theme "Effect of Mineral-Organic-Microorganism Interactions on Soil and Freshwater Environments" and includes five main topics:

- Reactivity and Transformations of Mineral Constituents and Metals at the soil-solution interfaces
- Nature, Dynamics and Transformations of Organic Compounds and Enzymes in Soils
- Microorganism-Colloid Interactions and their Effect on Bioavailability of Pollutants and Nutrients in Terrestrial and Freshwater Environments
- Effect of Microorganism-Colloid-Soil Interactions on Dynamics and Activity of Microbial Communities and Populations
- Integration of Mineral-Organic-Microorganism Interactions in the Evaluation of Soil and Freshwater Quality

These topics consider different aspects of the environmental behavior and fate of natural and anthropogenic organic and inorganic compounds in different soil systems and soil-surface water systems under the control of biotic and abiotic processes and parameters involved.

In this book, overlap of processes, constituents, etc., between the different topics underline the occurrence, importance, and interest of such interactions. In the near future such interactive phenomena will have to be much more defined to propose criteria for soil quality and for soil environmental functions that are needed to help the management of the whole terrestrial ecosystem.

CONTENTS

Part I. Reactivity and Transformations of Mineral Constituents and Metals at the Soil-Solution Interfaces

Part III. Microorganism-Colloid Interactions and their Effect on Bioavailability of Pollutants and Nutriments in Terrestrial and Freshwater Environments

Part IV. Effect of Microorganism-Colloid-Soil Interactions on Dynamics and Activity of Microbial Communities and Populations

SORPTION MECHANISMS AT THE SOLID-WATER INTERFACE

Philippe Behra,[1] Jamaâ Douch[1,2] and Frank Binde[1]

[1]Institut de Mécanique des Fluides – UMR 7507 Université Louis Pasteur - CNRS
2, rue Boussingault – 67000 STRASBOURG – France

[2]Ibnou Zohr University – Department of Chemistry
BP 28/S – 8000 AGADIR – Morocco

1. INTRODUCTION

Soils, solid phases or colloids have been for a long time considered as efficient barriers to trap pollutants during their migration. Meanwhile, this statement has been more and more questioned, and it becomes necessary not only to protect the resource « water » but also to prevent all deterioration of its quality. The cycling of water is a complex system which has to be better understood in order to predict the environmental impact of key parameters on the different compartments of the natural system.

It is thus important to bring new knowledge to the following questions: What is the fate and the role of micropollutants in the water cycle and in soils? What are the main mechanisms that allow a description of their transfer, transport and transformation in this cycle? Answering these questions should allow a better understanding and therefore a control of the fate of these compounds either in natural environments (by predicting the evolution of a pollution, by containing it and by remediating the site by suitable techniques), or during their anthropological use (elimination and control by proper processes). These studies are at the interface between the transfer of matter, water chemistry and the chemistry of the solid-water interface.

Effect of Mineral-Organic-Microorganism Interactions on Soil and Freshwater Environments
Edited by Berthelin, *et al*, Kluwer Academic / Plenum Publishers, New York, 1999

How can the transport of reactive solutes be modelled? The variability of hydrodynamic and bio-geochemical parameters makes illusive the elaboration of a « model » of fate applicable to every contaminated field site, for every kind of pollution and whatever the pollutant might be. In numerous case of pollution, several contaminants are simultaneously transported by water which can be considered both as a vector or a « scavenger » of solutes or particulate matter, and a reagent. As a function of the local conditions, there is practically no knowledge about the effects of interaction and competition resulting from the concerning substances or being modified by whether one or several of their characteristics (physical, chemical and biological nature).

The scientific approach, which is to work at different scales on experimental physical and/or complementary mathematical models, can by interactive simulation between experiments and modelling bring an answer to the mechanisms and the main parameters that rule the dispersion of different pollutants. One of the goals is also to validate the mathematical models and their applicability to natural systems.

The solid-water interfaces play a very important role in the fate and behaviour of major and trace elements present in atmospheric, marine and continental surface (soils…) or subsurface systems (Stumm and Morgan, 1996). The distribution of solutes between liquid and solid phase essentially depends on the chemical and physico-chemical properties of the two phases and the elements which are present there (Stumm, 1992). The important mechanisms which are involved in the fate of contaminants are on the one hand complexation, hydrolysis and redox reactions in the aqueous phase which influence the aqueous speciation, the sorption at liquid-solid interfaces which means surface complexation, and on the other hand precipitation or dissolution, fixation or release, diffusion processes, degradation of chemicals, flocculation or peptisation of colloids. Moreover scavenging of contaminants can be associated with precipitation or dissolution of (hydr)oxides, fixation or release of organic matter (non-living organic matter), either by flocculation or peptisation of colloids. The biological use of a solute can also play a non negligible role in its retention and accumulation in a biotope or its transfer to another.

Prior to their comparison the different mechanisms of sorption of a solute at the solid-water interface have to be defined. The term « sorption » covers all processes that contribute to the appearance or transformation of a surface by the presence of a solute or its environment as well as to the phase-transfer (liquid-solid) of the solute. These mechanisms are: (i) precipitation/dissolution; (ii) ion exchange; (iii) surface complexation; (iv) surface precipitation; (v) hydrophobic interactions; (vi) biological uptake; (vii) diffusion and absorption. The speciation of the considering solute will have to take into account these mechanisms as well in the aqueous phase as at the interfaces.

2. SORPTION AT THE SOLID-WATER INTERFACE

Several concepts permit to explain the sorption of cations and anions at the solid-water interface. After having considered the influence of the processes of precipitation and dissolution, we recall herein two approaches currently used (mechanisms (i) to (iv)) to describe the behaviour of non-conservative solutes in porous media.

2.1. Precipitation/Dissolution

The mechanisms of dissolution and precipitation are often associated with the processes of oxidation and reduction or aqueous complexation. The redox behaviour of Fe(III) or Mn(III, IV) (hydr)oxides might serve as an example. Under appropriate conditions, Fe(III) can be reduced leading to the dissolution of its oxides. Fe(II) appearing in the solution is stable as long as the reducing conditions are preserved. It can also precipitate with sulphide or carbonate forming iron sulphide (such as pyrite), or siderite, respectively. As oxidising conditions return, Fe(II) can be re-oxidised and then re-precipitated under formation of Fe(III) oxides. The consequences of dissolution, transport and re-precipitation are very important for the different ions present at the surface of the iron oxides, since the dissolution of the latter will result in the liberation of these ions. In the case of metal cations, the liberation might be followed by precipitation under formation of sulphides, but also by a more forward migration in the water column, the soil, the underground and the aquifer. Finally, they can be adsorbed again by ion exchange, complexation on the surfaces of oxides or co-precipitation with recently formed solid phases.

2.2. Ion exchange

The ion exchange describes the substitution of cations and anions on an ion exchange surface in the case of low energy sorption (electrostatic bond). It is represented by the cationic or anionic exchange capacity of the solid, N_E, which is a function of the intrinsic charge of the ion exchange surface due to crystallographic substitution in the lattice. In the case of two cations (M^{z+} and B^{w+}), the exchange can be described using the following reaction:

$$wM^{z+} + zB_s^{w+} \leftrightarrow wM_s^{z+} + zB^{w+} \qquad (1)$$

where M^{z+} present in solution is exchanged with a cation B^{w+} fixed on the solid, noted B_s^{w+} (the index s symbolises the cation fixed on the surface of the exchanger). The application of the law of mass action to this equilibrium reaction leads to a definition of the so-called selectivity coefficient where cations sorbed are given as molar fraction:

$$K_{M^{z+}/B^{w+}} = \frac{\overline{x_{M_s^{z+}}}^{w}\left[B^{w+}\right]^{z}}{\left[M^{z+}\right]^{w}\overline{x_{B_s^{w+}}}^{z}} \qquad (2a)$$

where $\overline{x_{M_s^{z+}}}$ and $\overline{x_{B_s^{w+}}}$ are the molar fractions for the sorbed M^{z+} and B^{w+} cations respectively:

$$\overline{x_{M_s^{z+}}} = \frac{z\left\{M_s^{z+}\right\}}{N_E} \text{ and } \overline{x_{B_s^{w+}}} = \frac{w\left\{B_s^{w+}\right\}}{N_E} \qquad (2b)$$

and N_E is the cationic exchange capacity defined as the electroneutrality condition for the solid phase by the relation:

$$N_E = z\left\{M_s^{z+}\right\} + w\left\{B_s^{w+}\right\} \qquad (3)$$

In natural media, the clays are the main minerals which exhibit ion exchange properties. Some oxides such as manganese oxides can also contribute to this process. Alkaline or earth alkaline cations and halogenide anions or organic ligands are often sorbed by ion exchange mechanism. For more details, see Bolt (1982). Transition metals exhibit stronger affinity with natural surfaces and other mechanisms are used to describe interactions.

2.3. The surface complexation model

The solid phases mostly present in natural media are composed of silica, feldspars, clays, metal (hydr)oxides, carbonates, sulphides, etc. These solids have ionisable functional groups (–OH, –COOH, $-NH_2$, or even $-OPO_3H$, –SH). In aqueous solution, the behaviour of these mineral surfaces (noted ≡S–OH) is similar to the one of a weak diacid (Figure 1).

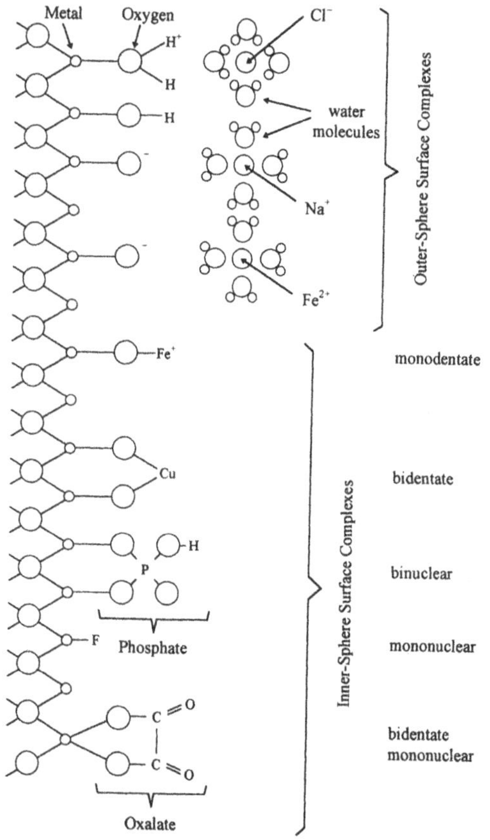

Fig. 1 Scheme of the bonding between cations or anions and a surface of oxides, (hydr)oxides, alumino-silicates or a surface comprising hydroxyl groups (after Sigg *et al.*, 1994).

$$\equiv S-OH_2^+ \leftrightarrow \equiv S-OH + H^+ \quad ; \quad K_{a1}^s = \frac{\{\equiv S-OH\}[H^+]}{\{\equiv S-OH_2^+\}} \quad (4)$$

$$\equiv S-OH \leftrightarrow \equiv S-O^- + H^+ \quad ; \quad K_{a2}^s = \frac{\{\equiv S-O^-\}[H^+]}{\{\equiv S-OH\}} \quad (5)$$

We can define a surface charge, q (in mol g^{-1}), which is a function of the pH of the solution, but also of the ionic strength, I, of the system as shown in Figure 2:

$$q = \{\equiv S-OH_2^+\} - \{\equiv S-O^-\} \qquad (mol\ g^{-1}) \qquad (6)$$

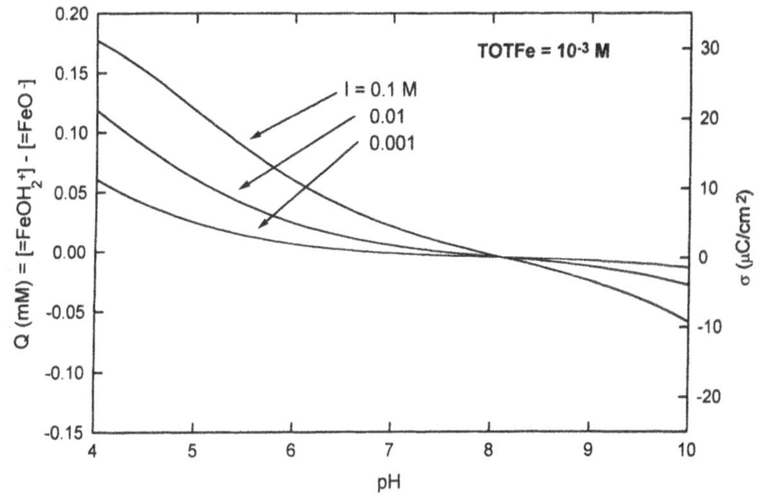

Fig. 2 Surface charge of FeOOH, Q (in mM) or σ (in μC cm^{-2}), as a function of pH and ionic strength, I (1:1 electrolyte). The common intersection point corresponds to the point of zero charge, i.e. pH_{pzc} = 8.11 for FeOOH. Conditions: 90 mg l^{-1} ($[\equiv Fe\text{–}OH]_{tot}$ = 1 mM, i.e. 200 μmol of reactive sites l^{-1}) suspension of FeOOH. Calculations from data given by Dzombak and Morel (1990); electrostatic corrections done with the diffuse double layer model (see section 3 « Structure of the Solid-Water Interface », and Figure 5).

The surface charge can be also called Q, in mM, or σ, in μC cm^{-2}. Thus, the surfaces are positively charged at low pH ($pH < pK_{a1}^s$) and negatively charged at higher pH values ($pH > pK_{a2}^s$). The pH value which corresponds to a zero charge is a characteristic property of every solid phase. Some pH values at the point of zero charge (pH_{pzc}) of different minerals or solid phases are given in Table 1.

Table 1 Values of point of zero charge of some solids (after Stumm, 1992)

Material	pH_{pzc}	Material	pH_{pzc}
α-Al_2O_3 (corundum)	9.1	β-MnO_2 (birnessite)	7.2
α-$Al(OH)_3$ (gibbsite)	5.0	SiO_2 (am., quartz...)	2.0-3.0
γ-AlOOH (boehmite)	8.2	TiO_2 (anatase, rutile)	6.3
BeO	10.2	ZrO_2	6.4
CuO (tenorite)	9.5	$ZrSiO_4$	5.0
Fe_3O_4 (magnetite)	6.5	Albite	2.0
α-FeOOH (goethite)	7.8	Chrysolite	>10
α-Fe_2O_3 (hematite)	8.5	Feldspars	2-2.4
FeOOH (amorphous)	8.1	Kaolinite	4.6
MgO	12.4	Latex	8.0
δ-MnO_2 (vernadite)	2.8	Montmorillonite	2.5

Note: The reported values are not absolute ones. They depend on the model used to estimate them (see section 3 « Structure of the Solid-Water Interface », and Figure 5).

The surface sites, in the form of $\equiv S-O^-$, can play the role of ligands for a given metal ion (M^{z+}). The mechanism of sorption is then analogous to a complexation reaction at the surface between the cation and surface sites (Schindler and Stumm, 1987), with an exchange of protons (Figure 1). For example, in the case of monodentate (n = 1) or bidentate (n = 2) surface complexes, the reactions and conditional constants, $*\beta_n^s$, associated to the equilibria are the following:

$$n(\equiv S-OH) + M^{z+} \leftrightarrow (\equiv S-O)_n M^{(z-n)+} + nH^+ \quad (7)$$

$$*\beta_n^s = \frac{\{(\equiv S-O)_n M^{(z-n)+}\}[H^+]^n}{[M^{z+}]\{\equiv S-OH\}^n} \quad (8)$$

Therefore the surface complexation will be favoured by an increasing pH of the solution (Figure 3a). The mass balance equation for the surface sites is defined as:

$$\begin{aligned}\{\equiv S-OH\}_{tot} &= \{\equiv S-OH_2^+\} + \{\equiv S-OH\} + \{\equiv S-O^-\} \\ &+ \{\equiv S-OM^{(z-1)+}\} + 2\{(\equiv S-O)_2 M^{(z-2)+}\}\end{aligned} \quad (9)$$

where $\{\equiv S-OH\}_{tot}$ (respectively $[\equiv S-OH]_{tot}$) represents the concentration of surface sites given in mol g^{-1} (respectively mol l^{-1}). Figure 3a exemplifies the difference in affinity of several mono, di- or trivalent cations for surface complex formation on an amorphous iron oxide (FeOOH). For divalent cations, the affinity for a (hydr)oxide surface which is often observed for increasing pH is: Hg(II) > Pb(II) > Cu(II) > Zn(II) > (Cd(II) > Ni(II) > Ca(II).

In the case of anion adsorption (L^{y-}), the surface site ($\equiv S-^+$) will behave like a metal centre towards the ligand L^{y-} (Fig. 1). This expresses in an exchange of the $-OH^-$ ion and the formation of mononuclear (n = 1) and/or binuclear (n = 2) complex(es):

$$n(\equiv S-OH) + L^{y-} \leftrightarrow (\equiv S-)_n L^{(y-n)-} + nOH^- \quad (10)$$

As in the case of cations, the according surface complex formation constant, $*\beta_n^s$, is defined from the law of mass action:

$$*\beta_n^s = \frac{\{(\equiv S-)_n L^{(y-n)-}\}[OH^-]^n}{[L^{y-}]\{\equiv S-OH\}^n} \quad (11)$$

Contrary to cations, the surface complex formation with anions will occur preferably in the domain of low pH but can be high even for alkaline solutions (Figure 3b). The aqueous speciation of anions which behave as weak acids affect a lot the sorption behaviour (see for example As(III)). On the other hand, the surface affinity of anions also depends on the oxidation state of element: the higher the oxidation state, the lower the affinity (case of S(VI)/S(IV) and Se(VI)/Se(IV)). The mass balance equation for the surface sites is given as:

$$\begin{aligned}\{\equiv S-OH\}_{tot} &= \{\equiv S-OH_2^+\} + \{\equiv S-OH\} + \{\equiv S-O^-\} \\ &+ \{\equiv S-L^{(y-1)-}\} + 2\{(\equiv S-)_2 L^{(y-2)-}\}\end{aligned} \quad (12)$$

The surface charge can be generalised in the case of different cations or anions adsorbing on the surface (mol g^{-1}):

$$\begin{aligned}q &= \{\equiv S\text{-}OH_2^+\} + (z-1)\{S-OM^{(z-1)+}\} + (z-2)\{(S-O)_2 M^{(z-2)+}\} \\ &- \left(\{\equiv S-O^-\} + (y-1)\{\equiv S-L^{(y-1)-}\} + (y-2)\{(\equiv S-)_2 L^{(y-2)-}\}\right)\end{aligned} \quad (13)$$

It is also possible to consider other types of surface complexes such as ternary surface complexes (see Bourg and Schindler, 1978; Bourg *et al.*, 1979; and Tiffreau *et al.*, 1995):

$$\equiv S-OH + M^{z+} + L^{y-} \leftrightarrow \equiv S-OM-L^{(z-y-1)+} + H^{+} \qquad \text{type A} \qquad (14)$$

$$\equiv S-OH + M^{z+} + L^{y-} \leftrightarrow \equiv S-L-M^{(z-y+1)+} + OH^{-} \qquad \text{type B} \qquad (15)$$

In the first case (type A), the cation and the ligand L behave like a cation, whereas in the second case (type B) the cation behaves like an anion. Surface complexation constants can be also defined to the previously defined equilibria.

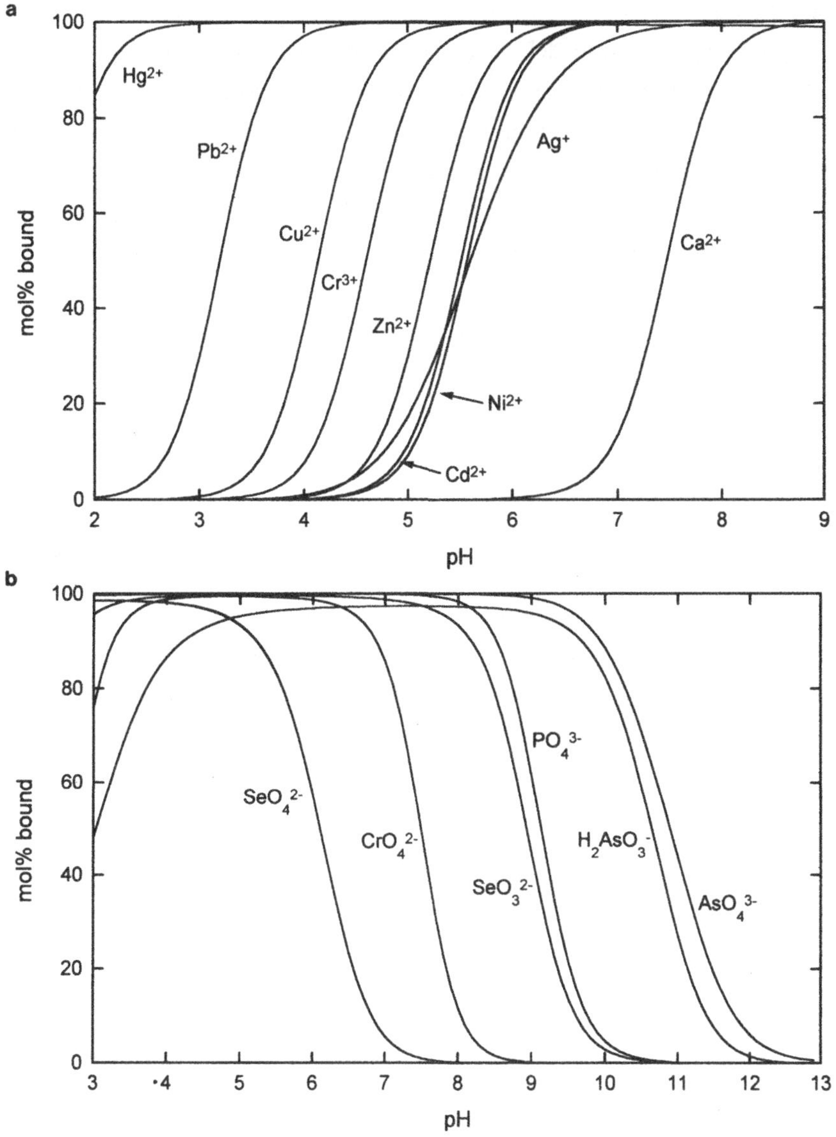

Fig. 3 Adsorption edge (plotted as mol % of the ions in the system adsorbed or surface bound) of different ions in the presence of FeOOH: (a) case of metal mono- to trivalent cations; (b) case of anions the aqueous speciation of which strongly depend on pH. $[\equiv Fe-OH]_{tot}$ = 1 mM (maximum 200 µmol of reactive sites l^{-1}, suspension 90 mg l^{1}), total ion concentration in the suspension = 0.5 µM, I = 0.1 M ($NaNO_3$). Calculations, from physico-chemical characteristics and intrinsic sorption constants given by Morel and Hering (1993), take into account hydroxo complex formation in solution; electrostatic corrections done with the diffuse double layer model (see section 3 « Structure of the Solid-Water Interface », and Figure 5).

The main advantage of the surface complexation model is that it takes into account the dependence of cations, anions and the surface on the pH of the system. Thus it appears evident that the sorption of an ion is a function of (i) the parameter which influences the charge of the surface such as pH or I, (ii) the speciation of the considered element in the presence of ligands, but also (iii) the type of solid phase the physico-chemical properties of which are very variable (see Table 1). Moreover, it is important to point out that the sorption of cations (respectively anions) performs well underneath (respectively above) the point of zero charge of a (hydr)oxide, when the solid is positively (respectively negatively) charged (see Figure 3).

The difference between the models of ion exchange and surface complexation is due to the nature of the interactions which are involved in the mechanism of fixation. The first model is considered to arise if weak interactions (dipole-dipole) occur, the surface charge being due to the substitution of ions in the crystal lattice, whereas the surface complexes are made up from the entirety of the chemical entities under formation of covalent bonds, the surface charge being caused by the presence of adsorbed ions as a function of the pH and ionic strength. As seen in Figure 3 for the surface complexation model, cations (respectively anions) can be strongly sorbed in a pH domain where the surface is positively charged, i.e. below pH_{pzc} (respectively negatively charged, i.e. above pH_{pzc}), what is consistent with the formation of covalent bonds. The law of mass action as well as the kinetic equations are applicable, that is assuming reversibility of system reactions. In the second case the role of the pH but also of the ionic strength appears more clearly.

2.4. Surface precipitation

The surface precipitation is a mechanism that can occur even though the saturation of surface sites has not been observed (Farley *et al.*, 1985). The composition of the surface is then modified after the appearance of a new phase the composition of which continuously varies between those of the original solid and those of the solute precipitated on the surface. In the case of a divalent cation and Fe hydroxide like ferrihydrite, the precipitation of the hydroxide of the divalent cation as well as the precipitation of Fe(III) hydroxide can be considered:

$$\equiv Fe-OH + M^{2+} + 2H_2O \leftrightarrow Fe(OH)_{3(s)} + \equiv M-OH_2^+ + H^+ \qquad {}^*\beta^s_{\equiv Fe-OM^+} \qquad (16)$$

$$\equiv M-OH_2^+ + M^{2+} + 2H_2O \leftrightarrow \equiv M(OH)_{2(s)} + \equiv M-OH_2^+ + 2H^+ \qquad K^{-1}_{spM(OH)_2} \qquad (17)$$

$$\equiv Fe-OH_2{}^+ + Fe^{3+} + 3H_2O \leftrightarrow Fe(OH)_{3(s)} + \equiv Fe-OH + 3H^+ \qquad K^{-1}_{spFe(OH)_3} \qquad (18)$$

It is important to notice that the formed solid phase is not pure, but consists of solid solution with two precipitates. Thus the activities of the new components will not be unity. The characteristic isotherms include three domains: at very low concentrations, the solute shows a Henry-type behaviour depending on the number of surface sites and the properties of the sorbed solute, in an intermediate concentration range the behaviour is close to that of a Langmuir-type (tending to the saturation of the sites), and finally the isotherm displays a concave curvature, the upper concentration limit being the solubility concentration of the considered cation as exemplified in Figure 4 (Lützenkirchen and Behra, 1996).

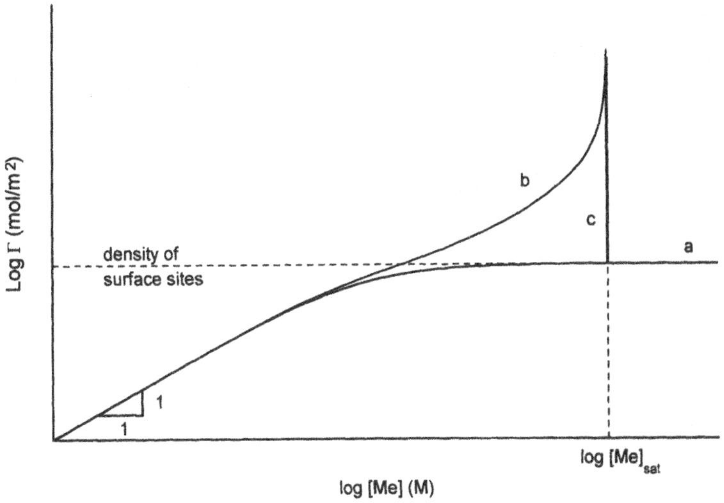

Fig. 4 Schematic sorption isotherms of a metal ion (Me) on the surface of an oxide at constant concentration of ligands (in solution and on surface sites) and constant pH: (a) adsorption only, for example by surface complexation; in the first part the behaviour is of Henry-type (slope of 1), in the second part the sites are saturated after formation of a mono-layer (Langmuir-type behaviour); (b) sorption (by surface complexation) and surface precipitation via formation of a solid solution; the isotherm is of Farley-Dzombak-Morel-type and looks like a BET-type isotherm; (c) adsorption and heterogeneous nucleation in the absence of a free energy barrier ($\Delta G^* \rightarrow 0$). Two other cases can occur: (i) adsorption and heterogeneous nucleation of a metastable precursor; (ii) same mechanism as in (i) but with transformation of the precursor into the stable phase (after Stumm, 1992). The arrows show the isotherm evolution for continual addition of dissolved Me. Γ_{Me} is the density of the solute fixed at the surface (in mol m^{-2}), $[Me]_{sat}$ the solubility concentration of Me for the stable metal (hydr)oxide. Calculations after Lützenkirchen and Behra (1996).
Note: To estimate the effect of ligands on sorption such as the formation of ternary surface complexes, the abscissa have to be presented as $\log[Me^{2+}]$ instead of log[Me].

Analysis of the surface by spectroscopic methods has given evidence for the formation of a precipitate at the solid-water interface while the conditions for precipitation in the solution were not reached. Numerous works are on the way to explore the different concentration ranges (Charlet and Manceau, 1993; Moulin, 1996).

3. STRUCTURE OF THE SOLID-WATER INTERFACE

The surface complexation model allows the description of interactions between the water and solid phases, as well as the behaviour of a solute at the solid-water interface. Nevertheless, at the surface of a charged solid an electrical field is created between the electrically neutral aqueous phase and the interface. With respect to the Gibbs free energy of adsorption, this charge contributes as an additional energy to the intrinsic free energy which is valid for an uncharged surface and corresponds to the chemical contributions. The total free energy corresponding to the conditions of the system, ΔG_{ads}, might therefore be assumed to be the sum of intrinsic and coulombic free energy:

$$\Delta G_{ads} = \Delta G_{int} + \Delta G_{coul} \tag{19}$$

The intrinsic free energy, ΔG_{int}, is obtained from the law of mass action applied to the corresponding equilibrium:

$$\Delta G_{int} = -RT \ln K_{int} \quad (20)$$

The coulombic free energy, ΔG_{coul}, is a function of the surface potential, Ψ:

$$\Delta G_{coul} = \Delta z F \Psi \quad (21)$$

where Δz is the variation in the charge at the interface for the adsorption reaction under consideration, F the Faraday constant, and Ψ the surface potential at the solid-water interface at which the reaction happens. Since,

$$\Delta G_{ads} = -RT \ln K_{ads} \quad (22)$$

the conditional equilibrium constant can be expressed as:

$$K_{ads} = K_{int} \exp\left(-\Delta z F\Psi / RT\right) \quad (23)$$

With respect to Eq. 4 and the amphoteric property of the surface species, ≡S–OH, the relation between the activity of protons at the surface, $\{H_s^+\}$, and that in solution, $\{H^+\}$, is given as:

$$\{H_s^+\} = \{H^+\} \exp\left(-F\Psi / RT\right) \quad (24)$$

Expressed as a function of pH, this relation turns into:

$$pH_s = pH + F\Psi / (2{,}303RT) \quad (25)$$

From relationship (25) it clearly appears that the pH value at the surface is different to the one measured in the solution except if the surface potential is zero. The corrective term, which must be taken into account to estimate the intrinsic constant of the exchanges, permits to correct the non-stoichiometric effects observed from experimental results. Different models, more or less complex, have been developed to estimate the surface potential as a function of the surface charge.

The most currently used models are: (i) the constant capacitance model *(CCM)*, in which the potential is a linear function of the surface charge, the capacitance is constant at a given ionic strength and all surface complexes are being considered as inner-sphere; (ii) the diffuse double layer model *(DLM)*, in which the surface potential exponentially decreases, the effect of the ionic strength is taken into account and all surface complexes are being considered as inner-sphere; (iii) the *Stern model*, which can be seen as a combination of both the previous models and which permits to distinguish between inner- and outer-sphere complexes (by describing the accumulation of counter ions in the intermediate layer), the effect of the ionic strength and background electrolyte being taken into account too; (iv) the three layer model *(TLM)*, which permits to connect the calculations with electrokinetic measurements and which requires a second capacitance for describing the acid-base properties of the surface, the other properties being comparable to those of the Stern model, the number of adjustable parameters meanwhile being very high (see Figure 5).

Other models known from the literature are: the model of James and Healy (1972), which does not consider the fixation on discrete sites; the model at one pK_a (van Riemsdijk *et al.*, 1986), which is defined with respect to the point of zero charge, in which surface species with a charge of 0.5 or -0.5 appear to describe the acid-base properties; the models MUSIC (multisite complexation model) of Hiemstra *et al.* (1989a,b) and CD-MUSIC (charge distribution multisite complexation model) of Hiemstra and van Riemsdijk (1996), which take into account the presence of different sites of coordination at the surface from crystallographic considerations; in the second case the charge which appears during ion adsorption is no longer considered as punctual; and finally Sverjensky's model (1993) in which contributions due to the solvatation at the interface are also included.

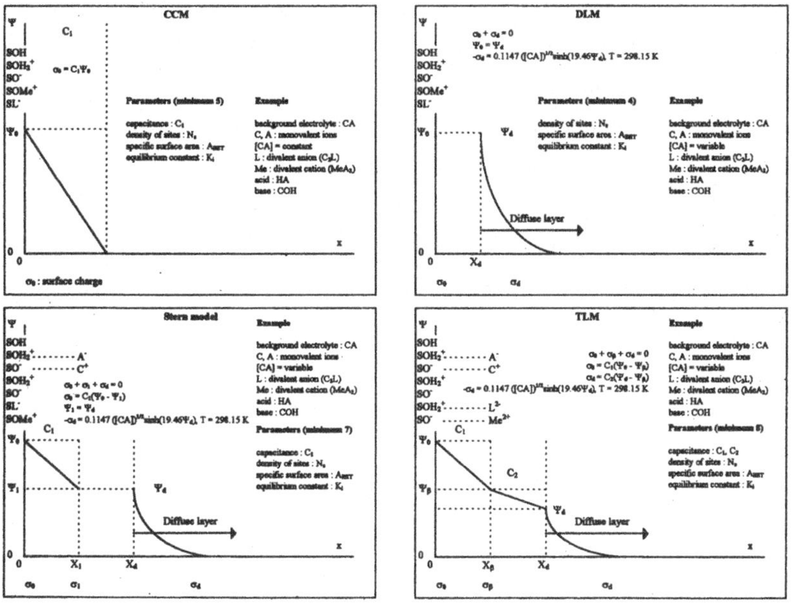

Fig. 5 Scheme of the development of the surface potential, Ψ, as a function of the distance, x, for: (a) the constant capacitance model (*CCM*), (b) the diffuse double layer model (*DLM*), (c) the Stern model, and (d) the three layer model (*TLM*). The parameter x_d indicates the head end of the diffuse layer (after Lützenkirchen, 1996).

It is important to recall that the estimation of the overall parameters (acid-base constants, surface complexation constants, site density, surface capacitance, activity coefficients in the case of surface precipitation, etc.) has to be done with only one of these electrostatic models. Among the computational programs used for these parameter estimations, the most used is certainly FITEQL (Westall, 1982; Westall and Herbelin, 1994). Taking into account the variations of the speciation in the studied system, the unknown parameters are estimated by the method of least squares using a classical iterative calculation (Newton-Raphson method) in order to solve the non-linear equation systems. For further details, see Davis and Kent (1990) and Lützenkirchen (1996).

4. CONCLUSIONS

The solid-water interfaces play a very important role in the fate and behaviour of major and trace elements present in groundwater quality (Stumm and Morgan, 1996). The sorption of solutes, *i.e.* their distribution between liquid and solid phase, essentially depends on the chemical and physico-chemical properties of the two phases and the elements which are present there (Stumm, 1992). We tried to summarise some sorption mechanisms, the term « sorption » covering all

mechanisms that express a change of phase between solid and liquid phases, that is the ion exchange, the surface complexation and precipitation, the precipitation/dissolution processes. We did not come up to (i) the sorption processes due to the hydrophobic properties of either the surface or the considered compound, the heterogeneity of the surface (Cases and Villiéras, 1992; Lecarme-Théobald, 1998), (ii) the absorption phenomena in which a solute is incorporated into the solid matrix, as well as (iii) the diffusion of an ion into the solid phase (Stipp *et al.*, 1994).
By studying the natural solid phases, it turns out that if adsorption is considered as an extremely fast mechanism, desorption is often kinetically limited. The question about the accessibility of ions and thus the reversibility of the exchanges raises. Moreover, the natural solid phases, contrary to those commonly studied presenting heterogeneity at the site level (Cases and Villiéras, 1992), are not pure and show numerous heterogeneities of surface and composition. Several minerals can coexist at the surface and thus modify the sorption properties towards a solute. Therefore the classical methods of potentiometric titration or electrophoretic mobility will only provide average information, which has to be completed by spectroscopic studies of the surface in order to well characterise the solid surface (Bonnissel-Gissinger *et al.*, 1996; Bonnissel-Gissinger *et al.*, 1998; Lützenkirchen *et al.*, 1996).
The spectroscopic studies of the surface appear to be essential in the cases where the processes could be described considering different possible reactions at the interfaces, which are consistent with thermodynamic considerations and coordination chemistry. While the actual tendency is to design models that fit very well the experimental results from the ensemble of numerous imaginable reactions like in a « meccano » game, it is fundamental to study and to validate them at different scales of observation, especially at the molecular and atomic scales. This knowledge will assure a greater credibility to the models, which must be predictive. Some limits appear here: the up-scaling (what does this notion mean?), the detection of compounds at the interfaces by surface spectroscopic techniques and thus their characterisation, the reversibility of the exchanges, the kinetics of the different processes and the effect of the temperature. The studies at field scale finally need to take into account the spatial and time variation of exchanges and heterogeneities in complex systems, which can only be reached by experiments considering the movement of the water and reactive solutes (Behra, 1994).

Acknowledgement

F.B. is supported by the Deutsche Forschungsgemeinschaft, DFG (GRK 366/1-97). This work was supported by the French « Ministère de l'Environnement » (Grant DRAEI/92159), by the « Action Intégrée Franco-Marocaine » (Grant 95873) and the DFG.

REFERENCES

Behra, Ph., 1994. Scale effects in the transport of contaminants in natural media. In: Chemistry of Aquatic Systems: Local and Global Perspectives. G. Bidoglio and W. Stumm (Editors). Kluwer Academic Publishers, Dortrecht, 433-463.

Bolt, G.H. (Editor), 1982. Soil Chemistry B. Physico-Chemical Models. Elsevier, Amsterdam.

Bonnissel-Gissinger, P., M. Alnot, Ph. Behra and J.J. Ehrhardt, 1996. Sorption du mercure (II) sur de la pyrite. In: Mécanismes de Sorptions aux Interfaces Solide-Liquide. Ph. Behra and J.J. Ehrhardt (Editors). Compte-rendu des Journées Scientifiques du GDR PRACTIS 1115 (2-3 mai 1996), LCPE, Villers-lès-Nancy, 79-84.
Bonnissel-Gissinger, P., M. Alnot, J.J. Ehrhardt and Ph. Behra, 1998. Surface oxidation of pyrite as a function of pH. Environ. Sci. Technol. 32, 2839-2845.
Bourg, A.C.M. and P.W. Schindler, 1978. Ternary surface complexes 1. Complex formation in the system silica-Cu(II)-ethylenediamine. Chimia 32, 166-168.
Bourg, A.C.M., S. Joss and P.W. Schindler, 1979. Ternary surface complexes 2. Complex formation in the system silica-Cu(II)-2,2' bipyridyl. Chimia 33, 19-21.
Cases, J.M. and F. Villiéras, 1992. Thermodynamic model of ionic and non-ionic surfactant adsorption-abstraction on heterogeneous surfaces. Langmuir 8, 1251-1264.
Charlet, L. and A. Manceau, 1993. Structure, formation and reactivity of hydrous oxide particles: Insights from X-ray absorption spectroscopy. In: Environmental Particles. J. Buffle and H.P. van Leeuwen (Editors). Vol.2. Lewis Pub., Boca Raton (Florida), 117-164.
Davis, J.A. and D.B. Kent, 1990. Surface complexation modeling in aqueous geochemistry. In: Mineral Water Interface Geochemistry. M.F. Hochella and A.F. White (Editors). Reviews in Mineralogy, Mineralogical Society of America, Washington.
Dzombak, D.A. and F.M.M. Morel, 1990. Surface Complexation Modeling. Hydrous Ferric Oxide. Wiley, New York.
Farley, K.J., D.A. Dzombak and F.M.M. Morel, 1985. A surface precipitation model of the oxide/water interface. J. Colloid Interface Sci. 106, 226.
Hiemstra, T., W.H. van Riemsdijk and G.H. Bolt, 1989a. Multisite proton adsorption modeling at the solid/solution interface of (hydr)oxides: a new approach. I. Model description and evaluation of intrinsic reaction constants. J. Colloid Interface Sci. 133, 91.
Hiemstra, T., J.C.M. de Wit and W.H. van Riemsdijk, 1989b. Multisite proton adsorption modeling at the solid/solution interface of (hydr)oxides: a new approach. II. Application to various important (hydr)oxides. J. Colloid Interface Sci. 133, 105.
Hiemstra, T. and W.H. van Riemsdijk, 1996. A surface structural approach to ion adsorption: The charge distribution (CD) model. J. Colloid Interface Sci. 179, 488.
James, R.O. and T.W. Healy, 1972. Adsorption of hydrolyzable metal ions at the oxide-water interface. J. Colloid Interface Sci. 40, 42.
Lecarme-Théobald, É., 1998. Comportement du tributylétain en milieu aqueux en présence d'une phase solide hététogène. Ph.D. thesis, Université Henri Poincaré, Nancy.
Lützenkirchen, J., 1996. Description des interactions aux interfaces liquide-solide à l'aide des modèles de complexation et de précipitation de surface. Ph.D. thesis, Université Louis Pasteur, Strasbourg.
Lützenkirchen, J., and Ph. Behra, 1996. On the surface precipitation model for cation sorption at the (hydr)oxide water interface. Aquatic Geochemistry 1, 375.
Lützenkirchen, J. J.P. Lickes, S. Contreras, J. Lambert, M. Alnot, F. Dumont and Ph. Behra, 1996. Sorption du cadmium sur un sable de cristobalite: Étude par spectroscopie de surface et modélisation. In: Mécanismes de Sorptions aux Interfaces Solide-Liquide. Ph. Behra and J.J. Ehrhardt (Editors). Compte-rendu des Journées Scientifiques du GDR PRACTIS 1115 (2-3 mai 1996), LCPE, Villers-lès-Nancy, 74-78.
Morel, F.M.M. and J.G. Hering, 1993. Principles and Applications of Aquatic Chemistry. 2nd edition. Wiley, New York.
Moulin, Ch., 1996. Apport de la spectroscopie optique pour les études aux interfaces. In: Mécanismes de Sorptions aux Interfaces Solide-Liquide. Ph. Behra and J.J. Ehrhardt (Editors). Compte-rendu des Journées Scientifiques du GDR PRACTIS 1115 (2-3 mai 1996), LCPE, Villers-lès-Nancy, 51-57.
Schindler, P.W. and W. Stumm, 1987. The surface chemistry of oxides, hydroxides and oxide minerals. In: Aquatic Surface Chemistry. W. Stumm (Editor). Wiley, New York.
Sigg, L., W. Stumm, and Ph. Behra, 1994. Chimie des Milieux Aquatiques. Masson, Paris.
Stipp, S.L., M.F. Hochella, G.A. Parks and J.O. Leckie, 1992. Cd^{2+} uptake by calcite, solid-state diffusion, and the formation of solid-solution: interface processes observed with near-surface sensitive techniques (XPS, LEED, and AES). Geochim. Cosmochim. Acta 56, 1941.
Stumm, W., 1992. Chemistry of the Solid-Water Interface. Wiley, New York.
Stumm, W. and J.J. Morgan, 1996. Aquatic Chemistry. 3rd edition. Wiley, New York.
Sverjensky, D.A., 1993. Physical surface-complexation models for sorption at the mineral-water interface. Nature 364, 776.
Tiffreau, Ch., J. Lützenkirchen and Ph. Behra, 1995. Modeling the adsorption of mercury (II) on (hydr)oxides. I. Amorphous iron oxide and α-quartz. J. Colloid Interface Sci. 172, 82-93.
Van Riemsdijk, W.H., G.H. Bolt, L.K. Koopal and J. Blaakmeer, 1986. Electrolyte adsorption on heterogeneous surfaces: adsorption models. J. Colloid Interface Sci. 109, 219.
Westall, J.C., 1982. FITEQL 2.1: A computer program for the determination of equilibrium constants from experimental data. Rep. 94-01. Department of Chemistry, Oregon State University, Corvallis (Oregon).
Westall, J.C., and A. Herbelin, 1994. FITEQL 3.1: A computer program for the determination of equilibrium constants from experimental data. Rep. 94-01. Department of Chemistry, Oregon State University, Corvallis (Oregon).

COMPARISON BETWEEN BACTERIAL AND CHEMICAL DISSOLUTION OF AL-SUBSTITUTED GOETHITE. INCIDENCE ON MOBILIZATION OF IRON

N. BOUSSERRHINE*, U . G . GASSER**,
E . JEANROY* AND J. BERTHELIN*

*Centre de Pédologie Biologique, U. P. R. 6831 du C. N. R. S. associée à l'Université Henri Poincaré, Nancy, 17, rue Notre-Dame des Pauvres, B. P. 5, F54501 Vandœuvre-lès-Nancy Cedex, France; ** College of Engineering ISW, Grüntal, P. O. Box 335, CH8820 Wädenswil, Switzerland.

INTRODUCTION

Iron oxides such as goethite, hematite and lepidocrocite are common products that form under various weathering conditions on the earth. Iron oxides may also form as a result of steel corrosion. While pure iron oxides may be obtained synthetically, natural iron oxides generally contain noniron metals (NIM), particularly Al. The solubility of iron oxides in pure water is generally extremely low (Schwertmann, 1991). However, Fe oxides dissolution is facilitated by strong acids, reducing and complexing agents (Cornell and Schwertmann, 1996).

In nature, bacterially mediated redox processes often strongly influence the cycle of iron bearing minerals (Lovley, 1991; 1993). The necessary energy for reduction is supplied by metabolic oxidation (or fermentation) of organic compounds with adenosine triphosphate (ATP) being the energy carrier. The products of metabolism are protons, electrons and metabolites. The latter may be mineralised to CO_2 and H_2O upon complete oxidation. Both electrons and protons hereby produced are transferred to the oxide surface, thus, causing formation and detachment of Fe II which is more soluble than Fe III under natural conditions. For goethite, a bulk reaction equation can be written as,

$$4\ FeOOH + CH_2O + 8H^+ \rightarrow 4\ Fe^{2+} + \text{metabolites} + CO_2 + 7H_2O$$

where CH_2O stands for the organic source of energy.

Effect of Mineral-Organic-Microorganism Interactions on Soil and Freshwater Environments
Edited by Berthelin, *et al*, Kluwer Academic / Plenum Publishers, New York, 1999

By reductively dissolving Fe oxides, bacteria may also release potentially toxic elements (NIM) from the solid in to the liquid phase (e. g. soil water, interstitial water of sediments) (Addy *et al.*, 1976; Cornwell, 1986, 1988; Singh & Subramanian, 1984; Todd *et al.*, 1988; Lovley, 1991; 1993). Although there is no doubt about the incidence of bacterial reductive dissolution (BRD) of NIM containing iron oxides in periodically anaerobic environments, very little is known about it. In contrast to BRD, some information is available about the chemical reductive dissolution (CRD) of Al or Co substituted iron oxides (Torrent *et al.*, 1987; Gasser *et al.*, 1996) and soil rich in trace metals containing iron oxides (Trolard *et al.*, 1995). The CRD rate of Al and Co containing iron oxides decreases with increasing NIM for Fe substitution of the Fe oxide (Gasser *et al.*, 1996; Torrent *et al.*, 1987).

There are a few cases where the dissolution of iron oxide by both types of processes (BRD and CRD) have been investigated. As no data have been reported for comparison between BRD and CRD of substituted goethite, this paper partly closes this gap.

MATERIALS AND METHODS:

Aluminium substituted goethite was synthesised following Schwertmann and Cornell (1991). The mole fraction x = NIM/NIM + Fe was 0.00 ² x ² 0.33. While goethites with an Al mole fraction x < 0.1 were synthesised from an Fe III system high Al-Gt (x = 0.33) was obtained from an Fe II system (Taylor and Schwertmann, 1978; Cornell and Schwertmann, 1996). Samples with x = 0.05 (A8), x = 0.33 (A11) (33 % Al substitution) and a pure goethite (S1) were selected for this study.

Goethite preparation : In brief, hundred mL each of 1M $Fe(NO_3)_3$ were used as an Fe(III) source (solution A). In case of Al, 250 mL 0.5M $Al(NO_3)_3$ were mixed with 150 mL 5M KOH to produce solution B; then, 80 mL of solution B were transferred to a 2 L polypropylene bottle, followed by 170 and 173 mL, respectively, of 5M KOH and solution A. Water was added to obtain a final suspension volume of 2 L and a KOH concentration of 2M. The suspensions were aged for 15 consecutive days at 63 ± 1 °C (336 ± 1 K). Bottles were opened daily for 5 min, recapped and shaken (by hand) end-over-end for approximately 15 seconds.

Sample A11 was obtained as follows : add 50 mL of freshly prepared 1 M $FeCl_2$ solution and 25 mL of 1 M $AlCl_3$ solution to a 2 L polyethylene bottle, dilute to approximately 1 L and adjust the pH to about 11.7 with KOH. Mix thoroughly. Slowly oxidise the suspension by opening the bottle daily and swirling the contents. Oxidation is completed after approximately 2-3 months when a dense, yellow product has been formed. Centrifuge the product and wash twice with 0.01 M KOH, to remove excess of Al. This system is thought to favour formation of Al-substituted goethite.

Amorphous materials were removed from the samples by a 2 h treatment with 3M H_2SO_4 at 50 °C (323 K) (Gasser *et al.*, 1996). Similar treatments were used by Schwertmann and coworkers to purify goethites substituted by Cr or V (Schwertmann *et al.*, 1989; Schwertmann and Pfab 1994).

Samples were washed 4 times with water prior to oven drying at 50°C overnight. After drying, samples were gently crushed in an agate mortar.

Total metal concentrations : Total metal concentrations of the goethites (Gt) were determined by wet chemical dissolution. A 30 to 50 mg sample of Gt was placed in a Teflon container, 5 mL nitric acid (65% wt) was added, and the container was placed in a microwave oven for 30 min. The completely dissolved sample was then transferred to and diluted to volume (water) in a 100 mL volumetric flask (Gasser et *al.*, 1996). The metal concentrations were determined with inductively coupled plasma atomic emission spectroscopy (ICP-AES ; Jobin Yvon JY 38 S).

Bacterial dissolution experiments were carried out with the bacterial strain 66/4-8 G isolated from a tropical soil (Oxisols). The isolate was subjected to broad range of physiological studies . It had a facultatively respiratory metabolism and acetic butyric fermentation type (Bousserrhine, 1995).

A modified Bromfield medium (A) was used for bacterial isolation and maintenance (Bromfield, 1954a; Bousserrhine, 1995). In 1000 mL, medium A contained 0.15 g yeast extract, 0.5 g each of KH_2PO_4 and $MgSO_4$, $7H_2O$, 1 g $(NH_4)_2SO_4$, 10 g glucose and 14 g of agar.

For reductive dissolution, the bacteria were grown in an agar free Bromfield medium (B) where ferric iron was provided in form of goethite (Bromfield, 1954 a, 1954 b). Bacteria were grown under standard anaerobic conditions. To a 250 ml serum glass bottle, 150 ml medium B and 0.8 g goethite were added mixed and autoclaved for 30 min at 110°C (383 K). Iron dissolution was determined at various times t (0 ^{2} t ^{2} 200 hours) by sampling suspension with a 5 mL syringe. Suspensions were filtered (0.45 µm MILLIPORE or SARTORIUS FILTER), and analysed for Fe II, total Fe, aliphatic acids and glucose. While Fe II was determined by spectral colorimetry using orthophenantroline reagent and Spectrophotometer at a wavelength λ = 490 nm (UV 1205, SHIMADZU) (Bousserrhine, 1995) , total Fe was measured by ICP-AES. Aliphatic acids were analysed by high performance liquid chromatography using a Gold Beckman instrument equipped with an UV detector set at λ 195 nm (Aminex column HPx87H and isocratic conditions) .

Inoculum of bacteria was prepared by culture of bacterial isolate on solid medium A in plate. After incubation at 28°C (301 K), bacteria were suspended in sterile physiological solution, washed twice, centrifuged at 48000 RCF (relative centrifugal force) and resuspended. A suspension of 10^7 bacteria per ml was used for inoculation. The medium was flushed with N_2 to remove O_2. and maintain anaerobic conditions.

Chemical dissolution The Chemical Reductive Dissolution (CRD) was investigated at 20°C (293 K) and 3°C (276 K) in a dithionite-citrate-bicarbonate (DCB) system following Gasser *et al.* (1996). This system was modified from Mehra and Jackson (1960), Homgren (1967) and Jeanroy *et al.* (1991).

The citrate-bicarbonate (CB) solution was 0.3M in trisodium citrate, and 0.1M in sodium bicarbonate and contained 100 g/L ethanol (Gasser *et al*, 1996). The final pH of the CB solution was 8.3. A 50 mg sample of Gt was placed in a polypropylene

Oak Ridge centrifuge tube, followed by 25 mL of CB and 250 mg of dithionite. Samples were placed on and end-over-end shaker for t hours (t= 0, 1, 2, 4, 8, 12, 24) and mixed at a rate of 30 revolutions per min (RPM). After shaking, samples were centrifuged at 48000 RCF for 10 min. Prior to Al and Fe analysis, aliquots of the supernatant were diluted 10 or 20 times with water.

Surface area (BET) and microporosity (μP) were determined from N_2 adsorption measurements (Boer et al. 1965; Brunauer et al. 1938). Nitrogen adsorption was measured at 6 to 8 different relative nitrogen pressures (P/P_o) between 0 and 0.3. Microporosity was obtained from M_N versus d_N plots (M_N and d_N are the mass and the thickness of the adsorbed N_2, respectively). The specific outer surface area was calculated OS= surface(BET)-μP.

X-ray diffraction (XRD) : In an agate mortar, 100 mg Gt was mixed with 25 mg $Pb(NO_3)_2$ as an internal standard. A CGR diffractometer equipped with an incident-beam monochromator was used to produce CoKα-radiation diffractograms between 18 and 65° 2θ at a scanning rate of 0.2° 2θ min^{-1}. Both XRD-line positions and unit cell dimensions were calculated as indicated earlier (Schwertmann *et al.*, 1989).

Expression of reducing equivalent production and of percentage of electron involved in Fe^{3+} reduction : The calculation of reducing equivalent production was done according to the global balance of acetic and butyric fermentation pathways with $C_6H_{12}O_6 \rightarrow CH_3COOH + 2CO_2 + 8e^-$ and $C_6H_{12}O_6 \rightarrow 2CH_3CH_2CH_2COOH + 2CO_2 + 4e^-$

The percentage (%) of electron production involved in Fe^{3+} reduction was calculated by the ratio between electrons involved in reduction of ferrous iron and electrons that are produced during fermentation processes.

RESULTS

Synthetic goethites characterisation

Goethite was the only mineral phase present in all samples. Unit-cell dimensions (Table 1) decreased with increasing Al concentration of the samples. Sample A11 showed a surface three times higher than those observed for A8 and S11 goethite.

Chemical dissolution

- **Kinetics of iron release**

After 8 hours of reaction with DCB more than 90 % (> 90 %) of total Fe were released from samples S1 and A8, whereas less than 40 % (< 40. %) were dissolved in

Table 1. Synthetic goethite analysis: specific surface (OS) (BET) and cristal parameters

	S1	A8	A11
Acetic acid (mM)	11.53	14.97	14.16
Butyric acid (mM)	14.86	15.16	14.58
Total acids (mM)	26.40	30.13	28.74
Acetic/Butyric	0.78	0.99	0.97
CO_2 produced (mM)	41.25	45.29	43,32
Glucose fermented (mM)	55	55	48.12
%	100	100	87,50
Calculated of the reducing equivalent production (meq e^-)	105	121	115
% of electron production involved in Fe^{3+} reduction	5.70	1.63	0.55
initial pH	6.2	6.2	6.2
final pH	4.00	3.40	3.50

sample A11 (Fig 1). After 10 hours, the total of Fe was solubilised from the two first goethites. Total solubilisation of A11 was obtained, however, after 72 hours. The initial reductive dissolution rates decreased with substitution (Fig. 1).

For the 4°C treatment, the initial Fe release rates were 2.29 nmol/m^2/s (sample S1) and 1.13 nmol/m^2/s (sample 18).

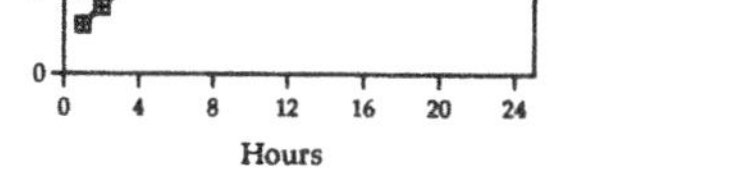

Figure 1. Release of Fe by chemical reductive dissolution (DCB) at 20° C from unsubstituted and substituted goethites. S1 (pure goethite), A8 (Goethite 5% A1), A11 (Goethite 33% Al).

• **Kinetics of release of substituted Al**

Released Al and Fe concentrations were normalised and expressed with respect to the percentage of total content in the mineral and of the total quantity of elements in solution at different times of chemical solubilization. Al was released congruently with respect to Fe (Fig 2 A and B) in samples A8 and A11..

Bacterial dissolution

• **Kinetics of iron release**

The goethites have been solubilized partially or completely during their contact with the ferri-reducing bacterial isolate. Like in chemical treatments, the fastest release rate was observed for unsubstituted goethite (Fig 3).

After 150 h., dissolution in the presence of bacteria, more than 90 % of Fe (> 90 % of Fe) were released from sample S1 (pure goethite). Only 40 % and 15 % respectively of A8 and A11 were dissolved. For all goethites, no solubilization occurred in the uninoculated assays (without bacteria). During this bacterial treatment, the ability to dissolve iron increased as follows : A11< A8 < S1. All the Fe solubilized was in ferrous form.

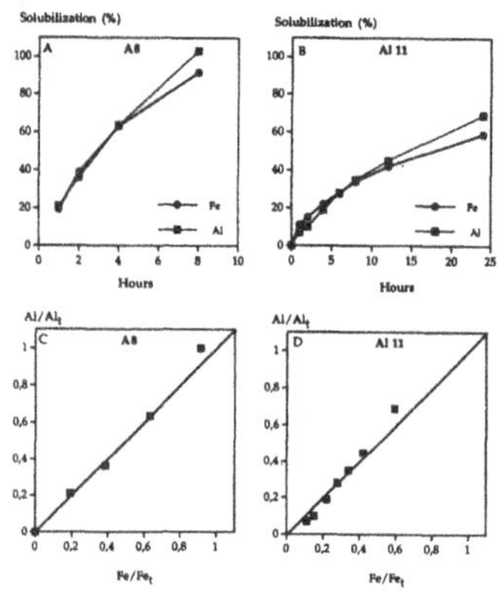

Figure 2. Metal release from samples A8 and A11 (0.5 and 033 mol. % Al, respectively) at 20° C. A and B = relative Fe and Al release by chemical reductive dissolution vs time; C and D = comparison of the corresponding relative Al vs Fe release.

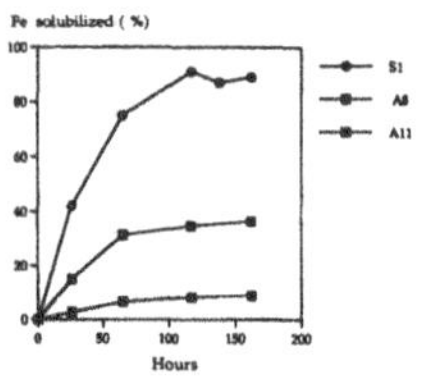

Figure 3. Release of Fe by bacterial reductive dissolution from unsubstituted and substituted goethites at 28% C S1 (pure goethite); A8 (Goethite 5% Al); A11 (Goethite 33% Al). The bars of standard deviation are included in spot on the curves.

• **Kinetics of release of Al**

During Al- goethites dissolution, Al and Fe were released simultaneously (Fig 4 A and B). Unlike chemical dissolution, however, dissolution was not congruent (Fig. 4). A deviation from congruency was observed as indicated, with in Figure 4C and 4D.

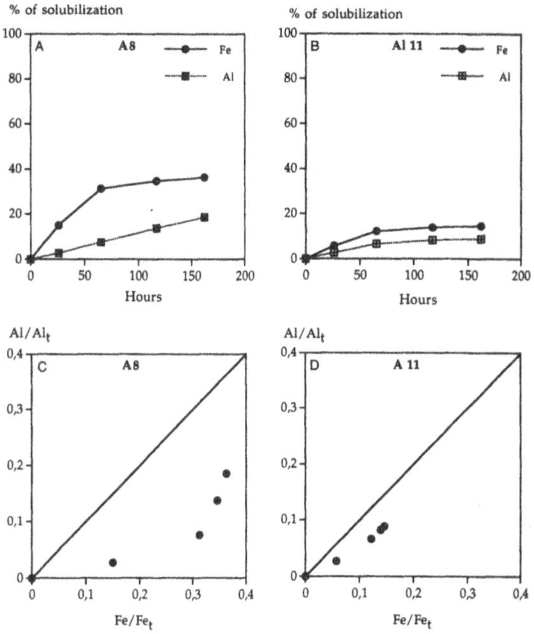

Figure 4. Metal release from samples A8 and A11 (0.05 and 0.33 mol. % Al, respectively) at 28° C. A and B: relative Fe and Al release under the influence of bacterial activity vs. time; C and D: comparison of relative Al vs. Fe release.

Fermentation products during growth on glucose

The results of glucose consumption and acid formation are shown in Table 2. In general, the global bacterial activity, expressed by CO_2 production and glucose consumption of the 66/4-8 G isolate was not affected by the presence of substituted goethites as compared to the pure one. Released aluminium did not induce any inhibition of the bacterial development and the fermentation processes. Table 2 shows complete glucose consumption at the end of fermentation corresponding to CO_2, organic acids, and biomass production. Acetic and butyric acids were the two acids produced (Table 2). In all cases, 66/4-8 G bacteria produced more butyric than acetic acid. The ratio of these metabolites were similar (Acetate/butyrate 0.8 - 1.0). It is worth noting that the total acid formation was higher during Al-substituted goethites bacterial reduction as compared to pure goethite.

Table 2. Fermentation balance at the end of the experiment in culture flasks for the bacterial isolate in presence of pure and substituted goethites. S1 (pure goethite); A8 (goethite 5% Al); A11 (goethite 33% Al)

			Unit cell dimension parameters				
Sample	x	BET (OS)	a	b	c	v	XRD lines (used for UCD calculation)
	mole/mole^{-1}	[m2/g^{-1}]	[nm]	[nm]	[nm]	[nm3]	
S1	0.000	46.0	0.46186 (6)	0.99532 (13)	0.30230 (5)	0.13897	11
A8	0.051	45.1	0.46144 (13)	0.99299 (26)	0.30169 (11)	0.13824	14
A11	0.33	135.1	0.45838 (6)	0.97930 (24)	0.29647 (5)	0.13308	6

DISCUSSION

Bacterial dissolution of sample A8 (5 % Al substitution), showed a curvilinear relation between Fe and time on the one hand and linear relation between Al and time on the other hand (Fig 4A). The continuous and constant release of Al observed during all the bacterial reduction indicates that Al solubilisation resulted both from goethite dissolution by bacterial iron reduction and from change in the pH of the medium (acid dissolution) under the effect of acid fermentation products. Acidification, i.e. a decrease in pH (from 6.2 to 3.4) constitutes a strong Al mobilizing process (Cornell and Schwertmann (1996).

The presence of structural Al (non-reducible element) in synthetic goethites decreased the rate of Fe bacterial release. The initial reductive dissolution rate dramatically decreased with increasing Al concentration in goethite : S1 (1.98 % Fe solubilised per hour) > A8 (0.51 % per hour) > A11 (0.1 % per hour). Norrish and Taylor (1961) reported similar results during chemical reductive dissolution (CRD) of soil Al-goethites : as Al substitution in soil goethites increased, the rate of reductive dissolution dropped (see also Jeanroy *et al.*, 1991).

Like BRD, CRD experiments showed a significant decrease of initial dissolution rate following an increase with Al concentration in goethites. Torrent *et al* (1987) obtained similar results and reported a good correlation for a second-order polynomial fit of the initial dissolution vs Al concentration of the goethites. Although the outer surface (OS) of A11 sample was much higher as compared to that of A8, initial dissolution is lower with the former sample. As indicated previously (Norrish and Taylor, 1961) noted that as Al substitution in soil goethites increased, the rate of reductive dissolution dropped (see also Jeanroy *et al.*, 1991).

During BRD, inhibition of both Fe III reduction and Fe II release may be facilitated by Al-precipitates at the goethite surface as proposed by Borggaard (1990) for the reductive chemical dissolution. The BRD of A8 shows (Fig. 4) that only 8 % of Al were released for 30 % of Fe released. In contrast, c.a. 30 % of Al and Fe were released with DCB. Reprecipitation of aluminium is prevented in CRD due to the complexing effect of citrate (Segal and Sellers, 1984). This may explain the higher Fe solubilization observed in CRD as compared to BRD. Selective chemical dissolution studies using DB (Dithionite-Bicarbonate) or DCB reported by Rajot (1992) support our hypothesis. Rajot showed that the amount of Fe released from Al-substituted goethite was the same during DB and DCB treatments for a short time. The DCB displayed, however, a higher efficiency for a long dissolution reaction. This confirms the important role of citrate to prevent Al precipitation and to promote Fe reduction (Borggaard, 1990 ; Jeanroy *et al.*, 1991; Peterschmitt, 1991). Further confirmation of our interpretation consist of the following : i) Sample A8 which was reduced once by ferri-reducing bacteria was washed 1 h in the dark by oxalate solution (pH 3) (Tamm, 1922) and submitted to an active ferri-reducing inoculum, exhibits an additional metal release (The total Fe released reached 60 % in contrast to 30 % obtained after the first reducing cycle). ii) Fe release remained below 38 % for non washed Al-goethite.

Carbon metabolism analysis shows that glucose fermentation seems a little modified by the presence of Al. Higher acetate production is observed during Al-goethite reduction and indicates that bacteria generated more ATP (Adenosine-Tri-Phosphate) and more NADH (Nicotinamide-adenine dinucleotide) and, consequently, more electrons that can be involved in reducing activity comparing to fermentation balance in presence of pure goethite (Jones and Woods, 1986). But, as mentioned previously, the reduction rate, however, is higher with the pure goethite sample (S1) than with the substituted ones (A8 and A11) due to the presence of Al that inhibits reduction process.

CONCLUSIONS

The chemical and bacterial reductive dissolutions of pure and Al substituted goethites lead to the following conclusions :

1) Structural Al (non reducible element) decreases the rate of goethite dissolution : initial dissolution rates decreased with increasing Al concentration of the samples.

2) Aluminium leads to an inhibition of the Fe (III) reduction by bacteria. Fe release may be restricted by an Al accumulation on to the goethite surface blocking access to sites necessary for iron reduction.

REFERENCES

Addy, S. K., Presley, B. J., and M. Ewing, 1976. Distribution of manganese, iron and other trace elements in a core from the northwest Atlantic. *J. Sed. Petrol.* 46 : 813-818.

Boer,J. H., De Linsen, B. G., and T. J. Osinga, 1965. Studies on pore systems in catalysts. VI. The universal t curve. *J. Catalysis.* 4 : 643-648.

Borggaard, O. K., 1990. Kinetics and mechanisms of soil iron oxide dissolution in EDTA, oxalate and dithionite. Proc. 9th Intern. Clay Conf., Strasbourg,1989, V.C.Farmer and Y.Tardy (Eds), *Sci. Géol. Mém., Strasbourg.* 85 : 139-148.

Bousserrhine, N., 1995. Étude des paramètres de la réduction bactérienne du fer et application à la déferrification de minéraux industriels. *Thèse Doc. Univ. Nancy I . France.*

Bromfield, S.M., 1954 a. Reduction of ferric compounds by soil bacteria. *J. Gen. Microbiol.* 11 : 1-6.

Bromfield, S. M., 1954 b. The reduction of iron oxide by bacteria. *J. Soil. Sci.* 5 : 129-139.

Brunauer, S., Emmet, P. H., and E. Teller, 1938. Adsorption of gases in multi-molecular layers. *J. Am. Chem. Soc.* . 60 : 309-319.
Cornell, R. M and U. Schwertmann, 1996. The iron oxides. Structure, properties, reactions occurence and use. VCH Publ., Weinheim, Germany.
Cornwell, J. C., 1986. Diagenetic trace-metal profiles in arctic lake sediments. *Environ. Sci. Technol.* 20 : 299-302.
Cornwell, J. C. 1988., Iron and manganese reduction in lacustrine sediments. *EOS Transact. Am. Geophys. Union.* 69 : 1106.
Gasser, U. G., Jeanroy, E., Mustin, C., Barres, O., Nüesch, R., Berthelin, J., and A. J. Herbillon, 1996. Properties of synthetic goethites with Co for Fe substitution. *Clay Miner.* 31 : 465-476.
Holmgren, G. G. S., 1967. A rapid citrate-dithionite-extractable iron procedure. *Soil Sci. Soc. Amer.* 31 : 210-211.
Jeanroy, E., Rajot, J. L., Pillon, P. and A.J. Herbillon, 1991. Differential dissolution of hematite and gœthite in dithionite and its implication on soil yellowing. *Geoderma.* 50 : 79-94.
Jones, D. T., and D. R.Woods, 1986. Acetone-butanol fermentation revisited. *Microbiol. Rev.* 50 : 484-524.
Lovley, D. R., 1991. Dissimilatory Fe(III) and Mn(IV) reduction. *Microbiol. Rev.* 55: 259-287.
Lovley, D. R., 1993. Microbial reduction of iron, manganese, and other metals. *Advances in agronomy.* 12 : 175-229
Mehra, O. P., and M. L. Jackson, 1960. Iron oxide removal from soils ans clays by dithionite-citrate system buffered with sodium carbonate. Proc. 7th Nat. Conf. *Clays and Clay Miner.* Washington, 1958, 317-327.
Norrish, K., and R. M. Taylor, 1961. The isomorphous replacement of iron by aluminium in soil goethite. *J. Soil Sci.* 12 : 294-306.
Peterschmitt, E., 1991. Les couvertures ferrallitiques des ghâts occidentaux (Inde du Sud). Caractères généraux sur l'escarpement et dégradation par hydromorphie sur le revers. *Thèse Doc. Univ. Nancy I. France.*
Rajot, J. L., 1992. Dissolution des oxydes de fer (hématite et gœthite) d'un sol ferralitique des Llanos de Colombie par des bactéries ferri-réductrices. *Thèse Doc. Univ. Nancy I.*
Schwertmann, U., 1984. The influence of aluminium on iron oxides. IX. Dissolution of Al-goethites in 6M HCL. *Clay Miner.* 19 : 9-19.
Schwertmann, U. 1991. Solubility and dissolution of iron oxides. *Plant and Soil.* 130 : 1-25.
Schwertmann, U. and R.M. Cornell, 1991. Iron oxides in the laboratory. VCH, Weinheim, 137 p.
Schwertmann, U., and G. Pfab, 1994. Structural vanadium in synthetique gœthites. *Geochim. Cosmochim. Acta.* 58 : 4349-4352.
Schwertmann, U., Gasser, U. and H. Sticher, 1989. Chromium-for-iron substitution in synthetic goethites. *Geochim. Cosmochim. Acta.* 53 : 1293-1297.
Segal, R. G., and R. M. Sellers, 1984. Redox reactions at solid-liquid interfaces. *Adv. Inorg. Bioinorg. Mechanisms,* 3 : 97-130.
Singh, S. K. and V. Subramanian, 1984. Hydrous Fe and Mn oxides-scavengers of heavy metals in the aquatic environment. *Environ. Contr.* 14 : 33-90.
Tamm, O. ,1922. Eine Methode zur Bestimmung der organischen Komponenten des Gelkomplexes im Boden. *Medd. Statens. Skogsförsok.* 19 : 385-404.
Taylor, R.M. and U. Schwertmann, 1978. The influence of aluminium on iron oxides. I. The influence of Al on Fe oxide formation from the Fe(II) system. *Clays Clay Min.,* 26: 373-383.
Todd, J. F., Elsinger, N. C., and W. S. Moore, 1988. The distibution of uranium and thorium isotopes in two anoxic fjords : Framvaren fjord (Norway) and Saanich Inlet (British Columbia). *Mar. Chem.* 23 : 393-415.
Torrent, J., Schwertmann, U., and V. Barron, 1987. The reductive dissolution of synthetic gœthite and hematite in dithionite. *Clay Miner.* 22 : 329-337.
Trolard, F., Bourrie, G., Jeanroy, E., Herbillon, A. J., and H. Martin, 1995. Trace metals in natural iron oxides from laterites: A study using selective kinetic extraction. *Geochim. Cosmochim. Acta.* 59 : 1285-1297.

PREPARATION AND THERMODYNAMIC EQUILIBRIA OF GREEN RUSTS IN AQUEOUS SOLUTIONS AND THEIR IDENTIFICATION AS MINERALS IN HYDROMORPHIC SOILS

J.-M. R. GÉNIN[1], G. BOURRIÉ[2,3], PH. REFAIT[1], F. TROLARD[2], M. ABDELMOULA[1], B. HUMBERT[1] AND A. HERBILLON[4]

1. Laboratoire de Chimie Physique pour l'Environnement, UMR 9992 CNRS - Univ. Henri Poincaré-Nancy 1, 405 rue de Vandoeuvre, F-54600 Villers-lès-Nancy, France.
Tel. : (33) 83 91 63 00 . Fax. : (33) 83 27 54 44
2. INRA - U.R. de Science du Sol et de Bioclimatologie
65 rue de Saint Brieuc, F 35042 Rennes Cedex, France
3. Géosciences Rennes, UPR 4661 CNRS-Univ. de Rennes 1
Campus de Beaulieu, F 35042 Rennes Cedex, France
4. Centre de Pédologie Biologique, UPR 6831 CNRS- Univ. Henri Poincaré
17, rue Notre-Dame des Pauvres, F 54501 Vandoeuvre-lès-Nancy, France

I. INTRODUCTION

Many natural minerals are made by stacking positively and negatively charged layers. Among them, the pyroaurite-sjögrenite group (Allmann et Lohse, 1966; Allmann, 1968) corresponds to the stacking of brucite-like layers carrying a positive charge and layers constituted of anions and water molecules. In this paper, we deal with the Fe(II)-Fe(III) compounds generally known by the generic name of Green Rusts, GRs, since the original classification by Bernal *et al.* (1959) and the major anions which are found in natural waters, *i.e.* Cl^-, SO_4^{2-} and CO_3^{2-} are concerned. From X-ray diffraction (XRD), two types of GRs are distinguished, Green Rust one (GR1) incorporating "planar anions", *e.g.* Cl^- and CO_3^{2-}, and Green Rust two (GR2) incorporating "three dimensional anions". The preparation of these compounds in the laboratory by controlled oxidation of a ferrous hydroxide precipitate in anion - containing aqueous solutions allows us, by monitoring the E_h and pH versus time curves to elaborate them at stoichiometry with the appropriate $[Fe^{++}]$ / $[OH^-]$ concentration ratio. The X-ray diffraction characterization of these "Green Rusts", and the corresponding Mössbauer spectra allow us to determine structural information and chemical formulae.

II. SYNTHETIC GREEN RUSTS

A. Preparation

The oxidation process for preparing GRs is achieved by stirring in the air ferrous hydroxide precipitated by mixing ferrous salt «183 \f "Symbol" \s 12»and caustic soda solutions of various concentrations ; the initial proportions of reactants are expressed by the factor $R = \{[Fe^{2+}]/[OH^-]\}$. The ferrous salt is chosen to provide the appropriate anions. GRs are commonly obtained for values of *R* in the range of 0.55 to 1, where an excess of ferrous salt is allowed. The experimental device has the major advantage that a continuous measurement of the redox potential E_h and pH of the solution with time can be done with a recorder. Saturated calomel electrode is used as reference for both E_h and pH, and the E_h value is measured by means of a platinum electrode. Figure 1 presents the E_h and pH *vs* time curves in the case of Cl^- ions. The overall trend for E_h is to increase whereas that of pH is to decrease. Both curves display several stages : two plateaus A and B are distinguished and two inflection points T_g and T_f designate the first and second stage, respectively. T_g corresponds to the complete formation of GR and T_f to that of the final product, in this case γ-FeOOH. Therefore, the first plateau is related to the equilibrium between $Fe(OH)_2$ and GR and the second one to the equilibrium between GR and γ-FeOOH.

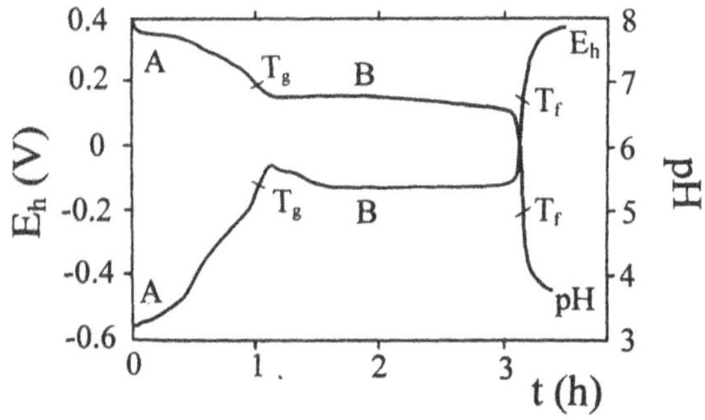

Figure 1. E_h and pH *vs* time curves obtained during the oxidation of $Fe(OH)_2$ in an aqueous solution with an excess of Cl^- ions at 25°C. E_h is referred to hydrogen standard electrode [Refait and Génin, 1993].

At a critical value of *R* a slight increase in the pH curve at the end of the first plateau disappears. The Mössbauer spectrum displays specificities of the peak intensities which infers the existence of a Fe(II)-Fe(III) stoichiometry. The formulae of synthetic GR's are :

GR1(Cl^-)	$[Fe^{II}_3 Fe^{III} (OH)_8]^+ \circ [Cl \circ nH_2O]^-$
GR2(SO_4^{2-})	$[Fe^{II}_4 Fe^{III}_2 (OH)_{12}]^{++} \circ [SO_4 \circ mH_2O]^{--}$
GR1(CO_3^{2-})	$[Fe^{II}_4 Fe^{III}_2 (OH)_{12}]^{++} \circ [CO_3 \circ pH_2O]^{--}$.

Departing from stoichiometry a general formula of GR is established ;

$$[Fe^{II}_{(1-x)}\ Fe^{III}_{x}\ (OH)_2]^{+x} \circ [\ x/n\ A^{-n} \circ m/n\ H_2O]^{-x}$$

where the positively charged hydroxide layers alternate with the layers constituted of anions A^{-n} and water molecules.

The stoichiometry determines the average degree of oxidation of iron for the various GRs, *e.g.* 2.25 for GR1(Cl^-) and 2.33 for GR2(SO_4^{2-}) and GR1(CO_3^{2-}). This difference is obviously connected to the fact that divalent anions are known to be preferred to monovalent ones by GRs and compounds of the pyroaurite group (Miyata, 1983).

B. Characterization

Three main methods are suitable to characterize GRs. The first one is X-ray diffraction (XRD). As pointed out, there exist two types of XRD patterns which correspond to *"Green Rust one"* (GR1) and *"Green Rust two"* (GR2). Most GRs are of the first type and there are no major differences among them. Thus XRD does not indicate which counter-anions lie in the interlayers of the structure. Some changes in lattice parameters, *e.g.* the *c* axis, can be detected from Cl^- to CO_3^{2-} ions, but departures from stoichiometry may also change lattice parameters. Moreover the small amount of GR and its dispersion in a mineral sample makes XRD irrelevant for characterisation in the field. Therefore we shall discuss the two other methods, Mössbauer and Raman spectroscopies.

Mössbauer spectroscopy is unmatchable to characterize GRs. Since it is only sensitive to Fe atoms ignoring all other phases which do not contain iron, a positive answer can be obtained. Moreover, using low temperature measurements, it is possible to stop any uncontrolled oxidation. Mössbauer spectra measured at 78 K of synthesised GRs *i.e.*GR1(Cl^-), GR2(SO_4^{2-}) and GR1(CO_3^{2-}) are displayed in Figure 2. Thus samples are precisely taken at the first inflection point, T_g, of the E_h or pH versus time curves in Figure 1. XRD analysis has revealed that the sample is constituted only of GR and no "final product" of oxidation, *e.g.*lepidocrocite or goethite is mixed with it. Lorentzian-shape lines are computer fitted (Table 1) with adjustable widths and identical Mössbauer f factor for all iron sites. Peak positions characterise each crystal iron site inside a solid phase and its degree of oxidation. Spectra display three quadrupole doublets D_1, D_2 and D_3 where D_1 and D_2 relate to two ferrous ion sites since isomer shift δ and quadrupole splitting ΔE_Q are about 1.3 and 2.9 mm s^{-1} respectively, and D_3 relates to a ferric ion site since δ and ΔE_Q are about 0.40 mm s^{-1}. Line intensities are proportional to the site abundances and vary from a GR to another (Table 1). In contrast, one cannot recognize one GR from another through the sole peak positions. The Fe(II) / Fe(III) ratio is precisely determined for stoichiometric conditions. The values of this ratio are 3, 2 and 2 for GR1(Cl^-), GR2(SO_4^{2-}) and GR1(CO_3^{2-}), respectively, a result consistent with the formula of each GR at stoichiometry as due to the charge balance between anion interlayers and ferric ions.

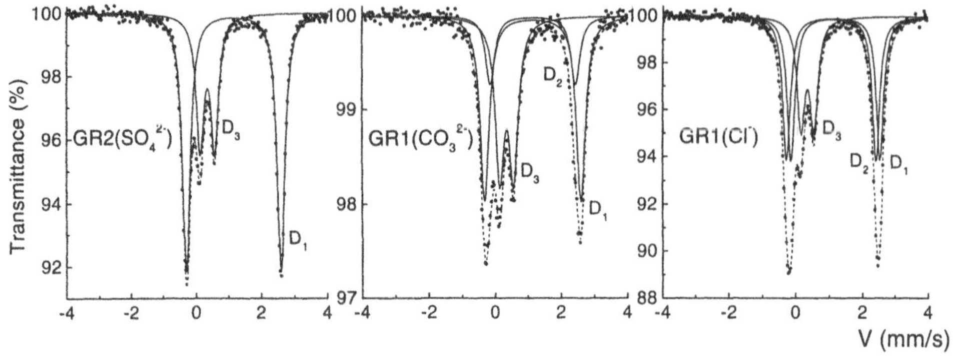

Figure 2. Mössbauer spectra measured at 78 K of GR2(SO_4^{2-}), GR1(CO_3^{2-}) and GR1(Cl^-).

Table 1 : Hyperfine parameters at 78 K of GR2(SO_4^{2-}), GR1(CO_3^{2-}) and GR1(Cl^-).

	GR2(SO_4^{2-})			GR1(CO_3^{2-})			GR1(Cl^-)		
Site	δ	ΔE_Q	*RA*	δ	ΔE_Q	*RA*	δ	ΔE_Q	*RA*
D_1	1.27	2.88	66	1.27	2.93	51	1.27	2.87	37
D_2	-	-	-	1.28	2.64	15	1.27	2.55	37
D_3	0.47	0.44	34	0.47	0.42	34	0.48	0.38	26

δ (mm s^{-1}) = isomer shift with respect to α-Fe, ΔE_Q (mm s^{-1}) = quadrupole splitting, RA (%) = relative abundance. Full widths at half maximum of Lorentzian-shaped lines are constrained to be equal and found to be 0.28 mm s^{-1} for all lines.

Raman spectroscopy was first used by Thierry *et al.* (1988) to characterize GRs in synthetic conditions and as corrosion products on steel (Boucherit *et al.* 1991). A special cell used for Raman experiments keeps the sample in an argon atmosphere without oxygen. The sample is set under a glove box and the cell is tightly closed by a glass window through which a laser beam is focused and the Raman backscattering collected. A 514.53 nm exciting radiation is used with a power lower than 1 mW to prevent sample degradation. The resolution is about 3 cm^{-1} and the precision on the wave number is 1 cm^{-1}. Two spectra are presented for calibration, one of GR1(CO_3^{2-}) and the other of GR1(Cl^-) (Fig. 3). Two bands are detected at 518 and 427 cm^{-1}. CO_3^{2-} anions are clearly detectable but one cannot ascertain that they are inside the GR phase. Thus, even though a GR is easily characterised, it is not possible to ascertain which type of anions it is related to. Raman effect is sensitive to the vibrational modes of the lattice as a consequence of the structure, not of the intercalated anion.

III. E_H-PH DIAGRAMS

The determination of thermodynamic data is of the utmost importance for equilibria. Moreover, these data can predict how iron can be thought of for water treatment and soil remediation. E_h and pH values for the various equilibria are measured from curves as those of

Figure 1. For instance, the values of the first plateau A correspond to the equilibrium between ferrous hydroxide and GR, whereas those of the second plateau B correspond to that between GR and the final oxyhydroxide. By varying the initial concentrations of reactants, *i.e.*, the initial ratio $R = [Fe^{++}] / [OH^-]$, the excess of anions, *e.g.*, Cl^-, SO_4^{2-} or CO_3^{2-} changes and the E_h *versus* pH variation corresponds to a straight line according to Nernst's law. Detailed explanation can be found in the original papers (Refait and Génin, 1993 ; Olowe and Génin, 1991 ; Drissi *et al.*, 1995 ; Génin *et al.*, 1996) and resulting E_h-pH diagrams are drawn accordingly. As an example, that of GR1(Cl^-) is presented in Figure 4 and Table 2. The domains of existence of GRs correspond to "triangles" between metallic Fe, $Fe(OH)_2$ and ferric oxyhydroxide. The allotropic form of FeOOH chosen in the diagram depends upon the form which is actually obtained experimentally, *e.g.* lepidocrocite for GR1(Cl^-). Several dissolved Fe species concentrations are displayed and the number *-n* on the lines means a molar concentration of 10^{-n}.

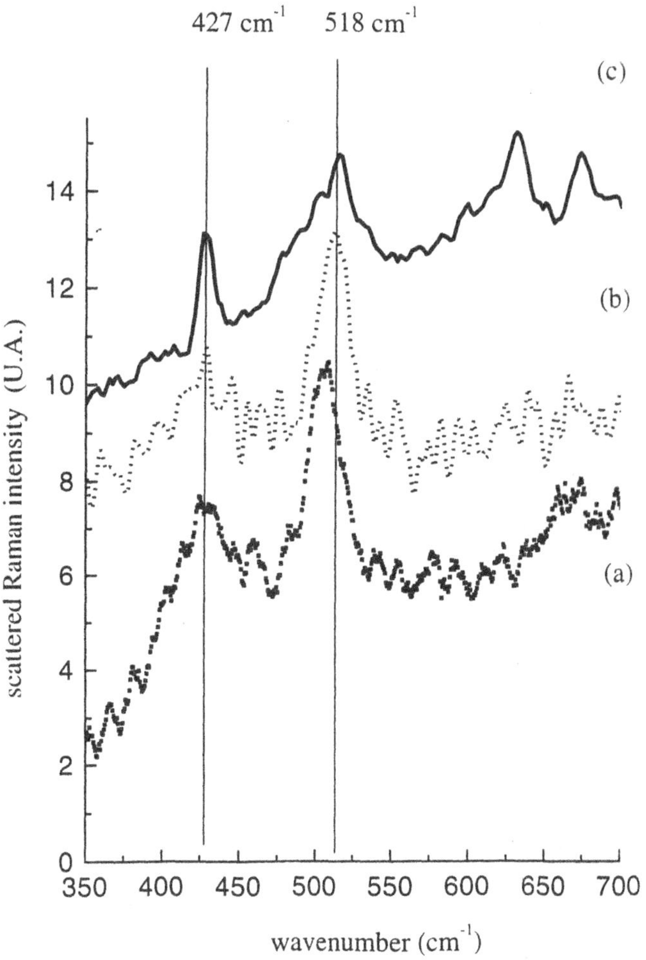

Figure 3. Raman spectra of two synthetic GR's: (a) GR1(CO_3^{2-}), (b) GR1(Cl^-) and (c) the mineral.

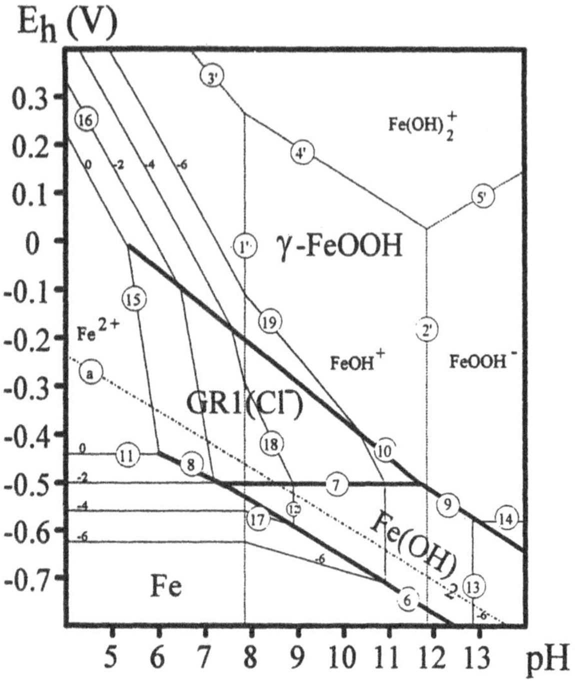

Figure 4. E_h *vs* pH diagram of GR1(Cl^-) [Refait and Génin, 1993]

Table 2 : Equilibrium Equations of E_h-pH Pourbaix Diagrams of GR1(Cl^-).

Water

(a) $H_2 \; 2\,H^+ + 2\,e^-$
$E_h = 0.000 - 0.0591$ pH

Iron species

(1) $Fe^{2+} + H_2O \; FeOH^+ + H^+$
$8.98 = \log[Fe^{2+}] - \log[FeOH^+] + pH$

(2) $FeOH^+ + H_2O \; FeOOH^- + 2\,H^+$
$12.10 = 0.5 \log[FeOH^+] - 0.5 \log[FeOOH^-] + pH$

(3) $Fe^{2+} + 2\,H_2O \; Fe(OH)_2{}^+ + 2\,H^+ + e^-$
$E_h = 1.324 - 0.0591 \log[Fe^{2+}] + 0.0591 \log[Fe(OH)_2{}^+] - 0.1182$ pH

(4) $FeOH^+ + H_2O \; Fe(OH)_2{}^+ + H^+ + e^-$
$E_h = 0.79 - 0.0591 \log[FeOH^+] + 0.0591 \log[Fe(OH)_2{}^+] - 0.0591$ pH

(5) $FeOOH^- + H^+ \; Fe(OH)_2{}^+ + e^-$
$E_h = -0.638 - 0.0591 \log[FeOOH^-] + 0.0591\,pH + 0.0591 \log[Fe(OH)_2{}^+]$

(6) $Fe + 2\,H_2O \; Fe(OH)_2 + 2\,H^+ + 2\,e^-$
$E_h = -0.080 - 0.0591$ pH

(7) *Chloride-containing media*
$4\,Fe(OH)_2 + Cl^- \; Fe_4(OH)_8Cl + e^-$
$E_h = -0.57 - 0.0591 \log[Cl^-]$

(8) *Chloride-containing media*

$4\ Fe + Cl^- + 8\ H_2O\ \ Fe_4(OH)_8Cl + 8\ H^+ + 9\ e^-$

$E_h = -0.13 - 0.0066 \log[Cl^-] - 0.0525pH$

(9) $Fe(OH)_2\ \ FeOOH + H^+ + e^-$

$E_h = E° - 0.0591$ pH, with $E° = 0.097$ V and 0.197 V for α- and γ-FeOOH, respectively

(10) *Chloride-containing media*

$Fe_4(OH)_8Cl\ \ 4\text{-}FeOOH + Cl^- + 4\ H^+ + 3\ e^-$

$E_h = 0.45 + 0.0197 \log[Cl^-] - 0.0788$ pH

(11) $Fe\ \ Fe^{2+} + 2\ e^-$

$E_h = -0.474 + 0.0296 \log[Fe^{2+}]$

(12) $Fe^{2+} + 2\ H_2O\ \ Fe(OH)_2 + 2\ H^+$

$13.33 = \log[Fe^{2+}] + 2$ pH

(13) $FeOH^+ + H_2O\ \ Fe(OH)_2 + H^+$

$4.35 = \log[FeOH^+] + pH$

(14) $Fe(OH)_2\ \ FeOOH^- + H^+$

$19.85 = -\log[FeOOH^-] + pH$

(15) $FeOOH^-\ \ FeOOH + e^-$

$E_h = E° - 0.0591 \log[FeOOH^-]$, with $E° = -1.08$ V and -0.98 V for α- and γ-FeOOH, respectively.

(16) *Chloride-containing media*

$4\ Fe^{2+} + Cl^- + 8\ H_2O\ \ Fe_4(OH)_8Cl + 8\ H^+ + e^-$

$E_h = 2.58 - 0.2364 \log[Fe^{2+}] - 0.0591 \log[Cl^-] - 0.4728$ pH

(17) $Fe^{2+} + 2\ H_2O\ \ FeOOH + 3\ H^+ + e^-$

$E_h = E° - 0.0591 \log[Fe^{2+}] - 0.1773$ pH, with $E° = 0.89$ V and 0.99 V for α- and γ-FeOOH, respectively.

(18) $Fe + H_2O\ \ FeOH^+ + H^+ + 2\ e^-$

$E_h = -0.21 + 0.0296 \log[FeOH^+] - 0.0296$ pH

(19) *Chloride-containing media*

$4\ FeOH^+ + Cl^- + 4\ H_2O\ \ Fe_4(OH)_8Cl + 4\ H^+ + e^-$

$E_h = 0.46 - 0.2364 \log[FeOH^+] - 0.0591 \log[Cl^-] - 0.2364$ pH

(20) $FeOH^+ + H_2O\ \ FeOOH + 2\ H^+ + e^-$

$E_h = E° - 0.0591 \log[FeOH^+] - 0.1182$ pH, with $E° = 0.35$ V and 0.45 V for α- and γ-FeOOH, respectively.

From the reactions, the lines in the E_h-pH diagrams are computed by using the standard chemical potentials or standard free enthalpy of formation of the various compounds. The determination of chemical potentials of GRs is easy and since there are more equations than variables it is possible to check the reliability of the determined values. They are compiled in Table 3. The values of $\mu°$(GR) are given for anhydrous GRs since the number of intercalated water molecules is not ascertained. As explained elsewhere (Refait and Génin, 1994), this does not matter since the water molecules do not intervene in the redox reactions.

Table 3 : Thermodynamic constants used for calculations.

Species	Average oxidation n° of Fe	$\Delta G°_f$ (cal mol^{-1})	Ref
Solid			
Fe	0	0	
$Fe(OH)_2$	+2	-116,300	(Wagman *et al.*, 1982)
$Fe^{II}_3Fe^{III}(OH)_8Cl$	+2.25	-509,500	(Refait and Génin, 1993)
$Fe^{II}_4Fe^{III}_2(OH)_{12}CO_3$	+2.33	-852,900	(Drissi *et al.*, 1995)
$Fe^{II}_4Fe^{III}_2(OH)_{12}SO_4$	+2.33	-901,100	(Génin *et al.*, 1996)
γ-FeOOH	+3	-112,100	(Schwartz, 1972)
α-FeOOH	+3	-114,800	(Schwartz, 1972)
Liquid			
H_2O		-56,690	(Wagman *et al.*, 1982)
Dissolved			
Cl^-		-31,350	(Bard *et al.*, 1985)
CO_3^{2-}		-126,170	(NBS, 1952)
HCO_3^-		-140,260	(NBS, 1952)
H_2CO_3		-148,940	(NBS, 1952)
SO_4^{2-}		-177,929	(Pourbaix *et al.*, 1990)
Fe^{2+}	+2	-20,300	(Pourbaix, 1966)
$FeOH^+$	+2	-66,300	(Wagman *et al.*, 1982)
$FeOOH^-$	+2	-90,627	(Pourbaix, 1966)
$Fe(OH)_2^+$	+3	-106,200	(Pourbaix, 1966)

IV. FOUGERITE, A NATURAL MINERAL

GRs have been already synthesized or found as corrosion products of steel in a natural environment. We have recently demonstrated from field evidence that GRs exist in nature as a mineral in hydromorphic soils in Brittany (France) at Fougères under forest (Trolard *et al.*, 1996; 1997). Here, additional data are given from the same and two other sites : at Quintin under grassland and at Naizin under intensive farming. The profile at Fougères consists of (i) a histic horizon from 0 to 15 cm, (ii) a granitic saprolite which presents a gleyed horizon from 15 to 70 cm overlying an oxidized horizon with accumulation of Fe(III) oxides. Large soil samples were taken in the gleyed horizon at 20 cm depth and kept in anoxic conditions in a container with the local ground water solutions and analysed by Mössbauer and Raman spectroscopies. Once exposed to the air, the color changes rapidly from a homogeneous bluish-grey color (5BG 6/1 Munsell Soil Color Chart) to greenish-grey (5GY 6/1) after about 10 minutes, then to pale olive (5Y 6/4) after one hour and eventually turns grey with light olive scattered spots (2.5Y 5/6) after one day of exposure. These color changes were already observed by Girard and Chaudron (1935) and Schwertmann and Fechter (1994) during the transformation of GRs into lepidocrocite. Such colours and colour changes are diagnosis features of gleyed horizons in Soil Classification Systems (Avery, 1973; Duchaufour *et al.*, 1976). Similarly gleyed saprolites were sampled in hydromorphic soils at Quintin on granite at 150 cm depth in a mid-slope and at Naizin on shales at 50 cm in the alluvial zone.

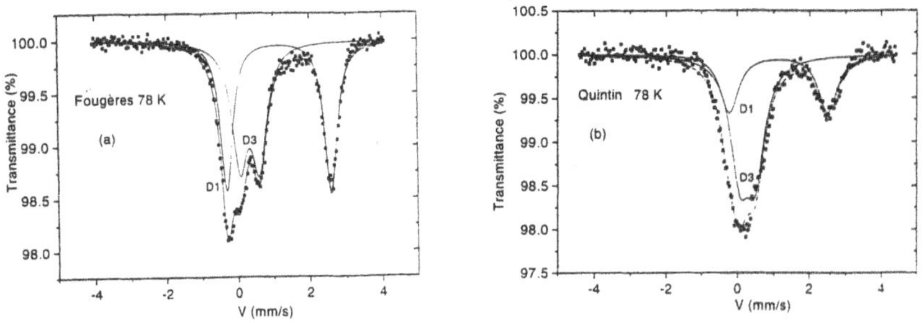

Figure 5. Mössbauer spectra measured at 78 K of the mineral denominated "fougerite" extracted from (a) Fougères and (b) Quintin.

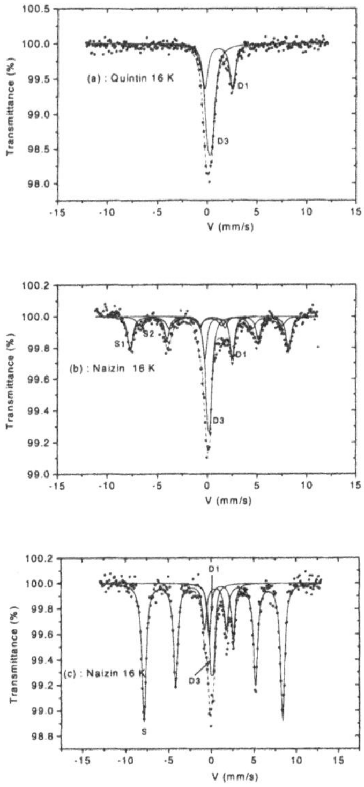

Figure 6. Mössbauer spectra measured at 16 K of the mineral denominated "fougerite" extracted from (a) Quintin and (b & c) Naizin.

Table 4. Hyperfine Parameters of soil samples measured at 78 K or 16 K and extracted from Fougères, Quintin or Naizin in Brittany (France).

		δ	ΔE_Q	H	W	*RA (%)*
78 K						
Fougères	D_1	1.25	2.87	-	0.41	51
	D_3	0.45	0.54	-	0.51	49
Quintin	D_1	1.26	2.71	-	0.62	36
	D_3	0.31	0.42	-	0.62	64
16 K						
Quintin	D_1	1.3	2.74	-	0.7	36
	D_3	0.42	0.41	-	0.7	64
Naizin 1	D_1	1.31	2.82	-	0.5	13
	D_3	0.26	0.33	-	0.5	14
	S_1	0.47	-0.26	503	0.6	73
Naizin 2	D_1	1.31	2.65	-	0.67	21
	D_3	0.27	0.28	-	0.67	33
	S_1	0.46	-0.26	491	0.67	36
	S_2	0.40	0.05	441	1.6	10

δ (mm s^{-1}) = isomer shift taking α iron as a reference, ΔE_Q (mm s^{-1}) = quadrupole splitting, H (kOe) = hyperfine field, W (mm s^{-1}) = full width at half maximum, RA (%) = relative abundance.

The Mössbauer spectra measured at 78 K or 16 K (Fig. 5 and 6 and Table 4) display in all cases the same characteristic quadrupole ferrous (isomer shift δ = 1.3 mm s^{-1}, quadrupole splitting ΔE_Q = 2.9 mm s^{-1}) and ferric (δ = 0.5 mm s^{-1}, ΔE_Q = 0.5 mm s^{-1}) doublets as synthetic GRs. The mole fraction ratio x = Fe(III) / $Fe_{tot.}$ observed ranges from 1/3 to 1/2 at Fougères, 1/2 to 2/3 at Naizin and is equal to 2/3 at Quintin. In this latter case, the presence of lepidocrocite is observed too. By microprobe Raman spectrometry, spectra obtained on fine particles of about 5 mm (Fig. 3c) show the same characteristic bands at 518 and 427 cm^{-1} as the synthetic GRs (Boucherit *et al.*, 1991), suggesting that the identification of a GR as a natural mineral is definitively established in various natural environments.

From the above spectroscopic data, the exact nature of the interlayer anion cannot be specified. The soil solution analyses show that chloride, sulphate and carbonate activities are very small : log [Cl^-] ≅ -3 ; log [SO_4^{2-}] @ -4 ; log [CO_3^{2-}] @ -8, so that soil solutions are largely undersaturated with respect to chloride, sulphate and carbonate GRs. We suggest that the natural GR incorporates OH^- anions. Such a solid phase was previously envisioned by Arden (1950) and Ponnamperuma *et al.* (1967) under the name of "ferroso-ferric hydroxide", and its bulk formula was written as $Fe_3(OH)_8$. It was considered to control Fe(II) in soil solution by Ponnamperuma *et al.* (1967), Lindsay (1979), Maître (1991) and Bourrié and Maître (1994). We assume that it is the GR1(OH^-), *i.e.* the Fe(II)-Fe(III) hydroxide with pyroaurite-like structure, the general formula being $[Fe^{II}{}_{1-x}Fe^{III}{}_x(OH)_2]^{+x}.[(OH^-)_x]^{-x}$ ^{o} $Fe(OH)_{2+x}$, x ranging from 1/3 to 2/3. The name "fougerite" is proposed, from the name of the city of Fougères (Brittany, France), and submitted to the International Mineralogy Association.

V. Conclusion

Synthetic GRs are prepared by oxidation of $Fe(OH)_2$ incorporating Cl^-, SO_4^{2-} or CO_3^{2-} anions. The existence of GRs can be predicted from the E_h-pH diagrams drawn previously. A GR incorporating OH^- ions is suspected to exist even though this is not confirmed by synthesis, since a Ni(II)-Fe(III) GR incorporating OH^- ions has been effectively observed (Refait *et al.*, 1994). Mössbauer and Raman spectroscopies characterize a GR in the samples extracted from hydromorphic soils. The mineral we propose to name "fougerite", has a Fe(III)/ $Fe_{tot.}$ ratio which varies from 1/3 to 2/3. Assuming that the GR "fougerite" is *GR1(OH^-)*, since other anions are not present, the formula $[Fe^{II}{}_{1-x}Fe^{III}{}_x(OH)_2]^{+x}.[(OH^-)_x]^{-x}$ ^{o} $Fe(OH)_{2+x}$, is proposed, *i.e.*, a Fe(II)-Fe(III) hydroxide. The aim is now to find the mechanism by which "fougerite" forms, in particular in view of the suspected role of bacteria in reductomorphic soils.

References

ALLMANN, R. and H.H. LOHSE, 1966. Die Kristallstruktur des Sjögrenits und eines Umwandlungsproduktes des Koenenits. *Neues Jb. Min., Monatsh.* 161-181.

ALLMAN, R. 1968. The crystal structure of pyroaurite. *Acta Cryst.* **B24**, 972-977.

ARDEN, T.V., 1950. The solubility products of ferrous and ferrosic hydroxides. *J. Chem. Soc.* **24**, 882-885.

AVERY, B.W., 1973. Soil classification in the Soil Survey of England and Wales. *J. Soil Sci.* **24**, 324-338.

BARD, A.J., R. PARSONS and J. JORDAN, Editors, 1985. *Standard potentials in aqueous solution*, Marcel Dekker Inc., New York.

BERNAL J.O., D.R. DASGUPTA and MACKAY A.L., 1959. The oxides and hydroxides of iron and their structural inter-relationships. *Clay Min. Bull.* **4**, 15-30.

BOUCHERIT N., A. HUGOT LE GOFF and S. JOIRET, 1991. Raman studies of corrosion films grown on Fe and Fe-6Mo in pitting conditions. *Corros. Sci.* **32**, 497-507.

BOURRIÉ G and V. MAÎTRE, 1994. Iron control in solutions from hydromorphic soils under temperate climate by equilibrium with a mixed Fe(II) - Fe(III) mineral, ferroso-ferric hydroxide. *15th International Congress of Soil Science*, Acapulco, Mexico, 143-144.

DRISSI H., PH. REFAIT, M. ABDELMOULA and J.-M.R. GÉNIN,1995. Preparation and thermodynamic properties of Fe(II)-Fe(III) hydroxide-carbonate (green rust one), Pourbaix diagram of iron in carbonate-containing aqueous media. *Corros. Sci.* **37**, 2025-2041.

DUCHAUFOUR PH., P. FAIVRE and M. GURY, 1976. *Atlas écologique des sols du monde,* Masson, Paris, 178.

GÉNIN J.-M.R., A.A. OLOWE, B. RÉSIAK, N.D. BENBOUZID-ROLLET, M. CONFENTE and D. PRIEUR, 1993. Identification of sulphated green rust two compound produced as a result of microbially induced

corrosion of steel sheet piles in a harbour. In *Marine corrosion of stainless steels : chlorination, and microbial effects,* European Federation of Corrosion by the Institute of Materials, London, **10**, 162-166.
GÉNIN J.-M.R., A.A. OLOWE, PH. REFAIT and L. SIMON, 1996. On the stoichiometry and Pourbaix diagram of Fe(II)-Fe(III) hydroxy-sulphate or sulphate-containing green rust 2: an electrochemical and Mössbauer spectroscopy study. *Corros. Sci.* **38**, 1751-1762.
GIRARD A. and G. CHAUDRON, 1935. Sur la constitution de la rouille. *Compt. Rend. Acad. Sci.*, **200**, 127-129.
INGRAM L. and H.F.W. TAYLOR, 1967. The crystal structures of sjögrenite and pyroaurite. *Min. Mag.* **36**, 465-479.
LINDSAY W.L., 1979. *Chemical equilibria in soils,* Wiley Interscience, New York, 449.
MAITRE V., 1991. *Géochimie des eaux libres extraites de sols hydromorphes sur granite dans le Massif Armoricain . Mobilité du fer et dynamique saisonnière.* Ph. D. thesis, Univ. Paris 6, INRA Rennes.
MIYATA S., 1983. Anion-exchange properties of hydrotalcite-like compounds. *Clays Clay Miner.* **31**, 305-311.
NBS, 1952. *Selected values of thermodynamic properties*, Circular of National Bureau of Standards 500, US Government Printing Office, Washington DC.
OLOWE A.A. and J.-M.R. GÉNIN, 1989. Potential-pH equilibrium diagrams for the iron-water system in the presence of sulphate ions : Domain of existence of green rust 2. *Proc. Int. Symp. Corros. Sci. Engng.*, CEBELCOR RT297, 363-390.
OLOWE A.A.and J.-M.R. GÉNIN, 1991. Mechanism of oxidation of ferrous hydroxide in sulphated aqueous media : importance of the initial ratio of the reactants. *Corros. Sci.* **32** 965-984.
PONNAMPERUMA F.N., E.M. TIANCO and T. LOY, 1967. Redox equilibria in flooded soils : I. The iron hydroxide system. *Soil Sci.* **103**, 374-382.
POURBAIX M., 1966. *Atlas of electrochemical equilibria in aqueous solutions*, Pergamon Press Ltd., Oxford.
POURBAIX M. and A. POURBAIX, 1990. Potential-pH diagrams for the system S-H_2O, from 25 to 150°C. Influence of access of oxygen in sulphide solutions. *Rapports Techniques CEBELCOR* **159**, RT299.
REFAIT PH. and J.-M.R. GÉNIN, 1993. The oxidation of ferrous hydroxide in chloride-containing aqueous media and Pourbaix diagrams of green rust one. *Corros. Sci.* **34**, 797-819.
REFAIT PH. and J.-M.R. GÉNIN, 1994. The transformation of chloride-containing green rust one into sulphated green rust two by oxidation in mixed Cl^- and SO_4^{2-} aqueous media. *Corros. Sci.* **36**, 55-65.
REFAIT PH., H. DRISSI, Y.MARIE and J.-M.R. GÉNIN, 1994. The substitution of Fe^{2+} ions by Ni^{2+} ions in green rust one compounds. *Hyp. Int.* **90**, 389-394.
SCHWARTZ H., 1972. Über die Wirkung des magnetits beim atmosphärischen Rosten und beim Unterrosten von Anstrichen. *Werkst. Korros.* **23**, 648-663.
SCHWERTMANN U. and H. FECHTER, 1994. The formation of green rust and its transformation to lepidocrocite. *Clay Miner.* **29**, 87-92.
THIERRY D., D. PRESSON, C. LEYGRAF, D. DELICHÈRE, S. JOIRET, C. PALLOTTA and A. HUGOT LE GOFF, 1988. *In situ* Raman spectroscopy combined with X-ray photoelectron spectroscopy and nuclear microanalysis for studies of anodic corrosion film formation on Fe-Cr single crystals. *J. Electrochem. Soc.* **135**, 305-310.
TROLARD F., M. ABDELMOULA, G. BOURRIÉ, B. HUMBERT and J.-M. R. GÉNIN, 1996. Evidence of the occurrence of a "Green Rust" component in hydromorphic soils. Proposition of the existence of a new mineral : "fougerite". *Compt. Rend. Acad. Sci.*, **323 IIA**, 1015-1022.
TROLARD F., J.-M.R. GÉNIN, M. ABDELMOULA, G. BOURRIÉ, B. HUMBERT and A.HERBILLON, 1997. Identification of a green rust mineral in a reductomorphic soil by Mössbauer and Raman spectroscopies. *Geochim. et Cosmochim. Acta* **61**, 1107-1111.
WAGMAN D.D., W.H. EVANS, V.B. PARKER, R.H. SCHUMM, I. HALOW, S.M. BAILEY, K.L. CHURNEY and R.L. NUTALL, 1982. The NBS tables of chemical thermodynamic properties. Selected values for inorganic and C_1 and C_2 organic substances in SI units. *J. Phys. Chem. Ref. Data* **11** (suppl. 2).

AN XPS AND AFM COUPLED STUDY OF AIR AND BIO-OXIDIZED PYRITE SURFACES

Toniazzo V.[1], Mustin C.[1], Vayer-Besancon M.[2], Erre R.[2] and Berthelin J.[1]

[1] Centre de Pédologie Biologique, UPR 6831 du CNRS
B.P.5, 54501 Vandœuvre-les-Nancy Cedex, France
[2] Centre de Recherche sur la Matière Divisée UMR 131 du CNRS
45071 Orléans Cedex 02

1. INTRODUCTION

In most sulfide ore deposits, the main gangue mineral is pyrite, which is often considered a nuisance in mining enterprises and natural environments due to acid mine drainage and soil and water pollution by heavy metals. Many chemical reactions, including natural metal leaching and industrial metal recovery, involve bacteria: sulfide oxidation by *Thiobacillus ferrooxidans* is a typical example. Frequently, bio-oxidation of sulfides is described in terms of pure biological and chemical processes that do not take into account the variability of mineral properties. In fact, biooxidation processes are complex and strongly dependent on the crystallographic and electrochemical characteristics of the mineral, (MARION *et al.,* 1991 ; MUSTIN *et al.,* 1992a). These physical properties also seem to govern both development of corrosion pores and oxidized phases at the sulfide surface (MUSTIN *et al.,* 1993a, MONROY-FERNANDEZ *et al.* 1995). The physico-chemical modification of the mineral interface, more often neglected, could be of primary importance for the biooxidation of sulfides.

Several authors (BRION *et al.,* 1980, DE DONATO *et al.,* 1993, NESBITT and MUIR 1994, SASAKI *et al.,* 1995) have demonstrated the presence of various amounts of oxidized iron and sulfur species ($FeSO_4$, $Fe_2(SO_4)_3$, FeOOH, S_8) at the pyrite surface by combining different qualitative spectroscopies (x-ray photoelectron spectroscopy, raman and infrared) and quantitative methods (chromatographic desorption). These oxidized compounds are thought to form non-continuous structures with various vertical extensions depending on their nature (DE DONATO *et al.,* 1993). However, this hypothesis has not been clearly demonstrated.

The quantification of these surface compounds during bioleaching of pyrite by *Thiobacillus ferrooxidans* has also been investigated (MUSTIN *et al.,* 1993a ; 1993b ; DE DONATO *et al.,* 1991). According to these previous studies, the superficial oxidized species (mainly sulfates and elemental sulfur) seems to be fundamental for bacterial oxidation of pyrite, and variation in their amount is a good tracer to follow a bioleaching cycle step-by-step (MUSTIN *et al.*, 1993c). Oxidized species could also be considered a bioavailable substrate for *Thiobacillus*; therefore the following global equations (Eq. 1 and Eq. 2), generally accepted for direct mineral biooxidation, must be discussed :

$$\text{(Eq. 1)} \qquad 2FeS_2 + 7.5\,O_2 + H_2O \xrightarrow{bact} 2Fe^{3+} + 4\,SO_4^{2-} + 2\,H^+$$

$$\text{(Eq. 2)} \qquad FeS_2 + Fe_2(SO_4)_3 \xrightarrow{bact} 3FeSO_4 + 2S^\circ$$

Unfortunately, techniques (x-ray photoelectron spectroscopy, raman and infrared spectroscopies, selective desorption...) already used to characterize the pyrite interface give global quantitative and qualitative information. Thus, it is essential to obtain precise topographic data of the bioleached pyrite surfaces in order to relate locally the setting of oxidized species, corrosion patterns and bacterial adsorption sites. Recently, near-field techniques such as atomic force microscopy (AFM) have been extensively developped to investigate mineral surfaces (LAMONTAGNE *et al.*, 1995). However, previous studies were often limited to simple observations.

The aim of this study is then to use AFM to complement XPS and compare the spatial distribution of oxidized species present on air-oxidized pyrite and on pyrite bioleached by *Thiobacillus ferrooxidans*. A new methodology is presented to gain the best advantage from the images and to relate these microscopic data to physico-chemical characteristics of the minerals.

This present work is the first step in a research project concerning the synergetic effect of surface mineral properties and bacteria on mineral weathering.

2. MATERIAL AND METHODS

2.1. Sulfide

The pure pyrite used for this study comes from a hydrothermal Peruvian deposit and contains minor impurities (less then 0.5%). Flakes of pyrite of about 2x2 mm^2 were selected and analyzed by x-ray photoelectron spectroscopy after natural oxidation in air or after 45 days of bioleaching by *Thiobacillus ferroxidans*.

2.2. Bacteria and nutrient medium

The chemiolithotrophic, Gram negative acidophilic bacteria *Thiobacillus ferrooxidans* was provided by the Deutsche Sammlung von Mikroorganismen (DSM 583). The nutrient medium used was made of 0.4 g KH_2PO_4, 1 g $(NH_4)_2SO_4$ and 0.4 g $MgSO_4.7H_2O$ per liter adjusted to pH 1.8 with 50 ml H_2SO_4 (1N). After sterilization, a pyrite fraction of 32-53 µm was added to obtain a 2% weight pulp. Then the reactor containing 200 ml of pulp was inoculated with an actively growing culture to have an initial population of 2×10^7 bacteria per ml. Several flakes of pyritewere enclosed in a polyamide woven bag (pore size of 20 µm to allow contact with the bacteria) and put in the reactor for bioleaching. The flakes are analysed by XPS and AFM after 45 days and compared with other flakes chemically oxidized by atmospheric oxygen (ambient wet air).

2.3. X-ray electron spectroscopy measurements

XPS measurements were carried out with a ESCALAB Mk II spectrometer (VG instruments) using $AlK\alpha_{1/2}$ excitation at an analyzer chamber pressure of about 10^{-7} Pa. The C(1s) core level of saturated hydrocarbons at 284.6 eV was used for energy calibration. Using these experimental conditions, the precision was ± 0.2 eV on the energy values and ± 0.6 eV on area ratios.

After a non-linear background subtraction, the line shapes of individual peaks were fitted to Gaussian spectral line functions using a non-linear least-squares fitting method. Several fundamental core levels such as C(1s), O(1s) and S(2p) were curve-fitted using this procedure. The fundamental $Fe(2p)_{3/2}$ core level is very sensitive to the effect of the chemical surroundings and allows a good determination of the degradation products. Thus, it was fitted using the method of Doniach and Sunjic (1976), which takes the asymmetry of the peak into account.

2.4. Atomic Force Microscopy

AFM images were obtained on a nanoscope (Park Scientific Instrument) operating in air and in contact mode with a Si_3N_4 cantilever tip and a laser beam deflection system. Naturally oxidized and bioleached flakes of pyrite were examinated several times (>10) over different areas (5 x 5 μm^2) to provide a statistical representation of the whole surface.

3. RESULTS AND DISCUSSION

On the assumption that oxidized superficial compounds are fundamental for the bioleaching processes of pyrite by *Thiobacillus ferrooxidans* (and probably good substrates for bacterial growth), it is important to compare the nature of iron and sulfur oxidized species present on the mineral surface after natural oxidation in air and after biooxidation. Table 1 summarizes the main results concerning the $S(2p)_{3/2}$ and $Fe(2p)_{3/2}$ core levels obtained on pyrite flakes using x-ray photoelectron spectroscopy.

Table 1 : Comparison of the main XPS results concerning naturally air oxidized and bio-oxidized flakes of pyrite.

	Air oxidized surface		Biooxidized surface		
Core Level	**BE**	**% Peak**	**BE**	**% Peak**	**Assignment**
$Fe(2p)_{3/2}$	**707.7**	**24**	**707**	**40**	**FeS_2**
	710.4	**53**	**709.5**	**25**	**Fe Oxides / Oxihydroxides**
	713	**23**	**711.8**	**35**	**Ferric Sulfate**
$S(2p)_{3/2}$	**162.1**	**49**	**162.4**	**27**	**Disulfide $S\text{-}S^{2-}$**
	167.7	**36**	**169**	**28**	**Ferric Sulfate**
	163.9	**15**	**163.5**	**45**	**Elemental sulfur S_8**

BE : Binding energy of core level (eV), calibrated with the C(1s) orbital of saturated hydrocarbons at 284.6 eV
% Peak : Part of the considered species in the total peak area of $Fe(2p)_{3/2}$ or S(2p).

On chemically air-oxidized pyrite, the $Fe(2p)_{3/2}$ core level decomposes into three components. The peak centered on 707.7 eV corresponds to pure pyrite (FeS_2); the peak at 710.4 eV is due to ferric oxidized species, probably ferric oxihydroxide (FeOOH) or oxides (Fe_3O_4 or Fe_2O_3. The contribution of ferric sulfates is detected at 713 eV. The study of the S(2p) core level confirms the presence of sulfates (peak at 167.7 eV), but it suggests also a minor contribution of elemental sulfur ($S(2p)_{3/2}$, BE = 163.9 eV).

The results concerning the decomposition of the $Fe(2p)_{3/2}$ and S(2p) core levels for bioleached pyrite are quite identical. Besides the signal of pure pyrite ($Fe(2p)_{3/2}$, BE = 707 eV ; $S(2p)_{3/2}$, BE = 162.4 eV), various oxidized compounds are detected: ferric oxides and oxihydroxides ($Fe(2p)_{3/2}$, BE = 709.5 eV), ferric sulfates ($Fe(2p)_{3/2}$, BE = 711.8 eV ; $S(2p)_{3/2}$, BE = 169 eV) or elemental sulfur ($S(2p)_{3/2}$, BE = 163.5 eV). No major differences concerning the nature of oxidized compounds on natural air-oxidized pyrite and on biooxidized pyrite, are detected by XPS measurements. Both are partially covered by similar oxidized species. However, shifts (of about 1 eV) observed for the peaks of ferric oxi-hydroxi-species and ferric sulfates in the $Fe(2p)_{3/2}$ core level of biooxidized pyrite could indicate that the chemical surroundings of iron are quite different for the two methods. Moreover, variation in the contribution of oxidized species to the global signal of iron (represented by the part of the peaks areas between 709.5 and 713 eV in the total peak area of $Fe(2p)_{3/2}$) for the two samples suggests some differences in the structure, the relative abundance or in the distribution of superficial phases at the interface.

In this way, AFM measurements have allowed more precise study of the distribution of these oxidized phases on chemically oxidized pyrite surfaces and have given us an idea of their geometric characteristics (shape, volume, ...).

The same AFM investigations were then performed on pyrite oxidized by *Thiobacillus ferrooxidans* to compare the effect of bio-oxidation and chemical oxidation on the surface morphology. For this, pyrite flakes exposed to biological attack in the little woven bag were removed from the bioleaching reactor after 45 days and analyzed by AFM.

The first observation is that all the surfaces are highly heterogeneous (Figure 1). Various structures are observed: well-rounded or angular bumps, "islands" (bumps with a large horizontal extension), holes (on bioleaching samples), smooth surfaces, etc. Moreover, one flake can present zonation with high structural diversity. As represented in examples in figure 1, some pictures show very smooth and regular surfaces with only a few little "bumps", while some others seem to be covered by a lot of piles of different vertical and horizontal extension.

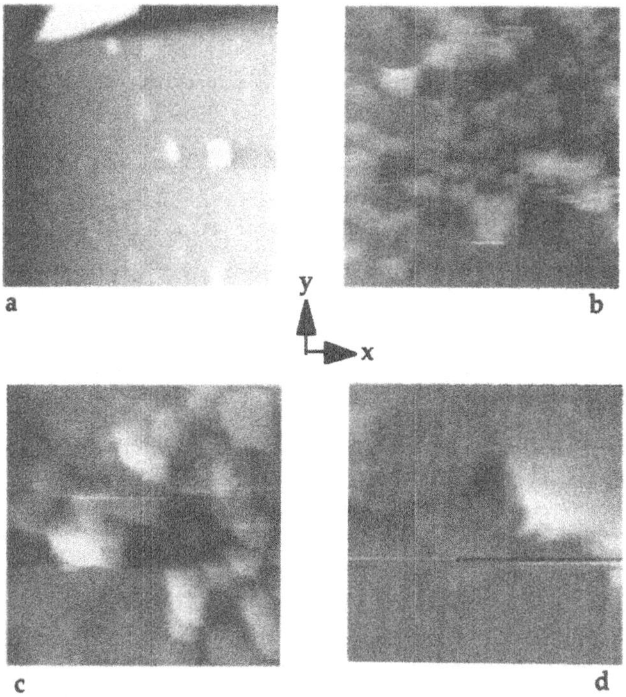

Figure 1 : AFM pictures (5 mm x 5 mm) of pyrite surfaces
a, b : two areas of the same chemically air oxidized pyrite flake
c, d : two examples of bioleached pyrite flakes
Gray levels(256) indicate the height of points : $scale_Z$ = 1500 Å (a) , $scale_Z$ = 5000Å (b,c,d)

However, the simple observation of the AFM images is not sufficient to differentiate natural oxidized surfaces from bioleached ones. Therefore, a new methodology to gain the best advantage of these pictures is necessary. This method uses geostatistical tools, statistical means analyzing geological data, such as, for example, metal abundance in mineral deposits. Let us consider an AFM picture as a regular and a two dimensional set of data points. In each point of the picture, the height z , expressed in gray levels or after correction in micrometers, could be defined as the value of a geostatistical variable (Z). If a step (h) and a direction in the xy plane are set, the variogram g_Z can be calculated at each point as follows:

$$2\gamma_Z(x,h) = E[\ \{ Z(x+h) - Z(x) \}^2\ ]$$

Then the plot of the variance (expressed in μm^2) versus an increasing step h (expressed in μm) gives the variogram of the Z variable in the chosen direction (Figure 2).
On pyrite images, the maximal variance g_{max} is representative of the heterogeneity of the interface and related to the maximal height of piles, while the range (maximum distance for points to be correlated) represents their mean horizontal extension.

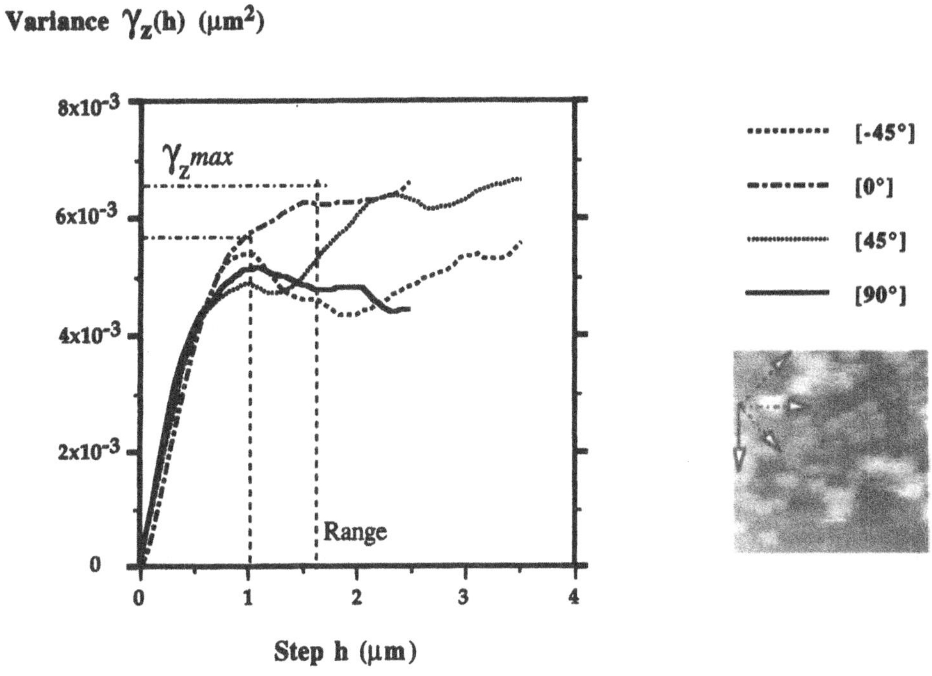

Figure 2: Example of variograms in four directions (-45, 0, 45, 90°) for the picture Fig. 1b

For all analyzed pictures, as shown by Figure 2, the first important result is the greater range in variograms for the x direction than for the ranges in the other directions. Because the x direction is the scanning one, this result puts the emphasis on the possible modifications of the surface due to the tip scan. Therefore, the piles' asymmetry is artificially increased. This problem and other difficulties concerning image analyses (noise, filters, sloping samples) will be discussed later, and methods to overcome these difficulties have already been established in the laboratory.

Another interesting point is related to the fluctuations at the variogram plate. They seem to indicate the existence of periodic structures at the surface of pyrite. Methods to reduce the noise of the images and to calculate the correct value of the periods in the different directions are also developed elsewhere.

In order to compare air-oxidized pyrite surfaces and bioleached ones, many AFM images have been studied using the variogram method in a defined direction (different from the scanning one). Each picture is represented by a (variance max, range) couple that can be plotted in a graph γ_Z=f(range) (Figure 3 and 4).

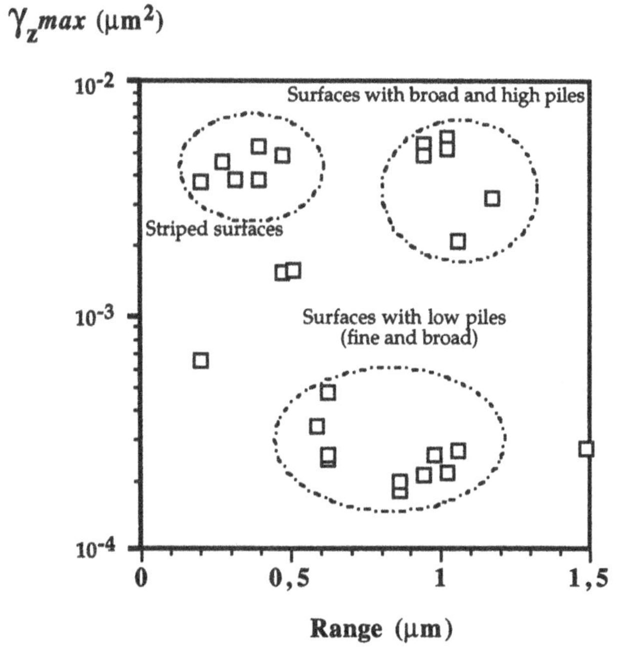

Figure 3: Variance/Range correlation for naturally air-oxidized pyrite surfaces

On natural oxidized pyrite (Figure 3), flake surfaces present oxidized phases that can be classified in two distinct groups: (i) the low piles (may be divisible into two categories: fine and broad) and (ii) the broad and high piles that have large ranges and variances. The striped surfaces can be separated from the other ones because they lead to a small range associated with a very high variance.

Results concerning bioleached surfaces are different from those obtained with naturally oxidized surfaces (Figure 4). A general and important increase of the variance is evident for more than 30% of the analyzed pictures. The pile distribution on the surface is much more continuous, thus it is not possible, in these cases to distinguish well-defined categories of piles. These increases of variance are not only due to the developement of pores during the bioleaching of pyrite (MUSTIN *et al.*, 1992b), but to the increase of the maximal height of piles. This increase indicates a real and global rearrangement of piles distribution at the interface, as suggested by the previous XPS results. This is related to an improvement of the surface heterogeneity induced by the bacterial oxidation.

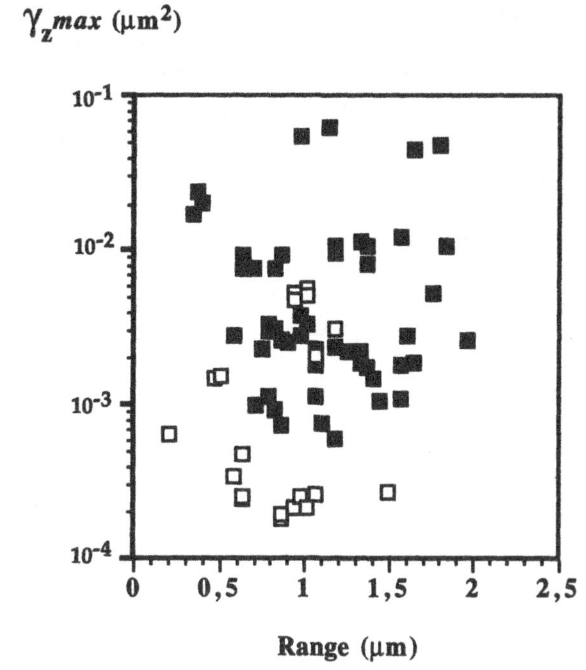

Figure 4 : Variance/Range correlation for bioleached pyrite in comparison with initial pictures.
□ Natural oxidized pyrites ■ Pyrites after 45 days of bioleaching

These results complement previous studies, in which only qualitative modifications of oxidized species could have been observed using global and "blind" spectroscopic techniques (XPS, IR ...). It is now established that the shape and distribution of the piles change during bioleaching. However, at this stage of the study, topographic structures could not yet be related to chemical data, corrosion patterns or adsorption sites for bacteria.

Even though this method seems to be very promising to evaluate the evolution of superficial phases distribution, various problems appeared during the study: the first one concerns the interpretation of the variogram in the x scanning direction. Because of the artifact of the tip scan, it is difficult to compare the x direction with the other ones. A second problem is related to the representativeness of AFM images for all complex surfaces (extended to a large scale). The high heterogeneity of the interface at the micrometer scale has been easily demonstrated by the examination of a few areas of the same flake. Although, the area investigated by the cantilever tip is quite small (5x5 μm^2), a complete analysis would require the acquisition of a very large number of images. To partially solve this problem, a new atomic force microscope, which can analyze 20x20 μm^2 ranges, will soon be available. Another disadvantage is the poor quality of some of the images, in which the variograms are then impossible to interpret and discuss. They present scratches or saturations due to the limitation of the vertical motion of the tip.

4. CONCLUSION

This work is the first step of a larger study related to the collection of chemical and topographic data on pyrite surfaces during bioleaching. In complement of some preliminary studies which have shown that pyrite surface is covered by various amounts of oxidized species, the present results prove that superficial phases are distributed in well-defined and separate piles. The geostatistical analysis of AFM pictures showed that the spatial distribution of piles on a pyrite surface is different after natural oxidation by air and after bioleaching and could lead to interesting results concerning the piles alignment, the periodicity of the structures and finally the isotropy or anisotropy of the interface.

Unfortunately, because of the presence of artifacts (slope, scratches, etc..), variograms and other parameters concerning the distribution of piles (periodicity, shape, surface coverage) could sometimes not be calculated. For example, variogram calculations are not available if flakes present a gentle slope in the scanned area. This is frequently observed on a random fresh cleaved surface of pyrite. In this case, the determination of surface coverage of piles for which one needs to know precisely the ground (i.e., the zero level) of the mineral surface, is impossible.

Thus, to minimize artifacts and to establish links between the abundance of oxidized surface species determined by spectroscopic techniques and the surface topography, a means to analyze AFM pictures is a necessity. Our focus in the future will be on the developpement of 3-D mathematical models of the interface, which will complete this statistical approach. A discrete method is also being developed to determine increasingly sophisticated surface properties including: roughness, local curvature, partition of piles and chemical composition. These models will provide an essential tool for realistic representation of the surface. In the future, tools could be used for the establishment of interfacial dynamic models during bioleaching of pyrite

6. REFERENCES

Brion D 1980. Etude par spectroscopie de photoelectrons de la dégradation superficielle de FeS_2, $CuFeS_2$, ZnS and PbS à l'air et dans l'eau.**Appl. Surf. Sci.** 5 133-152

De Donato, P., Mustin, C., Berthelin, J. and Marion, P. 1991. An infrared investigation of pellicular phases observed on pyrite by scanning electron microscopy during its bacterial oxidation. **C.R. Ac. Sci. Paris**, 312 série II:241-248.

De Donato, P., Mustin, C., Benoit, R. and Erre, R., 1993. Spatial distribution of iron and sulfur species on the surface of pyrite. **Appl. Surf. Sci., 68** : 81-93.

Lamontagne, B., Guay, D., Roy, D., Sporken, R. and Caudano, R., 1995. AFM and XPS characterization of the Si(111) surface after thermal treatment. **Appl. Surf. Sci., 90**: 481-487.

Marion, P., Mustin, C., Monroy, M. and Berthelin, J., 1991. Effect of auriferous sulfide minerals structure and composition on their bacterial weathering. **Source, Transport and deposition of Metals. Proceedings of the 25 years anniversary meeting. 30August- 3 September 1991.** 561-564.

Monroy-Fernandez, M. G., Mustin, C., de Donato, P., Berthelin, J., and Marion, P., 1995. Bacterial behavior and evolution of surface oxidized phases during arsenopyrite oxidation by *Thiobacillus ferrooxidans*. **Biohydrometallurgical Processing**, 57-66.

Mustin C., Berthelin J., Marion P. and de Donato, P. 1992a. Corrosion and electrochemical oxidation of a pyrite by *Thiobacillus ferrooxidans*. **Appl. Env. Microbiol. 58**: 1175-1182.

Mustin, C., de Donato, P. and Berthelin., J., 1992b. Quantification of the intergranular porosity formed in bioleaching of pyrite by *Thiobacillus ferrooxidans*. **Biotech. and Bioeng., 39** : 1121-1127.

Mustin, C., de Donato, P. and Berthelin, J., 1993a. An approch to spatial distribution of iron and sulfur species present on pyrite surface. **Biohydrometallurgical technologies, I, Biological Processes**, Torma, A.E., Wey, J.E. et Lakshmanan, V.I., éd., 163-175.

Mustin, C., de Donato, P. and Berthelin, J., 1993b. Investigation of superficial oxidized species development during the bacterial oxidation of pyrite by *Thiobacillus ferrooxidans*. **Biohydrometallurgical technologies, I, Biological Processes**, Torma, A.E., Wey, J.E. et Lakshmanan, V.I., éd., 175-184.

Mustin, C., de Donato, P., Berthelin, J., and Marion, P. 1993c. Surface sulphur as promoting agent of pyrite leaching by *Thiobacillus ferrooxidans*. **FEM'S Microb. Rev., 11** : 77-78.

Nesbitt, . H. W. and Muir, I.J., 1994. X-Ray Photoelectron Spectroscopic study of a pristine pyrite surface reacted with water vapour and air. **Geochim. Cosmochim. Acta., Vol. 58, No 21**: 4667-4679.

Sasaki, K., Tsunekawa, M., Ohtsuka, T. and Konno, H., 1995. Confirmation of a sulfur rich layer on pyrite after oxidative dissolution by Fe(III) ions around pH 2. **Geochim. Cosmochim. Acta., Vol. 59, No 15**: 3155-358.

TRANSFORMATION OF IRON-CONTAINING MINERALS IN KAOLIN DURING GROWTH OF A MIXED BACTERIAL CULTURE DERIVED FROM KAOLIN

E. S. Shelobolina, Z. A. Avakyan, and G. I. Karavaiko

Laboratory of Chemolithotrophic Microorganisms,
Institute of Microbiology, Russian Academy of Sciences
Prosp. 60-Letiya Octyabrya, 7/2, Moscow, Russia, 117811

1. INTRODUCTION

Kaolins are clay rocks composed mainly of kaolinites (not less than 50%), the rest being other silicates. They also contain iron and titanium minerals as impurities. Iron in kaolins occurs in the form of goethite, hematite, hydrogoethite, magnetite, and pyrite, or as iron oxide films on the surface of the silicate minerals, or it may be included in their crystalline lattices (Avgustinik, 1975).

There are some data in the literature on the leaching of iron from silicate rocks and minerals by aerobic heterotrophic microorganisms which are known to be producers of hydroxy and keto acids (Berthelin, 1988; Ogurtsova *et al.*, 1989).

Groudev removed iron from kaolins with spent medium from an *Aspergillus niger* culture supplemented with HCl (Groudev, 1989). In his experiments, the iron oxide and mineral films in kaolins were dissolved by citric and oxalic acids contained in the spent medium.

Facultative and obligate anaerobic microorganisms are known to be involved in the processes of iron reduction and dissolution, formation of soluble organo-iron complexes (Lovley, 1991), and biomineralization (Lovley, 1990).

The aim of this work was to study the dynamics of growth of the natural flora from kaolin and its role in the transformation of iron minerals in kaolin.

2. MATERIALS AND METHODS

2.1. Characterization of kaolin

Kaolin was available from the Prosyanov deposit (Ukraine) and had the following mineral composition (mass%): kaolinite, 90; mica, 1; illite, 3-4; quartz, 5. In this kaolin, iron is present in crystalline and amorphous forms and may also be a component of silicate structures. The iron minerals are hematite (α-Fe_2O_3), magnetite (Fe_3O_4), goethite (α-FeOOH), and lepidocrocite (γ-FeOOH). The chemical composition of this kaolin was as follows (%): SiO_2, 46.9; Al_2O_3, 36.0; TiO_2, 0.53; Fe_2O_3, 0.71; CaO, 1.18; MgO, 0.40; Na_2O, 0.17; K_2O, 0.88; calcination loss,13.21. Particle sizes of 86% (w/w) of kaolin were less than 0.01 mm.

Effect of Mineral-Organic-Microorganism Interactions on Soil and Freshwater Environments
Edited by Berthelin, *et al*, Kluwer Academic / Plenum Publishers, New York, 1999

2.2. Experimental procedure

The experiments were carried out at 30°C in a 5.5-l glass vessel (Fig.1) 90% full of a suspension containing nonsterile kaolin (3 kg), sterile nutrient medium (1 liter) composed of (g/l): glucose, 5; $(NH_4)_2SO_4$, 0.2; K_2HPO_4, 1; $MgSO_4*7H_2O$, 0.2, and sterile tap water (to a total of 5 liters). The medium and tap water were sterilized in an autoclave at 0.5 atm for 30 min. The relative volume of inoculum was 5%. The gas phase was not replaced. The bottle necks were closed with rubber stoppers fitted with tubes for sampling solid, liquid, and gaseous phases. The solid phase samples were withdrawn with a pipet through the glass tube connecting the side neck of the vessel with the kaolin sediment and bypassing gaseous and liquid phases. In a control experiment, sterile kaolin suspension, nutrient medium, and tap water were used. Within the first two days of the experiment, a separation of kaolin suspension into a dense 16-cm thick sediment and an 8-cm thick supernatant occurred. The experiment was continued for 38 days.

2.3. Inoculum. Inoculation was performed with a community of aerobic and anaerobic bacteria isolated earlier from Prosyanov kaolins. The microbial community was maintained by passages in a sterile kaolin suspension supplemented with the above-mentioned nutrient medium.

2.4. Bacterial enumeration. The main physiological groups of bacteria were counted by serial dilution to estimation at ten-fold intervals. The following media were used: 1:10 diluted nutrient broth for aerobic microbial flora; Vinogradsky agar medium (Kuznetsov and Dubinina 1989) for fermentative bacteria; Postgate medium (Postgate, 1966) for sulfate-reducing bacteria; Lovley medium (Lovley and Phillips, 1986) for iron-reducing bacteria; and Gil'tai medium (Romanenko and Kuznetsov, 1974) without asparagine for denitrifying bacteria.

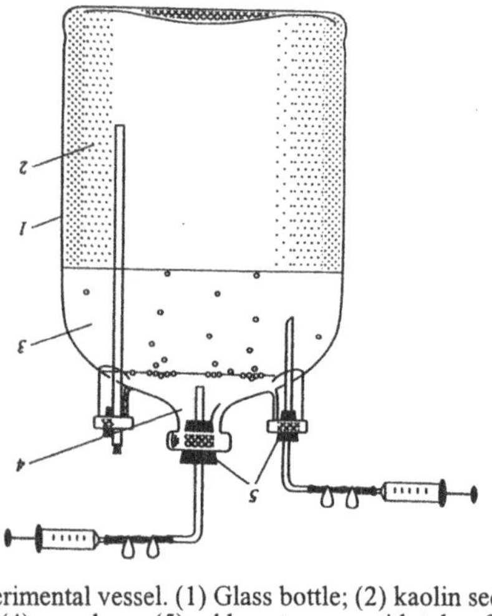

Fig.1. Experimental vessel. (1) Glass bottle; (2) kaolin sediment; (3) supernatant; (4) gas phase; (5) rubber stoppers with tubes for sampling.

2.5. Analytical methods. The fatty acid composition of the microbial community in kaolin samples was studied with an HP-5985B chromatograph-mass-spectrometer (Hewlett-Packard, USA). Lipids were extracted from kaolin samples by the Folch method using a chloroform-methanol-water system (Folch *et al.*, 1957). Methyl esters of fatty acids were obtained through acid methanolysis with 4.5 M HCl in dry methanol for 1.5 h at 70°C. The methyl esters of fatty acids formed were extracted with hexane, the extract was dried and then treated with 20 μl of N,O-bis(trimethylsilyl)trifluoroacetamide to produce volatile derivatives of acids. An aliquot of reaction mixture (4 μl) was introduced into the chromatograph injector for analysis.

Iron minerals in the first and final kaolin samples were analyzed after their concentration and subsequent isolation in 4 magnetic fractions. The first fraction was extracted with the help of a hand-held magnet at a magnetic intensity H = 1 kOe (Oe, oersted); the second fraction was extracted with a magnet at H = 4 kOe; the third and fourth fractions were isolated with an EKL-2 high-gradient magnetic separator in magnetic fields at H = 10 kOe and H = 15 kOe, respectively. The composition of iron minerals was studied by transmission electron microscopy (using a JEM-100C microscope with a Kevex-5100 energy dispersion device) and by measuring magnetic susceptibility with a KLY-2 Kappabridge instrument (Czechoslovakia).
Values of pH and Eh were determined with a pH-121 millivoltmeter. The pH was measured with an ESL-11G-04 glass electrode and an EVL-1M3 chlor-silver electrode, the Eh - with an ETPL-01M platinum electrode. To determine the gradients of pH, Eh, and total iron, samples were withdrawn in the end of the experiment from various layers of the sediment column. The levels of volatile fatty acids were measured by gas-liquid chromatography (Biotronic HPLC). The content of iron in solution and in kaolin (after subjecting it to dissolution in HF) was measured with a Perkin-Elmer 3100 atomic-absorption spectrophotometer. Oxalate-extractable iron was determined by the method of Tamm.

3. RESULTS

3.1. Composition of the microbial community

Analysis of the structure of the kaolin-originated microbial community by conventional physiological and biochemical techniques and by study of the lipid composition of total biomass made it possible to determine its generic and, in some instances, species composition (Turova and Osipov, 1996). Autotrophic bacteria were represented by *Nitrobacter* sp., aerobic and facultatively anaerobic bacteria - by the genera *Bacillus, Pseudomonas, Burkholderia, Nocardia, Caulobacter, Deinococcus*, and *Arthrobacter*, and anaerobic microorganisms by the genera *Clostridium, Bacteroides, Desulfovibrio*, and *Desulfobacter*. An iron-reducing microorganism, strain FeRed, and a representative of microscopic fungi were also identified.

3.2. The dynamics of microbial growth

The development of the microbial community under study is characterized as follows (Fig. 2). During the first 11 days of the experiment, both liquid and solid phases were dominated by aerobic, facultatively anaerobic and fermentative microorganisms. They caused a pH decrease in the supernatant (from 8.2 to 5.1), whereas the pH value of the sediment remained virtually constant. The Eh value decreased (Fig.3). The uptake of glucose was accompanied by an accumulation of microbial metabolites: organic acids (butyrate, acetate, lactate and formate) (Fig.4) that are further utilized as substrates by anaerobic bacteria which become predominant in the second developmental stage of microbial community (Fig.2). The development of anaerobic bacteria leads to the production of propionic acid, carbon dioxide, hydrogen, and hydrogen sulfide in the liquid and gaseous phases and to the eventual release of methane at the final experimental stage. By the end of the experiment, a gradient in coloring, Eh, pH, and total Fe is noticeable down the kaolin sediment column (Fig. 5).
The development of the gradient was due to several causes, the major one being the contact of the supernatant with the uppermost layer of the sediment column and the absence of such contact at the sediment bottom. In the upper portion of the kaolin column pH dropped to 5.6. At the bottom of the kaolin column, where no exchange with the supernatant liquid was possible, kaolin had a high pH value (7.7) and the Eh dropped from +450 mV to -130 mV. The total iron concentration remained virtually constant along the entire kaolin column, except at the top. In the control vessel, microorganisms did not develop, hence, no changes were revealed in physicochemical parameters of liquid and solid phases.

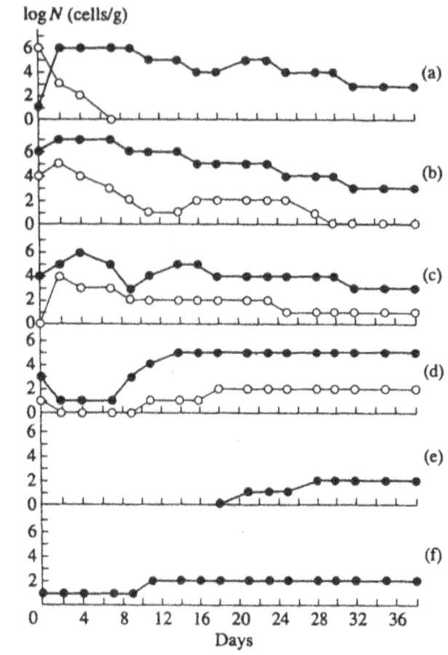

Fig.2. Dynamics of bacterial populations in the community: (a) denitrifying bacteria; (b) ae-robic and facultatively anaerobic heterotrophic bacteria; (c) fermentative bacteria; (d) sul-fate-reducing bacteria; (e) methanogens; (f) iron-reducing bacteria. (O) The number of microorganisms in supernatant; (●) the number of microorganisms in sediment.

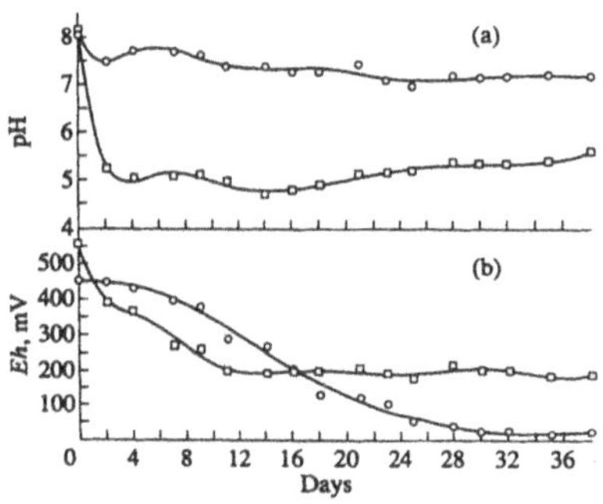

Fig.3. Changes in the pH and Eh during the development of the microbial community.
(a) pH; (b) Eh; (❑) supernatant; (❍) sediment.

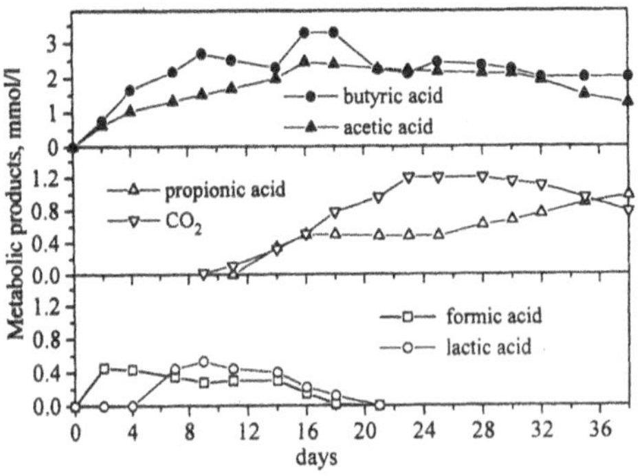

Fig.4. Dynamics of accumulation of metabolic products by microbial community

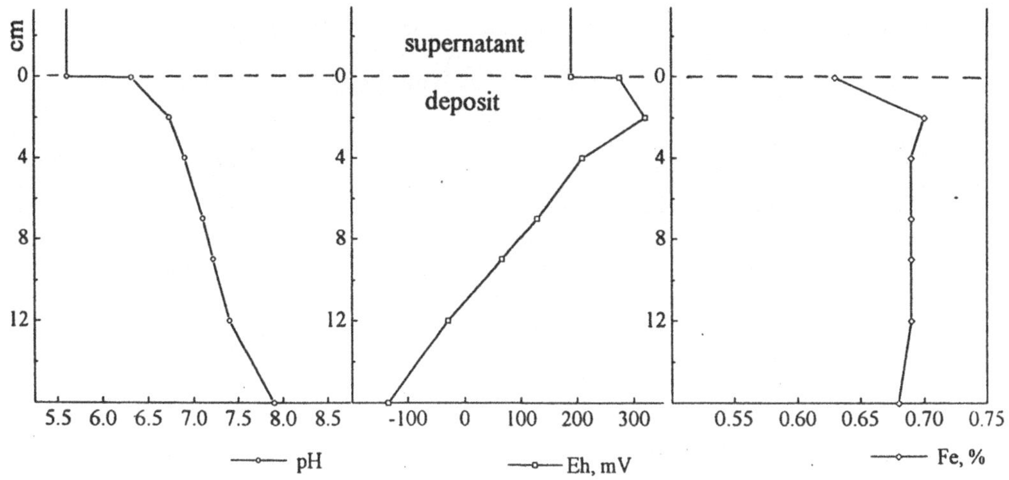

Fig.5. pH, Eh, and Fe gradients in the kaolin sediment column.

3.3. Transformation of iron minerals in kaolin

Owing to microbial activity, the portion of the oxalate-soluble iron in kaolin increased from 1.4 to 10% of the total iron, which is indicative of a decrease in the crystallinity degree of the iron mineral particles in kaolin.

A considerable increase in iron extractability by magnetic separation and changes in the mineral composition of magnetic fractions was also attributed to the microbial activity (Table 1).

Fractions	Initial kaolin		Final kaolin	
	Mineral*	Iron extraction**	Mineral*	Iron extraction**
Strong-magnetic (1+2)	Hematite, magnetite, goethite, lepidocrocite	1	Magnetite, hematite, ferrihydrite, lepidocrocite	38
Weak-magnetic (3+4)	Goethite, magnetite	11	Magnetite, goethite, lepidocrocite, hematite	14

* The iron minerals are listed in the decreasing order of their concentration in kaolin.
** Iron extractability is expressed as percentage of its total content in kaolin.

Table 1. Iron minerals extracted from kaolin by magnetic separation

3.4. Production of magnetite by the iron-reducing bacterial strain FeRed

A dramatic increase of the magnetite level in the magnetic fractions of the experimental kaolin sample suggests the possibility of biogenic synthesis of magnetite. To this end, we studied the composition of a black sediment formed during the cultivation of the iron-reducing bacterial strain FeRed in Lovley's medium in the presence of amorphous ferric hydroxide. Analysis of this sediment by transmission electron microscopy revealed the presence of dense oval magnetite particles (not more than 0.1-0.5 μm in diameter) (Fig. 6) differing by their morphology from the magnetite particles extracted from the initial and final kaolin samples from the experimental flask.

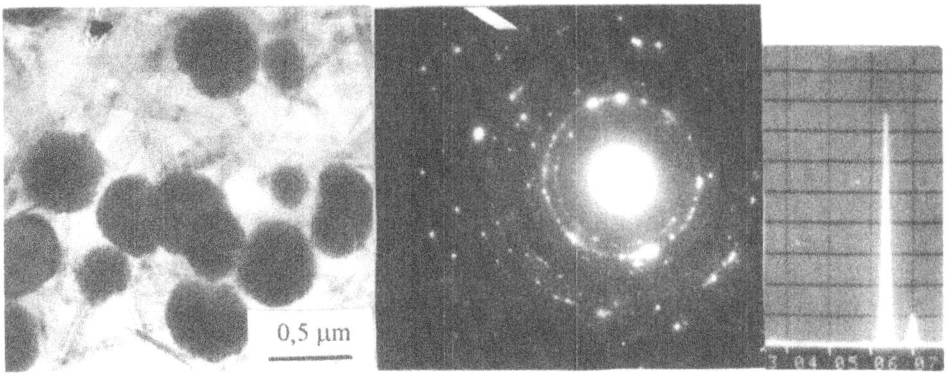

Fig.6. Electron image of magnetite particles formed during the cultivation of the strain FeRed

4. DISCUSSION

The development of a natural microbial community in kaolin involves different processes of iron transformation (reduction and oxidation, formation and degradation of organo-iron complexes, solubilization and formation of iron minerals) that strongly affect the content and composition of iron minerals in kaolin.

Dissimilative reduction of iron accompanied by magnetite synthesis may be brought about by several members of this microbial community. Experimental evidence has been presented for the ability of the iron-reducing bacterial strain FeRed, isolated from this community, to bring about iron reduction and magnetite neosynthesis. Similar processes may also be caused by certain sulfate-reducing bacteria, including some representatives of the genus *Desulfovibrio* (Coleman *et al.,* 1993). Bacteria which bring about enzymatic metabolism (in the community under study, these are *Clostridium sp.*, *Bacteroides sp., Bacillus sp.*) are known to be capable of transferring up to 5% of reductive equivalents from the substrate to ferric ions (Lovley, 1991). Nonspecific reduction and solubilization of iron, associated with strongly reductive metabolites (sulfide, slime, organic acids) released into the medium by members of the kaolin-derived microbial community, also may have been caused by representatives of the genera *Bacillus, Clostridium*, and *Desulfovibrio* (Lovley, 1991).

Changes in the iron concentration in the liquid phase resulted from iron complexation and, presumably, reprecipitation. Strains of *Bacillus sp.* and *Clostridium sp.* are known to be active producers of chelate-forming agents. Reprecipitation of iron in the form of insoluble compounds such as ferrous and ferric hydroxides, sulfides, and phosphates may result from the degradation of organo-iron complexes by the aerobic and facultatively anaerobic bacteria of this community which digest the organic moiety of such complexes.

Some heterotrophic bacteria, including representatives of *Arthrobacter sp.*, are known to be oxidizers of ferrous ions in neutral and alkaline media (Dubinina and Balashova, 1985). Production of ferrihydrite revealed in this study may be attributed to both microbial oxidation of iron and degradation of organo-iron complexes.

A considerable increase in the magnetite concentration to the magnetic fraction observed in our experiments may be the consequence of magnetite neogenesis as well as the result of dispersion and degradation of concretions of iron minerals and silicates.

5. CONCLUSION

The growth of a natural kaolin derived microbial community in kaolin supplemented with a nutrient medium was shown to decrease the extent of crystallization of particles of iron minerals , enhance the extractability of iron minerals by magnetic separation and modify the mineral composition of magnetic fractions. The magnetic fractions of the experimental kaolin sample proved to be enriched in magnetite. In addition, ferrihydrite was formed. The iron-reducing bacterial strain FeRed, isolated from the natural microbial community present in kaolin, has been shown to be capable of producing magnetite.

Acknowledgments

We thank Dr.G.A. Osipov for valuable advice and assistance in the analysis of the fatty acid composition of total biomass produced by the microbial community.

References

Avgustinik, A.I., 1975. Keramika (Ceramics), Leningrad. Stroiizdat.

Berthelin, J., 1988. Microbial weathering processes in natural environments. In: A.Lerman and M.Meybeck (Editors). Physical and Chemical Weathering in Geochemical Cycles. pp. 33-59.

Coleman, M.L., D.B.Hedrick, D.R.Lovley, D.C.White, Kenneth Pye, 1993. Reduction of Fe(III) in sediments by sulphate-reducing bacteria. Nature, 361, 436-438.

Dubinina, G.A. and V.V.Balashova, 1985. Microorganisms involved in the turnover of iron and manganese; their application in hydrometallurgy. In: G.I.Karavaiko *et al.* (Editors). Biogeotechnology of Metals. Moscow, pp. 145-161.

Folch, J., M.Lees and G.H.Sloan-Stanley, 1957. A simple method for the isolation and purification of total lipids from tissue. J. Biol. Chem., 226, 497-509.

Groudev, S.N., 1989. Microbial weathering of aluminum. In: Biogeotekhnologiya metallov. Prakticheskoe rukovodstvo (Handbook for Metal Biogeotechnology). Moscow, pp. 342-346.

Kuznetsov, S.I. and G.A.Dubinina, 1989. Metody izucheniya vodnykh mikroorganizmov (Methods of studying of water microorganisms) Moscow. Nauka.

Lovley, D.R. and E.J.P.Phillip , 1986. Organic matter mineralization with reduction of ferric iron in anaerobic sediments. Appl. Environ. Microbiol., 51, 1212-1222.

Lovley, D.R., 1990. Magnetite formation during microbial dissimilatory iron reduction. In: R.B. Frankel and R.P.Blakemore (Editors). Iron Biominerals. New York, Plenum Press, pp. 151-166.

Lovley, D.R., 1991. Dissimilatory Fe(III) and Mn(IV) reduction. Rev.Microbiol., 55, 259-287.

Ogurtsova, L.V., G.I.Karavaiko, Z.A.Avakyan, and A.A.Korenevskii, 1989. Activity of different microorganisms in leaching of elements from bauxite. Mikrobiologiya, vol. 58, no. 6, 956-962.

Postgate, J.R., 1966. Media for sulfur bacteria. Lab. Practice, vol. 15, no. 11, 1239-1244.

Romanenko, V.I. and S.I.Kuznetsov, 1974. Ekologiya mikroorganizmov presnykh vodoemov (Ecology of fresh water microorganisms). Leningrad. Nauka.

Turova (Shelobolina), E.S. and G.A.Osipov, 1996. Structure of microbial community transforming iron minerals contained in kaolin. Microbiology, 65, 597-603.

EFFECT OF SUCCINIC ACID PRODUCED BY MICROORGANISMS AND PLANT ROOTS ON COPPER SORPTION BY SOIL

Tatiana Pampura[1] and Mikhail Ustinin[2]

1 Institute of Soil Science and Photosynthesis RAS, Pushchino, Moscow Region, 142292, Russia
2 Institute of Mathematical Problems in Biology RAS, Pushchino, Moscow Region, 142292, Russia

INTRODUCTION

Copper mobility and bioavailability in soil are determined mainly by adsorption on the surface of soil particles (McLaren and Crawford, 1973; McBride, 1981, 1989). Sorption behavior of metals is dependent on interaction with various components of soil solution, especially chelating agents such as fulvic acid, aminoacids, organic acids, and phenolic compounds. Organic ligands are formed in soil during the decomposition of organic matter, root exudation, and microbial synthesis by rhizosphere microorganisms (McKeague *et al.*, 1986; Tan, 1986; Stevenson and Ardakani, 1972; McBride, 1989; Karpukhin *et al.*, 1993). Hodgson *et al.* (1965) found that as much as 99% of soluble Cu may exist in a complexed form in soil solutions from several mineral soils. Ponomareva (1972), Yashin and Kaurichev (1992) demonstrated an important specific role of water-soluble organic ligands for mobilization and migration of metals in soils and the landscape.
It was found that mobile ligands are quite effective in complexing and transporting metal ions to plant roots (Prasad *et al.*, 1976; Elawhary at al., 1970; Lindsay and Norvell, 1969).

In the presence of some nonadsorbing ligands, metal adsorption decreases due to the competition between the ligand and the surface for complexation of metal ions (Farrah and Pickering, 1976; Pampura, 1992; Harter and Naidu, 1996). On the other hand, adsorption of the dissolved ligand on the surface of a soil particle may enhance metal adsorption (Davis and Leckie, 1978; Huang et al, 1988). In some cases surface complexes exhibit an anionic character with decreasing adsorption with increasing pH (Huang *et al.*, 1988; Nowack *et al.*, 1996; Vulava *et al.*, 1997). The sorption behavior of heavy metals onto oxide surfaces in the presence of organic ligands depends on metal and ligand type, pH (Davis and Leckie, 1978; Elliott and Huang, 1979; Vuceta and Morgan, 1978), the individual and relative concentrations of metals and ligands (Elliott and Huang,1979; Pampura, 1992; Basta and Tabatabai, 1992 a,b; Vulava *et al.*, 1997), the order of application of metal and ligand to the adsorbing surface (Bryce, 1994; Vulava *et al.*, 1997) and the concentration of background electrolyte (Elliott and Huang, 1985; Vulava *et al.*, 1997).

Experimental investigations demonstrate that distribution of metal ions between the surface and solution is controlled by the interaction between metal ions, complexing ligands and solid surface. However, most experiments have been conducted with anthropogenic heavy metals and complexing agents such as NTA, EDTA and DTPA, which form much stronger complexes with (HM) than natural organic ligands. As a rule, selected oxides and clay minerals were used as the adsorbing phase (Linn and Elliott, 1988; Huang *et al.*, 1988; Nowack *et. al.*, 1996; Vulava *et al.*, 1997; McBride, 1981). Real soil systems are much more complicated then experimental ones. Unfortunately, information on the adsorption of metal on soil in the presence of natural organic ligands is sparse (for example, Chairidchai and Ritchie, 1990).

Effect of Mineral-Organic-Microorganism Interactions on Soil and Freshwater Environments
Edited by Berthelin, *et al*, Kluwer Academic / Plenum Publishers, New York, 1999

Information concerning HM adsorption behavior in soils is necessary to predict the fate of these dangerous pollutant in the ecosystems (De Vries and Bakker, 1996). Most experimental studies of metal adsorption behaviour have been conducted, however, in a simple electrolyte system in the absence of the ligand. The fact that some HM form strong complexes with dissolved organic compounds and therefore the activity of metal in solution is not equivalent to the concentration has often been neglected in published adsorption studied. From the thermodynamic standpoint, activity rather than concentration controls metal adsorption on solid surfaces (Reinds *et al.*, 1995). Moreover, it was shown that activity rather than concentration determines HM bioavailability and toxicity (Halvorson and Lindsay, 1977; Malzer and Barber, 1976; Brand *et al.*, 1986). However, only a few articles provide sorption curves as a function of copper activity (Sanders, 1982 and Cavallaro and McBride,1978) used ion selective electrodes and Lexmond (1980) used resin-method to determine Cu activity. These experiments were performed in the absence of organic ligands. McBride (1981) studied Cu adsorption isotherms on montmorillonite in terms of HM concentration and activity in the presence of different organic ligands. There is no information on the adsorption of metals on soil as a function of activity in the presence of natural organic ligands.

The objectives of the present study were :

1. experimental investigation of Cu adsorption isotherms on soil (Typical Chernozem) in terms of Cu concentration and activity in solution in the absence and presence of succinic acid. Succinic acid was used as a typical product of microbiological activity and a component of root exudates (Rovira, 1969; Smith, 1976). Mench *et al.* (1988) showed that 24% of maize root exudates consist of low molecular weight organic acids; succinic acid accounts for 40% of them.

2. computer simulation of the chemical equilibrium in the system soil-H_2O-NO_3-CO_2-Cu-Ca-Succinic acid.

MATERIAL AND METHODS

Soil

A sample of Typical Chernozem was collected in the Biosphere Station of the Geographic Institute of the Academy of Sciences of Russia in the Kursk region. Chemical characteristics of chernozem and its Ca-form are shown in Table 1. Mineralogical composition of water dispersible clay fraction (<2μm) was determined by x-ray diffraction using DRON-3 diffractometer (Cu-K_α radiation, Ni filter). The quantitative evaluation of major clay mineral groups was performed by the Biscay method (Biscay, 1965). The main components found were: irregular interstratified mica-smectite with high quantities of swelling phase (25%), hydromica of dioctahedral order (38%), kaolinite (11%), quartz (26%), plus traces of plagioclase and chlorite. Soil was powdered in an agate mortar, passed through 1-mm sieve and carried to the Ca-form by 10-fold treatment with 0.25 M $Ca(NO_3)_2$ solution. Then the excess of $Ca(NO_3)_2$ solution was removed by distilled water, the soil was dried, homogenised and used in the experimental study. Soil transformation into Ca-form resulted in displacement of exchangeable K, Na, and Mg by Ca (Table 1). It was done to simplify the calculation of chemical equilibrium in the system studied.

The transformation procedure did not change the content of <0.01 mm clay fraction. The difference inorganic matter content (on the order of 10 %) between Chernozem and Chernozem in Ca-form was most likely connected with the experimental error (organic matter content was analysed by the method of Tyurin (Arinushkina, 1970) (Table 1).

Table 1. Selected properties of soil used.

Soil type	pH H_2O	Organic matter %	Exchangeable cations, meq/100g soil				Clay fraction <0.01mm
			Ca	Mg	K	Na	
Chernozem	6.15	5.47	23.6	3.60	0.31	0.33	37.2
Chernozem Ca form	7.00	6.07	27.0	0.29	0.27	0.08	37.6

Adsorption Experiments

Copper sorption isotherms were experimentally obtained in the presence and absence of succinic acid (concentration of ligand in the initial solution was 5mM). In the copper sorption study, 2.5g duplicate samples of Ca-chernozem were treated with 25-mL aliquots of 5mM ($Ca(NO_3)_2$ + $Cu(NO_3)_2$) solution, at Cu/(Cu+Ca) ratio ranging from 0 to 1. To evaluate the effect of the organic ligand on Cu sorption, succinic acid was added to achieve the desired solute concentration. The pH drop caused by the addition of succinic acid was compensated by adding KOH in a special experiment. In this case suspension pH was adjusted to the value observed in the absence of ligand for the same amount of Cu added. Suspensions were shaken at room temperature for 15 min and let stand for 10 min. Final pH values and the activity of Cu and Ca were measured simultaneously by Ca, Cu ("ELIT-227", "NIKO", Moscow) and H-selective (ES-01.03.01...04) electrodes and reference electrode (ESr-00.07.01...0.9) (NPO "Measurement Technique", Moscow, and pH-meter OP-211/1("Radelcis", Hungary). Preliminary kinetic experiments indicated that steady-state values of pH, the activity of Cu and Ca and the concentration of ions were attained in a matter of minutes under the test conditions. The aliquots of solution were withdrawn and passed through 0.02-μm membrane filter. Filtrate was stored (if it was necessary) at a temperature of -18°C and then analysed for Ca and Cu by flame atomic absorption (Perkin Elmer 300 AAS spectrophotometer). Concentrations of succinic acid were determined by ion chromatography (chromatograph Biotronic). Eluent composition was 1.8 mM $NaHCO_3$+1.5 mM Na_2CO_3; the column height was 25 cm. The error of the Suc concentration determination (including procedure of aliquot withdrawal, filtration and dilution) was calculated using the formula

$$\varepsilon = \pm t_{\alpha n} S_x / \sqrt{n}\,, \tag{1}$$

where t_α is the Student coefficient, S_x is the standard deviation and n is the number of analysis replications. For a 95 % confidence interval and n = 5, the error was 8 %. The amounts of immobilised Cu, Ca and Suc were calculated by subtracting the final solution concentration from the initial concentration in control treatments without soil.

Description of Cu Sorption by Langmuir and Freundlich Equations

Experimental data were described by the Langmuir

$$S_{Cu} = S_m/(1+1/KC_{Cu}) \tag{2}$$

and Freundlich equations

$$S_{Cu} = kC_{Cu}^{\ p}, \tag{3}$$

where S_{Cu} is the amount of Cu sorbed (meq/100g of soil), C_{Cu} is the equilibrium Cu concentration (M), K is the Langmuir constant, S_m is sorption maximum and k and p are constants. Linear and non-linear parameters of the model were estimated by the root mean square method. The Fisher criterion was used to test regression coefficients for every model. Models were compared using the residual error (Pollard, 1982):

$$Error = \sqrt{\sum (y_i - f(x_i))^2}\,, \tag{4}$$

where y_i is the experimental value and $f(x_i)$ is the regression line.

Calculation of Ion Speciation and Adsorption in the System Soil-Ca-Cu-H_2O-NO_3-CO_2-Succinic acid

Special mathematical programs for the IBM PC/AT were developed to calculate:

1 - ion speciation in the equilibrium solution in the absence and presence of succinic acid;
2 - Cu adsorption isotherm in the presence of succinic acid.

Ion Speciation in the Equilibrium Solution in the Presence and in the Absence of Succinic Acid. Chemical equilibrium in the solution was described by a set of equations for:

- stability constants of Cu and Ca complexes with inorganic ligands (OH^-, HCO_3^-, CO_3^{2-}, NO_3);
- stability constants of Cu and Ca complexes with succinic acid;
- stepwise constants for Suc protonation (Table 2);
- mass balance of total Ca, Cu, Suc, NO_3.

Table 2. Stability constants of metal-ligand complexes (log K^0) (Sillen and Martell (1971), Lindsay (1979)

Ion species	OH^-	$2OH^-$	$3OH^-$	CO_3^{2-}	HCO_3^-	NO_3^-	$2NO_3^-$	Suc^{2-}	$HSuc^-$
Cu^{2+}	6	7.18	1.24	-11.4	-5.73	0.50	-0.40	2.93*	1.70*
$2Cu^{2+}$	-	-10.7	-	-	-	-	-	-	-
Ca^{2+}	1.3	-	-	-	-	-	-	1.20*	0.54*
H^+	14	-	-	-	-	-	-	5.28*	4*

*I = 0.1

$K_{CuCO3} = (CuCO_3)(H^+)/(Cu^{2+})pCO_2$; $K_{CuHCO3} = (CuHCO_3)(H^+)/(Cu^{2+})pCO_2$

$K_{Cu2(OH)2} = (Cu_2(OH)_2^{2+})(H^+)^2/(Cu^{2+})^2$ (Lindsay, 1979)

If the total concentration of ions in the solutions the pH and the pCO_2 are known (partial pressure of CO_2 (g) was taken to be equal to $10^{-3.52}$ (Lindsay, 1979)), we have a set of N equations with N unknowns (species concentrations). This set of equations was solved numerically with the Newton-Raphson iteration scheme. The stability constants used were corrected for a particular ionic strength of the equilibrium solution. These modified constants can be calculated from the thermodynamic stability constants and activity coefficients of the species by the formula:

$$K_{ML}=K^0_{ML}(\gamma_L\gamma_M\gamma_{ML}^{-1}), \quad (5)$$

were K_{ML} is the stability constant for a particular ionic strength, K^0_{ML} is the thermodynamic stability constant and γ_L, γ_M, γ_{ML} are the activity coefficients of ligand, metal and complex, respectively. Activity coefficients are calculated using the Davis equation (Stumm and Morgan, 1981):

$$\log \gamma_L = - 0.5\, z_i^2(I/(1+I) - 0.3I), \quad (6)$$

where I is the ionic strength and z_i is the valence of ion i. The ionic strength was calculated from the formula:

$$I = 1/2 \sum z_i^2 C_i, \quad (7)$$

where C_i is the concentration of the component i.

Modeling of Cu Adsorption Isotherm in the Presence of Succinic acid. The model treats the distribution of Cu between soil and solution as a result of competition between surface functional groups and ligands in the solution for complexation of the Cu ion. The model is based on the following facts established in the experiment:

1. $S_{Cu}=f(a_{Cu})$ is independent of the ligand presence if the suspension pH does not change with the addition of succinic acid;
2. activity of Ca is not significantly affected by ligand and is held nearly constant by soil Ca (Lindsay, 1979);

3. equilibrium solution contains 80% of initial amount of succinic acid.
It was also assumed that:
4. soil does not adsorb NO_3;
5. log pCO_2 is equal to -3.52.

The Freundlich equation, describing Cu adsorption as a function of its activity in soil solution in the absence of succinic acid, was included in the set of equations. Adsorbed (Cu) or desorbed (Ca) ions were treated as a species and were included in balance equations for the total ion concentrations in the soil-solution system. The set of equations was solved for known initial concentrations of Ca, Cu, NO_3, Suc in the solution, pH and pCO_2 and Ca activity in the equilibrium solution. Calcium activity was figured by two ways: it was taken equal to Ca activity in the absence of Suc, determined by Ca-selective electrode (model 1) or calculated using equilibrium concentration, determined by AAS-method (model 2).

RESULTS

Adsorption Experiments

The results of the copper adsorption experiment are presented in Table 3. It was found, that copper sorption resulted in a drop of suspension pH and the release of Ca. As the amount of Cu sorbed increased, the amount of Ca released became larger and the pH drop . This is consistent with the data of Kurdi and Donner (1983) and our recent investigation (Pampura *et al.*, 1993). In the absence of succinic acid the amount of Ca released was very close to the amount of copper sorbed for all metal ion concentrations (Table 3, Fig.1). As a result calcium concentration in the equilibrium solution was kept constant. The pH of the solution decreased linearly as the concentration of copper in the final solution increased (r^2=-0.87).

Table 3. Copper adsorption in the presence and absence of succinic acid.

Cu added meq/100g	Ion concentration Cu µM	Ca mM	solution Suc mM	Ion solution activity Cu µM	Ca mM	pH	Sorbed Cu meq/100g	Desorbed Ca meq/100g
				In the absence of Suc				
0	3.1	4.7	0	<1	2.6	5.86	0	0.4
1.5	3.9	4.6	0	<1	2.3	5.45	1.5	2.4
3.1	6.3	4.5	0	3.6	2.5	5.21	3.1	2.9
5.0	15	4.5	0	9.8	2.3	5.03	5.0	4.7
6.0	22	4.5	0	21	2.4	4.88	6.0	6.6
8.6	57	4.5	0	47	2.2	4.64	8.5	9.0
				In the presence of Suc				
0	-	-	-	<1	2.8	4.26	-	-
1.5	-	-	-	2.0	2.8	4.11	-	-
3.1	-	-	-	10	2.6	4.02	-	-
5.0	-	-	-	25	2.5	3.93	-	-
6.0	-	-	-	46	2.7	3.94	-	-
8.6	-	-	-	110	2.5	3.80	-	-
				In the presence of Suc (with compensation of pH changes)				
0	1.6	5.7	3.7	<1	2.2	5.82	0	2.4
1.5	3.2	5.5	4.5	<1	2.1	5.51	1.6	3.8
3.1	16	5.5	4.3	1.8	2.1	5.18	2.8	4.3
5.0	41	5.6	4.0	7.9	2.2	4.97	4.4	6.0
6.0	72	5.4	3.6	13	2.4	4.86	5.7	8.0
8.6	142	5.1	4.0	43	2.3	4.61	8.2	10.2

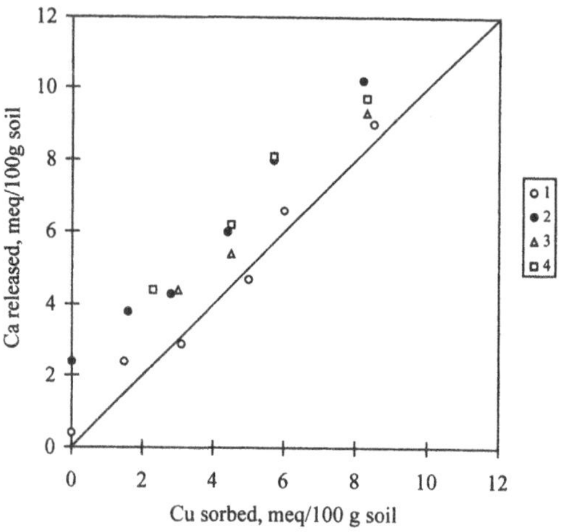

Figure 1. Calcium desorption as a function of copper sorption on soil. **Experiment**: 1 - in the absence, 2 - in the presence of succinic acid. **Model** (in the presence of succinic acid): 3 - a_{Ca} was taken equal to Ca activity in the absence of Suc, determined by ISE (model 1) (3), or calculated using equilibrium Ca concentrations, determined by AAS-method (model 2) (4).

On addition of succinic acid into the solution, its concentration was reduce by 20%. It was found that the concentration of Suc was not correlated with either Cu concentration (r^2=-0.32) or pH (r^2=0.18) in the final solution. The mean value of Suc final concentration in the solution was calculated (4 mM) and used in the model. The addition of succinic acid caused a pH drop in the final solution and, as in the case of ligand absence, pH decreased as copper concentration increased. Copper activity in the final solution increased in the presence of ligand (Table 3). In the pH range studied dissociation of succinic acid is suppressed (pK_1=5.28, pK_2=4), and the activity and concentration of copper are controlled by pH. The pH decrease caused Cu desorption as a result the metal concentration and activity in the final solution increased. In the case that the pH drop caused by the addition of succinic acid was compensated by KOH addition the copper activity was slightly less and the concentration was much higher than in the absence of succinic acid when the amount of copper added exceeded 1.5 meq/100 g soil (Table 3). When the amount of copper added was 1.5 meq/100g of soil or less, the addition of succinic acid was accompanied by a decrease in Cu concentration. However, because of lack of experimental data at low Cu concentration, this cannot be stated with certainty. Copper activities in the solution at this range of concentrations were below the detection limit (10^{-6}). Calcium activity in the presence of ligand hardly varied, whereas its concentration increased). The amount of Ca released exceeded the amount of copper sorbed (Table 3, Fig.1) in the presence of ligand.

Copper adsorption isotherms on Chernozem in the presence and in the absence of succinic acid expressed in terms of the equilibrium copper activity and concentration are shown in Fig.2. Experimental results indicated that the relationship between the amount of Cu sorbed (S_{Cu}) and its activity in the final solution did not vary with the addition of a ligand when pH dependence on the amount of Cu added was kept constant. By contrast, the adsorption isotherm, expressed in terms of copper concentration, shifted to higher concentrations corresponding to equal amounts of Cu sorbed. The exception was a short initial part of the curve, which shifted slightly to the lower concentrations.

Description of Cu Sorption by Langmuir and Freundlich Equations

Both Langmuir and Freundlich equations can adequately describe Cu sorption. The values of the copper adsorption maximum (S_m=12.4 meq/100 g) and Langmuir constant (K=40.9) in the absence of ligand were higher than when it was present (S_m=9.1 meq/100 g, K=29.40). The Freundlich equation describing the copper sorption in the absence of succinic acid in terms of copper activity (k=42.8, p=0.5) was then used to calculate the copper sorption isotherm in the presence of the ligand.

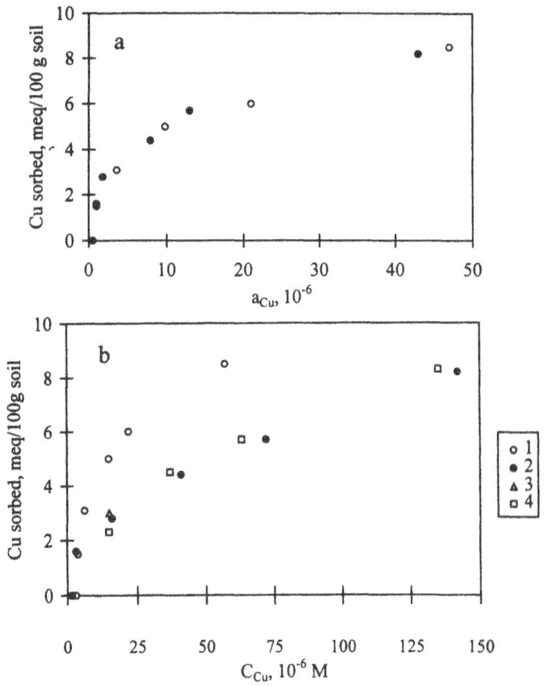

Figure 2. Experimental and calculated isotherms of Cu adsorption on Chernozem expressed in terms of equilibrium copper a) activity and b) concentration. Symbols are shown in Fig.1.

Calculation of Ion Speciation and Adsorption in the System Soil-Ca-Cu-H_2O-NO_3-CO_2-Succinic Acid

Ion Speciation in the Equilibrium Solution. Calculated ion speciation in the final solutions is presented in Table 4.

Calculations demonstrated that in the absence of succinic acid, the dominant Ca and Cu species in the final solutions were Ca^{2+} and Cu^{2+}. The portion of Cu^{2+} changed from 96.4 to 97.3% by varying the total Cu concentration and pH. Copper in the final solutions was present also in the forms complexed with inorganic ligands: $CuNO^{3+}$, $CuOH^+$ and others ($<n \cdot 10^{-2}$%). Concentration of $CuOH^+$ increased with increasing total Cu concentration and decreased with decreasing pH. These opposite effects were reflected in the value of $CuOH^+$ concentration in the final solutions (Figure.3) since Cu adsorption by soil was accompanied by a decrease in pH. The addition of succinic acid resulted in the decrease of Cu^{2+}, Ca^{2+}, and copper and calcium inorganic complex fractions at the expense of complexes with succinic acid (Table 4). The effect was more appreciable in the case of Cu than in the case of Ca because copper complexes with succinic acid were more stable than Ca ones.

Calculated activities of Cu^2 in the presence of succinic acid are in a good agreement with experimental (measured with ISE) copper activities. In the case when ligand is absent this agreement is somewhat worse (Table 5).

Table 4. Ion species in the equilibrium solutions (ranges).

	0 Suc	4 mM Suc
Cu added, meq/100g soil	0 - 8.6	0-8.6
pH	5.86 - 4.64	5.82 - 4.61
C_{Cu}, M	$3.1 \cdot 10^{-6}$ - $5.7 \cdot 10^{-5}$	$1.6 \cdot 10^{-6}$ - $1.4 \cdot 10^{-4}$
C_{Ca}, M	$4.7 \cdot 10^{-3}$ - $4.5 \cdot 10^{-3}$	$5.7 \cdot 10^{-3}$ - $5.1 \cdot 10^{-3}$
	Cu species, %	
Cu^{2+}	96.4 - 97.3	16.4 - 47.6
$CuSuc^0$	0	81.9 - 41.5
$CuHSuc^+$	0	1.3 - 10.1
$CuNO_3^+$	2.7	0.3 - 0.8
$CuOH^+$	0.7 - $4.2 \cdot 10^{-2}$	0.1 - $2.6 \cdot 10^{-2}$
$Cu(OH)_2^0$	$7.7 \cdot 10^{-2}$ - $2.8 \cdot 10^{-4}$	$6.7 \cdot 10^{-2}$ - $7.7 \cdot 10^{-4}$
$CuCO_3^0$	$5.7 \cdot 10^{-2}$ - $2.0 \cdot 10^{-4}$	$4.6 \cdot 10^{-3}$ - $5.0 \cdot 10^{-4}$
$Cu(NO_3)_2^0$	$2.4 \cdot 10^{-3}$ - $2.3 \cdot 10^{-3}$	$2.3 \cdot 10^{-3}$ - $6.4 \cdot 10^{-4}$
$CuHCO_3^+$	$4.4 \cdot 10^{-2}$ - $2.7 \cdot 10^{-3}$	-
$Cu_2(OH)_2^{2+}$	$4.1 \cdot 10^{-3}$ - $3.0 \cdot 10^{-4}$	-
	Ca species, %	
Ca^{2+}	100	91.1 - 97.0
$CaSuc^0$	0	8.5 - 1.4
$CaHSuc^+$	0	0.5 - 1.4
$CaOH^+$	$1 \cdot 10^{-7}$ - $1 \cdot 10^{-8}$	-
	Suc species, %	
Suc^{2-}	0	62.1 - 9.9
$Hsuc^-$	0	24.7 - 66.2
$CaSuc^0$	0	12.0 - 2.0
$CaHSuc^+$	0	0.7 - 1.8
H_2Suc^0	0	0.4 - 18.3
$CuSuc^0$	0	$3.2 \cdot 10^{-2}$ - 1.5
$CuHSuc^+$	0	$4.9 \cdot 10^{-4}$ - 0.4

Copper Adsorption Isotherm in the Presence of Succinic Acid. Calculated copper adsorption isotherms expressed in terms of copper solution concentration in the presence of succinic acid are presented in Fig.2b. Good agreement is observed between theory and the experimental data. The calculated adsorption isotherm was unaffected by Ca activity in the equilibrium solution because uncomplexed Suc was present in large excess and Ca and Cu did not compete for Suc (Table 4). The equivalence of Cu and Ca "exchange" breaks down in the presence of succinic acid. This may by attributable to the complexing of Cu and Ca with succinic acid in the solution. The formation of $CuHSuc^+$ and $CuSuc^0$ complexes resulted in the decrease of Cu^{2+} concentration in the solution and shifted the equilibrium (8) to the left:

$$2SOH + Cu^{2+} \leftarrow (SO)_2Cu + 2H^+ \quad (8)$$

In a similar manner, formation of $CaHSuc^+$ and $CaSuc^0$ complexes shifted the equilibrium (9) to the right:

$$(SO)_2Ca + 2H^+ \rightarrow 2SOH + Ca^{2+} \quad (9)$$

As a result, Cu sorption decreased and Ca desorption increased: $S_{Cu} < - S_{Ca}$. This explanation is supported by the model calculations, demonstrating good agreement with the experimental data (Fig.1).

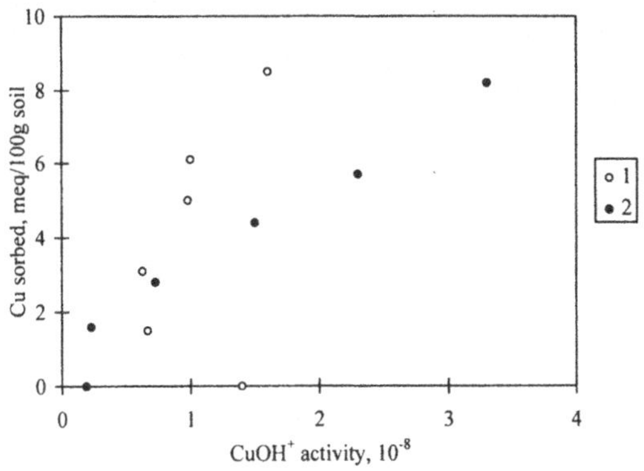

Fig.3. Copper adsorption as a function of $CuOH^+$ activity in the absence ( ) and presence ( ) of succinic acid

Table 5. Calculated and experimental activity of Cu in the equilibrium solutions in the presence and absence of succinic acid.

Cu added meq/100g soil	0	1.5	3.1	5.0	6.0	8.6
			in the absence of Suc			
a_{calc}	$1.9 \cdot 10^{-6}$	$2.4 \cdot 10^{-6}$	$3.9 \cdot 10^{-6}$	$9.2 \cdot 10^{-6}$	$1.4 \cdot 10^{-5}$	$3.5 \cdot 10^{-5}$
a_{exp}	$<10^{-6}$	$<10^{-6}$	$3.6 \cdot 10^{-6}$	$9.8 \cdot 10^{-6}$	$2.1 \cdot 10^{-5}$	$4.7 \cdot 10^{-5}$
			in the presence of Suc (with compensation for pH changes)			
a_{calc}	$1.5 \cdot 10^{-7}$	$3.6 \cdot 10^{-7}$	$2.4 \cdot 10^{-6}$	$7.8 \cdot 10^{-6}$	$1.6 \cdot 10^{-5}$	$4.1 \cdot 10^{-5}$
a_{exp}	$<10^{-6}$	$<10^{-6}$	$1.8 \cdot 10^{-6}$	$7.9 \cdot 10^{-6}$	$1.3 \cdot 10^{-5}$	$4.3 \cdot 10^{-5}$

DISCUSSION

There is much speculation that the hydrolised metal ions are active species to be adsorbed preferably to oxide surfaces and soil (James and Healy, 1972; Barrow, 1986). In this approach pH affects the adsorption through changing proportions of MOH^+, and, according to Barrow (1986), partly through varying electrostatic potential of variable-charge surfaces. These effects can be separated by plotting sorption against concentration of MOH^+. Barrow showed that for the soils studied most values fell near a common line when plotted against $ZnOH^+$, and he concluded that the simplest explanation was probably that $ZnOH^+$ dominated adsorption. Chairidchai and Ritchie (1990) showed that zinc adsorption in the absence of ligands was well correlated with $ZnOH^+$ in the final solution, but was not correlated with any of the other species in a way that made adsorption independent of pH. In contrast to this, zinc adsorption in the presence of organic ligands was not correlated with $ZnOH^+$.

In our experiments copper adsorption in the absence of organic ligand was not correlated with $CuOH^+$ activity in the solution (r^2=0.42). In the presence of succinic acid copper adsorption correlated with $CuOH^+$ activity (r^2=0.97), but the results obtained in the absence and presence of succinic acid did not fall near a common line when plotted against $CuOH^+$ (Fig.3). By contrast, adsorption isotherms in terms of Cu^{2+} activity fitted a common line, irrespective of whether succinic acid was present or not (Fig.2a). Thus, the simplest explanation is probably that adsorption was dominated by Cu^{2+} ions and succinic acid affected adsorption due to the changing concentration of Cu^{2+} ions.

Our calculations and experiments support the hypothesis that theadsorption isotherm expressed on terms of metal activity experimentally obtained in the absence of complexing ligands can be used to describe adsorption in the presence of ligands. However, the opposite conclusion may be deduced from the data of McBride (1981) for adsorption isotherms of Cu on montmorillonite in the absence and presence of different organic ligands. Only salicylic acid did not affect the Cu adsorption isotherm (at the amount of Cu adsorbed more then 40 meq/100g), whereas phthalic, fulvic and citric acids affected the adsorption isotherm expressed in terms of equilibrium Cu^{2+} activity. This disagreement may be connected with the difference of Cu adsorption mechanisms in the presence of different ligands and different adsorbing phases. In the case of phthalic, fulvic and citric acids and salicylic acid at low Cu concentrations, adsorption of copper on montmorillinite may be treated like adsorption of various Cu species (for example, free Cu^{2+} and metal-ligand complexes). In this case Cu adsorption is determined by the activity of adsorbing species.

Davis and Leckie (1978) demonstrated that the effect of ligands on metal uptake by soil depends on the ligand adsorption behaviour. Complexing ligands that are not adsorbed decrease metal adsorption, whereas the ligand adsorbing on the surface may create new adsorption sites for metal. It was found in our experiments that approximately 20% of initial succinic acid was removed from the solution regardless of the pH and copper solution concentration. Ligand removal may be caused by sorption. Uptake by microorganisms may also take place. These mechanisms may led to a different consequence. In the case of ligand adsorption, new adsorption sites for copper may be created and influence Cu distribution between soil and solution. In the case of microorganisms uptake ligand remaining in solution affects Cu adsorption on the soil surface. Unfortunately, the contribution of each mechanism can not be evaluated in our experiments and remains to be investigated.

If adsorption of succinic acid takes place the decrease in the copper concentration in the solution when a low amount of metal is added (Table 3) may be attributed to the increase in Cu adsorption due to complexing by adsorbed ligand, or alternatively, due to adsorption of metal-ligand complexes. As the copper concentration in the solution increases, new adsorption sites are completed and the adsorption mechanism changes. At higher metal concentrations independent of the mechanism of ligand removal, the decrease in the copper uptake may be explained in terms of the competition for Cu complexing between the surface sites and succinic acid remaining in the solution.

In the model calculations the presence of dissolved organic carbon (humic, fulvic and other organic compounds), which may be extracted from soil during mixing the soil and solutions, was ignored. Dissolved organic compounds can form complexes with Cu and affect copper activity in solution. This may lead to a significant discrepancy between calculated and measured Cu activity, especially at low Cu concentrations when the majority of dissolved Cu may be bound to DOC. The amount of DOC released from soil is a function of soil properties, it increases with an increase in soil:solution ratio, with an increase in the sodium and potassium concentration, and with decrease in the calcium concentration (Reinds *et al.*,1995). In Typical Chernozem the amount of humic acids (HA) dominates over more mobile fulvic acids (FA) (the ratio HA carbon: FA carbon is equal 1.5-1.6 (Orlov, 1992)), exchangeable Ca dominates over exchangeable K and Na, pH is close to neutral (table 1). The combination of these factors suppresses the DOC concentration. As a result, the amount of DOC in natural Chernozem soil solution is not high in spite of high content of organic carbon in this soil. Snakin *et al.*(1997) reported that mean DOC for 11 upper horizon Chernozem samples was 37.7±17.1 mg/L. For comparison, DOC ranged from 12 in a Grey Forest soil to more than 1000 mg/L in a Solonchaks (Snakin *et al.*, 1997). The relation between DOC and concentration of organic acids (in $\mu M_C/L$) was described by:

$$RCOO_{tot} = m\ DOC, \tag{10}$$

where $RCOO_{tot}$ denotes organic acids (Reinds *et al.*, 1995). Thalue of m is taken to be 5.5 μM_C/mg C in accordance with Henriksen and Seip (1980) and Bril (1994). Taking the soil moisture content to be 20%, soil:solution ratio 1:10, DOC concentration in soil solution 40 mg C/L, the estimated $RCOO_{tot}$ concentration in equilibrium solution will be on the order of 10^{-6} M. Soil transformation into Ca-form before experiments and the use of $Ca(NO_3)_2$ as a background electrolyte during adsorption experiments should result in additional significant decrease of DOC in the equilibrium solution. However, dissolved organic carbon released from the soil during adsorption experiments in the amount 10^{-6} M/L or less may influence Cu activity, especially at low Cu concentrations (concentration of copper in the equilibrium solutions ranged from $3.1 \cdot 10^{-6}$ to $5.7 \cdot 10^{-5}$, Table 4). As a result, calculated copper activity for a low amount of Cu added was higher ($1.9 \cdot 10^{-6}$ for no copper added and $2.4 \cdot 10^{-6}$ for 1.54 meq/100g Cu added) then copper activity determined by ICE (it was below the detection limit of 10^{-6}). At higher copper concentrations calculated Cu activities in the equilibrium solution agreed satisfactorily with experimental observations. The presence of DOC evidently can be ignored in the presence of succinic acid since the concentration of this ligand is much higher (5mM) then possible DOC concentration. This is supported by good agreement between experimental and calculated activities of copper in the presence of ligand (Table 5).

CONCLUSION

Experimental results show that in the presence of succinic acid pH of a soil suspension decreases, copper complexing is suppressed and copper adsorption behaviour is determined by solution pH. When the drop in pH caused by the introduction of succinic acid is compensated the relationship between copper activity in solution and amount of copper adsorbed is unaffected by the presence of succinic acid, whereas the relationship between copper adsorbed and total copper in solution shift to the higher copper concentration in solution in the presence of ligand. Experiments and chemical equilibrium calculations demonstrate that preferentially adsorbed copper species is most likely to be Cu^{2+}, both in the presence and absence of succinic acid. Copper sorption decreases and calcium desorption increases in the presence of succinic acid. As a result the equivalence of Cu and Ca "exchange" observed in the absence of ligand breaks down. Succinic acid affects Cu sorption and Ca desorption by changing the solution concentration of Cu^{2+} and Ca^{2+} due to the formation of soluble $CaSuc^0$ and $CaHSuc^+$ complexes for most studied concentrations. A mathematical model describing ion distribution in the soil-solution system in terms of competition between surface sites and succinic acid in a soil solution for Cu complexing, demonstrates good agreement with experimental data.

REFERENCES

Arinushkina, E.V., 1970. A Handbook of Soil Chemical Analysis. Moscow State University.

Barrow, N.J., 1986. Testing a mechanistic model. IV. Describing the effects of pH on Zn retention by soils. J. Soil Sci., 37, 295-302.

Basta, N.T. and M.A. Tabatabai, 1992 a. Effect of cropping systems on adsorption of metals by soils: I. Single-metal adsorption. Soil Sci., 153, 108-114.

Basta, N.T. and M.A. Tabatabai, 1992 b. Effect of cropping systems on adsorption of metals by soils: II. Effect of pH. Soil Science, 153, 195-204.

Biscay, P.E. 1965. Mineralogy and sedimentation of recent deep-sea clay in the Atlantic Ocean and adjacent seas and oceans. Geol. Soc. Amer. Bull., 76, 803-832.

Brand, L.E, W.G. Sunda and R.R.L. Guillard, 1986. Reduction of marine phytoplankton reproduction rates by copper and cadmium. J. Exp. Mar. Biol. Ecol., 96, 225-250.

Bril, J., 1994. The behaviour of dissolved humic substances in neutral soils. Report. DLO Institute for Agrobiological and Soil Fertility Research.

Bryce, A.L., W.A. Kornicker and A.W. Elzerman, 1994. Nickel adsorption to hydrous ferric oxide in the presence of EDTA: effects of component addition sequence. Environ. Sci. Technol., 28, 2353-2359.

Cavallaro, N. and M.B. McBride, 1978. Copper and Cadmium adsorption characteristics of selected acid and calcareous soils. Soil Sci. Soc. Amer. J., 42, 550-556.

Chairidchai, P. and G.S.P. Ritchie. 1990. Zinc adsorption by lateritic soil in the presence of organic ligands. Soil Sci. Soc. Amer. J., 54, 1242-1248.

Davis, J.A. and J.O. Leckie. 1978. Effect of adsorbed complexing ligands on trace metal uptake by hydrous oxides. Environ. Sci. Technol., 12, 1309-1315.

De Vries, W. and D.L. Bakker, 1996. Manual for calculating critical loads of heavy metals for soils and surface waters. Preliminary guidelines for environmental quality criteria, calculation methods and input data. Report 114. DLO Winand Staring Centre, Wageningen, The Netherlands.

Elgawhary, S.M., W.L. Lindsay and W.D. Kemper, 1970. Effect of complexing agents and acids on the diffusion of zinc to a simulated root. Soil. Sci. Soc. Amer. Proc., 34, 211-214.

Elliott, H.A. and C.P. Huang, 1979. The adsorption characteristics of Cu(II) in the presence of chelating agents. J. Colloid and Interface Sci., 70, 29-45.

Elliott, H.A. and C.P. Huang, 1985. Factors affecting the adsorption of complexed heavy metals on hydrous Al_2O_3. Water Sci. Technol., 17, 1017-1028.

Farrah, H. and W.F. Pickering. 1976. The sorption of zinc species by clay minerals. Aust. J. Chem., 29, 1649-1656.

Halvorson, A.D and W.L. Lindsay, 1977. The critical Zn^{2+} concentration for corn and nonadsorption of chelated zinc. Soil Sci. Soc. Am. J., 41, 531-534.

Harter, R.D. and R. Naidu, 1996. Role of metal-organic complexation in metal sorption by soils. Adv. Soil Sci., 55, 219-263.

Henriksen, A. and H.M. Seip, 1980. Strong and weak acids in surface waters in southern Scotland. Water Res., 14, 809.

Hodgson, J.F., H.R. Geering and W.A. Norvell. 1965. Micronutrient cation complexes in soil solution: Partition between complexed and uncomplexed forms by solvent extraction. Soil Sci. Soc. Amer. Proc., 29, 665-669.

Huang, C.P., E.A. Roads and O.J. Hao. 1988. Adsorption of Zn onto hydrous aluminosilicates in the presence of EDTA. Wat. Res., 22, 8, 1001-1009.

James, R.O. and T.W. Healy. 1972. Adsorption of hydrolisable metal ions at oxide-water interface. 3. A thermodynamic model of adsorption. J. Colloid and Interface Sci., 40, 65-81.

Karpukhin A.I., I.M. Yashin and V.A. Chernicov, 1993. Formation and migration of complexes of water-soluble organic substances with ions of heavy metals in taiga landscapes of European North. Izvestiya TSHA, Moscow, 107-125. In Russian.

Kurdi, F. and H.E. Donner. 1983. Zinc and copper sorption and interaction in soils. Soil Sci. Soc. Amer. J., 47, 873-876.

Lexmond, Th.M., 1980. The effect of pH on copper toxicity to forage maize grown under field conditions. Neth. J. Agric. Sci., 28, 164-183.

Lindsay, W.L. 1979. Chemical Equilibria in Soil. Wiley-Intersci. Publ., John Wiley and Sons.

Lindsay, W.L., and W.A. Norvell. 1969. Reactions of EDTA complexes of Fe, Zn, Mn and Co with soils. Soil Sci. Soc. Amer. Proc., 33, 86-91.

Linn, J.H. and H.A. Elliott, 1988. Mobilization of Cu and Zn in contaminated soil by nitrilotriacetic acid. Water, Air, and Soil Pollution, 37, 449-458.

Malzer G.L. and S.A. Barber, 1976. Calcium and stroncium adsorption by corn roots in the presence of chelates. Soil Sci. Soc. Amer.J., 40, 727-731.

McBride, M.B., 1981. Copper in solid and solution phases of soil. In: Y.F. Logeragan, Robson A.D., and K.D. Grahm (Editors). Copper in soils and plants. Academic Press, New York. pp. 24-43.

McBride, M.B., 1989. Reactions controlling heavy metal solubility in soils. Adv. Soil Sci., 10, 1-56.

McKeague, J.A., M.V. Cheshire, F. Andreux and J. Berthelin, 1986. Organo-mineral complexes in relation to pedogenesis. In: P.M.Huang and M. Schnitzer (Edirors). Interactions of Soil Minerals with Natural Organics and Microbes. Soil Sci. Soc. Amer. Inc., Madison, Wisconsin, pp. 549-592.

McLaren, R.G.and D.V. Crawford, 1973. Studies on soil copper. II. The specific adsorption of copper by soils. J.Soil Sci.,24, 4, 443-452.

Mench, M., J.L. Morel, A. Guckert and B. Guillet. 1988. Metal binding with root exudates of low molecular weight. J. Soil Sci., 39, 521-527.

Nowack, B., J. Lutzenkirchen, Ph. Behra and L. Sigg, 1996. Modeling the adsorption of metal-EDTA complexes onto oxides. Environ. Sci. and Technol., 30, 7, 2397-2405.

Orlov, D.S., 1992. Soil Chemistry. Moscow University Press.

Pampura, T.V., 1992. The influence of low molecular weight organic acids on sorption of Zn and Cu in Chernozem. Conference on Physical Chemistry and Mass Exchange processes in soils, Pushchino. Proceedings. Pushchino Research Centre. pp.85-89.

Pampura, T.V., D.L. Pinsky, V.G. Ostroumov, V.D. Gerchevich and V.N Bashkin, 1993. Experimental Study of the buffer capacity of Chernozem contaminated with copper and zinc. Eurasian Soil Sci., 25, 10, 27-38.

Pollard, J.H. 1982. A handbook of numerical and statistical techniques. Finances and statistics. Moscow. 344p. (in Russian).

Ponomareva V.V., 1972. Biogeochemical Processes in Podzols. Leningrad (In Russian).

Prasad, B., M.K. Sinha and N.S. Randhawa. 1976. Effect of mobile chelating agents on diffusion of zinc in soils. Soil Sci., 122, 5, 260-266.

Reinds, G.J., J. Brill, W. de Vries, J.E. Groenenberg and A. Breeuwsma, 1995. Critical loads and excess loads of cadmium, copper and lead for European forest soils. Report 96. DLO Winand Staring Centre, Wageningen, DLO Institute for Agrobiological and Soil Fertility Research, Haren, The Netherlands.

Rovira, A.D. 1969. Plant root exudates. Bot. Rev., 35, 35-57.

Sanders, J.R., 1982. The effect of pH upon the copper and cupric ion concentrations in soil solution. J. of Soil Sci., 33, 679-689.

Sillen, L.G. and A.E. Martell. 1971. Stability constants of metal ion complexes. Special publication N 25. Chemical Society. London.

Smith, W.H. 1976. Character and significance of forest root exudates. Ecology, 57, 324-331.

Snakin, V.V., A.A. Prisyazhnaya and O.V. Rukhovich, 1997. Soil liquid phase composition. Russian Academy of Sciences. Institute of Soil Science and Photosynthesis. REFIA House, Moscow.

Stevenson, F.J. and M.S. Ardakani. 1972. Organic matter reactions involving micronutrients in soils. In: J.J. Mortvedt, P.M. Giordano, and W.L. Lindsay (Editors). Micronutrients in agriculture. Soil Sci. Soc. of Am., Madison, Wisconsin. pp.79-114.

Stumm, W. and Morgan J.J. 1981. Aquatic Chemistry. John Wiley and Sons, New York.

Tan, K.H., 1986. Degradation of soil minerals by organic acids. In: P.M.Huang and M. Schnitzer (Editors). Interactions of soil minerals with natural organics and microbes. Soil Sci. Soc. Amer. Inc., Madison, Wisconsin, pp. 1-27.

Vuceta, J. and J.J. Morgan, 1978. Chemical modeling of trace metals in fresh waters: Role of complexation and adsorption. Environ. Sci. Technol., 12, 1302-1309.

Vulava, V.M., B.R. James and A. Torrents, 1997. Copper solubility in Myersville B horizon soil in the presence of DTPA. Soil Sci. Soc. Am. J., 61, 44-52.

Yashin, I.M and I.S. Kaurichev, 1992. Pedogenetic function of water soluble organic matter in taiga landscapes. Pochvovedenie, 10, 49-61 (In Russian).

INTERACTION OF IRON AND ORGANIC MATTER IN RELATION TO ITS UPTAKE BY PLANTS

A. M. ELGALA

Department of Soil Science,
Faculty of Agriculture,
Ain Shams University, Cairo, Egypt.

INTRODUCTION

Response to iron supply was observed for crops growing in newly reclaimed sandy and calcareous soils of Egypt. Fe-EDDHA was more effective as a soil application or foliar spray than $FeSO_4$ in correcting iron deficiency in these soils characterized by alkaline pH and high $CaCO_3$ content. Due to costs and irregular supply of the synthetic chelates in the market, crop production was affected. The addition of organic residues to these soils was found to reduce the incidence of iron deficiency in crops. Attempts were made to study the nature and role of humic substances formed under arid conditions with respect to the availability and uptake of iron by plants. The aim of this work is to present various studies dealing with humic and fulvic acids in relation to their chemical properties and the nature of complexes formed with Fe as a function of organic residue source and their role in solubilizing Fe and other nutrients from soils. A comparative study regarding the effect of synthetic versus natural chelates on Fe uptake by plants from soil or from nutrient solution having increasing concentrations of Fe will be presented.

MATERIAL AND METHODS

Soil Characteristics

The range of some of the physical and chemical characteristics of the studied soils are shown in Table 1. The reported physical and chemical properties (texture, $CaCO_3$, pH and CEC) of the alluvial, calcareous and sandy soils were determined by the methods described by Jackson (1958). Total iron by wet ashing with HNO_3, H_2SO_4 and $HClO_4$ and water soluble Fe were evaluated as described by Elgala and Hendawy (1972). The DTPA-extractable method described by Lindsay and Norvell (1978) was followed to evaluate the critical level values for Fe and other elements in alluvial, calcareous and sandy soils (Elgala *et al.*, 1986). The range of DTPA-extractable iron in cultivated areas of Egypt was reported by the Egyptian Academy for Scientific Research and Technology (ASRT, 1987).

Effect of Mineral-Organic-Microorganism Interactions on Soil and Freshwater Environments
Edited by Berthelin, *et al*, Kluwer Academic / Plenum Publishers, New York, 1999

Behavior of Iron in Soils

Determinations were made of the amounts of iron that remain soluble in water and that was removed with acidified NH_4OAc when 5 µg/g Fe in the form of ferrous sulfate or Fe-ethylene diamine di (0-hydroxyphenyl acetic "Fe-EDDHA") was added to 10 g soil samples. Bidistilled water was added to raise the solution volume to 50 ml. The mixture was shaken for one hour and then filtered. The soil sample on the filter paper was washed with 25 ml distilled water and then with 25 ml of 60% methanol. Ammonium acetate-extractable Fe was determined by passing 50 ml of NH_4OAc, pH 3 solution through the previously washed soil on the filter paper ; Fe was determined in the extract (Elgala and Hendawy, 1972).

Preparation of Humic and Fulvic Acid and Determination of Stability Constant

A laboratory experiment was conducted to study the nature of isolated and purified humic acid (HA) and fulvic acid (FA) formed after the enrichment of a calcareous soil with different sources of organic residues.

To five-hundred gram samples of calcareous soil (11.6% $CaCO_3$) various amounts of dried and ground clover residues (49.4% C, 0.84% N), wheat straw (50.2% C, 0.16% N) or farm manure (23.4 % C, 1.04% N) were added. These were incubated at 30°C and 50 % field capacity for 2 weeks, then organic matter was extracted with 0.5 N NaOH after pretreatment with 0.5 N $NaHCO_3$ and separated to humic acid HA and FA by acidifying the extract with 1 N HCl. The fractions were purified by electrodialysis and then dried at room temperature (Kononova, 1966).

In order to determine the stability constant formed between separated humic or fulvic acids and Fe, the ion exchange method that was first suggested by Schubert (1948) and used after modification by many authors was followed (Elgala *et al.*, 1976).

Sequestering Ability of Synthetic and Natural Chelating Agents

Natural and synthetic chelating compounds may differ in their abilities to solubilize various elements when present in soils. Thus the HA and FA previously prepared were compared with Na_2-EDIA and Na_2-EDDHA in their ability to bind certain cations from calcareous soil. Extractable metal cations from the soil were determined using 25 ml of various natural and synthetic chelates solutions. The concentration of the chelating agents in this study was 1000 µg/g for all material except FA,which was 400 µg/g.

Effect of Composted Material on Fe Availability

A sandy soil (pH 7.9, OM 0.5% and N 8%) collected from North-Tahreer section, and a calcareous soil (pH 7.5, OM 1.37% and $CaCO_3$, 35.5%) collected from El-Nobaria, Egypt, were used in this study. Separate composted heaps were prepared from corn stalk, clover residues and rice straw. The composted materials had C/N ratios of 18.6, 11.9 and 22.3, respectively. A greenhouse experiment was carried out using plastic pots with 8 kg soil. Composted materials were mixed thoroughly with the soil samples at the rate of 20 g/kg before planting. Five tomato seedlings were planted in each pot and these were thinned to 3 plants after 2 weeks. For one group of pots the soil was also enriched with iron in the form of $FeSO_4$ at the rate of 10 µg/g Fe. To compare the effects of natural chelating compounds with those of inorganic and synthetic chelating compounds on growth and yield of tomato plants, the following treatments were applied :

Control
$FeSO_4$, 10 µg/g Fe
Composted corn stalk
Composted corn stalk + $FeSO_4$
Composted clover residues
Composted clover residues + $FeSO_4$
Composted rice straw
Composted rice straw + $FeSO_4$

NPK fertilizers were added to each pot 15 and 45 days after planting. In each dose N was applied at the rate of 0.34 g/pot as ammonium sulfate, P at the rate of 0.24 g P_2O_5/pot as superphosphate and K_2O at the rate of 0.32 K_2O/pot (80 kg K_2O/feddan) as potassium sulfate. Each treatment consisted of three replicates in a complete randomized design. Soil moisture was maintained around 70-80% of the field capacity. The yield of fresh tomato fruits was measured throughout the period of 115 to 135 days from planting. The total fresh weight of tomato fruits was recorded. Fe content was determined in the dried tomato fruits using the atomic absorption spectro-photometer after wet ashing with 10 ml conc. HNO_3 and 4 ml of 1:1 conc. H_2SO_4 and $HClO_4$.

Effect of HA and Na_2-EDDHA on the Uptake of Fe

A pot experiment was carried out to study the effects of HA and NA_2-EDDHA on the uptake of Fe by barley (*H. vulgare var. baladi)* exposed to increasing concentrations of Fe. Plastic pots 5 cm in diameter and 5 cm in depth were each filled with 1000 g fine washed sand over gravel and coarse sand. 25 barley seeds were planted in each, then thinned to 20 after germination. Pots received Hogland solution with increasing concentrations of Fe. The Fe concentrations as $FeSO_4$ were 0.0, 30, 60, and 120 µg/g.

Plants were then cut, washed with distilled water, dried at 60°C and weighed. Amounts of 0.4 g of plant material was wet ashed and Fe was determined using atomic absorption spectrophotometer.

RESULTS AND DISCUSSION

Status and Behaviour of Iron in Arid Soils

Results of total iron tests (Table 1) indicate that in the alluvial soils, the amounts ranged between 1% and 6% depending on soil texture. The sandy soils contained 0.8% to 2.0% Fe, and the calcareous soils contained less than 0.6%, depending on $CaCO_3$ content (ASRT, 1989; Elgala and Hendawy, 1972). The amounts of water-soluble iron ranged from 1.0 to 3.3 ug/g depending on soil texture and humus content in the soil (Elgala and Hendawy, 1972). The DTPA soil test method suggested by Lindsay and Norvell (1978) was used to evaluate the critical level values of Fe, Mn and Zn in Egyptian soils using maize *(Zea mays, L.)* as a test plant. The suggested Fe critical levels in alluvial, sandy and calcareous soils were 5.6, 3.4 and 3.8 µg/g, respectively, (Table 1) (Elgala et al., 1989).

Table 1. Range and mean values of some of the physical and chemical characteristics in the main soil types

Soil properties	Alluvial typic (Torrets-Torrifluvents)	Sandy Typic Quartzipsamments	Calcareous Calciorthids
Texture	Sandy loam-clay	Sandy	Loamy
$CaCO_3$ (%)	2.0 - 4.0	1.0 - 4.0	9.0 > 90.0
pH	7.3 - 8.3	7.0 - 8.0	7.5 - 8.8
CEC meq/100 g	15.0 - 55.0	3.0 - 8.0	7.0 - 12.0
OM (%)	0.5 - 2.0	0.2 - 0.8	0.4 - 0.8
Total Fe (%)	1.0 - 6.0	0.8 - 2.0	< 0.6
Water-soluble Fe, µg/g	1.6 - 2.0	1.0 - 3.3	2.0 - 3.0
DTPA-extractable Fe, ug/g	2.5 - 15.0	1.5 - 4.4	2.8 - 4.8
Critical level, µg/g	5.6	3.4	3.8

* Range and mean values are based on analysis of hundreds of soil samples .

The Egyptian Academy for Scientific Research and Technology "ASRI" has initiated a research project dealing with "micronutrients and clay mineralogy of the soils of Egypt". The objective of the project was to evaluate ranges of micronutrient levels in the various soils and to make tentative classification of these soils with respect to their content of micronutrients (ASRI, 1987). A summary of the work reported by the project team for certain Governorates is shown in Table 2 . It seems that appreciable amounts of available Fe, Mn, Zn and Cu are present in most soils of the Nile Valley. Low DTPA-Fe (5 to 15%) can be found only in some soils belonging to the great Vertic Torrifluvents and Typic Torrerts; it is higher in the newly reclaimed soils of the Calciorthids. All the soils of Fayoum and most of the newly reclaimed soils west of the Nile Delta contained DTPA-Zn levels within the deficient ranges (95-100%).

Table 2. Percentage of samples taken from various Governorates that were deficient in extractable Fe, Mn, Zn and Cu

Governorates	Fe	Mn	Zn	Cu
Damanhour	2	0	23	0
Kafr El-Shaikh	5	0	30	5
Menofia	0	0	15	0
Qaliobia	0	0	15	0
El-Sharkeia	5	0	20	0
Giza	0	0	13	0
Beni Sweif	15	2	8	2
Fayoum	0	0	100	0
Newly Reclaimed Land W, Nile Delta	14	12	95	12

Results show that the application of iron sulfate did not cause pronounced change in the amounts of water-soluble forms of Fe in various soils, but Fe-EDDHA addition resulted in a significant increase in the amount of water-soluble and acidified ammonium acetate soluble-Fe in alluvial, sandy and calcareous soils (Fig. 1) (Elgala and Hendawy, 1972). Positive response in plant growth, iron content and yield was obtained when iron was applied in the form of Fe-EDDHA. The iron chelate proved to be more effective than $FeSO_4$ in correcting iron deficiency on crops grown in sand or in calcareous soils, particularly when applied as a foliar spray. Similar results were previously found by Elgala (1971) and Wallace (1966).

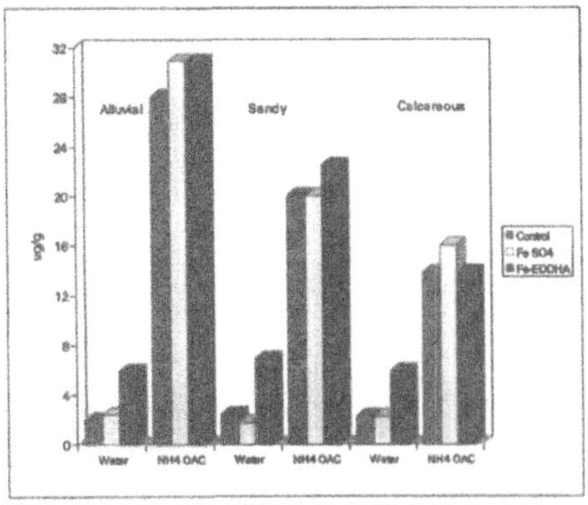

Figure 1. Amounts of iron soluble in water and in NH4 OAc (pH 3) from various soil samples previously treated with iron sulphate or iron chelate.

Effect of Organic Source on the Properties of Formed Humic Acids

Results in Table 3 reveal that the stability constants values were higher with humic acids than with fulvic acids. The stability constants of HA with Fe range from 6.65 to 7.59 and those for FA range from 4.94 to 5.80. This indicates that the stability constant values are not related to the sum of COOH and phenolic-OH groups, but possibly to the total number and location of phenolic groups attached to the organic compounds.

However, humic acid samples had more phenolic-OH groups, which revealed the importance of phenolic-OH groups in determining the stability with Fe. It has been reported that phenolic-OH groups have higher affinity for metal ions than COOH groups do, and that different types of carboxyl groups occur on soil humus substances with different acidic strengths (Chaberek and Martell, 1959; Elgala *et al.*, 1976; Theng and Posner, 1967). The effect of source of organic residues on the stability constant with Fe was slight and generally in the order : farm manure > wheat straw ≥ clover residues. This again agrees with the idea of the importance of phenolic-OH group attached to HA in complex reactions (El-Damaty *et al.*, 1979; Schnitzer and Skinner, 1966). Schnitzer (1992) indicated that soil inorganic and organic components are closely associated to form water-soluble and water-insoluble complexes of widely differing chemical and biological stabilities. These interactions include chelation and adsorption on external and internal surfaces and often involve low energy binding.

Table 3. Some chemical characteristics of humic and fulvic acids isolated from calcareous soils enriched with different organic residues and log stability constants of Fe complexed with HA and FA.

Type of plant residue	Humus fraction	C/N	COOH	Phenolic-OH	Total acidity	Log Stability constants values
			---------- meq/100 g acid	---------		
Clover	HA	14.5	218	716	934	6.68
	FA	12.5	527	626	1153	4.94
Wheat straw	HA	26.1	204	796	1000	6.65
	FA	29.7	650	682	1332	5.80
Farm manure	HA	12.5	290	800	1190	7.59
	FA	11.0	550	735	1285	5.18

Comparative Ability of Synthetic and Natural Chelates to Solubilize Fe, Zn, Mn and Ca

The data in Figures 2 and 3 indicate the higher affinity of Na_2-EDTA or Ca-EDTA than Na_2-EDDHA, so the solubilizing abilities for Fe, Zn and Mn were lower. More Fe, Zn and Mn particularly Fe and Mn) were extracted with Na_2-EDDHA as compared with Na_2-EDTA (Chaberek and Martell, 1959; Wallace, 1962). Except for HA from clover, HA samples generally removed greater amounts of Fe than Na_2-EDTA and lower amounts than Na_2-EDDHA. Zn solubility was lower but Mn solubility was higher with HA than with Na_2-EDTA. This could be explained by the low stability constants of HA with Zn (Table 3). In general, the differences among HA sources in removing Fe, Mn and Zn were also related to differences in their log stability constants with humic acids.

The average amounts of different elements solubilized by FA exceeded those removed by HA. The amounts of Fe removed by FA were comparable to those removed by Na_2-EDDHA. Mn solubility by FA was 2 to 4 times (depending on organic residue source) that extracted by Na_2-EDTA. It is worth noting that the highest amounts of soluble Ca were also caused by FA, reaching about 3 times that for FA of wheat straw as compared to that removed by Na_2-EDTA. Differences between HA and FA sources in the solubility of studied cations were related to differences in their molecular structure, total acidity, type of functional groups and consequently, the stability constants as well as their affinity for Ca (Elgala *et al.*,1976; Kononova and Alexandrova, 1973; Schnitzer and Skinner, 1966).

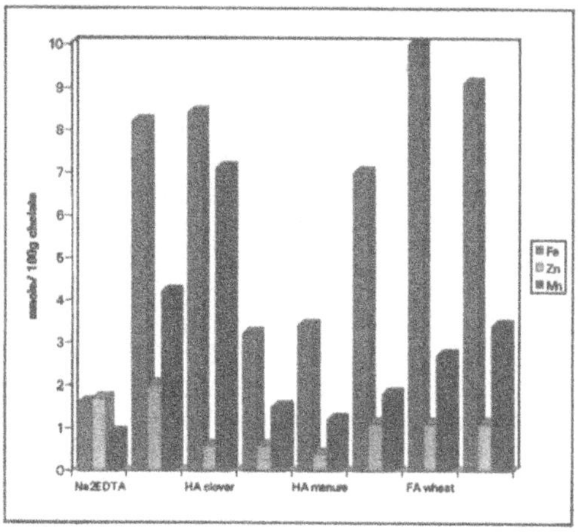

Figure 2. Sequestering ability of some chelating agents for Fe, Mn and Zn from calcareous soil.

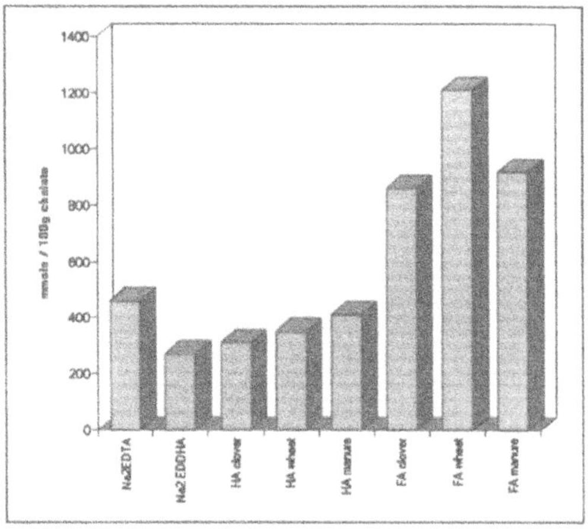

Figure 3. Sequestering ability of some chelating agents for Ca from calcareous soil.

Natural Chelates Versus Synthetic Fe Chelate or $FeSO_4$ for Supplying Iron to Tomato Plants

Results of the effect of different composted materials and Fe sources on yield and Fe content of tomato fruits are shown in figures 4 and 5. Date reveal that the application of different composted materials either alone or enriched with $FeSO_4$ caused a significant increase in yield of tomato fruits as compared to other treatments. The highest fruit yield nearly double the control, was obtained with the addition of clover residues to sandy and calcareous soils . No remarkable difference in yield was found when the composted materials were enriched with $FeSO_4$.

On the other hand, $FeSO_4$ treatment caused slight increase in fruit yield and Fe-EDDHA application to both soils caused pronounced increase in yield but less than the effect of the organic residues. The more favorable effect of organic residues on the tomato yield than that of supplying the inorganic or chelated form of Fe could be related to the role of humus in improving the physical, chemical and nutritional conditions, besides supplying the plants with available forms of iron. Parr *et al.* (Personal communication) indicated that organic wastes and residues offer the best possible means of restoring the productivity of severely eroded agricultural soils or of reclaiming marginal soils.

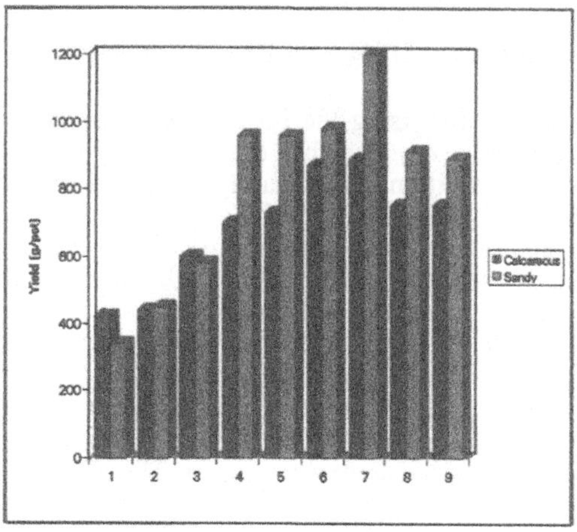

1 = Control
2 = $FeSO_4$
3 = Fe-EDDHA
4 = Composted corn stalk
5 = Composted corn stalk + Fe
6 = Composted clover
7 = Composted clover + Fe
8 = Composted rice straw
9 = Composted rice straw + Fe

Figure 4. Effect of different composted materials and Fe sources on fruit yield of tomato plants in calcareous and sandy soils.

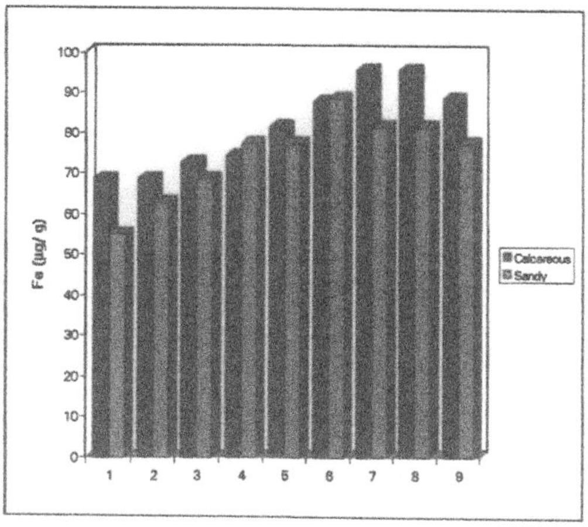

1 = Control
2 = $FeSO_4$
3 = Fe-EDDHA
4 = Composted corn stalk
5 = Composted corn stalk + Fe
6 = Composted clover
7 = Composted clover + Fe
8 = Composted rice straw
9 = Composted rice straw + Fe

Figure 5. Effect of different composted materials and Fe sources on Fe content of tomato plants in calcareous and sandy soils.

Concerning Fe concentration in fruits results indicate that the values were relatively lower in fruits of plants grown in sandy soil. In general, the composted materials increased Fe content by about 28% for calcareous soil and about 47% for the sandy soil. The highest values were found with clover in both soils.

On the other hand, the percentage increases in Fe content were 0 and 12% with $FeSO_4$ treatment and 6% and 24% with Fe-EDDHA in calcareous and sandy soils, respectively. Such results also indicate the role of humus in improving the yield quality of tomato fruits, by supplying available Fe and maintaining nutritional balance among nutrients (Ismail *et al.*, 1984).

Effect of Humic Acid and Na_2-EDDHA on Dry Matter and the Uptake of Fe

Figure 6 shows that increasing the Fe concentration up to 120 µg/g in the nutrient solution had no effect on dry weight, even though plants suffered in the highest concentration. Humic acid additions significantly increased the dry weight at 0.0 and 30.0 µg/g iron, but decreased it at 120 µg/g.

The iron concentration in plants progressively increased with increasing Fe concentration in the nutrient solution. Humic acid slightly increased Fe uptake only at 30 µg/g but decreased both iron uptake and iron concentration in plants exposed to high Fe concentrations. The decrease was almost 50% in the highest Fe treatment (120 µg/g).

With Na_2-EDDHA, Fe uptake and Fe concentration in plant initially increased over that of the control treatment when the concentration in the nutrient solution was 30 µg/g. With further increases in Fe concentration in the nutrient solution, Fe uptake and Fe concentration were much lower than those of the control treatment. The effect of Na_2-EDDHA on reducing Fe uptake was even larger than that of humic acid. Na_2-EDDHA is known to form a highly stable complex with iron and is an efficient chelate for supplying plant with iron (Wallace, 1962). This was the case at low iron concentration. At high iron concentration (120 µg/g); Na_2-EDDHA reduced Fe uptake to only 17 % and Fe concentration to 25 percent of their values in the control treatment competing very effectively with plant roots (Elgala *et al.*, 1978). It is evident that humus materials beside their role in supplying plants with Fe, may guard against possible toxicity.

It could be concluded that organic materials have more value than their nutritional content because of their beneficial effect on solubilizing Fe and other nutrients through chelation or adsorption to maintain balance and favorable nutritional conditions for plants. Organic material also affect the physical properties of soils.

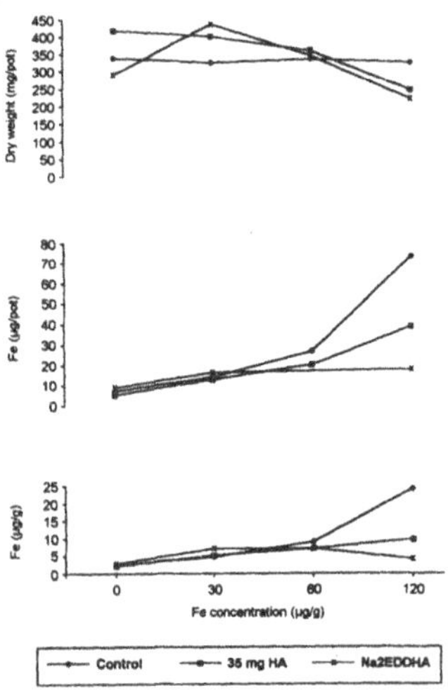

Figure 6. Effect of humic, Na_2-EDDHA and Fe concentration in nutrient solution on dry weight and Fe-content in barley plants

REFERENCES

ASRI, 1987. Egyptian Academy of Scientific Research and Technology, Project on "Micronutrients and clay mineralogy of soils of Egypt", Rep. No. 12.

Chaberek, S. and A. Martell, 1959. Organic sequestering agents. John Wiely and Sons Inc. pp. 416-496.

El-Damaty, A.; A.M. Elgala; M.H. Hilal and M. Abdel-Latif, 1979. Studies on humus of an arid region. II- Effect of organic residues and soil type on the chemical characterization of humic and fulvic acids. Egypt. J. Soil Sci. 15, 175-183.

Elgala, A.M., 1971. Studies on the need of sandy and calcareous soils to iron supply. UAR J. Soil Sci., 11, 39-46.

Elgala, A.M. and S. Hendawy, 1972. Studies on iron availability and behavior in some soils of Egypt. Egypt. J. Soil Sci., 12, 21-30.

Elgala, A.M.; A.H. El-Damaty and Abdel-Latif, 1976. Stability constants of complexes of humic and fulvic acids isolated from organic-enriched Egyptian soils with Fe, Mn and Zn cations. Z. Pflanzenern. Bodenk., 3, 233-300.

Elgala, A.M.; A.S. Ismail and M.A. Ossman (1986). Critical levels of iron, manganese and zinc in Egyptian soils. J. Plant Nutrition, 9, 267-280.

Elgala, A.M.; A.I. Metwally and R.A. Khalil, 1978. The effect of humic acid and Na_2-EDDHA on the uptake of Cu, Fe and Zn by barley in sand culture. Plant and Soil, 49, 41-48.

Ismail, A.S.; A.M. Elgala and S. Montasser, 1984. The use of plant residues in newly reclaimed soils. II- Effect of composted materials enriched with Fe and Zn on the availability of these elements to tomato plants. 9th Int. Cong. for Stat. Comp., Social and Demogr. Res., Ain Shams Press.

Jackson, M.L., 1959. Soil chemical analysis. Prentic-Hall, Inc., England Cliffs, N.J.

Kononova, M.M., 1966. Soil organic matter. Pergamon Press Oxford, London, N.Y., Paris.

Kononova, M.M. and I.V. Alexanderova, 1973. Formation of humic acids during the plant residue humification and their nature. Geoderma, 9, 157-164.

Lindsay, W.L. and W.A. Norvell, 1978. Development of DIPA soil test for zinc, iron, manganese and copper. Soil Sci. Soc. Am. J., 42, 421-428.

Schnitzer, M., 1992. Organic-inorganic interactions in soils and their effects on soil quality. In : P.M. Huang, J. Berthelin, J.-M. Bollag, W.B. McGill and A.L. Page (eds), Environmental Impact of Soil Component Interactions. Vol. 1. Natural and Anthropogenic Organics. Lewis Publ. London, pp.3-20.

Schnitzer, M. and S.I.M. Skinner, 1966. Organo-metallic interaction in soils. 5- Stability constants of Cu^{++}, Fe^{++}, Zn^{++} - fulvic acid complexes. Soil Sci., 102, 361-365.

Schubert, J., 1948. The use of ion exchangers for the determination of physical-chemical properties of substances particularly radiotrabers in solution. J. Phys. and Chem., 52, 340-356.

Theng, H.I. and A.M. Posher, 1967. Nature of carboxyl groups in humic acid. Soil Sci., 104, 191-201.

Wallace, A., 1962. A decade of synthetic cheating agents in inorganic plant nutrition, A. Wallace, Los Angeles.

Wallace, A., 1966. Current Topics in Plant Nutrition. Edward Brothers Inc., Ann Arbor, MI.

EFFECTS OF ORGANIC MATTER, IRON AND ALUMINIUM ON SOIL STRUCTURAL STABILITY

M. Arias, M.T. Barral and F. Díaz-Fierros

Departamento de Edafoloxía.
Facultade de Farmacia.
Universidade de Santiago de Compostela.
España.

1. INTRODUCTION

Soil structure is the result of the elementary soil particles being bound together at certain points on their surface by cementing substances, such as organic matter (OM), iron and aluminium oxides, colloidal silica or calcium carbonate (Baver et al., 1972). The relative importance of these stabilizing agents depends on their nature and abundance in the soil.

Although there is a general agreement on the fundamental role that OM, Fe and Al play in the formation and stabilization of soil aggregates, some questions deserve further research, namely, the opposing effects ascribed to OM, which sometimes increases aggregate stability (Emerson, 1983; Bartoli et al., 1992; Tarchitzky et al., 1993) and sometimes favours dispersion (Visser and Caillier, 1988), the forms of Fe and Al implicated (Arduino et al., 1989; Colombo and Torrent, 1991), the mechanisms of their aggregative effect or which of the two metals is the most efficient stabilizer (Bartoli et al., 1988).

In this work, we aim at clarifying the role of OM, Fe and Al in soil aggregation. We use statistical methods to examine the possible relationships between several indices of structural stability and the forms and contents of OM, Fe and Al of natural soils, and we also examine the changes in the particle-size distribution following the substraction of these potential stabilizing agents. These results are compared to those obtained in previous works on synthetic systems in which varying amounts of Fe or Al oxides, humic acids (HA), or Fe- or Al-HA associations were added to kaolin or quartz substrates (Arias et al., 1995; Arias et al., 1996). The results are discussed in relation to theoretical models of soil aggregation.

2. MATERIAL AND METHODS

Surface samples (0-20 cm) of 33 soils, developed over gabbros, amphibolites and schists in 12 sites of A Coruña province (NW Spain), were taken. The samples corresponded to Umbric or Ochric A horizons (FAO, 1988). In each site, soils representing three different soil uses were sampled: Temporary pasture and cropping, in a mixed rotation, and permanent pasture. The soils were air-dried and then sieved to isolate the <4 mm fraction. Particle size analysis and the determination of soil pH in water and of total carbon and total nitrogen contens, were all carried out using the methods described by Guitián and Carballas (1976). Soil texture was loamy or silty loamy. Organic carbon contents ranged between 4.2-24.5 %. Soils were acidic, with pH in water between 4.0 and 5.2. The selective extractants of iron and

Effect of Mineral-Organic-Microorganism Interactions on Soil and Freshwater Environments
Edited by Berthelin, *et al*, Kluwer Academic / Plenum Publishers, New York, 1999

aluminium used were: (p) 0.1 N sodium pyrophosphate at pH 10, extracting for 16 h (Bascomb, 1968), which is considered to dissolve all organic forms of Fe and Al; (d) dithionite-citrate, extracting for 16 h (Holmgren, 1967), which is considered to extract both amorphous and crystalline Fe oxides; (o) oxalic acid-ammonium oxalate at pH 3, extracting for 4 h in the dark (Schwertmann, 1964), considered to dissolve only amorphous Fe and Al oxides, and (s) 0.5 N NaOH, extracting for 16 h (Borggaard, 1985), capable of dissolving all Al outside the network of crystalline silicates. Soil data are summarized in Table 1.

Table 1. Summary of soil properties (%). Total organic carbon (C_t), total nitrogen (N), ratio carbon/nitrogen.

	C_t	N	C/N	C_p	Fe_d	Fe_o	Fe_p	Al_s	Al_o	Al_p	Si_o
Max.	15.95	1.25	24	4.12	6.46	2.41	1.29	2.31	1.69	1.56	0.31
Min.	2.46	0.20	6	0.72	1.60	0.50	0.21	0.32	0.27	0.20	0.01
Mean	7.26	0.52	14	2.04	2.99	1.19	0.65	1.01	0.82	0.57	0.10

p = sodium pyrophosphate.
o = oxalic acid-ammonium oxalate.
d = citrate-bicarbonate-dithionite.
s = sodium hydroxide.

2.1. **Structural stability tests**

We examined the aggregate stability to water in the form of simulated rainfall and as a dispersive medium, using mechanical agitation, a chemical dispersant or ultrasound to maximize dispersion. The following structural stability test were used:

2.1.1. *Simulated rainfall.* Fifty g of soil sieved between 4 and 0.25 mm was evenly spread on a sieve of mesh size 0.25 mm. The soil thickness on the sieve was ² 5 mm. The soil was subjected to 30 minutes of rainfall simulated by ejecting water at an exit pressure of 0.25 kg cm^{-2}. The simulated rain had mean drop diameter 1.25 mm, intensity 45 mm h^{-1}, and kinetic energy 14.38 J m^2 mm^{-1}. From the results of duplicate experiments, the mean proportion of soil lost from the sieve was calculated (SL in g m^{-2} min).

2.1.2. *Hénin and Monnier (1956) structural stability test (HM test).* A subset of 16 samples, covering the variation interval of iron content of the aforementioned samples, was selected. Air-dried samples were immersed in water and wet-sieved following a standardized procedure. The stability of untreated samples is compared to that of samples pretreated with alcohol or bencene. The treatment with alcohol or benzene affects the moistening of the samples and reduces the breakdown of the aggregates due to the pressure of the entrapped air. The mass of stable aggregates retained on the sieve (mesh 0.2 mm) was recorded. The proportion (% w/w) of stable aggregates for untreated, alcohol-treated and benzene-treated aggregates (Ag, Ag_a and Ag_b, respectively) were calculated, and the instability index, I_s, was obtained using the equation

$$I_s = \frac{(\% <20\ \mu m)_{max}}{[(Ag+Ag_a+Ag_b)/3 - 0.9(\%\ CS)]}$$

where $(\% <20\ \mu m)_{max}$ indicates the highest proportion of suspended particles <20 μm determined for the three sample treatments, and % CS is the highest proportion of coarse sand (the fraction 0.2-2 mm) forming part of the stable aggregates.

2.1.3. *Dispersion tests*. Nine surface samples were selected from among the samples studied before. Six samples are pasture/cropped soil pairs with the same location (1-2, 4-5 and 7-8). The other three samples (3, 6 and 9) were unpaired pasture soils (Table 2). Aggregates 3-2 mm were isolated by sieving. Their composition is indicated in Table 2. Three dispersion procedures were compared: a) mechanical agitation in water (20 g in 100 ml, 1 h), which represents conditions of moderate energy; b) mechanical agitation (20 g en 100 ml, 1 h) in a solution of sodium hexametaphosphate (0.36%) and sodium carbonate (0.08%) (HMP), where the dispersion is due to combined effect of the Na^+ cation, the phosphate anion and the increase in pH; c) agitation with ultrasounds (20 g in 100 ml, energy 300 J ml^{-1}) (US), which represents conditions of high energy.

Table 2. Soil use and 3-2 mm aggregate composition (g kg^{-1})

Soil	Use	Agregate composition			
		Fe_d	Al_d	C	N
1	Cultivation	53.6	7.5	31.7	2.5
2	Pasture	40.4	7.2	51.2	2.6
3	Pasture	33.0	6.7	78.1	3.9
4	Cultivation	30.6	9.2	65.4	9.8
5	Pasture	31.5	4.4	59.6	4.3
6	Pasture	26.2	10.2	135.0	12.4
7	Cultivation	17.2	3.1	21.0	1.7
8	Pasture	14.3	3.0	58.7	4.8
9	Pasture	17.3	4.4	83.1	13.0

2.2. Effect of OM, Fe and Al substraction in the particle-size distribution

The surface horizons of two soils dedicated to cultivation of maize were selected (samples 1 and 7, Table 2). Aggregates 3-2 mm were used in this experiment. The influence of OM, Fe and Al and on aggregate formation and stability was determined analytically, by selectively extracting these potential cementing substances, and comparing the particle-size distributions (PSD) of the soils before and after the extraction. The extractants used were: H_2O_2 (to fully oxidise organic matter); H_2O_2+oxalic acid-ammonium oxalate (to extract organic matter and amorphous Fe and Al); and H_2O_2+citrate-bicarbonate-dithionite (to eliminate organic matter and amorphous and crystaline Fe oxides).

3. RESULTS

3.1. Structural stability

Soil loss (SL) under simulated rainfall ranged between 2.2 and 25.2 g m^{-2} min^{-1} (mean 10.7, _= 6.0). These SL values are medium or high in comparison with soil losses measured by Benito and Díaz-Fierros (1989) for other soils of this country under similar experimental conditions. Because rainfall simulators are not standardized, these values cannot be compared with other data from the literature.

Multiple-regression analysis selected Fe_o, Al_o, C and Fe_d (in that order), as significant predictors (at $p < 0.005$) of SL. Fe_o and Fe_d were not significantly correlated. The regression equation had the form

$$SL = 21.4 - 5.4\,Fe_o + 3.6\,Al_o - 0.5\,C - 1.2\,Fe_d \qquad \text{(Eq. 1).}$$

Using the four parameters it was possible to explain up to 41% of the variation in SL. Fe_o explained 33% of the variation in SL.

Soil use is a qualitative variable that was not considered in regression. Nevertheless, it has a major influence on soil stability against simulated rainfall, as cultivated soils invariably had higher PS than the nearby pasture soils. Benito and Díaz-Fierros (1992) also observed a decline in structural stability of soils as a consequence of continuous cropping. Temporary and permanent pastures were not appreciably different as regards their structural stability. The influence of cultivation on structural stability is evident when we examine the quotient between the SL measured and that obtained from the regression equation (Fig. 1). The quotient for cultivated soils was higher than 1, which means that cultivated soils are less stable than the equation predicts. The reverse is true for pasture soils.

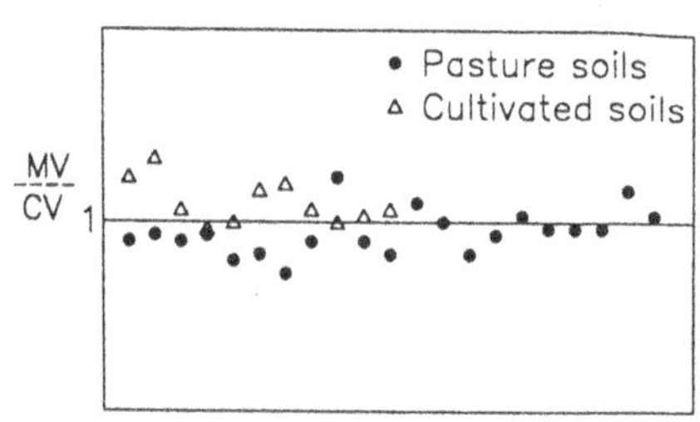

Figure 1. Quotient between the measured value (MV) of soil loss (SL) in the rainfall simulator and SL calculated (CV) from de equation reggression (Eq. 1).

The values of the instability index (I_s) for the 16 samples submited to the HM test ranged between 0.05 and 1.13 (mean 0.24). The highest values corresponded to cultivated soils. According to Monnier and Stengel's classification (1982), the soils are classified as very stable ($\log 10I_s < 1$). Nevertheless, this classification seems to be insufficient at a regional scale, because these soils had only a moderate stability against a simulated rainfall, when compared with other soils from the same area.

We did not found any significant lineal relationship betweeen Is or Ag (the proportion (% w/w) of stable aggregates) and soil properties, now including ESP (exchangeable Na/CEC) as a variable representative of the exchange complex. Nevertheless, there is a significant potential relationship between Fe_o and I_s (Fig. 2). The greatest variations in Is correspond to Fe_o values between around 0.5 y 1%. There is a significant ($p<0.01$) lineal relationship between Is and soil loss under simulated rainfall ($r = 0.767$). Pauwels et al (1976) compared the results of several aggregate stability tests with soil loss measurements on small erosion plots under simulated rainfall, and only the test of Hénin and Monnier (1956) was correlated with soil loss.

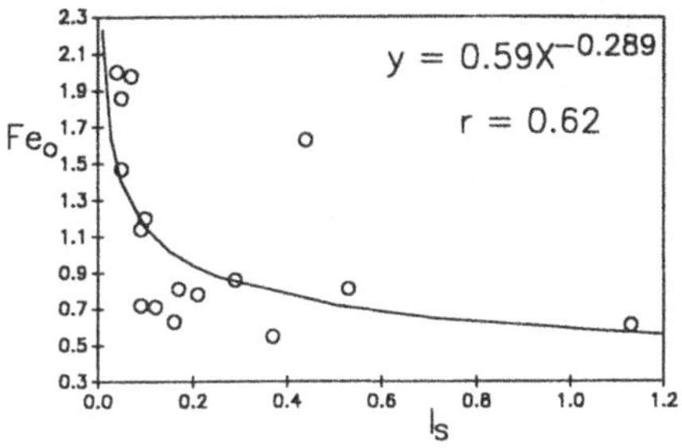

Figure 2. Relationship between oxalic-extractable iron (Fe_o) and the Hénin and Monnier I_s index.

The results from the dispersion tests showed that US was the most efficient procedure for both aggregate breakdown and clay dispersion, as it is shown by lower geometric mean diameter (GMD) values in Table 3. From the particle size distribution data a dispersion index I_d and an aggregation index I_a were obtained using the equations:

I_d= % fraction <0.002 mm dispersed in water/% fraction <0.002 mm dispersed with US.

I_a= % fraction 3-2 mm dispersed in water/% fraction 3-2 mm dispersed with US.

Table 3. Geometric mean diameter (mm) of samples dispersed in water (a), sodium hexametaphosphate-carbonate (b), and ultrasounds (c). I_d: dispersion index. I_a: aggregation index.

Sample	Water	HMP	US	I_d	I_a
1	0.8	0.7	0.6	28.1	1.3
2	1.1	0.8	0.4	17.3	2.8
3	1.4	1.3	0.2	16.9	8.7
4	0.8	0.8	0.7	24.8	1.0
5	0.8	0.6	0.1	16.1	5.8
6	1.9	1.7	0.5	22.7	15.2
7	0.2	0.3	0.2	37.5	1.5
8	1.2	0.9	0.8	15.8	25.9
9	1.4	1.0	0.3	12.6	5.5

Cultivated soils presented higher values for the I_d dispersion index and lower values for the I_a aggregation index, and so they could be considered as less stable than the nearby pasture soils (Table 3). I_d was significantly related ($p < 0.05$) to the I_s instability index obtained in the HM test ($r = 0.69$), and with the soil loss (SL) under simulated rainfall ($r = 0.76$). Nevertheless, neither I_d nor I_a were significantly related to the C, N, Fe, Al or clay contents, which is probably due to the reduced number of samples and to the fact that the relative importance of the stabilizing agents varies in every soil.

The dispersion method had little influence on the concentrations of C, N, Fe and Al in particle size fractions (Table 4). Usually, the fraction 3-2 mm remaining after dispersion with US of the aggregates of diameter 3-2 mm had lower contents of these elements than the same fraction remaining after dispersion in water or HMP. The C, N and Al concentrations were not appreciably different in particle size fractions for the same dispersion procedure. Iron concentrations were slightly higher in the coarser fractions (3-2 mm and 2-0.2 mm). Nevertheless, there were clear differences in the total amounts (T, mg kg^{-1}) of the elements in each particle size fraction, obtained by multiplying the element concentration in the fraction (C, mg g^{-1}) by the proportion in weight of the fraction in the original sample 3-2 mm (F, g kg^{-1}). In Fig. 3 we see that the percentage of the element associated to the fraction < 0.05 mm increases in the order water < HMP < US. For three samples, 25% of the Fe remains in the fraction 3-2 mm. This fact seems to indicate that although US is the most effective treatment for aggregate dispersion, a complete breakdown of the coarsest Fe-rich aggregates is not achieved. On the contrary, C almost completely disappears from the coarse fractions, thus indicating that US easily breaks interparticular bonds in which organic matter is concerned.

Table 4. Concentrations of C, N, Fe and Al (mg g^{-1})in particle-size fractions separated after the dispersion treatments.

Fraction		3-2 mm			2-0.2 mm			0.2-0.05 mm		
Treatment		Max.	Min.	Mean	Max.	Min.	Mean	Max.	Min.	Mean
	C	168	14	64	121	24	59	106	16	49
Water	N	13	0	5	7	1	4	6	1	4
	Fe	60	15	33	76	16	37	39	11	25
	Al	10	2	6	12	2	7	13	2	6
	C	151	6	57	120	13	60	96	12	47
HMF	N	13	0	4	8	1	4	6	0	3
	Fe	71	14	33	66	17	35	37	9	23
	Al	9	2	6	10	2	6	11	1	5
	C	249	6	51	209	11	63	150	11	58
US	N	28	0	4	25	1	5	4	1	3
	Fe	73	6	29	66	10	27	33	7	20
	Al	7	1	4	8	2	4	8	1	4

3.2. **Effect of OM Fe and Al substraction in the particle-size distribution**

The results of the experiments examining the stability to water of microaggregates (soil dispersion) and macroaggregates (simulated rainfall, HM test) indicated that Soil 1 comprised more stable aggregates. This stability is attributed to the high organic matter and iron contents of Soil 1; Al is present in much smaller amounts in this soil, and so it is not thought to play an important stabilizing role. The particle size fractions contributing the largest amount of OM, Fe and Al to the aggregates were the silt and the clay fraction.

Organic matter appears to be an important aggregating substance in both soils, since the oxidation of OM with H_2O_2 caused a roughly two-fold increase in the fraction of clay-sized particles, mainly at the expense of the silt-sized fraction (0.002-0.05 mm) and, to a lesser extent, the 3-2 mm fraction (Table 5). Extraction of H_2O_2-oxidized Soil 1 with oxalic acid - ammonium oxalate further increased the fraction of clay-sized particles, mainly at the expense of the silt-sized fraction. Extraction with dithionite-citrate-bicarbonate produced the same effect on the fine particles, but it also caused fragmentation of 3-2 mm particles into 0.2-

2 mm particles. Extraction of H_2O_2-oxidized Soil 7 with oxalic acid - ammonium oxalate had similar effects that H_2O_2 alone, whereas extraction with dithionite-citrate-bicarbonate slightly increased the fraction of clay-sized particles at the expense of the silt-sized fraction.

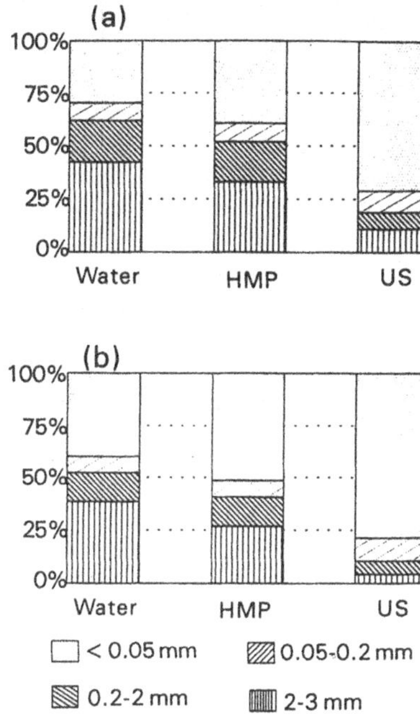

Figure 3. Iron contents (a) (% of total Fe) and carbon contents (b) (% of total carbon) in the particle-size fractions separated after dispersion in water, HMP and US.

Table 5. Particle size distribution (%) of aggregates 3-2 mm subjected to different dispersion treatments: HMP (a), HMP+H_2O_2 (b), HMP+H_2O_2+oxalic acid-ammonium oxalate (c), HMP+H_2O_2+ citrate-bicarbonate-dithionite (d).

Fraction	3-2 mm		2-0.2 mm		0.2-0.05 mm		0.05-0.002 mm		<0.002 mm	
Soil	1	7	1	7	1	7	1	7	1	7
Treatment										
a	24.6	9.0	14.8	5.5	13.6	13.3	33.3	60.9	13.6	11.0
b	22.3	7.9	14.0	7.0	14.1	13.2	29.5	51.0	20.1	20.8
c	22.0	6.2	13.8	6.7	14.1	13.4	25.0	52.2	25.1	21.5
d	17.8	8.5	17.5	7.6	14.5	12.3	25.8	47.9	24.4	23.6

From these results it appears that OM is the main aggregant in soil 7, whereas in soil 1, with higher iron contents, amorphous iron (extractable in oxalic acid - ammonium oxalate) is implicated, together with organic matter, in the formation of silt-sized pseudoparticles out of smaller soil particles, whereas the crystalline iron forms favour formation of coarser particles.

4. DISCUSSION

Many theoretical models of soil aggregation stress the role of the associations between clay, organic matter, and iron and aluminium components in the formation of soil aggregates (Edwards and Bremner, 1967; Mortland, 1970; Tisdall and Oades, 1982). The model proposed by Tisdall and Oades (1982) is particularly interesting as it attempts to illustrate the architecture of an aggregate and describes the effectiveness of the binding agents at different stages in the structural organization of the aggregates. In this model of soil organization, four levels of aggregation can be distinguished:

a) In soils with high contents (> 2 %) of organic carbon, water-stable aggregates >2000 μm in diameter consist of aggregates and particles held together mainly by a fine network of roots and hyphae. The stability is controlled by agricultural practices. Inorganic binding agents including highly disordered aluminosilicates and crystalline Fe oxides also stabilize these aggregates but to a lesser extent than organic materials.

b) Aggregates 20-250 μm are stable to rapid wetting and are not destroyed by agricultural practices. However, they can be destroyed by ultrasonic vibration. They consist largely of particles 2-20 μm bonded together by persistent organic material and inorganic cements.

c) Aggregates 2-20 μm consist of particles <2 μm bonded together so strongly by persistent organic bonds that they are not disrupted by agricultural practices.

d) Aggregates <2 μm are often floccules where individual clay plates come together to form a fluffy mass. Some particles <2 μm in diameter have been shown to be aggregates of very fine material held together by organic matter and iron oxides. Organic matter material is probably sorbed onto the surface of clays.

An idealized model can be drawn showing that an aggregate of soil is built up of structural units of various sizes held together by various binding agents (Fig. 4).

The results of our experiments confirm and clarify Tisdall and Oades' model. The rainfall simulator and the Hénin and Monnier's test refer to aggregates >200-250 μm. At this level, cropping clearly appears as a major factor that reduces structural stability. This effect can be attributed to the breaking of the network of hyphae and fine roots as a result of tillage, but also because a reduced return of organic residues and an increased rate of organic matter mineralization. Iron appears as the main inorganic binding agent of aggregates >200 μm.

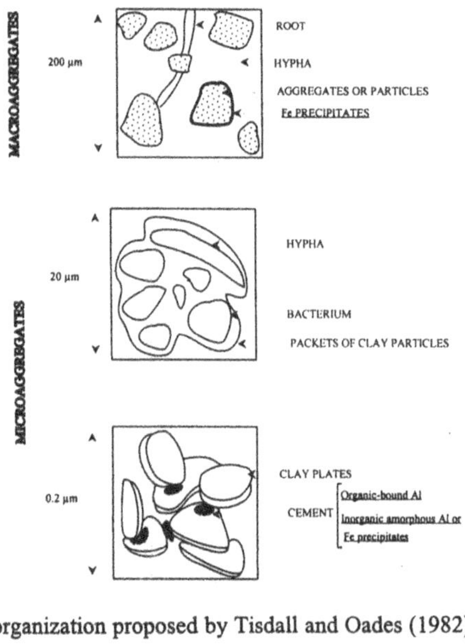

Figure 4. Model of aggregate organization proposed by Tisdall and Oades (1982) simplified. The main findings of this work are underlined.

The dispersion test integrates aggregate rupture and clay dispersion. The effects of cultivation are yet recognizable at this level. Coarse aggregates have low water stability, whereas silt-sized aggregates are very stable, even with ultrasonic dispersion. Organic matter and amorphous iron are responsible for the aggregation of clay-sized particles to form silt-sized pseudoparticles. Crystalline Fe compounds mainly contribute to the formation of sand-sized aggregates.

Previous experiments with inorganic Fe or Al precipitates, and Fe or Al-HA associations, added to mineral substrates (kaolin and quartz), showed that Al compounds (both organic and inorganic forms) were the most effective aggregating agents at the microaggregate level (<250 μm) (Arias *et al.* 1996; Arias *et al.*, 1997). Aluminium and iron oxides affect kaolin aggregation by neutralizing its surface charge, and thus decreasing its colloidal stability (Arias et al, 1995). The Al hydroxide seems to have a greater capacity for intimate coating, resulting in more effective masking of the negative charges of the substrate. The effects of HA-Al and HA-Fe associations on kaolin aggregation cannot be only attributed to surface charge modifications; other (e.g. steric) effects must be invoked, in order to explain the greater aggregating efficiency of HA-Al associations. Quartz aggregation is attributed mainly to cementation, especially at high doses of metal oxide or HA-metal associations.

The question arises why aluminium is a more effective binding agent than iron in experiments with synthetic samples, whereas iron appears as the main inorganic cementing substance in experiments with natural soils. A possible explanation is that natural soils are subjected to changes in the oxidation-reduction status which favour redistribution of iron towards more aerated oxidating surfaces of aggregates, thus reinforcing their cohesion and increasing their resistance to slaking when submerged in water. Aluminium is not affected by changes in oxidation-reduction conditions. Moreover Al concentrations are generally lower in soils than Fe, and selective extractions are less defined for Al forms than for Fe compounds.

5. CONCLUSIONS

The role of organic matter (OM), Fe and Al in the aggregation of natural soils has been investigated. Amorphous Fe and organic matter are the main soil constituents involved in the stability in water of microaggregates and macroaggregates. Comparison of particle-size distributions after OM oxidation and selective extraction of Fe confirmed that both are important soil aggregants. Soil management has also a major influence on structural stability. Cropped soils are less stable than nearby soil dedicated to pasture.

The results are consistent with the aggregation model proposed by Tisdall and Oades (1982).

6. REFERENCES

Arduino, E.; Barberis, E. and Boero, V., 1989. Iron oxides and particle aggregation in B horizons of some Italian soils. Geoderma, 45, 319-329.

Arias, M., Barral, M.T. and Díaz-Fierros, F., 1995. Effects of iron and aluminium oxides on the colloidal and surface properties of kaolin. Clays and Clay Minerals, 43, 406-416.

Arias, M., Barral, M.T. and Díaz-Fierros, F., 1996. Effects of associations between humic acids and iron or aluminium on the flocculation and aggregation of kaolin and quartz. European Journal of Soil Science, 47, 335-343.

Arias, M., López, E. and Barral, M.T., 1997. Comparison of functions for evaluating the effect of Fe and Al oxides on the particle size distribution of kaolin and quartz. Clay Minerals, 32, 3-11.

Bartoli, F.;Philippy, R. and Burtin, G., 1988. Aggregation in soils with small amounts of swelling clays. I. Aggregate stability. J. Soil Sci., 39, 593-616.

Bartoli, F.; Burtin, G. and Guérif, J., 1992. Influence of organic matter on aggregation in Oxisols rich in gibbsite or in goethite. II. Clay dispersion, aggregate strength and water-stability. Geoderma, 54, 259-274.

Bascomb, C.L., 1968. Distribution of pyrophosphate-extractable iron and organic carbon un soils of various groups. J. Soil Sci., 19, 251-268.

Baver, L.D.; Gardner, W.H. and Gardner, W.R., 1972. Soil Physics. 4th ed. John Wiley, New York.

Benito, E. and Díaz-Fierros, F., 1989. Estudio de los principales factores que intervienen en la estabilidad estructural de los suelos de Galicia. Anales de Edafología y Agrobiología, 48, 229-253.

Benito, E. and Díaz-Fierros, F., 1992. Effects of cropping on the structural stability of soils rich in organic matter. Soil and Tillage Research, 23, 153-161.

Borggaard, O.K., 1985. Organic matter and silicon in relation to the crystallinity of soil iron oxides. Acta Agric. Scand., 35, 398-406.

Colombo, C. and Torrent, J., 1991. Relationships between aggregation and iron oxides in Terra Rossa from southern Italy. Catena, 18, 51-59.

Edwards, A.P. and Bremner, J.M., 1967. Microaggregates in soils. Journal of Soil Science, 18, 64-73.

Emerson, W.W., 1983. Inter-particle bonding. In: Soils: An Australian Viewpoint, pp. 477-498. Division of Soils, CSIRO, Melbourne.

FAO, Food and Agriculture organization of the United Nations, 1988. Soil map of the world. Revised legend. World Soil Resources Report 6D, FAO, Roma.

Guitián, F. and Carballas, T., 1976. Técnicas de análisis de suelos. Pico Sacro (ed). Santiago de Compostela, España.

Hénin, S. and Monnier, G., 1956. Evaluation de la stabilité de la structure du sol. C.R. VIe Congrès AISS, Paris, 49-52.

Holmgrem, G.G.S., 1967. A rapid citrate-dithionite extractable iron procedure. Soil Sci. Soc. Amer. Proc., 31, 210-211.

McKeague, J.A., 1967. An evaluation of 0.1 N pyrophosphate-dithionite in comparison with oxalate as extractants of the accumulation products in podzols and some other soils. Canadian Journal of Soil Science, 46, 13-22.

Mehra, O.P. and Jackson, M.L., 1960. Iron oxide removal from soils and clays by dithionite-citrate system buffered with sodium bicarbonate. 7th Nat. Conf. Clay Minerals, 317-327.

Monnier, G. and Stengel, P., 1982. Structure et état physique du sol. Techn. Agri. Fasc. 1140.

Mortland, M.M., 1970. Clay-organic complexes and interactions. Advances in Agronomy, 22, 75-117.

Pauwels, J.M., Gabriels, D. and Eeckout, G., 1976. Evaluation of different criteria to assess the stability of the soil surface. Mededelingen van de Landbouwhogeschool Gent, 41, 135-139.

Schwertmann, U., 1964. Differenzierung der Eisenoxide des Bodens durch Extraktion mit Ammonium oxalat-Lösung. Z. Pflanzenernäh Dúng. Bodenk, 105, 194-202.

Tarchitzky, J., Chen, Y. and Banin, A., 1993. Humic substances and pH effects on sodium- and calcium-montmorillonite flocculation and dispersion. Soil Sci. Soc. A. J., 57, 367-372.

Tisdall, J.M. and Oades, J.M., 1982. The effect of crop rotation on aggregation in a red-brown earth. Australian Journal of Soil Research, 18, 423-434.

Visser, S.A. and Caillier, M., 1988. Observations on the dispersion and aggregation of clays by humic substances. I. Dispersive effects of humic acids. Geoderma, 42, 331-337.

INTERACTIONS OF MUGINEIC ACID WITH ALLOPHANE, IMOGOLITE, MONTMORILLONITE AND GIBBSITE

S. Hiradate* and K. Inoue

Faculty of Agriculture, Iwate University,
3-18-8 Ueda, Morioka 020-0066, Japan
(*Present address: National Institute of Agro-Environmental Sciences,
Ministry of Agriculture, Forestry and Fisheries, 3-1-1 Kan-nondai,
Tsukuba, Ibaraki 305-8604, Japan).

INTRODUCTION

Roots of rice (*Oryza sativa* L.) and oats (*Avena sativa* L.) in water culture under Fe-deficient conditions excrete an Fe-chelating compound identified as mugineic acid (MA; Fig. 1) (Takemoto *et al.*, 1978; Takagi, 1993). In our previous paper (Inoue *et al.*, 1993), we studied interactions between MA and Fe oxides in relation to the chemistry of Fe nutrition of graminaceous plants grown in Fe-deficient soil conditions.

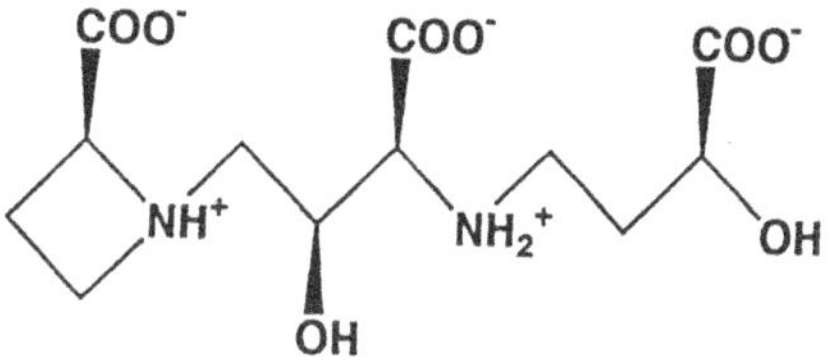

Fig. 1. Chemical structure of mugineic acid dominant between pH 3.4 and 7.8.

Adsorption of MA or MA-Fe complexes by Fe oxides, dissolution of Fe from Fe oxides by MA, and formation of MA-Fe complexes basically involve ligand exchange and nucleophilic substitution (Inoue *et al.*, 1993). Surfaces, other than those of Fe oxides, are also available for ligand exchange in soils. Especially, allophane and imogolite have abundant reactive Al and large specific surface areas (Wada, 1989). The adsorption interactions of humic substances and phosphate with Al atoms in the edges of aluminosilicate structures are similar to those with Fe atoms in Fe oxides. Therefore, it is likely that MA will react with aluminosilicates in the same way with Fe oxides (Fig. 2).

Takagi *et al.* (1988) studied the dissolution of Al, Fe, and Ca from a limed humic volcanic ash soil (in which the major clay minerals were allophane and imogolite) in the presence of ethylenediaminetetraacetic acid (EDTA), diethylenetriaminepentaacetic acid (DTPA), desferrioxamine B (FOB), MA, and 2'-dehydroxymugineic acid (DMA). EDTA and DTPA at pH <7 and FOB at pH 5-9 dissolved Al to a greater extent than Fe from the soils. On the other hand, MA and DMA dissolved significant amounts of Fe but only very small

Effect of Mineral-Organic-Microorganism Interactions on Soil and Freshwater Environments
Edited by Berthelin, *et al*, Kluwer Academic / Plenum Publishers, New York, 1999

amounts of Al from the soil at pH 7-8. Takagi *et al.* (1988), however, did not study the adsorption of MA by volcanic ash soils. The objective of this study is to investigate the adsorption of MA by allophanes (A), imogolite (Im), montmorillonite Mt), and gibbsites (Gb), and the dissolution of Al from these clay minerals by MA.

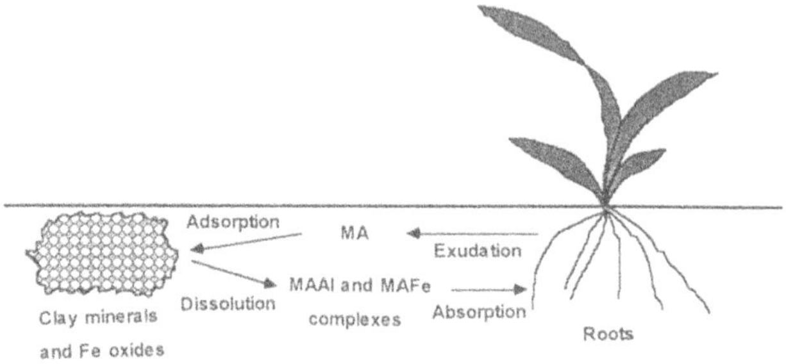

Fig. 2. Diagram showing interactions of MA exuded from graminaceous plant roots with aluminosilicates and Fe oxides in soils.

MATERIAL AND METHODS

Mugineic Acid

Mugineic acid (MA)was extracted from the roots of young seedlings of barley which were cultivated in a modified Hoagland and Arnon no. 2 nutrient solution (pH 5.5), and purified using Amberlite IR-120B (H+ form), Dowex 50W X2 (H+ form), and gel filtration (Sephadex G-10) (Inoue *et al.*, 1993). MA Al and MA Fe c orrespond to mugeneic acid-Al and mugeneic acid-Fe complexes respectively.

Aluminosilicates and Gibbsites

Allophanes (A-PA and A-1041) were prepared from weathered pumice and volcanic ash soils, respectively, Choyo, Kumamoto Pref.; and imogolite (Im-KiG) from gel films in weathered pumice bed, Kitakami, Iwate Pref. (Henmi and Wada, 1974). Montmorillonite (Mt) was prepared from Hojun bentonite, Gunma Pref, One gibbsite sample (Gb-R) was obtained from the Kanto Chemical Co. Ltd., and another (Gb-C) from concretions formed in volcanic ash soils at Mt. Cyokai, Akita Pref.

Clay fractions (<2 *m*m) of these samples were collected by repeated dispersion, sedimentation, and siphoning. Except for the Gb-R clay, the clay fractions were treated with H_2O_2 to remove organic matter. After washing with water, the clay samples were then successively treated with dithionite-citrate-bicarbonate (Mehra and Jackson, 1960) to remove Fe oxides and 2% $NaCO_3$ (Jackson, 1979) to remove citrate adsorbed. Clay samples were neutralized to pH 7 with 0.1 M HCl, dialyzed with deionized-water to remove free salts, freeze-dried, and stored in plastic containers.

Selected physicochemical properties of the clay samples are shown in Table 1. The A-PA and A-1041 clays are characterized by a predominance of allophane with Si/Al atomic ratios of 0.98 and 0.56, respectively. The A-1041 clay also contains a small amount of imogolite. Imogolite is predominant in the Im-KiG clay. The Mt clay contains a small amount of cristobalite.

Characteristics of Clay Minerals

Total and external surface areas of the clay samples were determined by the EGME method (Eltantawy and Arnold, 1973) and by adsorption of N_2 at -195Åé in conjunction with the BET equation using a Micrometritics AccuSorb physical adsorption analyzer (Model

2100D, Micrometritics Instrument Corp., Norcross, GA), respectively. Internal surface areas were calculated as the difference between total and external surface areas (Table 1).

One hundred mg of each clay mineral was treated with 200 mL of 0.2 M ammonium oxalate-oxalic acid (pH 3.0) for 4 h at 30Åé in the dark (Schwertmann, 1964). After centrifuging suspensions, concentrations of Al in the supernatants were determined by ICPAES (Maxim, Fison Instruments, Ecublens, Switzerland) (Table 1).

Table 1. Selected characteristics of clay minerals (oven-dry basis).

Sample	Dominant clay mineral	Si/Al atomic ratio	Oxalate-oxalic acid extractable Al	Specific surface area Total (a)	External (b)	Internal (a-b)
			$mg\ g^{-1}$		$m^2\ g^{-1}$	
A-PA	allophane	0.98¶	167	908	337	571
A-1041	allophane/ imogolite (10/1)	0.56¶	175	649	270	379
Im-KiG	imogolite	0.55-0.56¶	212	1397	258	1139
Mt	montmorillonite	-	8	773	50	723
Gb-C	gibbsite	-	10	164	30	134
Gb-R	gibbsite	-	25	146	17	129

¶ Henmi and Wada (1974)

Determinations of Mugineic Acid and Aluminum

Concentrations of MA were determined using a Shimadzu LC-7A high performance liquid chromatograph (HPLC, Shimadzu Corp., Tokyo) with a fluorescence spectromonitor (Shimadzu FLD-6A) after derivation with *o*-phthalaldehyde. Operating conditions were as used by Inoue *et al.* (1993) and Hiradate and Inoue (1996). Aluminum ions did not affect the determination of MA by HPLC. Concentrations of Al were determined using the ferron method (Davenport, 1949).

Adsorption and Dissolution Experiments

Ten mg portions of each clay sample (oven-dry) were placed in stoppered 10-mL centrifuge tubes; 0.5-mL of 1 M NaCl and deionized water (<3.5 mL) were added and the pH of suspensions was adjusted with 1.0 mL of 0 to 4 mM HCl or NaOH to an appropriate pH in the range 3.5 to 10. After pre-equilibrating for 24 h with occasional shaking, appropriate amounts of 1.00 mM MA were added. The volume of each suspension was 5.0 mL. The final NaCl concentration in each tube was 100 mM; concentrations of MA ranged from 10 to 250 *m*M. As MA-Al complexes, like MA-Fe complexes, may decompose photochemically, suspensions were shaken in the dark for 4 h at 25Å}0.5Åé. Clear filtrates were obtained by passing suspensions through 0.2-*mm* pore-size filter membranes. Amounts of Al dissolved by MA and those of MA adsorbed by clay minerals were calculated as the differences between amounts initially present in the suspensions and those in the filtrates.

RESULTS AND DISCUSSION

Dissolution of Aluminum by Mugineic Acid

Amounts of Al dissolved from clay minerals by MA depended on the kind of aluminosilicates and their crystallinity. Figure 3 shows influences of pH on concentrations of MA in ultrafiltrates and concentrations of Al dissolved from aluminosilicates (A-PA, A-1041, Im-KiG, and Mt) or gibbsites (Gb-C and Gb-R) with and without MA. In the allophane and imogolite systems, considerable amounts of Al were dissolved by 100 *m*M MA over a pH range of 5 to 10. At pH 5-7, Al was not detected in filtrates obtained by ultrafiltration in the absence of MA but was detected in filtrates obtained in the presence of MA, indicating the formation of a soluble MA-Al complex. At pH 3-10, amounts of Al dissolved from the samples are almost equal to the amounts of Al estimated from the solubility product of amorphous $Al(OH)_3$ (-log Ksp = 32.3) and the dissociation constant of amorphous Al-hydroxide (log K = 1.3). Aluminum was dissolved only from the A-PA, A-1041, and Im-KiG clays by MA; it was not dissolved from the Mt, Gb-C, and Gb-R clays. The amounts of Al dissolved by MA varied with the type of mineral and depended on crystallinity and, for the allophanes and imogolite, on the Si/Al atomic ratio as follows:

A-1041 > Im-KiG, A-PA >> Gb-C, Gb-R, Mt

At pH 5-7, dissolutions of Al by MA are apparent for allophane and imogolite (Fig. 3) and increased markedly with increasing MA concentration at pH 6 (Fig. 4). Amounts of Al dissolved from allophanes and imogolite by MA were comparatively smaller than that of Fe dissolved from ferrihydrite (Inoue *et al.*, 1993). Only a small amount of Al was dissolved from Gb-C even in a 200 *m*M MA solution.

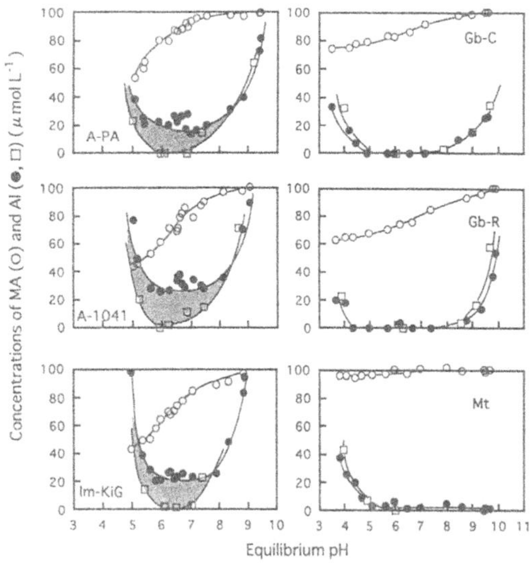

Fig. 3. Concentrations of total mugineic acid (MA) in filtrates and concentrations of Al dissolved from clay minerals as influenced by pH.
Symbols: ○, Concentrations of total MA (MA + MAAl) ; ●, Concentrations of Al in the presence of MA; □, Concentrations of Al in the absence of MA. Areas of dots represent concentrations of Al dissolved from each clay mineral by MA. Dominant clay minerals: A-PA, allophane (Si/Al = 0.98); A-1041, allophane (Si/Al = 0.56); Im-KiG, imogolite (Si/Al = 0.55-0.56); Gb-C and Gb-R, gibbsite; Mt, montmorillonite.

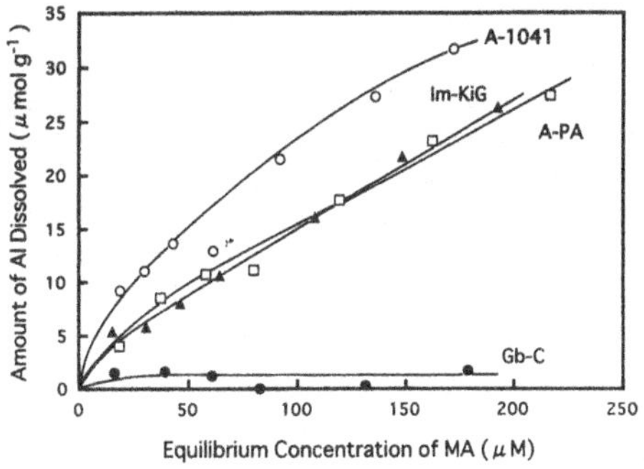

Fig. 4. Amounts of Al dissolved from selected clay minerals at pH 6.0Å ± 0.1 as influenced by concentrations of total mugineic acid. For legends of clay samples see Fig. 3.

Similarity between the dissolution of Al from allophanes and imogolite by MA and that of Fe from Fe oxides (Inoue *et al.*, 1993) leads us to suggest an analogous mechanism: (i) complexation of MA with Al exposed on surfaces of allophanes and imogolite by ligand exchange; and (ii) release of MA-Al complexes from adsorption sites by nucleophilic substitution.

Applied chemicals primarily interact with clay surfaces. Dissolutions of Al by MA were most significant in the A-PA, A-1041, and Im-KiG clays, which have large external surface areas (Table 1). Although the Mt clay also has a large specific surface area, Al was not dissolved from it. Consequently, dissolutions of Al from clay minerals by MA depend on not only specific surface areas but also the characteristics (e.g. crystallinity and structure).

Adsorption of Mugineic Acid by Aluminosilicates and Gibbsites

Figure 3 also indicates amounts of MA adsorbed by the samples. The adsorption of MA was significant at pH <7. Amounts of MA adsorbed increased with decreasing pH in the following order:

Im-KiG, A-1041 > A-PA >> Gb-R, Gb-C >> Mt

The A-PA, A-1041, and Im-KiG clays, which had considerable amounts of oxalate-oxalic acid extractable Al and large external surface areas (Table 1), adsorbed large amounts of MA. Montmorillonite, which had a small amount of reactive Al and small external surface area, did not adsorb MA at all. Even in the presence of 100 mM NaCl, MA was specifically adsorbed by ligand exchange on reactive Al sites of allophanes and imogolite. The adsorption of MA by Im-KiG and A-1041 (Si/Al = 0.55-0.56) was greater than by A-PA (Si/Al = 0.98) and the adsorption of MA by A-1041 was greater than by Im-KiG, suggesting that the proportions of reactive Al and external surface area in the sample are important factors in determining adsorption characteristics.

The shapes of the primary particles are probably another important factor in MA adsorption. Allophane consists of "hollow spherules" with diameters of 3.5-5.5 nm (Wada, 1987) and with pores of 0.3-0.5 nm (Henmi, 1988) whereas imogolite has a tubular structure with an external diameter of about 2.2 nm and with an inside diameter of 0.64 nm (Parfitt, 1980). Allophane particles are highly aggregated and the spaces between particles in aggregates is not accessible to nitrogen gas. Allophane and imogolite therefore have both large 'internal' surface areas and significant interparticle volumes. The chemical structures of allophanes with Si/Al = 1 and Si/Al = 0.5 may be different to some extent. Each MA molecule is about the same size as three molecules of acetic acid with a diameter of 0.90 nm. MA molecules or MA-Al complexes are probably unable to penetrate into the internal surfaces of primary particles and into the interparticle spaces of aggregates because of their large molecular size. Consequently, it seems likely that MA molecules and MA-Fe complexes (MAFe) are adsorbed only on the external surfaces of allophanes and imogolite aggregates.

X-ray diffraction analysis for a structurally analogous Co(III) complex of MA-Fe(III) complex indicates that azetidine nitrogen, secondary amine nitrogen, and both terminal carboxylate oxygens coordinate as basal planar donors, and hydroxyl oxygen and intermediate carboxylate oxygen bind as axial donors in a nearly octahedral configuration (Mino *et al.*, 1983; Sugiura and Nomoto, 1984). We assume, therefore, that Al^{3+} can also form six-fold coordinated complexes (MAAl) with MA.

Figure 5a shows adsorption isotherms of MA by A-PA, A-1041, Im-KiG, and Gb-C at pH 6. The adsorption of MA approached maximum in A-PA and Gb-C, suggesting saturation of the binding sites with MA. On the other hand, maximum adsorption of MA on A-1041 and Im-KiG was not approached in our experiments. Except for A-PA, these adsorption trends are approximately consistent with the order of the external surface areas of the samples (Table 1). The interaction of MA with ferrihydrite was a complicated reaction involving adsorption of MA and/or MAFe by Fe oxides and MAFe desorption from Fe oxides (Inoue *et al.*, 1993). In the same way, interactions of MA with allophanes and imogolite probably involve adsorption of MA and/or MAAl and dissolution of MAAl.

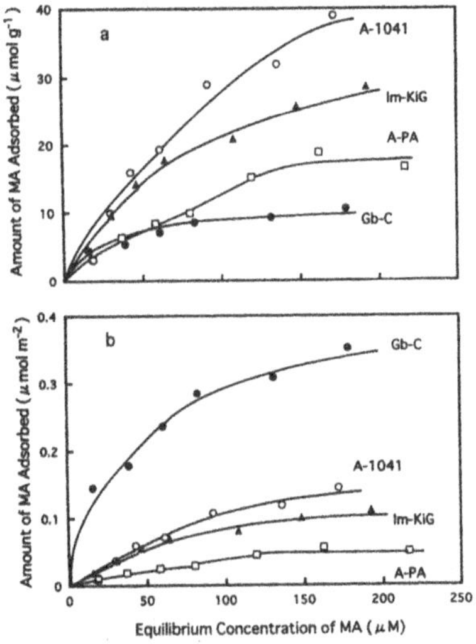

Fig. 5. Adsorption isotherms of mugineic acid on selected clay minerals at pH 6.0Å $\pm$ 0.1. For legends of clay samples see Fig. 3.

The adsorption isotherms do not follow the Langmuir model. This may be because the different chemical forms of MA and MAAl present each have different affinities for the adsorption sites. Amounts of MA adsorbed on the samples at a given equilibrium solution concentration were in the following order (Fig 5a):

A-1041 > Im-KiG > A-PA >> Gb-C

The A-PA, A-1041, Im-KiG, and Mt clays have larger total and internal surface areas than the Gb-C and Gb-R clays (Table 1). The amounts of MA adsorbed, however, did not correlate with the total and internal surface areas of the samples but did correlate with their external surface areas and Si/Al atomic ratios. The amounts of MA adsorbed per an unit surface area of samples decreased with increasing the Si/Al atomic ratio (Fig. 5b), suggesting that the adsorption of MA is related with the reactive Al of the clay samples. The A-1041 and Im-KiG clays with low Si/Al atomic ratios adsorbed greater amounts of MA than the A-PA clay with a high Si/Al atomic ratio. Allophanes and imogolite adsorbed greater amounts of MA than the Gb-C clay. Based on amounts of MA adsorbed per unit external surface area, however, the amounts of MA adsorbed by A-1041, Im-KiG, and A-PA (0.14, 0.11, and 0.05 *m*mol m^{-2}, respectively) were smaller than that of Gb-C (about 0.35 *m*mol m^{-2}). Although allophanes and imogolite had external surface areas (258-337 m^2g^{-1}, Table 1) of the same order as ferrihydrite (268 m^2g^{-1}, Inoue *et al.*, 1993), amounts of MA adsorbed by A-PA, Im-KiG, and A-1041 at pH 6 were less than that by ferrihydrite (>0.45 *m*mol m^{-2}, Inoue *et al.*, 1993) at pH 7.

Implication to Agriculture

Allophane, imogolite, and allophane-like constituents (Wada, 1989) are commonly present in volcanic ash soils. Since most volcanic ash soils are generally also rich in ferrihydrite and have pH values <7, Fe-deficiency of graminaceous crops such as barley, oats, wheat, rye, sorghum, and rice is unlikely to be a problem. In central African countries such as Rwanda and Tanzania, however, Andic Hapludolls, Lithic Mollic Vitrandepts, and Humic Nitosols formed in volcanic ash soils generally have pH >7 (Mizota *et al.*, 1988; Mizota and Chapelle, 1988). In such soils, interactions of MA with the reactive Al present in allophane and imogolite are likely to have a significant effect on the extent of dissolution of Fe from Fe oxides.

CONCLUSION

Mugineic acid interacts with allophane and imogolite in a similar manner as it does with ferrihydrite. MA is adsorbed only on the external surfaces of allophane and imogolite and dissolves Al as MAAl from them at pH 5-8. Adsorption of MA and dissolution of Al by MA are controlled by the crystallinities and structural characteristics of clay minerals. In volcanic ash soils with pH >7, allophane and imogolite are likely to strongly influence the extent of dissolution of Fe from ferrihydrite by MA.

Acknowledgment: The authors are grateful to Dr. C.W. Childs, Victoria University of Wellington, New Zealand, for critical reading and valuable discussion of the manuscript.

REFERENCES

Davenport, W.H.Jr. 1949. Determination of aluminum in presence of iron. Anal. Chem., 21, 710-711.

Eltantawy, I.M. and P.M. Arnold. 1973. Reappraisal of ethylene glycol mono-ethyl ether (EGME) method for surface area estimation of clays. J. Soil Sci., 24, 232-238.

Henmi, T. 1988. Chemical structure of wall of allophane hollow spherical particle. Soil Phys. Cond. Plant Growth, 56, 47-50. (in Japanese).

Henmi, T. and K. Wada. 1974. Surface acidity of imogolite and allophane. Clay Miner., 10, 231-245.

Hiradate, S. and K. Inoue. 1996. Determination of mugineic acid, 2'-deoxymugineic acid, 3-hydroxymugineic acid, and their iron complexes by ion-pair HPLC and colorimetric procedures. Soil Sci. Plant Nutr., 42, 659-665.

Inoue, K., S. Hiradate and S. Takagi. 1993. Interaction of mugineic acid with synthetically produced iron oxides. Soil Sci. Soc. Am. J., 57, 1254-1260.

Jackson, M.L. 1979. Soil Chemical Analysis--Advanced Course, 2nd edition, Published by the author, Madison, WI, 895pp.

Mehra, O.P. and M.L. Jackson. 1960. Iron oxide removal from soils and clays by a dithionite-citrate system buffered with sodium bicarbonate. Clays Clay Miner., 7, 317-327.

Mino, Y., T. Ishida, N. Ota, N. Inoue, K. Nomoto, T. Takemoto, H. Tanaka and Y. Sugiura. 1983. Mugineic acid-iron(III) complex and its structurally analogous cobalt(III) complex: Characterization and implication for absorption and transport of iron in graminaceous plants. J. Am. Chem. Soc., 105, 4671-4676.

Mizota, C., I. Kawasaki and T. Wakatsuki. 1988. Clay mineralogy and chemistry of seven pedons formed in volcanic ash, Tanzania. Geoderma, 43, 131-141.

Mizota, C. and J. Chapelle 1988. Characterization of some Andepts and Andic soils in Rwanda, central Africa. Geoderma, 41, 193-209.

Parfitt, R.L. 1980. Chemical properties of variable charge soils. In: B.K.G. Theng (Editor). Soils with Variable Charge. New Zealand Soc. Soil Sci., Lower Hutt. pp. 167-194.

Schwertmann, U. 1964. Differentzierung der Eisenoxide des Bodens durch Extraktion mit Ammoniumoxalat-Loesung. Z. Pflanzenernaehr. Dueng. Bodenkd., 105, 194-202.

Sugiura, Y. and K. Nomoto. 1984. Phytosiderophore Structure and properties of mugineic acids and their metal complexes. In: M.J. Clarke, J.A. Ibers, D.M.P. Mingos, G.A. Palmer, P.J. Sadler and R.J.P. Williams (Editors). Siderophores from Microorganisms and Plants. Structure and Bonding 58, Springer-Verlag, Berlin, pp. 107-135.

Takagi, S. 1993. Production of phytosiderophores. In: L.L. Barton and B.C. Hemming (Editors). Iron Chelation in Plants and Soil Microorganisms, Academic Press, Inc., San Diego, CA. pp. 111-131.

Takagi, S., S. Kamei and M.-H. Yu. 1988. Efficiency of iron extraction from soil by mugineic acid family phytosiderophores. J. Plant Nutr., 11, 643-651.

Takemoto, T., K. Nomoto, S. Fushiya, R. Ouchi, G. Kusano, H. Hikino, S. Takagi, Y. Matsuura and M. Kakudo. 1978. Structure of mugineic acid, a new amino acid possessing an iron-chelating activity from roots washings of water-cultured Hordeum vulgare L. Proc. Japan Acad., 54, Ser. B, 469-473.

Wada, K. 1987. Minerals formed and mineral formation from volcanic ash by weathering. Chem. Geol., 60, 17-28.

Wada, K. 1989. Allophane and imogolite. In: J.B. Dixon and S.B. Weed (Editors). Minerals in Soil Environments (2nd ed.). Soil Sci. Soc. Am., Madison, WI. pp. 1051-1087.

ALUMINUM SPECIATION, TOXICITY AND TRANSFER FROM SOILS TO SURFACE WATERS IN TWO CONTRASTING WATERSHEDS EXPOSED TO ACID DEPOSITION IN THE VOSGES MOUNTAINS (NORTHEASTERN FRANCE)

Ouafae Maitat, Jean-Pierre Boudot, Denis Merlet and James Rouiller

Centre de Pédologie Biologique, UPR 6831 du CNRS Associée a l'Université Henri Poincaré, Nancy I, 17 Rue Notre-Dame des Pauvres, B. p. 5, f-54501 - Vandœuvre-lès-Nancy Cedex - France

INTRODUCTION

Despite considerable research in the past decades, acid deposition remains a serious environmental problem for many areas in the northern hemisphere, particularly in the northeast United States (Driscoll and Van Dreason, 1993; David and Lawrence, 1996) and Central and Northern Europe (Ulrich *et al.*, 1980; Nilsson *et al.*, 1983; Rosseland and Henriksen, 1990; Matzner and Prenzel, 1992; DeVries *et al.*, 1995), including the north-east of France in the Vosges mountains (Probst *et al.*, 1990; Boudot *et al.*, 1994a). Acid inputs may have negative impacts on the chemistry and the biological quality of soils and surface waters, particularly through the appearance of soluble inorganic Al species (some of them being toxic to plants and aquatic communities), nutrient imbalance and low pH. Both forest decline and freshwater impoverishment may be related to acidification processes and its related nutrient imbalance as well as the occurrence of soluble Al (Baker and Schofield, 1982; Boudot *et al.*, 1994a).

In aqueous systems, aluminum persists as various chemical species. The toxicity of Al is strongly dependent on the prevalent species, so that the concentration of total Al alone cannot be used as a reliable index of Al toxicity. Current evidence and reinterpretation of early literature suggest that both monomeric Al such as Al^{3+}, hydroxy-Al [$AlOH^{2+}$, $Al(OH)_2^+$, $Al(OH)_4^-$] and $AlSO_4^+$ species are the most toxic forms to plants, together with the (doubtful for natural waters) Al_{13} polymer (Blamey *et al.*, 1983; Li *et al.*, 1989; Kinraide, 1991; Barcelo *et al.*, 1993; Baylis *et al.*, 1994; Boudot *et al.*, 1994a). Aluminum complexed either with fluoride, phosphate or organic matter may be regarded as non-toxic to plants (Suthipradit *et al.*, 1990; MacLean *et al.*, 1992; Baylis *et al.*, 1994; Boudot *et al.*, 1994a).

Effect of Mineral-Organic-Microorganism Interactions on Soil and Freshwater Environments
Edited by Berthelin, *et al*, Kluwer Academic / Plenum Publishers, New York, 1999

Toxic species of Al with respect to fish are believed to include AlF^{2+}, as well as the same toxic species as those mentioned above for plants (Driscoll *et al.*, 1980; Baker and Schofield, 1982; Birchall *et al.*, 1989; Lydersen *et al.*, 1990; Wilkinson *et al.*, 1990; Parkhurst *et al.*, 1990; Poléo, 1995; Harris *et al.*, 1996).

In addition, cations such as Ca and Mg, and to a lesser extent K and Na, have been shown to mitigate aluminum toxicity to plants (Rhue and Grogan, 1977; Cameron *et al.*, 1986; Kinraide and Parker, 1987; Blamey *et al.*, 1992). Ca alone must be retained in this respect for aquatic organisms (Brown, 1983; Sadler and Lynam, 1988; Wood *et al.*, 1990). Therefore a detailed knowledge of both the overall water composition and Al speciation is required to assess Al toxicity on biota, as well as to understand Al translocation mechanisms within the soil and from the soil to surface waters.

Acidification occurring in watersheds on acidic bedrocks should be regarded as resulting from both natural pedogenetic processes and anthropogenic H^+ input through atmospheric pollutants such as H_2SO_4, HNO_3 and HCl. The respective influence of these two kinds of proton sources depends on both the features of the soil and the intensity of the pollution.

The present paper is devoted to the assessment of the respective influence of both natural pedogenetic processes and acid anthropogenic inputs on Al translocation and toxicity in an acidic watershed in the Vosges mountains, through Al speciation analysis in the soil solutions as well as in surface waters.

MATERIAL AND METHODS

The present study was conducted in the Vosges mountains (NE-France), in the Mortagne catchment, which falls at an altitude 450 to 700 m a.s.l.. On average, this area receives an annual rainfall of 1100 mm and moderate atmospheric inputs reaching 10-30 kg SO_4-S.ha^{-1}.$year^{-1}$, 1-2 kg NH_4-N.ha^{-1}.$year^{-1}$, 6-20 kg NO_3-N.ha^{-1}.$year^{-1}$, 11–33 kg Cl.ha^{-1}.$year^{-1}$, 0.9 Kg F.ha^{-1}.$year^{-1}$ and 0.4–1 kg free H^+.ha^{-1}.$year^{-1}$, depending on the year.

We investigated both a Triassic sandy sandstone watershed with quartz-rich podzols (Typic Haplorthods) and a Triassic silty sandstone watershed with acid brown soils (Typic Dystrochrepts). Soils and vegetation have been described elsewhere (Becquer, 1991; Boudot *et al.*, 1995).

Soil solutions were continuously sampled with zero-tension PVC plate-lysimeters at the bottom of each soil horizon (4 to 7 replicates). Throughfalls were obtained using collectors settled 1 m above the forest floor in each stand, in order to assess quantitative inputs of cations and anions to soils through precipitation. Throughfalls, soil leachates and spring waters were collected at the end of each rainfall event. Samples were filtered with pre-rinsed cellulose nitrate filters with 0.45 μm pore diameters. Both soil solutions and surface waters were analyzed for major cations, anions, dissolved organic carbon and Al speciation according to Boudot *et al.* (1994b, 1995). Total Al was determined on pre-acidified samples by a sequential Jobin-Yvon ICPAES apparatus (Al_{ICP}). Total monomeric Al (both inorganic and organic) was obtained by ion chromatography (Al_{IC}). An 8-hydroxyquinoline flash extraction (5 s) at pH 5 ($Al_{8\text{-}HQ}$) allowed the determination of the inorganic monomeric Al species, except for the Al-F ones (Boudot *et al.*, 1994b). Therefore:

$Al_{ICP} - Al_{IC}$ = polymeric + colloidal Al (namely "non-monomeric Al" hereafter)

$Al_{IC} - Al_{8\text{-}HQ}$ = organic Al + F-bound Al

$Al_{8\text{-}HQ} = Al^{3+} + Al\text{-}OH + Al\text{-}Si + Al\text{-}PO_4 + Al\text{-}SO_4 + Al\text{-}Cl$

Speciation in these two latter Al pools was assessed with the MINEQL$^+$ V 3.01 equilibrium program (Schecher and McAvoy, 1992, 1994), of which the database has been previously carefully updated (Boudot *et al.*, 1994; Pokrovski *et al.*, 1996; Stumm and Morgan, 1996).

A relevant Al Toxicity Index (ATI), either for the regional plants or for the local fish in rivers (namely, brown trout), was calculated as the following (Boudot *et al.*, 1994a):

$$ATI_{plants} = [9\{Al^{3+}\} + 4\{AlOH^{2+}\} + \{Al(OH)_2^{+}\} + \{AlSO_4^{+}\} + 9\{Al(OH)_4^{-}\} + 117\{Al_{13}^{7+}\}] / [4\{Ca^{2+}\} + 4\{Mg^{2+}\} + 0.02\{K^{+}\} + 0.02\{Na^{+}\}]$$

$$ATI_{brown\ trout} = [\{Al^{3+}\} + \{AlOH^{2+}\} + \{Al(OH)_2^{+}\} + \{Al(OH)_4^{-}\} + \{AlSO_4^{+}\} + 0.5\{AlF^{2+}\}] / 1/6\{Ca^{2+}\}$$

In these expressions, brackets denote molar activities and each element is weighted by a coefficient intended to reflect its relative beneficial or detrimental effect when this is known (Grauer and Horst, 1991). In both expressions, an ATI value higher than 1 denotes Al toxicity.

RESULTS

Podzol area

We will present hereafter data related both to lysimetric waters collected at the bottom of the A_2 horizon (similar data were obtained at the bottom of the A_0 horizon) and to one spring draining this area (Fig. 1 to 3).

Soil solutions from the podzol eluvial horizons showed low pH values (in the range 3–4), low nitrate content and medium Al concentrations (usually 300–1000 µg.L^{-1}) (Fig. 1a and 2a). In this soil, Al was leached predominantly as monomeric organic Al and F-bound Al together with a very variable non-monomeric Al pool (Fig. 2a). As none of these species are toxic to plants, no Al toxicity occurred (Fig. 1a).

In the spring waters draining the podzol, pH was usually in the range 4–5 (with occasional higher values during baseflow periods), nitrate remained very low and Al concentrations were often abnormally high for a stream (usually 90–600 µg.L^{-1}) (Fig. 1b and 2b). Maximal Al content was observed during high runoff periods, minimal Al concentrations during baseflow periods. During highflow events and compared to the lysimetric waters collected at the bottom of the eluvial horizons, additional inorganic Al species occurred, among which ionic Al^{3+} dominated the hydroxy-Al and the $AlSO_4^+$ species (Fig. 2b). Because most of these species are toxic to fish, Al toxicity occurred in these periods (Fig. 1b).

With respect to anion distribution in both soil and stream waters in this non-nitrifying area, it can be emphasized that NO_3, Cl, F and SO_4 are exogenous anions originating from acid deposition and that organic molecules are endogenous anions related to natural pedogenetic processes. The ratio of the exogenous anions to the endogenous free acidity [= (NO_3 + Cl + F + SO_4 free acidity in meq.L^{-1}) / free organic acidity (in meq.L^{-1}) at the sample pH, determined by equilibrium calculation using the triprotic model of Driscoll *et al.* (1994)] was on average around 1 in the soil solutions at the bottom of the eluvial horizons, then increased dramatically to 10–100 in spring water (Fig. 3). This strong increase of the contribution of the exogenous anions to the actual acidity of waters between the bottom of the eluvial horizons and the springs, a zone where plant nutrition can be neglected, demonstrates the anthropogenic origin of the acidity responsible for the transfer of Al from soil to surface water.

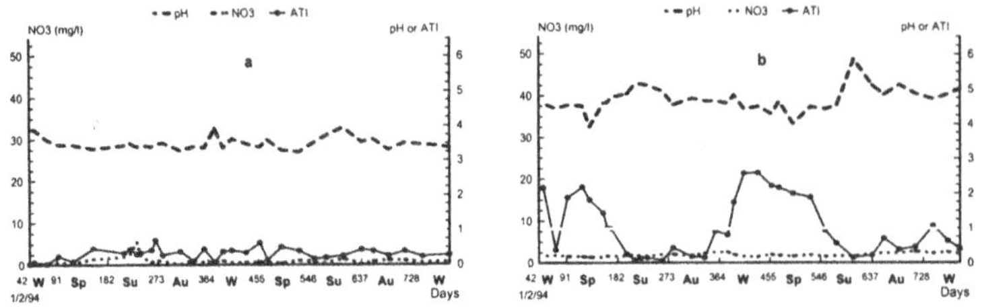

Figure 1 - pH, nitrate content and Al Toxicity Index (ATI) in the leaching waters collected at the bottom of the podzol A_2 horizon (a) and in a spring draining from the podzol area (b) (W, Sp, Su and Au denote the seasons). Al toxicity occurred when ATI was higher than 1.

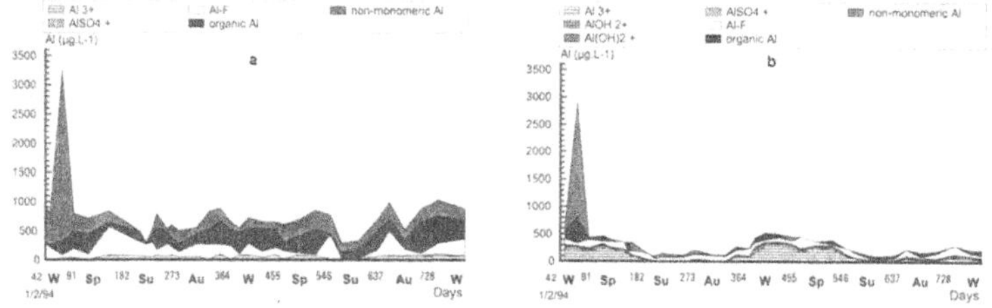

Figure 2 - Al speciation in the leaching waters collected at the bottom of the podzol A_2 horizon (a) and in a spring draining from the podzol area (b) (W, Sp, Su and Au denote the seasons).

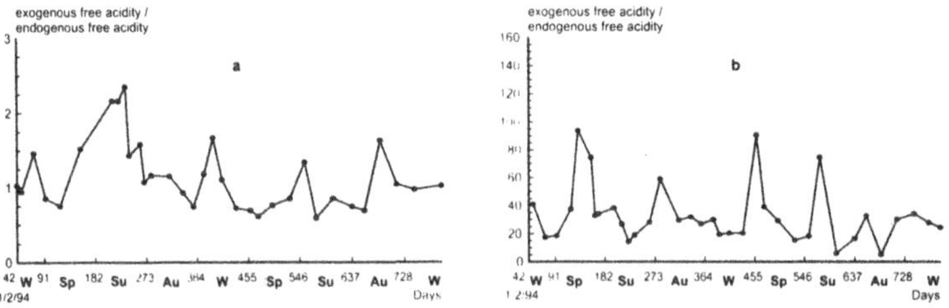

Figure 3 - Exogenous free acidity to endogenous free acidity ratio in the leaching waters collected at the bottom of the podzol A_2 horizon (a) and in a spring draining from the podzol area (b) (W, Sp, Su and Au denote the seasons).

Acid brown soil area

Soil solutions collected at the bottom of the A_1 horizon of the acid brown soil had low pH (from 3.5 to 4.7), rather high nitrate concentrations (from 0 to 60 $mg.L^{-1}$) and high Al content [usually 500–2000 (3500) $\mu g.L^{-1}$)] (Fig. 4a and 5a). According to its speciation pattern, Al was leached mainly as inorganic monomers dominated by Al^{3+} (Fig. 5a). As a consequence, Al toxicity occurred, as attested by rather high ATI values (Fig. 4a). In addition, a strong positive correlation ($R^2 = 0.85$) occurred between NO_3^- and monomeric (inorganic and organic) Al concentrations (Fig. 6).

Conversely, spring waters draining from the acid brown soil area showed a lower acidity than the A_1 soil solutions (pH 4.7 to 6.5), a lower nitrate concentration, and a very low monomeric Al content (Fig. 4b and 5b). Consequently, no Al toxicity occurred with respect to aquatic organisms (Fig. 4b)

In this ecosystem, NO_3 originated mostly from endogenous nitrification. Partitioning nitrate into soil nitrification-derived nitrate and pollution-derived nitrate (settled at the same level than in the podzol watershed), the exogenous to endogenous free acidity ratio can be calculated as [(exogenous NO_3 + Cl + F + SO_4 free acidity) / (endogenous NO_3 + free organic acidity) ($meq.L^{-1}$)]. As for the podzol area, it was found that the contribution of the exogenous free acidity to the actual acidity of waters increased dramatically (20 to 45 times) from the soil solutions to spring water (data not shown). This emphasizes the predominance of the man-made acidity over that derived from the pedogenetic processes in the actual acidity of surface waters in this area too.

DISCUSSION

In the podzol area of the watershed studied, Al translocation within the eluvial horizons occurred mainly as non-toxic organic and F-bound Al, as well as under a very variable non-monomeric Al pool. In this catchment, the mean annual concentration of F remained constant from throughfalls to the bottom of the eluvial horizons and to the springs, so that F should be ascribed to the atmospheric pollution. Therefore the Al-F species should be regarded as the shift of the original Al speciation due to atmospheric deposition. The spring waters studied in this area suffer currently from a harmful contamination of Al, responsible for the occurrence of a strong Al toxicity to aquatic biota during heavy rainfall / snow-melt / high runoff periods (but some other springs have a high Al content and a high Al Toxicity Index throughout the year, irrespective of the hydrologic conditions). The organic Al and the Al-F species in the spring waters, as well as their non-monomeric Al pool, can be regarded as inherited from the upper soil horizons, mostly through lateral flow during storm events. The new inorganic Al species in the springs may originate either (i) from the decomplexation of the pre-existent soluble organic and F-bound Al from the eluvial horizons into ionic Al, or (ii) from the remobilization of the previously flocculated Al of the spodic horizons, although the organic Al species remained flocculated within the B_h horizon. To answer this question, it must be emphasized that the pH shift between the bottom of the eluvial horizons (3.3–3.9) and the spodic horizons (3.7–4) is too low to support the decomplexation hypothesis, so that the remobilization one appears to be likely. Because the contribution of the pollution-derived free acidity increased strongly from the bottom of the eluvial horizons to the springs, the main source of acidity available was related to acid deposition. In other words, Al translocation from the soil to the springs must be ascribed to the redissolution of the spodic horizons under the predominant influence of the acidic atmospheric pollution.

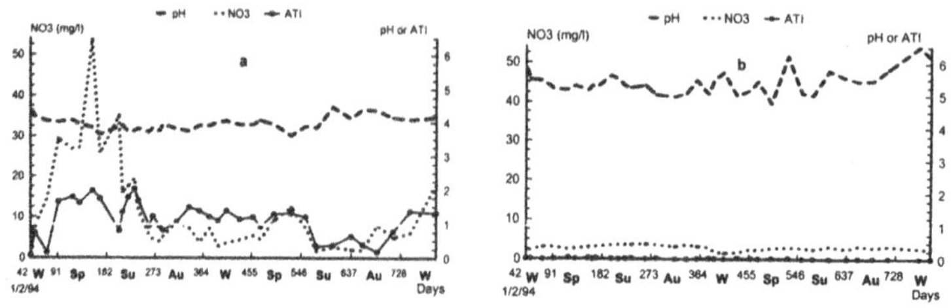

Figure 4 - pH, nitrate content and Al Toxicity Index (ATI) in the leaching waters collected at the bottom of the acid brown soil A_1 horizon (a) and in the spring draining from the acid brown soil area (b) (W, Sp, Su and Au denote the seasons). Al toxicity occurred when ATI was higher than 1.

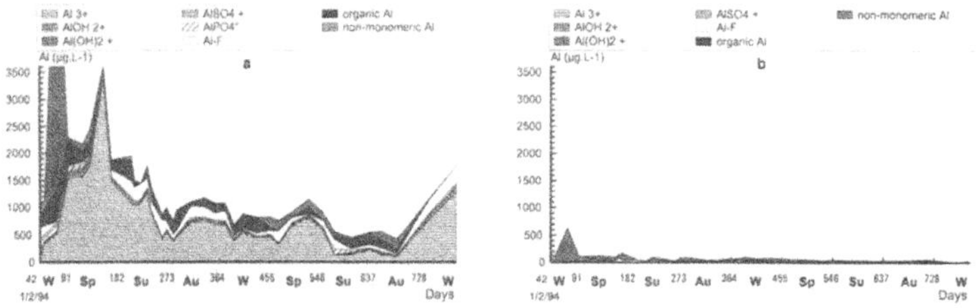

Figure 5 - Al speciation in the leaching waters collected at the bottom of the acid brown soil A_1 horizon (a) and in the spring draining from the acid brown soil area (b) (W, Sp, Su and Au denote the seasons).

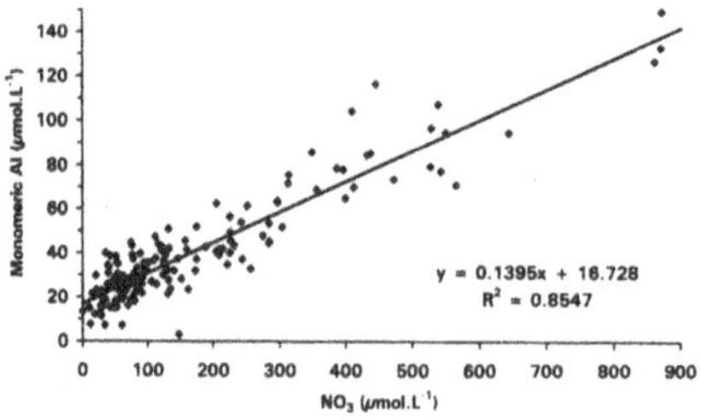

Figure 6 - Relationship between NO_3^- and monomeric (organic + inorganic) Al in the leaching waters from the acid brown soil A_1 horizon.

In the acid brown soil, Al mobilization mostly occurred as inorganic monomeric species, among which Al^{3+} dominated. This process was controlled mostly by nitrate in excess of biological uptake, that is under the dependence of the natural biological cycles (nitrification, microbial uptake, plant nutrition, etc.) rather than determined by external input. Because the prevalent Al species are toxic to plants and because this soil is considerably depleted in exchangeable Ca and Mg, Al toxicity occurs repeatedly during the year. All the monomeric Al appears to be retained between the soil and the springs draining from this and similar areas because the mineralogical composition of the regolith is richer in weatherable minerals than that of the podzol. As a consequence, no Al toxicity occurred in these springs. Nevertheless, as the acid neutralizing capacity (ANC) of these springs [given the mean ionic charge of each element at the equivalence point, ANC ($\mu eq.L^{-1}$) = {(2Ca + 2Mg + K + Na + NH_4 + 2Al + Fe + 2Mn) - (2 SO_4 + NO_2 + NO_3 + Cl + F + PO_4 + 1.2 fulvic acid) (μM/l)}] was often very low, even in baseflow conditions (usually 0–120 $\mu eq.L^{-1}$), they clearly remain open to acid stress during heavy rainfalls, as well as to a progressive deterioration in the future if acid deposition will not regress sufficiently.

ACKNOWLEDGMENTS

This work was partly a contribution to the former "Science and Technology for Environmental Protection/European Network of Catchments Organised for Research on Ecosystems" programme of the Commission of the European Communities (1991–1994).

REFERENCES

Baker, J.P. and C.L. Schofield, 1982. Aluminum toxicity to fish in acidic waters. Water Air Soil Pollut., 18, 289–309.

Barcelo, J., P. Guevara and C. Poschenrieder, 1993. Silicon amelioration of aluminum toxicity in teosinte (*Zea mays* L. ssp. *mexicana*). Plant and Soil, 154, 249–255.

Baylis, A.D., C. Gragopoulou, K.J. Davidson and J.D. Birchall, 1994. Effects of silicon on the toxicity of aluminum to Soybean. Comm. Soil Sci. Plant Anal., 25, 537–546.

Becquer, T., 1991. Production endogène de protons par les cycles de l'azote et du soufre dans deux sapinières vosgiennes : bilans saisonniers et incidence sur la toxicité de l'aluminum. Thèse de doctorat de l'Université de Nancy I, 154 pp.

Birchall, J.D., C. Exley, J.S. Chappell and M.J. Phillips, 1989. Acute toxicity of aluminum to fish eliminated in silicon-rich acid waters. Nature, 338, 146–148.

Blamey, F.P.C., D.G. Edwards and C.J. Asher, 1983. Effets of aluminum, OH:Al and P:Al molar ratios, and ionic strength on soybean root elongation in solution culture. Soil Sci., 136, 197–207.

Blamey, F.P.C., D.C. Edmeades and D.M. Wheeler, 1992. Empirical models to approximate calcium and magnesium ameliorative effects and genetic differences in aluminum tolerance in wheat. Plant and Soil, 144, 281–287.

Boudot, J.-P., T. Becquer, D. Merlet and J. Rouiller, 1994a. Aluminum toxicity in declining forests: a general overview with a seasonal assessment in a silver fir forest in the Vosges mountains (France). Ann. Sci. For., 51, 27–51.

Boudot, J.-P., D. Merlet, J. Rouiller and O. Maitat, 1994b. Validation of an operational procedure for aluminum speciation in soil solutions and surface waters. Sci. Tot. Environ., 158, 237–252.

Boudot, J.-P., O. Maitat, D. Merlet and J. Rouiller, 1995. Occurrence of non-monomeric species of aluminum in undersaturated soil solutions and surface waters. Consequences for the determination of mineral saturation indices. J. Hydrol., 177, 47–63.

Brown, D.J.A., 1983. Effect of calcium and aluminum concentrations on the survival of brown trout (*Salmo trutta*) at low pH. Bull. Environ. Cont. Toxicol., 30, 582–587.

Cameron, R.S., G.S.P. Ritchie and A.D. Robson, 1986. Relative toxicities of inorganic aluminum complexes to Barley. Soil Sci. Soc. Am. J., 50, 1231–1236.

David, M.B. and G.B. Lawrence, 1996. Soil and solution chemistry under red spruce stands across the northeastern united states. Soil Sci., 161 (5), 314–328.

De Vries, W., J.J.M. Van Grinsven, N. Van Breemen, E.E.J.M. Leeters and P.C. Jansen, 1995. Impacts of acid deposition on concentrations and fluxes of solutes in acid sandy forest soils in the Netherlands. Geoderma, 67, 17–43.

Driscoll, C.T. and R. Van Dreason, 1993. Seasonal and long-term temporal patterns in the chemistry of Adirondack lakes. Water Air Soil Pollut., 67, 319–344.

Driscoll, C.T., J.P. Baker, J.J. Bisogni and C.L. Schofield, 1980. Effect of aluminum speciation on fish in dilute acidified waters. Nature, 284, 161–164.

Driscoll, C.T., M.D. Lehtinen and T.J. Sullivan, 1994. Modelling the acid-base chemistry of organic solutes in Adirondack, New York, lakes. Water Resour. Res., 30 (2), 297–306.

Grauer, U.E. and W. Horst, 1991. Comments on the Calcium-Aluminum Balance (CAB). Soil Sci. Soc. Am. J., 55, 897–898.

Harris, W.R., G. Berthon, C. Exley, T.P. Flaten, W.F. Forbes, T. Kiss, C. Orvig and P.F. Zatta, 1996. Speciation of aluminum in biological systems. J. Toxicol. Environ. Health, 48, 543–568.

Kinraide, T.B., 1991. Identity of the rhizotoxic aluminum species. Plant and Soil, 134, 167–178.

Kinraide, T.B. and D.R. Parker, 1987. Cation amelioration of aluminum toxicity in wheat. Plant Physiol., 83, 546–551.

Li, Y.C., A.K. Alva and M.E. Sumner, 1989. Response of cotton cultivars to aluminum in solutions with varying silicon concentrations. J. Plant Nutrit., 12, 881–892.

Lydersen, E., A.B.S. Poléo, I.P. Muniz, B. Salbu and H.E. Bjornstad, 1990. The effects of naturally occurring high and low molecular weight inorganic and organic species on the yolk-sack larvae of Atlantic salmon (*Salmo salar* L.) exposed to acidic aluminum-rich lake water. Aquat. Toxicology, 18, 219–230.

MacLean, D.C., K.S. Hansen and R.E. Schneider, 1992. Amelioration of aluminum toxicity in wheat by fluoride. New Phytol., 121, 81–88.

Matzner, E. and J. Prenzel, 1992. Acid deposition in the German Solling area: effects on soil solution chemistry and Al mobilization. Water Air Soil Pollut., 61, 221–234.

Nilsson, S.I. and B. Bergkvist, 1983. Aluminum chemistry and acidification processes in a shallow podzol on the Swedish west coast. Water Air Soil Pollut., 20, 311–329.

Parkhurst, B.R., H.L. Bergman, J. Fernandez, D.D. Gulley, J.R. Hockett and D.A. Sanchez, 1990. Inorganic monomeric aluminum and pH as predictors of acidic water toxicity to brook trout (*Salvelinus fontinalis*). Can. J. Fish. Aquat. Sci., 47, 1631–1640.

Poléo, A.B.S., 1995. Aluminum polymerisation – a mechanism of acute toxicity of aqueous aluminum to fish. Aquat. Toxicol., 31, 347–356.

Probst, A., E. Dambrine, D. Viville and B. Fritz, 1990. Influence of acid atmospheric inputs on surface water chemistry and mineral fluxes in a declining spruce stand within a small granitic catchment (Vosges massif, France). J. Hydrol., 116, 101–124.

Pokrovski, G.S., Schott, J., Harrichoury, J.C. and A.S. Sergeyev, 1996. The stability of aluminum silicate complexes in acidic solutions from 25 to 150°C. Geochim. Cosmochim. Acta, 60 (14): 2495–2501.

Rhue, R.D.G. and C.O. Grogan, 1977. Screening corn for Al tolerance using different Ca and Mg concentrations. Agron. J., 69, 755–760.

Rosseland, B.O. and A. Henriksen, 1990. Acidification in Norway – loss of fish populations and the 1000-lake survey 1986. Sci. Total Environ., 96, 45–56.

Sadler, K. and S. Lynam, 1988. The influence of calcium on aluminum-induced changes in the growth rate and mortality of brown trout, *Salmo trutta* L.. J. Fish Biol., 33, 171–179.

Schecher, W.D. and D. McAvoy, 1992. MINEQL+: A software environment for chemical equilibrium modeling. Comput. Environ. Systems, 16, 65–76.

Schecher, W.D. and D. McAvoy, 1994. MINEQL+, a chemical equilibrium program for personal computers. User's manual, version 3.0, 112 pp. Environmental Research Software, Hallowell, ME, USA.

Stumm, W. and J.J. Morgan, 1996. Aquatic chemistry (3rd ed.). John Wiley & sons, New York, 1022 pp.

Suthipradit, S., D.G. Edwards and C.J. Asher, 1990. Effects of aluminum on tap-root elongation of soybean (*Glycine max*), cowpea (*Vigna unguiculata*) and green gram (*Vigna radiata*) grown in the presence of organic acids. Plant and Soil 124, 233–237.

Ulrich, B., R. Mayer and P.K. Khanna, 1980. Chemical changes due to acid precipitation in a loess-derived soil in central Europe. Soil Sci., 130, 193–199.

Wilkinson, K.J., G.C. Campbell and P. Couture, 1990. Effect of fluoride complexation on aluminum toxicity towards juvenile Atlantic Salmon (*Salmo salar*). Can. J. Fish. Aquat. Sci., 47, 1446–1452.

Wood, C.M., D.G. McDonald, C.G. Ingersoll, D.R. Mount, O.E. Johannsson, S. Landsberger and H.L. Bergman, 1990. Effects of water acidity, calcium and aluminum on whole body ions in brook trout (*Salvelinus fontinalis*) continuously exposed from fertilization to swim up : a study by instrumental neutron activity analysis. Can. J. Fish. Aquat. Sci., 47, 1593–1603.

ULTRAFILTRATION AS A MEANS TO INVESTIGATE COPPER RESISTANCE MECHANISMS IN SOIL BACTERIA

I. LAMY[1], S. LOYS[1], L. COURDE[2], T. VALLAEYS[2] and R. CHAUSSOD[2]

1 INRA, Unité de Science du Sol, Route de Saint-Cyr,
78026 VERSAILLES Cedex, FRANCE
2 INRA-CMSE, Laboratoire de Microbiologie des Sols,
BV 1540, 21034 DIJON Cedex, FRANCE

1. INTRODUCTION

Copper is a trace element of major concern for agricultural soils. It exhibits a high toxicity against microorganisms and is widely introduced into soils as a component of pesticide treatments or urban wastes such as sewage sludges or refuse composts. In most French vineyards, "Bordeaux mixture" (copper sulfate) has been applied for more than a century, sometimes leading to copper concentrations in soils much higher than the threshold values (100 mg kg^{-1} in France). Microorganisms exposed to high concentrations of copper or other trace elements are known to develop resistance mechanisms and represent a suitable material for the study of such processes, both at a physiological and molecular level (Capasso et al., 1996).

Heavy metal contamination has been shown to affect the soil biomass (Brookes and McGrath, 1984) and microbial activities (Lighthart et al., 1983; Obbard and Jones, 1993; Yeates et al., 1994), as well as the species composition of bacterial communities (Bååth, 1989; Doelman et al., 1994; Frostegård et al., 1996), reflecting a differential sensitivity to copper among components of the soil microflora (Huysman et al., 1994).

Various mechanisms of metal resistance have already been described in the literature. These include for example the alteration of the membrane transport system responsible for metal accumulation, the extracellular chelation or precipitation of metal by a number of secreted metabolites, the intracellular sequestration of metal by the cell wall, or energy-dependent efflux of metal ions (Brown et al., 1992; Cooksey, 1993; Brady et al., 1994).

However, resistance mechanisms have been accurately determined only in a relatively small number of microbial strains. In addition these determinations mainly result

Effect of Mineral-Organic-Microorganism Interactions on Soil and Freshwater Environments
Edited by Berthelin, *et al*, Kluwer Academic / Plenum Publishers, New York, 1999

from a time-consuming molecular approach involving hybridization and restriction analysis (Cooksey, 1990), cloning and transposon mutagenesis (Bender and Cooksey, 1987) or sequencing (Kim et al., 1995). The latter describes the purification of a copper-binding protein using anion exchange columns and acrylamide gel electrophoresis. Alternative methods based on centrifugation and metal ion affinity chromatography columns have been used in order to purify copper-induced proteins (Harwood-Sears and Gordon, 1990). However, such methods appear unsuitable for the study of copper resistance mechanisms mediated by non-specifically produced extracellular organic products that have been described in a number of bacterial species (Bitton and Freihofer, 1978; Mittelman and Geesey, 1985).

The aim of our work was therefore to develop a general purpose scheme which could be used for screening a variety of copper resistance mechanisms developed by soil bacteria. The method should allow the determination of the respective parts of the exomolecules produced in response to the presence of copper, the role of the cell wall surfaces or the part played by copper-binding organic metabolites produced in the cells. Therefore the proposed scheme is based on the comparison of bacterial behavior in growth media with and without copper and the study of the three compartments where resistance mechanisms could take place and therefore be identified (namely the intracellular medium, the cell walls and the extracellular medium). Recently ultrafiltration or gel filtration chromatography techniques have been used to characterize soluble microbial products by molecular weight distribution (Kuo and Parkin, 1996). However the combination of a fractionation/ultrafiltration scheme has never been used, to our knowledge, for screening bacterial resistance mechanisms. In this study, the molecular weight distribution of organic carbon and copper was estimated by the ultrafiltration of both the extracellular and intracellular compartments of a bacterial strain which had been isolated from a copper-enriched soil and cultivated on different growth media. In addition, the role of the cell wall compartment in copper retention was investigated using potentiometric titrations.

2. MATERIALS AND METHODS

2.1. Bacterial Strain

The bacterium used in this study was isolated as previously described by Vallaeys et al. (1996), from an agricultural experimental site located at Hagetmau near Bordeaux (France). Repeated applications of copper sulfate (used as a monometallic contaminant) over 8 years resulted in experimental plots containing various concentrations of copper, and the most contaminated plot (90 mg Cu kg^{-1} soil) was used. The strain was selected for its growth characteristics and high level of copper resistance, as well as for its representativeness in the contaminated soil. It has been identified as a Gram-positive bacterium related to the *Arthrobacter* genus (Courde et al., 1996).

2.2. Growth Medium

Bacterial cells were cultured in the following minimal medium: 0.17 M K_2HPO_4; 0.07 M KH_2PO_4; 5 mg l^{-1} yeast extract (Difco, USA); 37.9 μM $(NH_4)_2SO_4$; 840 μM $MgSO_4$, $7(H_2O)$; trace elements (6.96 μM $ZnSO_4$, $7(H_2O)$; 0.809 μM H_3BO_3; 1.05 μM $CoCl_2$, $6(H_2O)$; 6.26 μM $CuSO_4$; 0.21 μM $NiCl_2$, $6(H_2O)$; 9.00 μM $FeSO_4$, $7(H_2O)$; 4.28 μM EDTA), and 6.00 mM L-asparagine (Sigma, France). Asparagine was used in this medium both as a source of carbon and as a copper ligand (Zevenhuizen et al., 1979). In order to induce a potential resistance mechanism, the minimal medium was supplemented when necessary with copper sulfate (2.00 mM). The pH of the medium was 6.5. Precultures were grown in the absence of copper sulfate at 28°C for 48 hours in the medium described above. Two 400 ml flasks of fresh culture medium (with and without copper) were then inoculated with 5 ml of the preculture and placed on a rotary shaker (200 rpm) at 28°C until optical density at 600 nm reached 0.5.

2.3. Preparation of Cell Fractions

The three different compartments were separated using following procedures. Cultures were centrifuged 10 min. at 6000 *g* and 4°C. The supernatant containing the growth medium and the bacterial exoproducts filtrated at 0.2 μm was defined as the extracellular compartment. The cell pellets were washed once with water, centrifuged 10 min. at 6000 *g* and 4°C, and resuspended in a minimum volume of water (about 20 ml or 40 ml). Lysis was performed using a French press (Model M S/N 12000-002, Fred and Carver Inc., Wabash, USA). The lysates were centrifuged (17500 *g*, 30 min., 4°C) in order to separate the cell wall compartment from the soluble intracellular compartment. The cell walls were lyophilized, and the solution including intracellular constituents was frozen.

2.4. Ultrafiltration Separations

In order to estimate the molecular weight distribution of the soluble bacteria products, the extracellular and intracellular compartments were fractionated using three Amicon ultrafilters (Amicon Inc., Beverly, USA). The YM10, YM3 and YC05 filters with nominal molecular weight cut-offs (MWCO) of 10 000, 3 000 and 500, respectively, were used in sequence in an Amicon model 8050 stirred cell, pressurized with nitrogen at 370 kPa. Fifty ml of sample was ultrafiltrated with the first YM10 membrane up to 15 ml remaining in the cell, which were kept for further analysis. The filtrate was applied to the second YM3 ultrafiltration membrane under identical conditions. Similarly, 15 ml of the second retentate obtained were kept for further analysis, while the filtrate was further applied to the last YC05 ultrafiltration membrane, to finally collect one retentate and one ultrafiltrate. Four fractions were thus obtained : > 10 000; 3 000 - 10 000; 500 - 3 000; and < 500. In a separate experiment, another 50 ml of intracellular medium was ultrafiltrated using an YM30 ultrafilter membrane (MWCO 30 000) in order to obtain two fractions: > 30 000 and < 30 000.

Blank solutions prepared from asparagine with or without copper were ultrafiltrated and analyzed in the same way as the extracellular compartments. The retention behavior of the ultrafilter membranes for the fractionation was thus checked with respect to asparagine and its copper complexes, and these blanks were used to interpret the ultrafiltration results in the presence of bacteria.

The high organic carbon content of the intracellular compartments required 5-fold dilution before ultrafiltration. This choice was made as a compromise between the risk of ultrafiltration membrane plugging and a possible change of copper speciation as a result of the dilution.

2.5. Compartment Analyses

Total copper concentrations were determined in both the defined compartments and the ultrafiltrated fractions with a flame IL251 atomic absorption spectrometer (Instrumentation Laboratory Inc., Lexington, MA, USA). Values are obtained +/- 0.2 mg Cu l^{-1}. For the cell wall compartment, aliquots of lyophilized cell walls (50 mg) were calcined 4h at 450°C then resuspended with HCl 0.1 M and made up to volume with water before copper analysis.

The organic carbon was determined in the extracellular and intracellular compartments as well as in their respective ultrafiltrated fractions by catalytic oxidation and IR CO_2 determination with a Dohrmann DC 190 (Rosemount Analyt. Inc., Santa Clara, CA, USA). Mean values are given with a standard deviation lower than 2%.

Potentiometric titrations were performed at a constant ionic strength of 0.1 M $NaClO_4$ under nitrogen flow. Solution temperatures were maintained at 25°C using a water cell with a thermostat. About 50 mg of lyophilized cell walls were resuspended in water to obtain a 3 g l^{-1} initial suspension. Aliquots of this stock cell walls suspension were titrated with 0.1 M NaOH in the presence of a known amount of $HClO_4$, and in absence or presence of different concentrations of $Cu(ClO_4)_2$. The pH was monitored with a Metrohm model 60102100 glass electrode and a Tacussel model XR110 calomel reference electrode including a salt bridge filled with 0.1 M $NaClO_4$ connected to a Tacussel TTProcessor pH meter (Radiometer, Copenhagen, Denmark). In the presence of copper, the concentration of free copper in solution was determined using a copper-selective electrode (Tacussel XS290) which was calibrated with freshly prepared metallic buffers and with blank titrations without cell walls at the same ionic strength.

2.6. Data Calculations

A concentration factor was calculated for each ultrafiltration by the ratio: R = Vi/Vf, where Vi is the initial volume introduced into the ultrafiltration cell, and Vf is the final volume of the retentate in the cell. During the ultrafiltration procedure, molecules that do not pass through the membrane concentrate in the ultrafiltration cell, while molecules that pass through the membrane are found again at the same concentration in the filtrate and in the retentate. Therefore after analysis, the total organic carbon (TOC) of each fraction was calculated as described below :

TOC>10000=[TOCmeas>10000 - TOC3000-10000 - TOC500-3000 - TOC<500]/R10000

TOC3000-10000=[TOCmeas3000-10000 - TOC500-3000 - TOC<500]/R3000

TOC500-3000=[TOCmeas500-3000 - TOC<500]/R500

TOC<500=TOCmeas<500

Where TOCmeasX = total organic carbon determined in the corresponding fraction X, and TOCX = effective content of organic carbon in the fraction X. Copper contents were obtained with the same calculations from the copper determinations in the isolated fractions.

TOC and copper recoveries were calculated. Over 95% of the TOC and copper applied were recovered, except for the intracellular fractionations for which only 85% were recovered.

3. RESULTS

3.1. Copper Distribution Between Compartments

Results of copper distribution among the various isolated compartments are shown in Table 1.

Table 1. Copper distribution in mg and % of the available copper (taken here as the extracellular amount, see text) in the different operationally defined compartments for the bacteria cultured in the absence and presence of copper (respectively Cu as micronutrient, and Cu as pollutant).

Compartment	Cu-micronutrient medium (mg)	Cu-enriched medium (mg)
extracellular	1.2	86.4
intracellular	0.096 (8%)	1.1 (1.2%)
cell walls	0.03 (2.5%)	1.2 (1.4%)

3.1.1. Micronutrient Copper Medium

When copper was initially present in the growth medium at a micronutrient amount of 1.2 mg (0.5 mg l^{-1} in 2.4 l), the extracellular copper concentration after bacterial growth was similar to the initial copper concentration. The intracellular compartment displayed an increased copper concentration of 2.4 mg l^{-1} (in 40 ml), thus represented 0.096 mg Cu corresponding to 8% of the total initial copper (Table 1). The copper concentration in the cell wall compartment was 0.2 mg Cu g^{-1} of cell walls (for a total of 150 mg harvested). Therefore, about 0.03 mg Cu corresponding to 2.5% of the total available copper occurred in the cell wall compartment.

3.1.2. Copper-Enriched Medium

Conversely, when copper was introduced at a total concentration of 127 mg l^{-1}, a decrease of copper content was observed in the intracellular compartment which exhibited a copper concentration of only 53.3 mg l^{-1}. This global Cu distribution confirms that at trace levels bacterial cells concentrate Cu and that the intracellular Cu concentration rises with the external concentration.

The total initial copper amount in the copper-enriched growth medium was 304.8; only 86.4 mg were found in the extracellular compartment. The main difference between total initial and after culture copper amounts could be accounted for by the presence of a blue precipitate observed in this case. This precipitate was identified by XR diffraction as composed of only asparagine and copper. Its presence did not correspond to a bacterial resistance mechanism, since it could also be obtained by mixing asparagine 6 mM and copper 2 mM. However, this precipitate was not predicted in the solubility products compilations (Sillen and Martell, 1971). Most of this precipitate was removed from the bacterial cells during the first centrifugation step requiring a few supplementary washings compared to the minimal non-enriched copper medium. It must be noted that this artificial step may be avoided by changing the metal/ligand ratio or by using a more appropriate growth medium.

For the purpose of calculations, the precipitate was considered to be an unavailable form of copper for the bacteria. Therefore the total soluble amount in the extracellular compartment (86.4 mg) was taken as the amount of available copper. 1.2% of the amount of so-defined available copper was in the intracellular compartment and 1.4% in the cell wall compartment (Table 1). Since some matter could have been lost during the few supplementary washings, as already mentioned, the values given in Table 1 can be viewed as slightly underestimated. Nevertheless, results clearly demonstrate a decrease in the net copper flux into the cells in the copper-enriched medium.

3.2. Extracellular Compartment: Molecular Weight Distributions of Organic Carbon and Copper

Results of copper and organic carbon calculations in the various ultrafiltrated fractions of the extracellular compartments are given in Figure 1. To help interpret these results, copper and organic carbon contents of the various fractions are compared with the results of the corresponding ultrafiltrated blanks (asparagine solutions with corresponding organic carbon and copper contents).

It is clear from the blank fractionations presented in Figure 1, that retention of asparagine occurs with the MWCO 500 membrane, even though asparagine molecular weight is 132.12. This shows the importance of blanks for direct interpretation of ultrafiltration results which depend on the nature, the charge, and the conformation of the molecules.

For the copper-micronutrient growth medium, most organic carbon of the extracellular compartment was found to pass through the MWCO 500 membrane. Copper was found exclusively in the < 500 fraction, indicating its presence as a free or mineral-complexed ion or in the complexed form with low molecular weight organic ligands. By comparison with the blank, organic carbon with nominal molecular weight above 3 000 was also revealed in the extracellular compartment. This high molecular weight organic carbon could be attributed to exobacterial products. Therefore our method could constitute the basis for further investigation on the nature and properties of this material.

For the copper-enriched growth medium, organic carbon of the extracellular compartment was found to be divided between the two lowest fractions, < 500 and 500 - 3 000, but higher molecular weight organic carbon was also found (Fig. 1). Copper was mainly found in the < 500 fraction. By comparison with the blank and with the distribution obtained in the case of the copper-micronutrient medium, the differences in carbon and copper contents of the 500 - 3 000 fractions (Figure 1) indicated the formation of new bacterial products, which could be the consequence of the expression of a resistance mechanism.

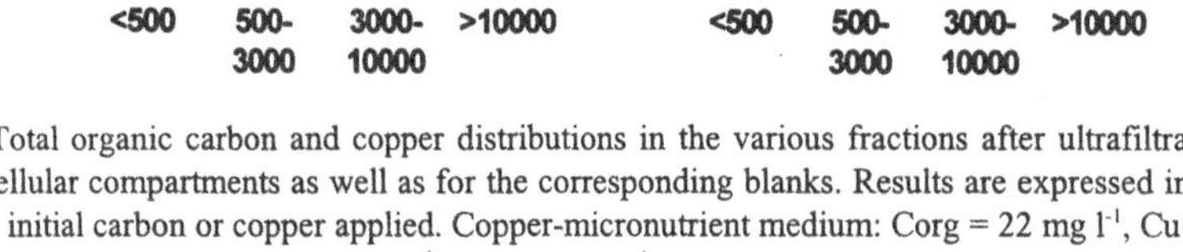

Figure 1. Total organic carbon and copper distributions in the various fractions after ultrafiltration for the two extracellular compartments as well as for the corresponding blanks. Results are expressed in percentage of the total initial carbon or copper applied. Copper-micronutrient medium: Corg = 22 mg l^{-1}, Cu = 0.5mg l^{-1}; copper-enriched medium: Corg = 79 mg l^{-1}; Cu = 33 mg l^{-1}.

3.3. Intracellular Compartment: Molecular Weight Distributions of Organic Carbon and Copper

Total organic carbon concentrations of the ultrafiltrated bacterial intracellular compartment were 3532 mg l^{-1} and 1531 mg l^{-1} for the minimal growth medium and the copper-enriched medium, respectively. This difference could mainly be attributed to the technical difficulties encountered in the sample preparation in presence of high copper concentrations. As already mentioned, the multiple washings for the first separation step of extracellular medium and cells could lead to loss of cells and/or loss of intracellular medium due, for instance, to bacterial lysis.

Results of the ultrafiltration of these two intracellular compartments are shown in Figure 2. The main feature is that most organic carbon was found in the > 10 000 fraction for both copper media. A separate experiment with a MWCO 30 000 membrane fractionation was thus investigated. The results are also presented in Figure 2. Copper distributions were different depending on the growth medium used. For the copper micronutrient medium, intracellular copper was found associated partly with high molecular weight organic carbon and partly with low molecular weight organic carbon in the < 500 fraction. But the 30 000 fractionation revealed that copper was mainly associated with organic carbon < 30 000. On the contrary, for the copper-enriched growth medium, intracellular copper was found distributed among the various fractions, and the 30 000 fractionation revealed the predominant association of copper with > 30 000 molecular weight molecules. This observation led to the hypothesis of a resistance mechanism including intracellular production of copper-binding high molecular weight molecules.

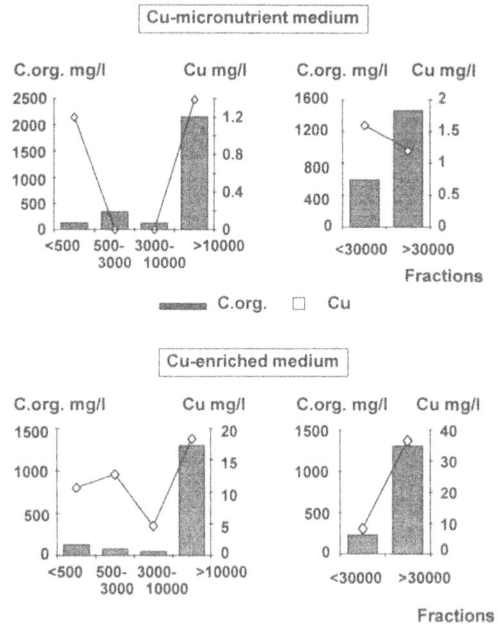

Figure 2. Total organic carbon and copper distributions in the various fractions after ultrafiltration for the two intracellular compartments. Results are shown also for the ultrafiltrations with the MWCO 30 000 membrane.

3.4. Cell Wall Compartment

Potentiometric titrations of the cell wall compartment in the absence and presence of copper are shown in Figure 3. These titrations were used here to test the proton and copper binding capacities of this compartment. The titration curve in absence of copper reflects the sequential binding of protons by the cell wall surfaces (Figure 3a). Whether the functional groups originated from surface sites and/or by-products adsorbed on the surfaces can not be distinguished here. Therefore, the titration curve can be viewed as a whole response of the titratable functional groups for this operationally defined compartment. The protonation curves do not produce distinct equivalence points that could characterize the neutralization of well identified reactive groups. Such potentiometric titration curves are generally interpreted by the presence of carboxylic acid-type groups which protolyze around pH 4-5, and phenolic and/or amino groups that become deprotonated when pH reaches 7 (Gonçalves et al., 1987; Lamy et al., 1988; Deneux-Mustin et al., 1994).

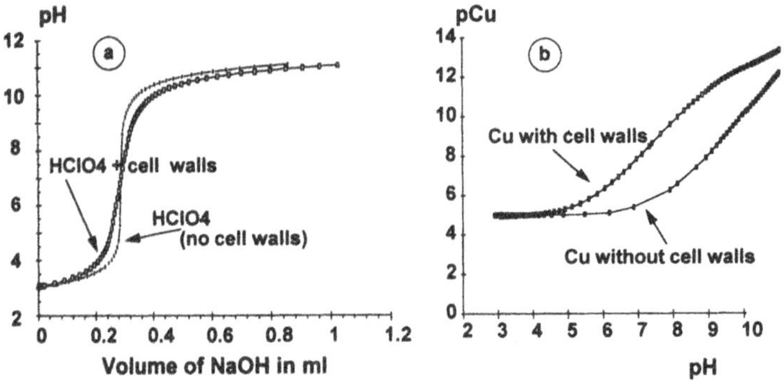

Figure 3. (a) Titration curve of the cell walls (0.5 g l^{-1}) given pH as a function of base added, and (b) Results of the titration curves of Cu(II) 10μM as a function of pH in the absence and presence of cell walls (0.5 g l^{-1}), expressed as pCu = -log[Cu^{2+}] vs pH.

Release of protons during surface complexation or sorption was seen by comparison of the cell walls titration curves in the absence and presence of copper. Therefore, interactions of copper with the potential ligands correspond to competition between the metal ion and protons for complexing sites. Figure 3b shows data from titration curves of copper solutions in the absence or presence of cell walls, where simultaneous determination of Cu(II) and H+ were made; the concentration of free copper was measured using a Cu(II)-selective electrode. In the absence of cell walls, the decrease of Cu^{2+} at high pH is due to the hydrolysis of copper. In the presence of cell walls, the difference between the two curves gives a measure of copper bound to the cell walls. This suggests that surface complexation or sorption is pH dependent, but occurs effectively at pH > 4. The copper maximum binding ability was not quantified in this study, but it is obvious that the cell wall compartment is able to remove copper from solution. Results presented in Table 1 show, however, that such a mechanism does not play a major part in the copper resistance mechanism.

4. DISCUSSION

The aim of this study was to develop a method enabling rapid screening of mechanisms involved in copper bacterial resistance and an evaluation of the mechanisms' relative importance. Therefore, the bacteria were grown in conditions enhancing potential resistance mechanisms, without leading to the death of the cells. A balance between the free copper ion, the most toxic (Zevenhuizen et al., 1979), and the other copper-complexed species whose toxicity depends on the lability and the formation constants of the species (Angle and Chaney, 1989), was therefore sought. In this study, an organic compound was used both as carbon source and copper ligand. This resulted in the free copper concentration being controlled at $t = 0$ in our particular initial conditions, but varying with time according to bacterial growth, the substrate utilization and the release of exomolecules influencing the Cu^{2+} activity in the growth medium. In order to control more accurately Cu^{2+} activity during growth, it seems desirable to use a medium in which the free metallic cation is buffered (Angle and Chaney, 1989; Knight and McGrath, 1995).

Cell walls, exoproducts, and intracellular ligands have been described in the literature as means of bacterial resistance strategies. In our study, the total copper distribution among the various compartments gave a first but incomplete idea of the processes. In the presence of high levels of copper, our results showed a relative decrease in the net copper flux in the cells but no indication was obtained to share between a Cu-efflux strategy or a reduction of the Cu transfer (Cervantes and Gutierrez-Corona, 1994).

Sequential ultrafiltrations were used to further investigate the extracellular and intracellular media and to characterize their copper and organic carbon molecular weight distributions. Ultrafiltration technique has been shown to be a useful tool in separation processes (Geckeler and Volchek, 1996). However this technique must be used with caution because of the several factors influencing the separation of different substances, such as adsorption to the ultrafilter, pH, ionic strength or even the molecular shape and charge (Kuo and Parkin, 1996). Ultrafiltration was used here in a standardized procedure, including comparison between the ultrafiltrated fractions of the compartments with those of blanks to obtain a relative estimation of a molecular weight distribution. This comparison showed the different bacterial behavior when the strain was grown on two different media (i.e., copper-micronutrient or -enriched media) and allowed us to hypothesize about major resistance mechanisms. Our results suggested both a synthesis of copper-binding high molecular weight molecules in the intracellular compartment and, in a lesser extent, a secretion of low molecular weight molecules in the extracellular compartment. In this latter case, much of the copper and organic carbon was found in the < 500 fraction. These findings agree with the various mechanisms of bacterial metal resistance described in the literature (Brown et al., 1992; Cervantes and Gutierrez-Corona, 1994). However, simultaneous processes have rarely been proposed as a strategy of bacterial resistance. In this study, the combination of different processes including extracellular binding of metal by low molecular weight molecules and intracellular sequestration by high molecular weight molecules was simultaneously seen.

In the absence of added copper, the isolated cell wall compartment was checked for its ability to bind copper. With the bacterial strain used, no binding role could be attributed to the cell wall compartment in the presence of copper. The part played by the cell wall compartment in excess copper conditions has been described in the literature for fungal mycelia (Huang et al., 1991) and algal surfaces (Xue et al., 1988; Crist et al., 1990; Gonzalez-Davila et al., 1995), but more rarely for bacteria.

The method described here can give a rapid answer to whether the resistance mechanism is copper-dependent or is part of an intrinsic process of the bacteria. In our conditions, the resistance mechanism seems to be induced by copper introduced in excess concentrations into the growth medium.

Discrimination between the various mechanisms involved in copper resistance can be performed using this approach, as well as a focus on one or more different processes if simultaneously involved. The method described here is simple to apply and involves only the ultrafiltration of the intra- and extracellular compartments after their separation. Because of the low volumes involved, sequential ultrafiltration of one compartment is easy and rapid to achieve in less than 3 hours including washings and dryings of the ultrafiltration cell if only one such apparatus is available. Therefore the fractionation/ultrafiltration method allows us to rapidly determine which resistance mechanism is dominating, before conducting further investigations on a particular compartment. In spite of the difficulties encountered, which can be well improved in particular concerning the growth medium, the proposed approach seems suitable for screening bacterial resistance mechanisms and can easily be adapted for routine work.

5. ACKNOWLEDGMENTS

Part of this work was supported by the National Institute of Agronomical Research, INRA, AIP Ecosol, France.

6. REFERENCES

Angle, J.S. and R.L. Chaney, 1989. Cadmium resistance screening in nitrilotriacetate-buffered minimal media. Appl. Environ. Microbiol., 55, 2101-2104.

Bååth, E., 1989. Effects of heavy metals in soil on microbial processes and populations (a review). Water Air Soil Poll., 47, 335-379.

Bender, C.L., and D.A. Cooksey, 1987. Molecular cloning of copper resistance genes from *Pseudomonas syringae* pv. *tomato*. J. Bacteriology, 169, 470-474.

Bitton, G. and V. Freihofer, 1978. Influence of extracellular polysaccharides on the toxicity of copper and cadmium toward *Klebsiella aerogenes*. Microbial Ecology, 4, 119-125.

Brady, D., D. Glaum and J.R. Duncan, 1994. Copper tolerance in *Saccharomyces cerevisiae*. Lett. Appl. Microbiol., 18, 245-250.

Brookes, P.C. and S.P. McGrath, 1984. Effects of metals toxicity on the size of the soil microbial biomass. J Soil Sci., 35, 341-346.

Brown, N.L., D.A. Rouch and B.T.O. Lee, 1992. Copper resistance determinants in bacteria. Plasmid, 27, 41-51.

Capasso, C., F. Nazzaro, F. Marulli, A. Capasso, F. La Cara and E. Parisi, 1996. Identification of a high-molecular-weight cadmium-binding protein in copper-resistant *Bacillus acidocaldarius* cells. Res. Microbiol., 147, 287-296.

Cervantes, C. and F. Gutierrez-Corona, 1994. Copper resistance mechanisms in bacteria and fungi. FEMS Microbiology reviews, 14, 121-138.

Cooksey, D.A., 1993. Copper uptake and resistance in bacteria. Molecular Microbiol., 7, 1-5.

Cooksey, D.A., 1990. Plasmid-determined copper resistance in *Pseudomonas syringae* from Impatiens. Appl. Environ. Microbiol, 56, 13-16.

Courde, L., T. Vallaeys, G. Laguerre, M.C. Breul and R. Chaussod, 1996. Effets d'applications répétées de cuivre sur la diversité des populations microbiennes du sol. In : Proceedings of the « Congrès de la société française de microbiologie: biodiversité et fonctionnement des sols », Lyon, France.

Crist, R.H., J.R. Martin, P.W. Guptill, J.M. Eslinger and D.L.R. Crist, 1990. Interaction of metals and protons with algae. 2. Ion exchange in adsorption and metal displacement by protons. Environ. Sci. Technol., 24, 337-342.

Deneux-Mustin, S., J. Rouiller, S. Durecu, C. Munier-Lamy and J. Berthelin, 1994. Détermination de la capacité de fixation des métaux par les biomasses microbiennes des sols, des eaux, et des sédiments: interêt de la méthode du titrage potentiométrique. C. R. Acad. Sci. Paris, série II, 319, 1057-1062.

Doelman, P., E. Jansen, M. Michels and M. van Til, 1994. Effects of heavy metals in soil on microbial diversity and activity as shown by the sensivity-resistance index, an ecologically relevant parameter. Biol. Fert. Soils, 17, 177-184.

Frostegård, Å., A. Tunlid and E. Bååth, 1996. Changes in microbial community structure during long-term incubation in two soils experimentally contaminated with metals. Soil Biol. Biochem., 28, 53-63.

Geckeler, K.E. and K. Volchek, 1996. Removal of hazardous substances from water using ultrafiltration in conjunction with soluble polymers. Environ. Sci. Technol., 30, 725-734.

Gonçalves, M.L.S., L. Sigg, M. Reutlinger and W. Stumm, 1987. Metal ion binding by biological surfaces: voltametric assessment in presence of bacteria. Sci. Total Environ., 60, 105-119.

Gonzalez-Davila, M., J.M. Santana-Casiano, J. Perez-Pena and F.J. Millero, 1995. Binding of Cu(II) to the surface and exudates of the alga *Dunaliella tertiolecta* in seawater. Environ. Sci. Technol., 29, 289-301.

Harwood-Sears, V. and A. Gordon, 1990. Copper-induced production of copper-binding supernatant proteins by the marine bacterium *Vibrio alginolyticus*. Appl. Environ. Microbiol, 56, 1327-1332.

Huang, C., C.P. Huang and A.L. Morehart, 1991. Proton competition in Cu(II) adsorption by fungal mycelia. Wat. Res., 11, 1365-1375.

Huysman, F., W. Verstraete and P.C. Brookes, 1994. Effect of manuring practices and increased copper concentrations on soil microbial populations. Soil Biol. Biochem., 26, 103-110.

Kim, B.-K., T.D. Pihl, J. N. Reeve and L. Daniels, 1995. Purification of the copper response extracellular proteins secreted by the copper-resistant methanogen *Methanobacterium bryantii* BKYH and cloning, sequencing, and transcription of the gene encoding these proteins. J. Bacteriol., 177, 7178-7185.

Knight, B. and S.P. McGrath, 1995. A method to buffer the concentrations of free Zn and Cd ions using a cation exchange resin in bacterial toxicity studies. Environ. Toxicol. Chem., 14, 2033-2039.

Kuo, W.C. and G.F. Parkin, 1996. Characterization of soluble microbial products from anaerobic treatment by molecular weight distribution and nickel-chelating properties. Wat. Res., 30, 915-922.

Lamy I., M. Cromer and J.P. Scharff, 1988. Comparative study of copper(II) interactions with monomeric ligands and synthetic or natural organic materials from potentiometric data. Anal. Chim. Acta, 212, 105-122.

Lighthart, B., J. Baham and V.V. Volk, 1983. Microbial respiration and chemical speciation in metal-amended soils. J. Environ. Qual., 12, 543-548.

Mittelman, M.W., and G.G. Gessey, 1985. Copper-binding characteristics of exopolymers from freshwater-sediment bacterium. Appl. Environ. Microbiol, 49, 846-851.

Obbard, J.P. and K.C. Jones, 1993. The effect of heavy metals on dinitrogen fixation by *Rhizobium*-white clover in a range of long-term sewage sludge amended and metal-contamined soils. Environ. Poll., 79, 105-112.

Sillen, L.G. and A.E. Martell, 1971. Stability constants of metal-ions complexes. The Chemical Society, London.

Vallaeys, T., L. Courde, V. Brenac, M. C. Breuil, M. Linères, R. Chaussod and G. Laguerre, 1996. Effects of repeated applications of the fungicide "Bordeaux mixture" (copper sulfate) on the diversity of bacterial populations as estimated by PCR-RFLP analysis of 16S rDNA. Proceed. 2nd International Symposium on environmental aspects of pesticide microbiology, 7-11 July, Beaune, France.

Xue, H.-B., W. Stumm and L. Sigg, 1988. The binding of heavy metals to algal surfaces. Wat. Res., 22, 917-926.

Yeates, G.W., V.A. Orchard, T.W. Speir, J.L. Hunt and M.C.C. Hermans, 1994. Impact of pasture contamination by copper, chromium, arsenic timber preservative on soil biological activity. Biol. Fert. Soils, 18, 200-208.

Zevenhuizen, L.P.T.M., J. Dolfing, E.J. Eshuis and I.J.Scholten-Koerselman, 1979. Inhibitory effects of copper on bacteria related to the free ion concentration. Microbial Ecology, 5, 139-146.

APPLICATION OF ORGANIC GEOCHEMISTRY TECHNIQUES TO ENVIRONMENTAL PROBLEMS.

P. FAURE, P. LANDAIS, M. ELIE,
M. KRUGE, E. LANGLOIS & O. RUAU

C.N.R.S. - C.R.E.G.U.
BP 23, 54501 Vandœuvre-lès-Nancy Cedex, France.

INTRODUCTION

A major concern in environmental studies is the estimation of the impact of anthropogenic activities on natural systems. Organic compounds are among the more abundant and the more various pollutants. Because they can display significant adsorption capacities and thus transport other (non organic) pollutants, they should be carefully analyzed. The oil industry has developed advanced organic geochemical techniques in order to improve knowledge of the structure and evolution of natural organic matter. Most of the advanced techniques that are required for the characterization of organic compounds can be directly used in environmental studies. Three major problems regarding organic pollutant impacts on the environment must be addressed: (i) the characterization of the source of the organic pollutants, (ii) their migration and dispersion in water, soils and sediments and (iii) their stability during degradation processes such as biodegradation, oxidation, etc. Those problems require an approach similar to that frequently used in the petroleum field. In petroleum exploration, it is important to characterize the source rocks for oil, to study its migration and dispersion in the reservoir rock and to estimate the changes induced by different alterations. The aim of this paper is to present applications of oil exploration analytical techniques to environmental problems.

Effect of Mineral-Organic-Microorganism Interactions on Soil and Freshwater Environments
Edited by Berthelin, *et al*, Kluwer Academic / Plenum Publishers, New York, 1999

MATERIAL AND METHODS

Petroleum field examples documented in this paper are supported by samples collected in various locations : type II kerogens and oils (Lodève, France), coals (Mahakam Delta, Indonesia), type II and III kerogens (Paris basin, France).

The sewage muds selected for the present study came from the city of Marseille (France), and the agricultural soils treated with these muds were collected near Aix-en-Provence (France) (Pierrisnard, 1996).

The recent sediments analyzed in this study came from three Moselle Rivers (Lorraine Basin, France): (i) la Fensch, (ii) la Rosselle and (iii) la Bièvre. These sediments were sampled by the Rhin-Meuse Water Basin Agency.

The coarser fraction of the sediments (> 250 μm) was used in order to isolate the different organic fractions under a microscope (higher plant seeds, leaves and woods, coals, coke residues, etc.).

The different samples were extracted using hot chloroform for 45 minutes. Liquid chromatography of the extracts on alumina and silica gel columns allowed the separation of the aliphatic and aromatic hydrocarbons and the polar fraction.

Gas chromatography-Mass spectrometry (GC-MS). Aliphatic and aromatic hydrocarbons were analyzed by gas chromatography-mass spectrometry (HP 5890 Serie II GC coupled to a HP 5971 mass spectrometer), using an on-column injector, a 60 m DB-5 J&W, 0.25 mm i.d, 0.1μm film fused silica column. The temperature program was 40 to 300 °C at 3°C/min followed by an isothermal stage at 300°C for 15 min (constant helium pressure of 16.1 psi)

Flash pyrolysis-Gas Chromatography-Mass spectrometry (Py-GC-MS). Pyrolysis of kerogens, global sediments and isolated particles was performed with a CDS 2000 pyroprobe. Samples were loaded in quartz tubes and heated at 620°C for 15 seconds. The GC-MS characteristics and temperature program were as described above. Flash pyrolysis yields were estimated by mass balance.

Fourier Transform Infrared Microspectroscopy (FTIR). Powdered sediment particles were analyzed using a special sample preparation technique (Ruau *et al.*, 1995) thereby avoiding drawbacks usually encountered when using bulk infrared on KBr pellets e.g. contamination by water adsorbed on the highly hygroscopic KBr (Painter *et al.*, 1981). Sediment aliquots (m < 0.5 mg) were placed between the two diamond windows (2 mm diameter, 1 mm thickness) of a compression cell (7.5 cm X 5.1 cm, Spectra-Tech int.). After compression of the sample by screwing, the top window was removed. The visible light X 10 objective gave an enlarged image of the sample and allowed the selection of the area to be analyzed. Spectra were recorded on the sample placed on top of the bottom window. The micro-FTIR analysis was performed on a Brucker IFS-88 equinox spectrometer coupled with a Brucker multipurpose infrared microscope which was fitted with a 250 μm narrow band MCT detector cooled to 77K. The standard analytical conditions were X 15 infrared objective, 40-60 μm diameter infrared spot, 200 scans (100 seconds), spectral resolution of $4cm^{-1}$. Spectra were ratioed to the background collected on a clean diamond window in the same analytical conditions. The assignments of the main I.R. bands were determined by reference to previous works (Bellamy, 1975; Landais and Rochdi, 1990).

RESULTS AND DISCUSSION

SOURCE CHARACTERIZATION

The chemical structure of an immature kerogen depends on the original mixture of organic sources (marine, lacustrine and terrestrial) and on the physical, chemical and biochemical conditions of deposition. Kerogens are currently classified in three families. Type I kerogens derive from bacterial and algae deposited in large lakes, whereas type II kerogens come from the accumulation of planktonic biomass (algae) in shallow epicontinental seas. The accumulation of continental organic matter (higher plants) in sediments characterizes type III kerogens (Durand and Monin, 1980).

During burial of the sediments, the increase in temperature induces a progressive rearrangement of the kerogen (Tissot and Welte, 1984). First, defunctionalization leads to the generation of CO_2, water and some light hydrocarbons during diagenesis. Then follows the catagenesis stage, during which the kerogen yields a chloroform-soluble heavy liquid (organic extract that can lead to an expellable oil). Then, the residual kerogen begins to aromatize. During the last stage, called metagenesis, aromatization proceeds while methane is generated. Chemical characterization of kerogens facilitates the identification of the original organic source, its degree of thermal degradation and the evaluation of its oil potential.

Py-GC-MS is useful for the identification of kerogen types because it provides detailed molecular information on the kerogen (Larter and Douglas, 1980). Two different kerogens (A and B) were pyrolyzed by Py-GC-MS (Figures 1 and 2) and analyzed to determine their composition and therefore identify their respective origins. The pyrogram obtained with kerogen A (Figure 1) is characterized by a high abundance of light n-alkanes (C_{15}-C_{25}) and alkyl-benzene compounds. The pyrogram of kerogen B (Figure 2) displays a more complex molecular composition with mono and di-aromatic hydrocarbons, alkyl-phenols and heavy n-alkanes. N-alkanes of the medium molecular weight range, C_{12}-C_{20} derive from the accumulation of marine organic matter (planktonic origin) (Tissot and Welte, 1984; Sicre *et al.*, 1994). Kerogens of terrigenous (higher plants) organic source, yield pyrograms dominated by aromatic compounds (Giraud, 1970), and contain long-chain n-alkanes (Sicre *et al.*, 1994). The distribution of the latter displays a characteristic odd predominance (during the immature stage) (Bray and Evans, 1961; Tissot and Welte, 1984). The pyrogram obtained with kerogen B is a typical example of continental origin, whereas kerogen A is typical of a marine source.

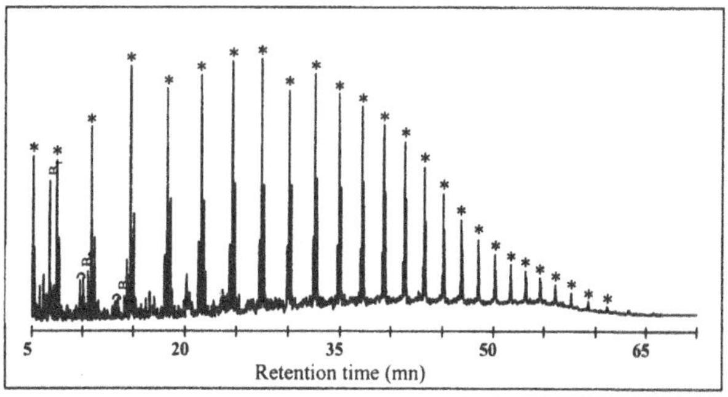

Figure 1 : Kerogen A : Pyrogram (Total Ion Chromatogram : TIC) of an alginite (marine origin) pyrolysed at 620 °C. Peak symbols refer to compounds listed in table 1.

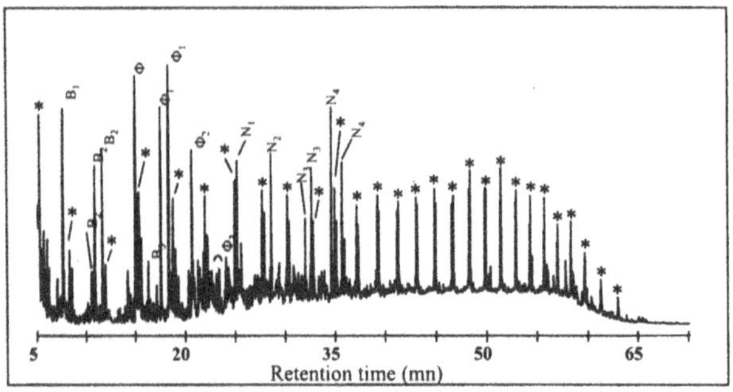

Figure 2 : Kerogen B : Pyrogram (TIC) of a vitrinite (terrestrial origin) pyrolysed at 620 °C. Peak symbols refer to compounds listed in table 1.

Symbol	Assignment
B_i	Alkyl-benzenes
N_i	Alkyl-naphtalenes
Φ_i	Alkyl-phenols
*	n Alkanes
B-diol	Benzene-diol
$M\Phi_i$	Alkyl-methoxy-phenol
FCA_i	Alkyl-furane-carboxaldehyde
Py	Pyrol

Table 1: Identification of peaks labeled in Figs 1 and 2 (i = number of carbon of the alkyl substituents).

By using the same methodology, it is possible to characterize particles isolated from recent sediments. The pyrograms of higher plant seeds and a coke residue isolated from the coarse fraction of La Rosselle river sediments illustrate such data. The higher plant seeds pyrogram (Figure 3) mainly exhibits peaks attributed to oxygenated compounds such as alkyl-phenols and alkyl-methoxy-phenols. The coke residue pyrogram (Figure 4) shows the predominance of oxygenated compounds (alkyl-phenols), aromatic (alkyl-benzenes and alkyl-naphtalenes) and aliphatic hydrocarbons. The obvious difference observed between the two pyrograms suggests that Py-GC-MS is an efficient tool for investigating the molecular signature of organic material. The presence of oxygenated compounds in the seed as well as the coke residue pyrogram, may be due to weathering processes (such as oxidation and biodegradation) occurring during transportation and deposition in sediment. Trace amounts of alkyl-methoxy-phenol in the seed pyrogram, can be also related to the contribution of lignin or altered lignin (Saiz Jimenez and de Leeuw, 1986). The high aromatic hydrocarbon contribution in the coke residue underlines its origin in combustion processes.

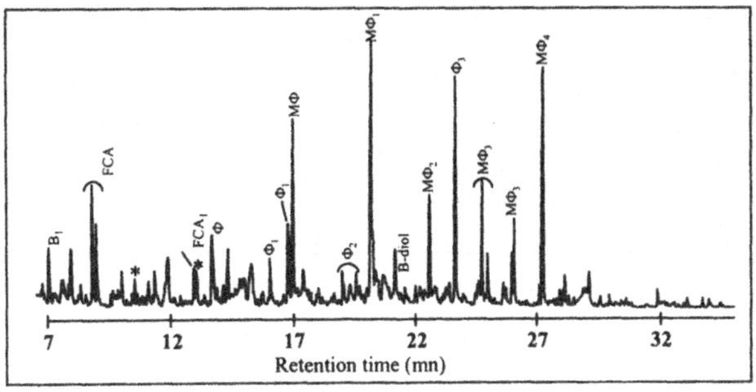

Figure 3 : Pyrogram (TIC) of a higher plant seed (natural origin) pyrolysed at 620 °C. Peak symbols refer to compounds listed in table 1.

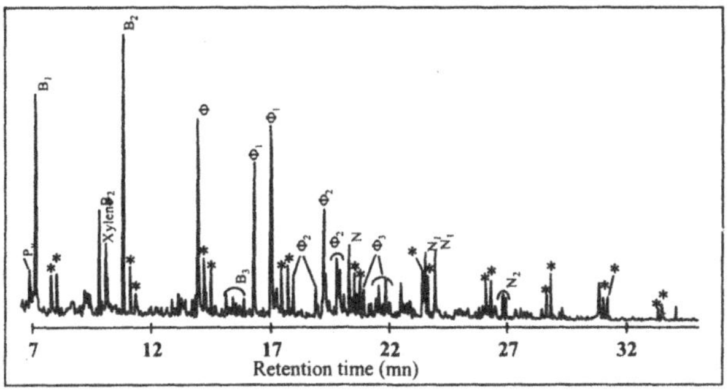

Figure 4 : Pyrogram (TIC) of a coke residue (anthropic origin) pyrolysed at 620 °C. Peak symbols refer to compounds listed in table 1.

Micro-FTIR analysis was carried out (Figures 5 and 6) in order to improve the organic structure characterization of the isolated particles. The coke residue spectrum (Figure 5) shows only oxygenated (hydroxyl and carboxyl) and aromatic bands. The oxygenated bands for higher plant seeds (Figure 6), are relatively important, and N,S,O bonding can be identified in the 1500-1100 cm^{-1} spectral range. The micro-FTIR data are in agreement with Py-GCMS results. The global oxygenated composition of these two particles is confirmed by micro-FTIR. Functional group information, especially for oxygenated species, is essential to evaluate the complexation potential of the organic compounds.

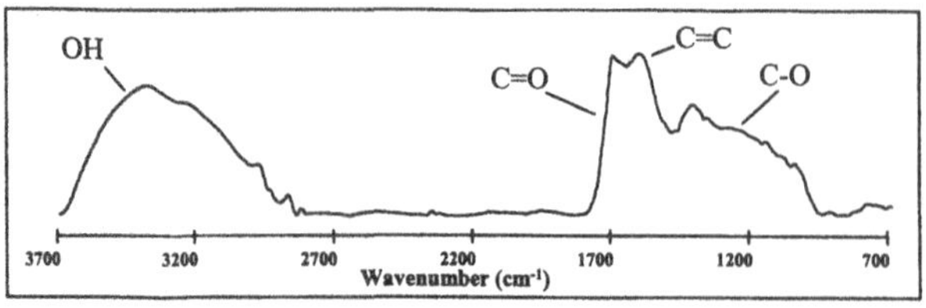

Figure 5 : Spectrum of a coke residue (.anthropic origin) obtained by μFTIR

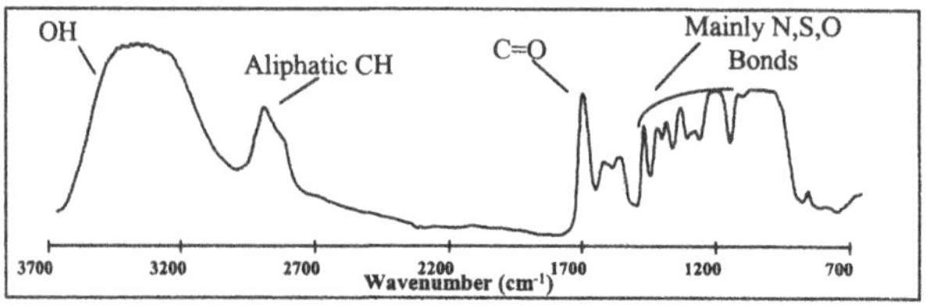

Figure 6 : Spectrum of a higher plant seed (natural origin) obtained by μFTIR.

By combining Py-GC-MS and micro-FTIR data, it is possible to deduce the geochemical signature of the selected particles as well as a gross characterization of the main chemical functional groups. This combination makes it possible to identify and classify the particles contained in a sediment and relate them to their respective origins. It also becomes possible to distinguish the natural organic background signal from those introduced by human activities. Information concerning the structures present and their reactivity can be obtained and used to evaluate the opportunities for the formation of organic-metal complexes.

Migration-dispersion

The migration and dispersion of oils generated during thermal maturation of kerogens is a major problem in petroleum exploration. The source rock and the related oil must be identified even if the chemical composition of the oil has been severely modified during migration and trapping. The saturated hydrocarbon distribution of a crude oil and its suspected source rock analyzed by GC-MS are presented in Figure 7. The chromatograms of the total saturates fraction do not allow a genetic correlation to be established. Indeed, the source rock chromatogram shows large amounts of heavy and light normal-alkanes, whereas the oil chromatogram displays only peaks corresponding to heavy compounds. The lack of light hydrocarbons (mainly n-alkanes) in the oil is probably related to an alteration process (biodegradation). However, the presence of fossils such as polycyclic hydrocarbons are observed in these two chromatograms. Geochemical fossils are biological markers that frequently convey genetic information about the type of organisms contributing to the sedimentary organic matter. They are used for correlation (oil-oil and oil-source rock), for the reconstruction of depositional environments, and also as indicators for maturation (Tissot and Welte, 1984). The biomarkers belonging to the hopanoid family are characterized by their m/z=191 mass fragment. GC-MS allows the single m/z=191 distribution to be plotted in order to study its characteristic distribution. As observed on Figure 7, this distribution is identical in both saturated fractions, strongly suggesting a genetic relationship between the oil and the source rock.

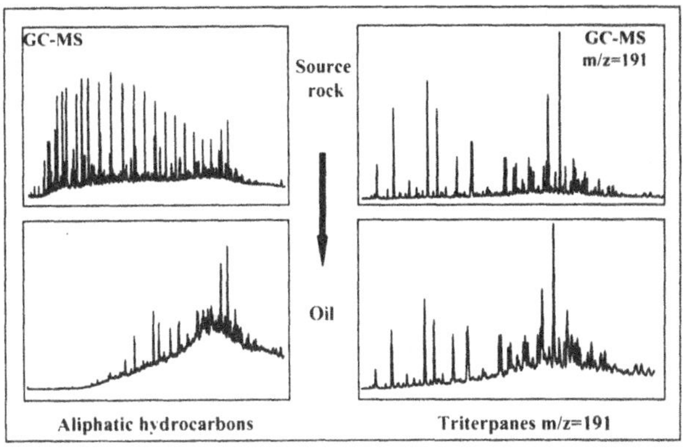

Figure 7 : Correlation between a source rock and an oil by using the triterpanes distribution.

The same approach can be used in the environmental field to establish a relationship between a pollutant source and a pollutant hosting sediment. Because of new legislation and economic interest, sewage muds are currently used in agricultural soils. The organic composition of sewage sludge should be characterized and the soil contamination should be evaluated in order to estimate the environmental impact (Pierrisnard, 1996).

The Figure 8 shows the correlation between a sewage mud and agricultural soil using the GC-MS analysis. The total saturates chromatograms are relatively different and therefore do not support a possible contamination of the soil by the sewage mud. By using the distribution of the naphthalene group (isolation of the m/z=142, 156, 170 mass chromatograms from the total ion current chromatogram of aromatic compounds), it is possible to identify the contamination from the sewage sludge in the organic matter of the agricultural soil. This example suggests that the distribution of the diaromatic compounds as well as other families (benzene, phenanthrene, etc.) can provide a good tool for the detection of organic contamination. However, in order to improve the quality of such genetic correlation, it should be useful to compare the distributions of different biomarkers such as triterpanes, steranes, and the different hydrocarbons families (benzene group, naphthalene group, phenanthrene group, etc.).

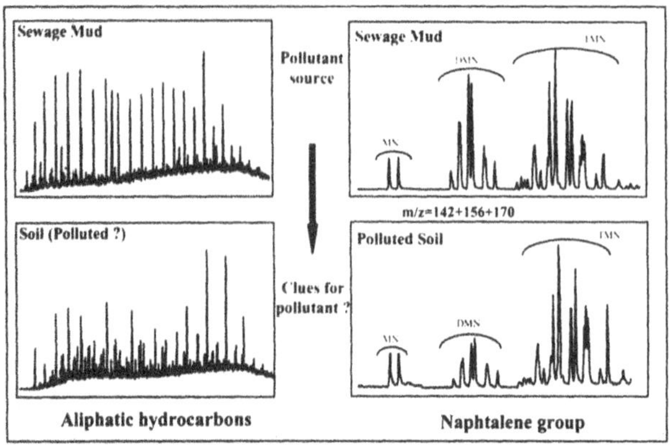

Figure 8 : Correlation between a sewage mud and a soil by using the naphtalene group distribution. (MN: Methyl-Naphtalene, DMN: Dimethyl-Naphtalene, TMN: Trimethyl-Naphtalene).

Organic matter degradation

The composition of petroleum may be strongly modified by alteration processes (such as oxidation biodegradation) during and after accumulation. Crude oil alteration influences its quality and economic value and adversely affects crude oil - source rock correlations. Aliphatic distributions (GC-MS) and micro-FTIR spectra of two crude oils, coming from the same source rock, but showing different alteration levels are presented in Figure 9. The molecular and structural changes induced by alteration can be assessed by comparison of these two oils. The original oil shows an abundance of normal-alkanes on the total saturates chromatogram and a strong aliphatic band absorption on the micro-FTIR spectrum. The total saturates chromatogram of the altered oil is characterized by the drastic removal of normal alkanes and the occurrence of a hump of unresolved complex compounds. The micro-FTIR spectrum presents a less intense aliphatic band whereas oxygenated functions (hydroxyl and carboxyl) are noticed.

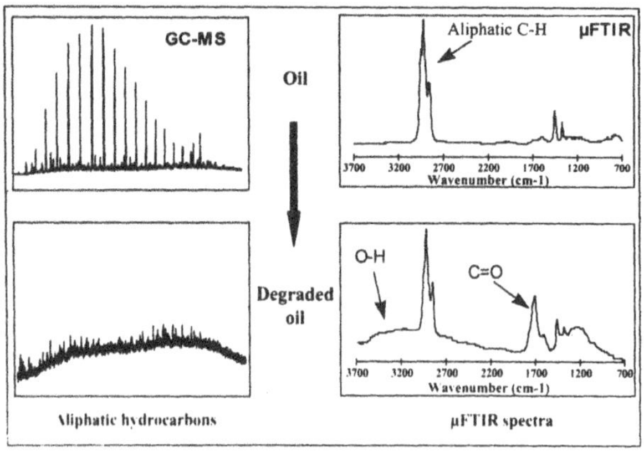

Figure 9 : Total saturates distribution obtained by GC-MS and micro-infrared spectra of a preserved and degraded oil.

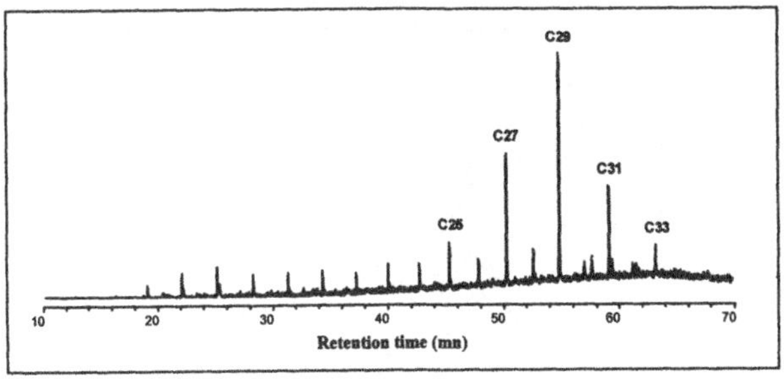

Figure 10 : Total saturates distribution obtained by GC-MS of a preserved river sediment.

Biodegradation by aerobic and/or anaerobic microorganisms results in a partial or total removal of normal-alkanes and slightly branched alkanes (Tissot and Welte, 1984). The remaining aliphatic distribution chiefly constitutes highly branched and/or cyclic hydrocarbons that are not resolved by classic chromatography techniques and so result in an unresolved complex mixture (UCM) (Gough and Rowland, 1990). Such degradation is commonly associated with surface-derived meteoric formation waters, which can induce oxidation processes. The oxidation of organic substances leads to the depletion of reactive aliphatic C-H moieties and the formation at these sites of carbonyl, carboxyl, ether and hydroxyl functionalities. As a consequence, the oxygen functional group distribution is expected to be one of the main structural differences between unoxidized and weathered organic matter (Jakab *et al.*, 1989).

Recent deposited organic matter in sediments can also be altered. The alteration processes can modify the molecular composition and the functional group distribution (formation of complexes). The total saturates distributions of two river sediments (Figure 10 and 11) allow two different alteration levels to be distinguished. The chromatogram of the less altered sediment (Figure 10) emphasizes a marked predominance of odd normal-alkanes in the C_{25}-C_{33} range. A similar predominance is also observed in the chromatogram of the altereted sediment (Figure 11), but an unresolved complex mixture of compounds as well as a molecular sulfur peak appear. The presence of UCM, as well as the decrease of the n-alkanes / iso-cyclo-alkanes ratio, suggests that the river sediment has undergone biodegradation.

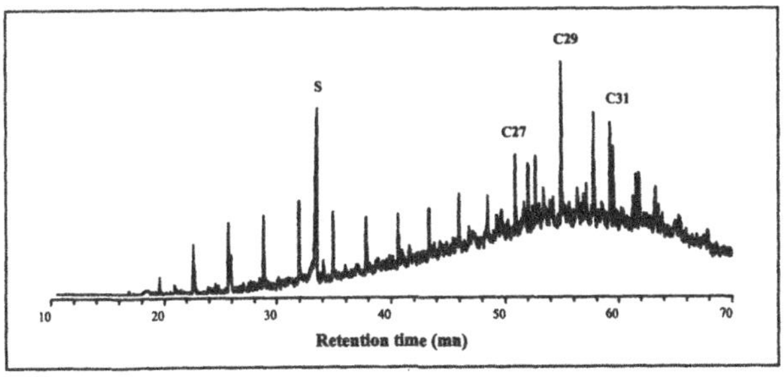

Figure 11 : Total saturates distribution obtained by GC-MS of a degraded river sediment.

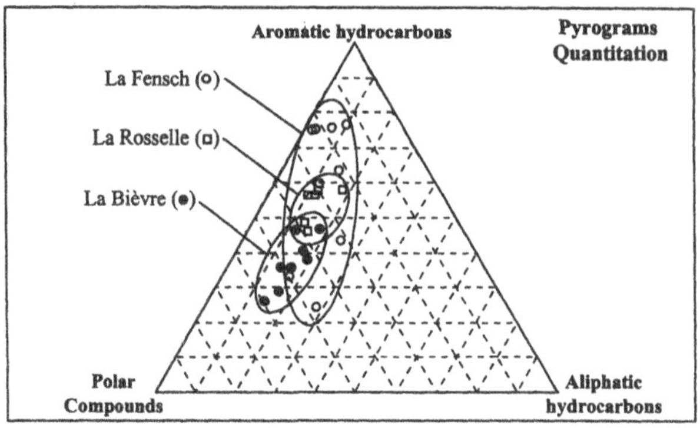

Figure 12 : Triangular diagram of the organic constitution of three Moselle rivers sediments.

Application to regional river sediments

Three Moselle river sediments (La Fensch, La Bièvre and La Petite Rosselle) were analyzed by pyrolysis-GC-MS. All the compounds identified in the pyrograms were divided into three families : (i) polars, (ii) aromatic hydrocarbons and (iii) aliphatic hydrocarbons. In the triangular diagram of Figure 12, each sample, characterized by the relative abundance of each group of compounds, allows a typical river behavior to be identified. La Bièvre and La Petite Rosselle rivers sediments show homogeneous compositions (low dispersion of the data on the triangular diagram). La Petite Rosselle sediments are principally composed of polar and aromatic compounds. La Briève sediments present an identical chemical composition but the aromatic amount is less important. The same regional localization of these two rivers explains the relative homogeneity of the sediment compositions because the natural source (higher plant, algae, bacteria,...) are identical. Fensch river sediments show a larger variability of their chemical compositions deduced from Py-GC-MS analyses (Figure 12) and an enrichment in aliphatic and heavy aromatic hydrocarbons compared to the other sediments. The other Fensch sediments compositions are well grouped and are slightly more aromatic than the Bièvre and the Petite Rosselle sediments. The dispersion of several points underlines an anthropogen contribution to the sediment composition and the other well-grouped points correspond to the natural background of the Fensch river.

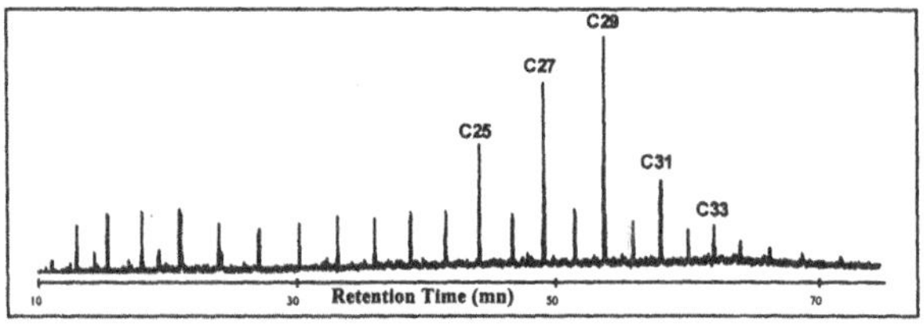

Figure 13 : Total saturates chromatogram of the global sediment of la Rosselle sediments (Moselle, France).

Figure 13 shows the saturates chromatogram obtained after chloroform extraction of the global Rosselle river sediment. It is characterized by a slightly bi-modal n-alkane distribution (C14-C20 and C20-C35 groups) with a very strong odd predominance in the C25-C34 range. The microscopic study of the Petite Rosselle sediments reveals a large amount of coal particles. After separation of the coal particles from the sediment, the aliphatic composition analysis was carried out for the coal particles and for the residual sediment. The isolated coal presents a bi-modal distribution of the normal-alkanes with a slight odd predominance for the heavy normal-alkanes (Figure 14), typical of a type III kerogen. The residual sediment saturates distribution (Figure 15) presents only some heavy normal-alkanes, showing a very strong odd predominance typical of recent higher plants input.

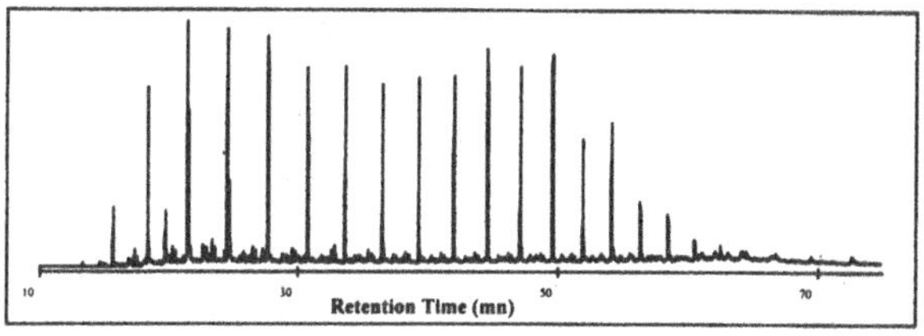

Figure 14 : Total saturates chromatogram of coal particles isolated from sediments of la Rosselle sediments (Moselle, France).

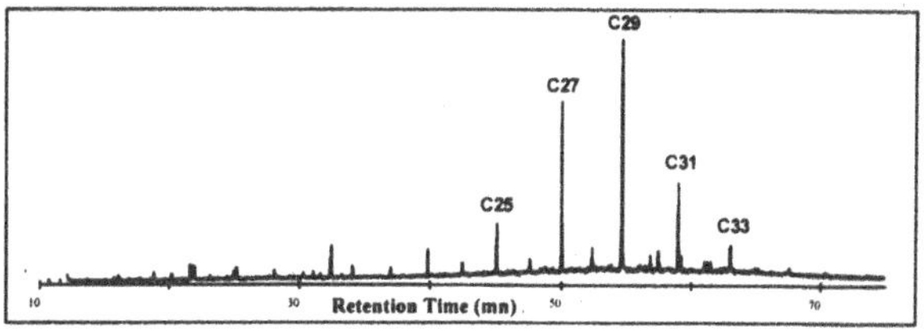

Figure 15 : Total saturates chromatogram of la Rosselle sediments residual fraction after separation of the coal particles (Moselle, France).

Comparison of Figures 13 to 15 reveals that the saturates distribution of the global sediment can be explained by a mixture of the saturates derived from the coal and from the residual fraction. The n-alkanes content of the global sediments thus results from a mixture between an autochthonous organic input (recent higher plants) and an allochthonous input (coal), which can be considered here as a pollutant because no coal-bearing strata are exposed in or close to the river bed. This approach is therefore able to distinguish the natural background input of hydrocarbons in the sediment from that brought by human activity.

CONCLUSION

Organic geochemistry technologies facilitate studies of fossilized organic matter. Molecular and functional group analysis are currently performed in order to characterize the origin of the kerogens, to study migration and dispersion and to estimate the efficiency of degradation processes. The direct transfer of this technology to environmental problems is mainly successful. Indeed, the organic characterization and the identification of possible anthropogen contributions can be assessed by both the molecular and functional group approaches. By using compound family distributions (e.g., biomarkers), it is possible to correlate a pollutant source with the contaminated targets. The evolution of the organic matter by different alteration processes can be estimated by studying molecular and functional group modifications.

However, this technology transfer is not fully achieved. Some organic compounds such as polyaromatic hydrocarbons (PAH) can come from natural sources as well as anthropoogen inputs. The determination of carbon isotope composition is then the only way to distinguish these two potential sources. Gas chromatography - isotopic ratio - mass spectrometry (GC-IR-MS) determines isotopic compositions at the molecular level and may facilitate the correlation between polluted sediments and suspected sources.

Organic compounds have the ability to form complexes with metals. These complexes can mobilize and transport metals to targets such as aquifers. It is desirable to identify which organic compounds are the metal carriers. Gas chromatography - atomic emission detection allows the analysis of the elemental composition of each organic compound and the identification of those that are chemically bonded to metallic species.

Gas chromatography techniques coupled with different detectors such as a mass spectrometer, an isotope ratio analyzer, an atomic emission spectrometer, and a μFTIR are very efficient techniques that can be used in order to solve environmental problems. The ongoing technological transfer from petroleum geochemistry to the environment field is promising, especially when studying source, dispersion, and correlation aspects of pollution problems.

REFERENCES

Bellamy, L.J., 1975. The Infrared Spectra of Complex Molecules. Third Edition, Chapman and Hall, Ltd.,London.

Bray E.E. and E.D. Evans, 1961. Distribution of n-paraffins as a clue to recognition of source beds. Geochim. Cosmochim. Acta, 22, 2-15.

Durand B. and J.C. Monin, 1980. Elemental analysis of kerogens (C, H, O, N, S, Fe). In : B. Dyrabd (Editor), Kerogen.Technip, Paris, pp. 113-142.

Giraud A., 1970. Application of pyrolysis and gas chromatography to the geochemical characterisation of kerogen in sedimentary rocks. Bull Amer. Assoc. Petrol. Geol., 54, 439-455.

Gough M.A. and S.J. Rowland, 1990. Characterization of unresolved complex mixtures of hydrocarbons in petroleum. Nature, 344, 648-650.

Jakab E., Y. Yongseung, and H.L.C. Meuzelaar, 1989. Effects of weathering on the molecular structure of coal. In : C.R. Nelson (Editor), Chemistry of Coal Weathering, pp 61-82.

Landais P. and A. Rochdi, 1990. Reliability of semiquantitative data extracted from transmission microscopy Fourier transform infrared spectra of coal. Energy and fuels, 4, 290-295.

Larter S.R. and A.G. Douglas, 1980. A pyrolysis-gas chromatography method for kerogen typing. In : AG. Douglas and J.R. Maxwell (Editors), Advances in Organic Geochemistry, Pergamon, London, 579-583.

Pierrisnard F., 1996. Impact de l'amendement des boues résiduaires de la ville de Marseille sur des sols à vocations agricole : comportement du Cd, Cr, Cu, Ni, Pb, Zn, des hydrocarbures et des composés polaires. Thèse de l'université Aix Marseille III.

Painter P.C., M.M. Coleman., R. W. Snyder., O. Mahajan., M. Komatsu and P.L., Walker JR., 1981. Low temperature air oxidation of caking coals : Fourier transform infrared studies. Appl.Spectroscopy, 35, 106-110.

Ruau O., L. Mansuy. and P. Landais., 1995. Mise au point d'une nouvelle technique de préparation de la matière organique pour analyse en microspectroscopie infrarouge en mode transmission. C.R. Acad. Sci. Paris, t. 321, série II a, 201-208.

Saiz Jimenez C. and J.W. De Leeuw., 1986. Chemical characterization of soil organic matter fractions by analytical Pyrolysis-Gaz Chromatography-Mass Spectrometry. Journal Analytical and Appl.Pyrolysis, 9, 99-119.

Sicre M-A, S. Peulve, A. Saliot, J.W. De Leeuw and M. Baas, 1994. Molecular characterization of the organic fraction of suspended matter in the surface waters and bottom nepheloïd layer of the Rhône delta using analytical pyrolysis, Org. Geochem., Vol 21, No 1, 11-26.

Tissot B. P. and D.H. Welte, 1984. Petroleum Formation and Occurence. 2nd edition. Springer-Verlag, Berlin.

IN SITU ATR-FTIR CHARACTERIZATION OF ORGANIC MACROMOLECULES AGGREGATED WITH METALLIC CATIONS

F. QUILES, A. BURNEAU and K. KEIDING*

Laboratoire de Chimie Physique pour l'Environnement (LCPE)
UMR 9992 CNRS-Université Henri Poincaré, Nancy I
405, rue de Vandoeuvre 54600 Villers-lès-Nancy, France
Tel. (33) (0)3 83 91 63 00 - Fax (33) (0)3 83 27 54 44

*Environmental Engineering Laboratory
Aalborg University, Sohngårdsholmsvej 57, DK-9000 Aalborg, Denmark
Tel. (45) 9815 8522 ext 6510 - Fax (45) 9814 2555

INTRODUCTION

Interactions of organic matter with metallic cations in natural aqueous environments influence its transport by their participation in its "precipitation". Flocculation of organic macromolecules is of importance in drinking water production, water recycling and waste water treatment. Metallic cations (alkaline-earth, Cu(II), etc.) act as flocculating ions for the organic matter in suspension in aqueous environments.

Interactions of metallic cations with natural macromolecules are usually studied in aqueous media by titration methods with the aid of computerised theoretical models (Sposito *et al.*, 1982; Tipping, 1993; Kinniburgh *et al.*, 1996). Infrared spectroscopy has been mainly used by numerous authors for structural characterisations of lyophilised humic substances (Hernandez *et al.*, 1993; Relan *et al.*, 1993). The problem is that since the degree of hydration of these samples is not the same as in "natural" conditions, these data can not describe the structure of the complex formed. To our knowledge, only MacCarthy *et al.* (1975) has achieved the acquisition of transmission spectra of sodium humate (with a very high concentration). More recently some authors have used ATR-FTIR to characterize solutions of sodium humates (Morra *et al.*, 1989).

In this study, the flocs of humic acids, activated sludge extracellular polymeric substances (EPS) and alginate gels have been characterized by *in situ* Fourier transform infrared (FTIR) spectroscopy with the attenuated total reflectance (ATR) technique to describe features of their flocculation or gelation with alkaline-earth cations and Cu(II) and Na^+ for comparison. The goal is to show the effects of ion-hydration in flocculation by using the series Mg^{2+}, Ca^{2+}, Sr^{2+} and Ba^{2+} (because these represent ions with the same valence, no acid-base properties in the tested pH and concentration range, but a decreasing radius of hydration from Mg^{2+} to Ba^{2+}). The ATR-FTIR technique allows the study of structures at the interface between the hydrated flocs and a ZnSe crystal, with a penetration depth (that is on the distance from the surface at which the electric field strength is divided by e) of the order of 1.5 m at 1000 cm^{-1}. As no pre-treatment, especially no drying of the sample, is required, not only information on the infrared characteristics of the organic macromolecules, but also on the ion-polymer interaction is obtained. Strong interactions between naturally occurring polyelectrolytes and metal may a priori involve the COO^- -groups, with some possible secondary effect from the phenolic hydroxyl groups. The interactions between the flocculating cations and the macromolecules should primarily show in the infrared spectra as a shift of peak positions of the absorbance bands related to COO^- -groups. This paper is thus mainly devoted to the description of infrared absorptions related to carboxylate groups in organic macromolecules.

METHODS

Sample preparation

For infrared analysis, all the solutions were prepared with deionised and boiled water.

Two g of the commercial Aldrich Na-humate was dissolved in 100 ml of deionized water with addition of 0.4 g NaF and left over night under magnetic agitation in order to dissolve silicate impurities. This humate Na-salt was then dialyzed with a Spectra/Por cellulosic membrane (12-14000 D, before wash hand) against deionized water until the test with $AgNO_3$ (to detect residual Cl^-) was negative. The sample was then freeze dried. Finally, 0.2 g of the purified humate Na-salt was dissolved in 100 ml water and the pH was adjusted to 8 or 12. Four ml of a 0.1 molar alkaline-earth chloride (M= Mg, Ca, Sr or Ba) solution molar were added to 10 ml of the mother solution. The Cu-humate floc was obtained by dissolving 20 mg of the Na-humate salt in 10 ml of water, the pH was then adjusted to 6.5. 0.125 g of [$CuSO_4$, 5 H_2O] (Prolabo, Normapur) was added and the whole was briefly shaken. After flocculation, the samples were centrifuged (at 30 000 g for 15 min) and the supernatant eliminated.

The EPS were extracted from activated sludges sampled from the treatment plant station of Malzeville (France) and kept in a polycarbonate container at 4°C. Before the extraction of the EPS, the supernatant of the sludges was eliminated by centrifugation at low speed. The supernatant was replaced by an identical volume of ultra-pure water. The EPS were then extracted from 50 ml of these activated sludges by a combination of a treatment with ultrasound and a Dowex 50X8 (Na^+, 20-50-mesh, Aldrich-Fluka, with a strong affinity for Ca^{2+} and Mg^{2+}, time of contact one hour under agitation). The resin was next eliminated by filtration and the polymers separated from the suspended matter by centrifugation (20 000 g, 30 min.). The extracted EPS were dialyzed against a solution of NaCl to yield the Na-salt in a pure form and were finally lyophilized. Seven ml of alkaline-earth chloride 0.1 M solution were added in each of four solutions prepared with 30 mg EPS (Na-salt) dissolved in 5 ml water. The obtained flocs were then centrifuged in the same conditions as for humate flocs.

A solution of Na-alginate (Acros, used without any further purification) was prepared by dissolving 0.251 g of the product in 100 ml of water. 4.4 ml of the alkaline-earth chloride 0.1 M solution were added to 20 ml of the preceding solution. No gel formation was observed by addition of Mg^{2+}, but this solution was also analyzed by infrared spectroscopy.

Infrared spectroscopy

Attenuated total reflection Fourier transform infrared (ATR-FTIR, Kortüm, 1969) spectra were obtained on a Perkin Elmer 2000 spectrometer equipped with a TGS thermal detector. The detector and sample compartments were purged with a current of dry and decarbonated air provided by a compressor (Balston). The ATR accessory used was a flat horizontal ZnSe crystal prism manufactured by Specac (6 internal reflections on the upper surface, angle of incidence: 45°). The centrifuged wet flocs and gels were deposited on the crystal and 40 bi-directional, double-sided scans were recorded with a resolution of 4 cm^{-1}. The optical path difference velocity was 0.1 cm/s, the total acquisition time was about 8 min. ATR spectra are shown with an absorbance scale corresponding to $\log(R_{reference}/R_{sample})$, where R is the internal reflectance of the device. In order to eliminate the strong absorbance of liquid water, a reference spectrum of this last was recorded and subtracted from the sample spectrum. This was sufficient to obtain a good signal-to-noise ratio spectrum in most cases. However, it was also necessary to eliminate the contribution of water vapor for soluble samples spectra with very weak absorbance. Integrated intensities of humate spectra have been measured by using the "area" procedure of the Perkin Elmer program IRDM.

RESULTS

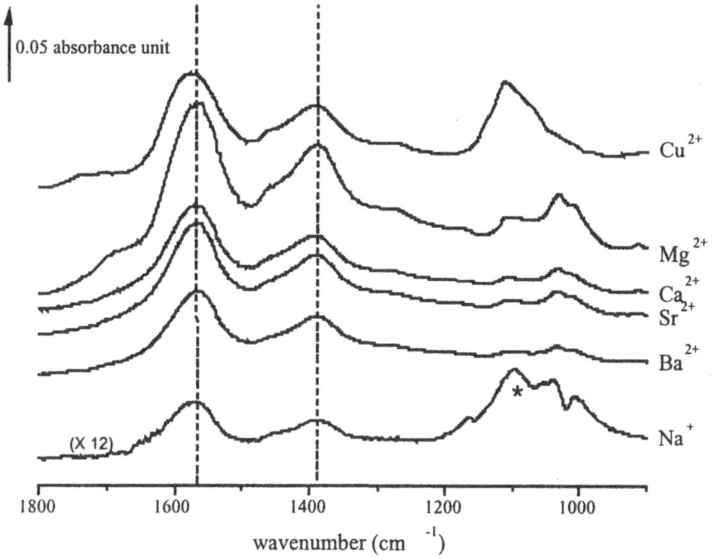

Figure 1. ATR-FTIR spectra of Cu-, Mg-, Ca-, Sr-, Ba- and Na-humate at pH 8. Alkaline -earth chlorides and sulphate copper were used to flocculate Na-humate solutions. * Sample still containing silicate impurities.

Figure 1 compares the ATR-FTIR spectra of a sodium humate solution to those of flocs obtained by addition of alkaline earth salts and copper(II). The attribution of the various peaks is achieved in accordance with the literature (Niemeyer *et al.*, 1992). The results are summarized in Table 1.

The spectra of sludge EPS flocs (figure 2) are more complex since humic acids (and/or their salts) are only one component of EPS. According to Thomas and Kyogoku (1977), we can see from the spectra (Figure 2 and Table 1) the presence of the characteristic amide bands of proteins (1649, 1548 and 1316 cm^{-1}). The vibrations at 1240 and 1078 cm^{-1} are attributed to phosphate esters from lipids and nucleic acids. Sugars can be recognised through the lines at 1119, 1078 and 982 cm^{-1}. These constituents have effectively been identified and titrated by Frolund *et al.* (1994) in EPS extracted from activated sludge.

Another result of aggregation by a metallic cation is the formation of a gel, as for alginates (Skjåk-Braek *et al.,* 1989). Figure 3 shows that it is also possible to obtain good ATR-FTIR spectra of these alkaline-earth and copper-alginate gels. The lines at 1596 and 1414 cm^{-1} are attributed to the asymmetric and symmetric stretching vibrations of the carboxylate functions, respectively. The alcoholic bending and stretching vibrations can be seen at 1414 and 1287 cm^{-1}. Finally, the lines between 1200 and 970 cm^{-1} are characteristic of the vibrations of ether functions, C-O and C-C stretching in sugar cycles. The attributions of each peak are summarised in Table 1.

Table 1. Attribution in the region 1800-900 cm^{-1} of infrared vibrational bands of the ATR-FTIR spectra of humate, EPS and alginate alkaline-earth wet flocs or gels.

Compound	Wavenumbers (cm^{-1})	Assignment
Humate salts	1700	C=O stretching (esters, cetones, residual carboxylic acids)
	1567 (1579*)	COO- asymmetric and aromatic ring stretching modes (*value for Cu-humate)
	1458	C-H bending
	1390	COO^- symmetric stretching
	1288	Aromatic ring stretching
	1105; 1032; 1010; 913	C-O, C-C vibrations of sugars
	1104	asymmetric stretching of SO_4^{2-} (for the floc obtained with Cu^{2+})
EPS salts	1733	C=O stretching (esters)
	1649	Amide I band (C=O, N-H, C-N)
	1631	Aromatic ring stretching
	1548	COO^- asymmetric stretching Amide band II (C-N, CNH)
	1455	C-H bending (lipids), aromatic ring stretching
	1404	COO^- symmetric stretching
	1316	Amide band III (C=O, CN-H, C-N)
	1240	P=O stretching in phosphate esters (nucleic acids, lipids), aromatic ring stretching
	1149; 1119; 1078; 1047; 982	C-O-C, C-O, C-C stretchings, CH_2 (sugars)
		PO_2^- stretching (nucleic acids)
Alginate salts	1596	COO- asymmetric and aromatic ring stretching modes
	1414	COO^- symmetric stretching
	1144; 1126; 1100; 1085; 1035; 1004; 995; 950	C-O, C-C vibrations of sugar cycles

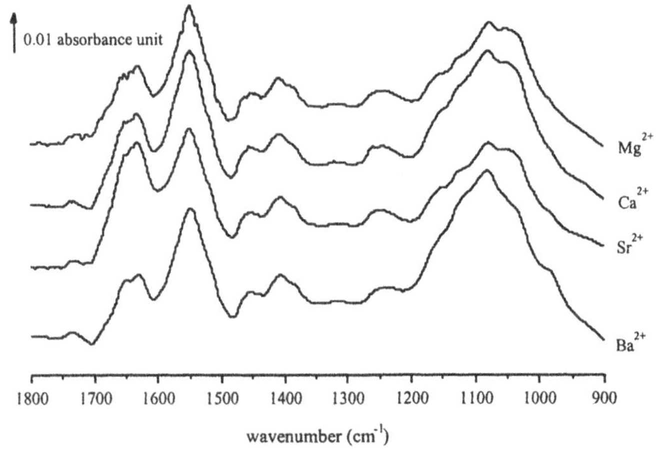

Figure 2. ATR-FTIR spectra of Mg-, Ca-, Sr- and Ba-EPS.

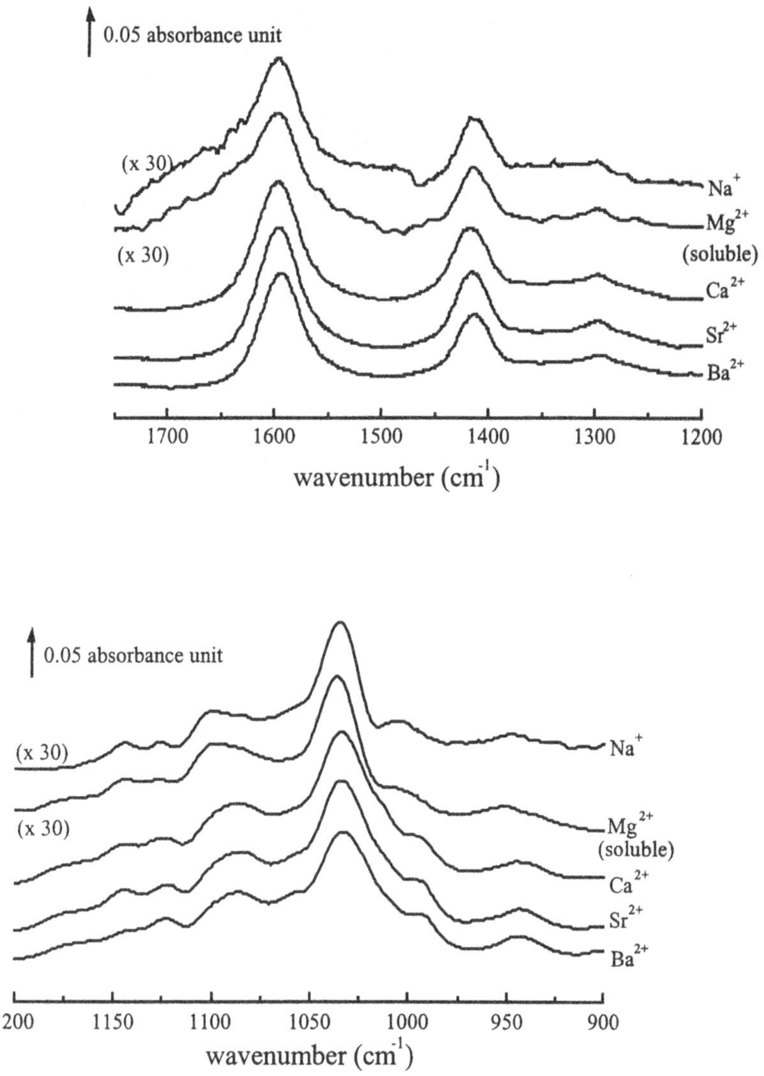

Figure 3. ATR-FTIR spectra of Na-, Mg-, soluble alginates and Ca-, Sr- and Ba-alginate gels.

DISCUSSION

The discussion of these data needs references to the spectra of isolated carboxylate anions interacting with cations in water. Whereas many spectra of solid carboxylate salts have been obtained (Bellamy, 1975), the studies of dilute solutions in water are scarce. The main features of these spectra are tentatively summarized as follows:

(1) The structure of a carboxylate ion in a dilute solution of the sodium salt in water is taken as a reference, in which the anion is considered as "free" with respect to the counter-ion. In this state, specific interactions appear only between carboxylate ion and water. This view is supported by the fact that the vibrational spectra of sodium and tetrabutylammonium benzoate salts are similar (spectra not shown here).

(2) On the basis of a supposed similarity between the $-CO_2^-$ and $-NO_2$ groups, which are isoelectronic, a free carboxylate group is often considered symmetric because of a

resonance between the two CO bonds with equal partial double bond characters. For such a configuration, the two CO stretching vibrations are described as symmetrical, $_{s}(COO^{-})$, and antisymmetrical, $_{a}(COO^{-})$, modes. This notation is used throughout the paper although there is no decisive evidence of the symmetry of the free anion (Spinner, 1964).

(3) Most free carboxylate ions display at least two bands between 1300 and 1610 cm^{-1}. The band with the lowest wavenumber, which is the most intense in Raman spectra, is assigned to $_{s}(COO^{-})$. Although there are no definite values for both modes of free ions, which are more or less mixed with other molecular vibrations, three main kinds of spectra are observed: (i) The wavenumber values of the purest $_{a}(COO^{-})$ and $_{s}(COO^{-})$ modes are probably those of the formate ion, respectively 1585 and 1351 cm^{-1}, corresponding to a splitting of 234 cm^{-1} (Ito and Bernstein, 1956). (ii) Aqueous aliphatic carboxylate solutions display $_{a}(COO^{-})$ between 1551 and 1541 cm^{-1} and $_{s}(COO^{-})$ between 1416 and 1408 cm^{-1} (Babaniss and McVey, 1995), with a mean splitting of 134 cm^{-1}. With respect to formate, the $_{s}(COO^{-})$ increase corresponds to a coupling of this mode with the C-C stretching vibration (Spinner, 1964). (iii) Sodium benzoate derivatives display still different spectra. For 41 substituted sodium benzoates, the $_{a}(COO^{-})$ and $_{s}(COO^{-})$ ranges are not very large:1575-1533 and 1406-1377 cm^{-1}, respectively, with a mean splitting of about 157 cm^{-1} (Dunn and McDonald, 1969). Thus a frontier between aliphatic and aromatic compounds appears at 1407 cm^{-1} for $_{s}(COO^{-})$.

(4) Relative to the free state of a given carboxylate anion, the $_{a}(COO^{-})$ mode shifts to a higher wavenumber on formation of a unidendate complex, because of its increasing (C=O) character, while the $_{s}(COO^{-})$ shifts in the opposite direction because it tends to a (C-O) mode. The mixing with other molecular vibrations may indeed perturb this trend. In contrast, bidendate complexation decreases the splitting between the two carboxylate stretching vibrations. In solid acetato complexes, separation values superior to 200 cm^{-1} are generally indicative of unidendate structure, whereas a difference of less than 100 cm^{-1} suggests a bidendate chelate and a separation of 150 cm^{-1} points to either a bridging complex or a "free" ion (Nakamoto, 1978).

The absorptions corresponding to carboxylate vibrations of humates are located at 1567 cm^{-1} and 1390 cm^{-1} (Figure 1). According to the above analysis of literature, this last value points to a dominant contribution of benzoate derivatives rather than aliphatic carboxylate, in agreement with the usual representation of humic substances (Filella *et al.*, 1995). From figure 1, it is seen that the spectra in the region of COO^{-} bands of Mg-, Ca-, Sr- and Ba-humates are similar both amoong themselves and with Na-humate. In contrast, a shift of 12 cm^{-1} to high wavenumbers is observed with Cu-humate. This shift is assigned to inner-sphere complexes of some carboxylate groups with Cu^{2+} (Tackett, 1989). The lack of any shift for alkaline-earth flocs suggests mainly outer-sphere complexes involving weak interactions between carboxylate groups and these cations, which cannot be differentiated by infrared spectra. The same conclusions can be drawn for the EPS flocs spectra (Figure 2) (although one carboxylate band is overlapped with amide band II) confirming that the interactions with alkaline-earth cations are the same in the flocs for each macromolecule family. Since the flocculent power depends, however, on the nature of the alkaline-earth elements (decreasing from Ba to Mg), the difference must be assigned to the hydrated states of the ions. It is indeed known that the hydration strength decreases while the naked ion radius increases and that the hydrated ion radius increases from Ba to Mg (Gregory, 1989).

It is noteworthy that the absorption coefficient at 1390 cm^{-1} is smaller than at 1567 cm^{-1} for every humate floc at pH 8 (Figure 1). This contrasts with the spectra observed for the same flocs at pH 12, where the absorption coefficients become nearly the same for both wavenumbers (spectra not shown). In order to quantify this pH effect, integrated intensities have been measured in two ranges (1320-1485 and 1485-1700 cm^{-1}) with respect to a linear baseline between 1220 and 1720 cm^{-1}. The ratio of the two values is reported in Figure 4 for every spectrum and is 40.3 2.7 at pH 8 and 52.4 4.8 at pH 12. Although the uncertainty may be on the order of the pH effect for Sr-humate, the increase of the relative intensity of the low frequency band by pH increase is found regularly and is statistically established (Figure 4). The effect appears even though absorptions of other parts than carboxylate groups (ring, CH bending) are integrated in the intensity measurements shown in figure 4. This intensity variation is still not understood but it cannot be due to the carbonatation of the solutions (Nakamoto, 1978) because the samples were prepared with boiled water. Since many phenol groups (except ortho hydroxybenzoates) are ionised when pH increases from 8 to 12, one might wonder if the absorption enhancement around 1400 cm^{-1} is due to the formation of phenolate groups. Ionisation of C_6H_5OH indeed induces an intensity increase of the (C-O) band, but with a shift of only 30 cm^{-1} from 1241 cm^{-1} to 1271 cm^{-1} (spectra not shown here). If this effect seems to be insufficient in this simple case, it would be interesting to know what would append with hydroxybenzoic acid compounds. Figure 4 could also be analysed as a characteristic feature of the interaction of humic acids with alkaline-earth ions. It then suggests the presence of an outer-sphere complex (since the intensities of the peaks of the carboxylate group vary with the pH and with the ionic strength) (Stumm and Morgan, 1995).

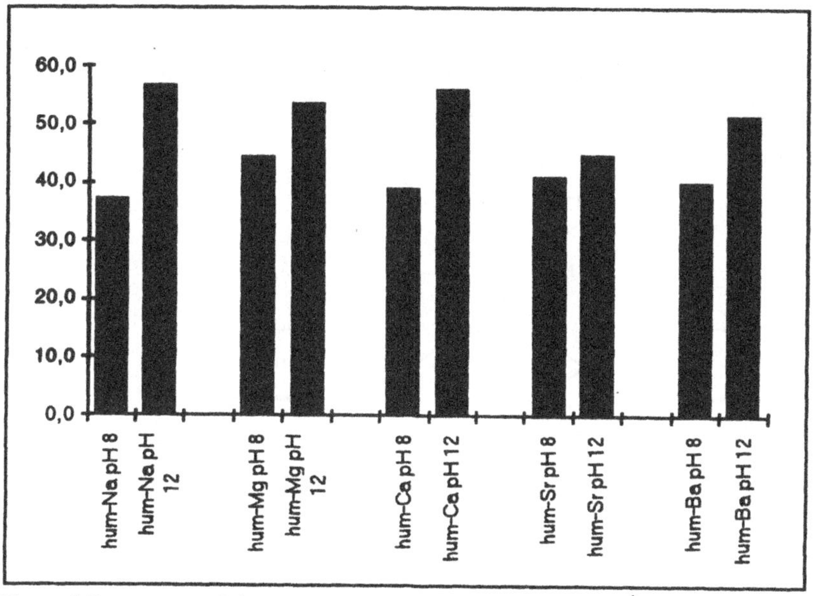

Figure 4. Percentages of the peaks area between 1320 and 1485 cm^{-1} with respect to the total area between 1320 and 1700 cm^{-1}, at pH 8 and 12 of ATR-FTIR spectra of humates. Relative uncertainty is estimated at 10% on each result.

The values of the alginate (COO^-) frequencies (1596 and 1414 for the asymmetric and symmetric stretchings, respectively) point to the aliphatic character of their carboxylic groups in contrast to those of humate salts. We can see from Figure 3 and the similar general shape of the spectra in the region of carboxylate frequencies that the conclusions given for the preceeding examples are the same: the formation of outer-sphere complexes seems to be confirmed. In this state, the polymer chains may organize more or less regularly as shown in Figure 5, tending to organize the alginate gel into a sheet-like structure (Yokoyama *et al.*, 1992). Meanwhile, we can see a difference in the intensities and frequencies between the gels (Ca^{2+}, Sr^{2+} and Ba^{2+}) and the soluble salts (Na^+ and Mg^{2+}) in the region of C-O and C-C stretching vibrations. Particularly, we note two intensity inversions for the peaks at 1144 and 1126 cm^{-1} and 1100 and 1085 cm^{-1}. A shift of about 11 cm^{-1} is also observed for the peak initially positioned at 1004 cm^{-1} for the Na^+ and Mg^{2+} soluble salts. All these spectral features could be attributed to a modification of the conformation of the sugar cycles in the gel as previously observed with poly-oxyethylene chains (Kalnins and Lyubinova, 1976; Ovsepyan *et al.*, 1978).

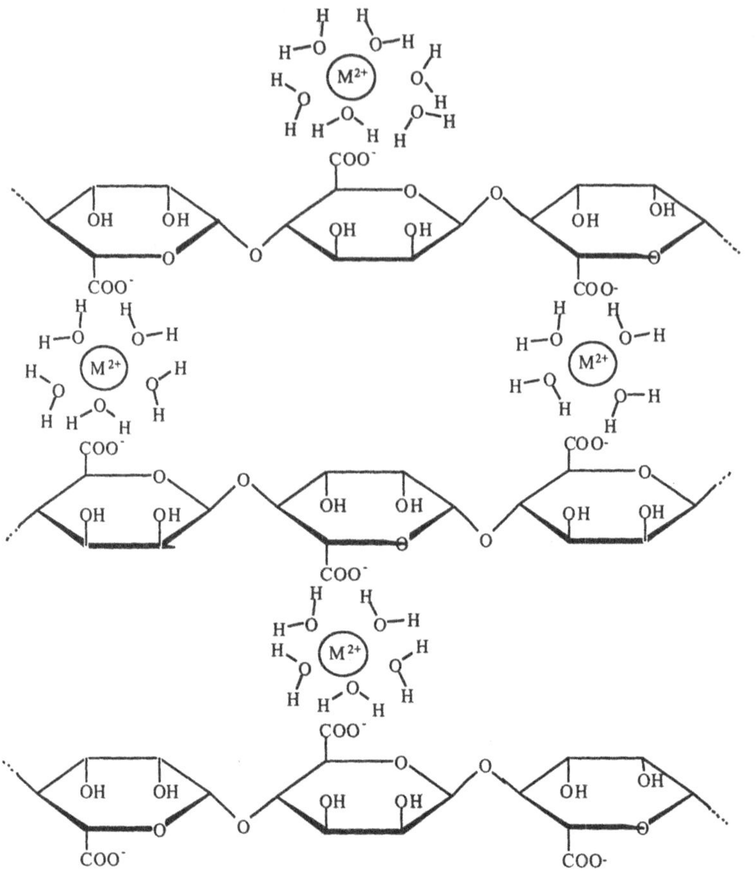

Figure 5. Schematic representation of the outer-sphere complexes inducing a sheet-like structure of alginate gels.

Supported by the calculation of Kinniburgh *et al.* (1996), we gather that the metal-polyelectrolyte interaction is purely electrostatic, i.e., not involving changes of the orbital structure of the organic molecule. From a theoretical point of view, these observations correspond to an effect of ion-hydration or solvation forces (Israelachvili, 1992). From polymer chemistry, they are described by the Flory treatment of gelation (Axelos *et al.*, 1994).

CONCLUSIONS

From an engineering point of view, two important consequences occur:
a) it may be gathered as a general rule that Mg^{2+} is a poorer flocculent than is Ca^{2+}, hence a priori the two ions are not interchangeable. Exceptions to this rule may be found when more diverse systems are studied.
b) the effect of a strong flocculent is to provide flocculation at sub-stoichiometric concentration. Thus, such flocs must possess an open structure, with negatively charged surfaces. On the contrary a weak flocculent will produce weaker flocs with a closed structure and fairly low surface charge. In mixed systems of such flocs, removal of the strong flocculent will lead to significant dissolution of the floc, whereas the removal of a weak flocculent will lead to some disintegration, forming negatively charged particles. This view is in accordance with data presented on activated sludge floc systems, based on Fe^{3+}/Ca^{2+} flocculation (Caccavo *et al.*, 1996).

Concerning infrared studies, the complexity of even the simplest system is too great to allow detailed interpretations. There is still a need for fundamental research to determine the structures of interaction between carboxylate groups and cations in aqueous media as a function of the nature of the cation. Work is in progress in this field by using ATR-FTIR and Raman spectroscopy.

ACKNOWLEDGMENTS

The authors thank Fréderic Jorand (Laboratoire Santé-Environnement, Université H. Poincaré, Nancy I) for the separation and the purification of the EPS.

REFERENCES

Axelos M.A.V., M.M. Michele and J. Francois, 1994. Phase diagrams of aqueous solutions of polycarboxylates in the presence of divalent cations. Macromolecules 27, 6594.

Bellamy L. J., 1975. The Infra-Red Spectra of Complex Molecules. Chapman and Hall Ltd, London.

Babaniss S.E. and I.F. McVey, 1995. Aqueous infrared carboxylate absorbances: aliphatic monocarboxylates. Spectrochimica Acta Part A. 51, 2385.

Caccavo F. Jr, B. Frolund, F. Vanommen Kloeke and P.H. Nielsen, 1996. Defloculation of Activated Sludge by the Dissimilatory (Fr(III). Reducing Bacterium *Shewanella alga* BrY. Appl. Environ. Microbiol. 62, 1487-1490.

Dunn, G.E. and R.S. McDonald, 1969. Infrared spectra of aqueous sodium benzoates and salicylates in the carboxy-stretching region: chelation in aqueous sodium salicylates. Can. J. Chemi. 47, 4577.

Filella M., N. Parthasarathy and J. Buffle, 1995. Humic and fulvic compounds. Encyclopedia of Analytical Science, 2017.

B. Frolund, 1994. Qualitative and quantitative characterization of activated sludge exopolymers. Ph.D. Thesis, Aalborg University, Denmark.

J. Gregory 1989. Fundamentals of flocculation. Fundamentals in Environmental Control, 3, 185.

Hernandez T., J. I. Moreno and F. Costa, 1993. Infrared spectroscopic characterization of sewage sludge humic acids. Evidence of sludge organic matter-metal interactions. Agrochimica, 37(1-2), 12.

Israelachvili J.,1992. Intermolecular & Surface Forces. Chapter 13. Academic Press, London.

Ito K. and H.J. Bernstein, 1956. The vibrational spectra of the formate, acetate, and oxalate ions. Can. J. Chem., 34, 170.

Kalnins K. and G.V. Lyubinova, 1976. Vibrational spectra and conformations of poly(ethylene oxide) oligomers. Zh. Prikl. Spectrosk. 25, 269.

Kinniburgh D. G., C. J. Milne, M. F. Benedetti, J. P. Pinheiro, J. Filius, L. K. Koopal and W.H. van Riemsdik, 1996. Metal ion binding by humic acid: Application of the NICA-Donnan model. Envir. Sci. & Technol., 30, 1687.

G. Kortüm, 1969. Reflectance spectroscopy. Principles, Methods, Applications. Chapter VIII. Springer Verlag, Berlin,.

MacCarthy P., H B. Mark Jr and P.R. Griffiths, 1975. Direct measurement of the infrared spectra of humic substances in water by Fourier transform infrared spectroscopy. J. Agric. Food Chem., 23, 600.

Morra M.J., D.B. Marshall and C.M. Lee, 1989. FT-IR analysis of Aldric humic acid in water using cylindral internal reflectance. Commun. Soil Sci. Plant Anal. 20, 851.

K. Nakamoto, 1978. Infrared and Raman Spectra of Inorganic and Coordination Compounds. (3rd Edition), John Wiley and Sons, London.

Niemeyer J., Y. Chen and J.-M. Bollag, 1992. Characterization of humic acids, composts and peat by diffuse reflectance spectroscopy. Soil Sci. Soc. Am. J., 56, 135.

Ovsepyan A. M., V.V. Kobyakov and R.G. Zhbankov, 1978. IR spectra of aqueous solutions of some polymers. Zh. Prikl. Spectrosk., 29, 62.

Relan P.S., V.K. Garg and S. Kumar, 1993. Infrared studies of the complexes of bivalent metal cations with humic materials. Intern. J. Trop. Agric., 9, 227.

Skjåk-Braek G., H. Grasdalen and O. Smidsrod, 1989. Carbohydr. Polym. 10, 31.

Spinner E., 1964. The vibration spectra of some substituted acetate ions. J. Chem. Soc., 4217.

Sposito G., F.T. Bingham, S.S. Yadav and C.A. Inouye, 1982. Trace metal complexation by fulvic acid extracted from sewage sludge: II. Development of chemical models. Soil Sci. Soc. Am. J., 46, 51.

Stumm W. and J. J. Morgan, 1995. Aquatic chemistry. Third edition. John Wiley and Sons, London.

Tackett J.E., 1989. FTIR characterization of metal acetates in aqueous solution. Appl. Spectrosc. 43, 483.

Thomas G.J. Jr and Kyogoku, 1977. Infrared and Raman spectroscopy. In: E.G. Brame Jr and J.G. Grasselli (Editors), Vol. I, Chapter 11, Biological Science.

Tipping E., 1993. Modelling ion binding by humic acids. Colloids and Surfaces A: Physicochemical and Engineering Aspects, 73, 117.

Yokoyama F., C.E.Achife, K. Takahira, Y.Yamashita and K. Monobe, 1992. Morphologies of oriented alginate gels crosslinked with various divalent metal ions. J. Macromol. Sci.-Phys., B31(4), 463.

THE STRUCTURE OF ORGANIC NITROGEN IN PARTICLE SIZE FRACTIONS DETERMINED BY ^{15}N CPMAS NMR

Heike Knicker, Michael W. I. Schmidt & Ingrid Kögel-Knabner

Department of Soil Science
Technische Universität München
85350 Freising-Weihenstephan, Germany

1. INTRODUCTION

Nitrogen availability in soils plays an important role for the productivity of agricultural systems. After entering the soil, material of biogenic origin experiences decay and microbial reworking. During these processes, the labile compounds are quickly mineralized into inorganic nitrogen forms, directly available for the production of new biomass. The more stable compounds and metabolic products accumulate to form the refractory organic pool of soils (Kelly and Stevenson, 1996). Their nitrogen will be sequestered from the overall nitrogen cycle and therefore from bioproductivity over a expended time range.

In soils, it was shown that refractory organic matter is intimately associated with the silt and clay fractions in soils (Christensen, 1992), although some young organic material is present in these fractions (Balesdent *et al.*, 1987; Magid *et al.*, 1996). An understanding of the chemical structure and composition of the N-functionality in these fractions may give some insight on the mechanism involved in the stabilization of Soil Organic Nitrogen (SON).

With common analytical methods only 21 to 40 % (Christensen and Bech-Andersen, 1989) of the total nitrogen of soil organic matter were identified, mostly as amino acids. The structure of the remaining nitrogen, however, is still a subject of controversy. In a recent study, Schulten *et al.* (1993) investigated the composition of organic matter in particle-size fractions of an agricultural soil by means of analytical pyrolysis. They suggested, that the non-hydrolyzable organic nitrogen is present in form of N-heterocyclic compounds, specifically alkyl-substituted pyrroles, pyridines and indoles. However, such compounds were not identified in soils in amounts, necessary to form the major building block of the refractory N-pool. It has also to be born in mind, that these compounds may be formed during the pyrolysis experiment.

Solid-state ^{15}N NMR spectroscopy represents a milder method and an alternative to common analytical approaches. Since this technique does not depend on the solubility of the sample, insoluble soil fractions, such as bulk soils or particle size separates, can be analyzed. Unfortunately, ^{15}N NMR spectroscopy is a very insensitive technique. Solid-state ^{15}N NMR spectra of soils with natural ^{15}N with an acceptable signal-to-noise ratio abundance can only be obtained within reasonable measurement time, if their organic N-content is higher than 1 % (Knicker *et al.*, 1993; Knicker & Lüdemann, 1995). Such nitrogen concentrations, however, are

Effect of Mineral-Organic-Microorganism Interactions on Soil and Freshwater Environments
Edited by Berthelin, *et al*, Kluwer Academic / Plenum Publishers, New York, 1999

rarely found in mineral soils and particle size fractions, excluding them from ^{15}N NMR spectroscopic investigations.

One possibility to increase the concentration of organic nitrogen in soil fractions rich in mineral matter represents the removal of mineral matter with hydrofluoric acid (HF). Recent studies by Skjemstad *et al.* (1994) and Schmidt *et al.* (1997) showed that applying this technique to bulk soils and particle size samples leads to major improvements of solid-state ^{13}C NMR spectra and may even allow their ^{15}N NMR spectroscopic examination.

A series of HF-treated fine particle size separates of a Haplic Podzol were subjected to solid-state ^{15}N NMR spectroscopy. The objective of the present work was to obtain insight in the nature of refractory nitrogen associated with fine particle size fractions of soils. Possible alterations of the organic-N composition due to HF-treatment were estimated using ^{15}N-enriched degraded algal material.

2. MATERIAL AND METHODS

A mixed algal culture of strains of *Chlamydomonas, Chlorella, Closterium and Scenedesmus* was grown in liquid medium containing ^{15}N-enriched potassium nitrate (Zelibor *et al.*, 1988). The freeze dried algae were mixed with quartz sand (6:100), inoculated with 1 ml of an aqueous extract from a natural compost and incubated at 25°C (Knicker *et al.*, 1996). After two months, the sample mixture was harvested and freeze dried. An aliquot was subjected to HF-treatment (10%) (Schmidt, *et al.*, 1997).

The soil material was obtained from the Aeh horizon of a Haplic Podzol (FAO, 1990) developed under a coniferous forest (Faesheim, Westfalia, Germany). Soil samples were sieved (< 2 mm) and subjected to particle-size fractionation. Fractionation was performed by a combination of wet sieving and sedimentation after dispersion with ultrasonic treatment according to Christensen (1992), modified as described by Schmidt *et al.* (1996). The ultrasonic energy was applied with an ultrasonic probe (Labsonic U, Braun Melsungen, FRG), resulting in 440 J/ml of applied ultrasonic energy. Separates of silt (medium silt 20 to 6.3 µm, fine silt 6.3 to 2 µm) and clay fractions (0.45 to 2 µm) were subjected to solid-state ^{15}N NMR spectroscopy.

Table 1: Carbon (C) and nitrogen (N) content of the clay, fine silt and medium silt fraction of a Haplic Podzol before and after treatment with 10% hydrofluoric acid (HF). The relative enrichment factors E_C and E_N were calculated from the ratio of the C and N content before and after HF-treatment. The percentage of α-amino-N was determined after acid hydrolysis with 6N HCl.

	Before HF-Treatment			After HF-Treatment			Relative Enrichment Factor		% of total N identified as α-amino-N
Size Fraction:	C (g/kg)	N (g/kg)	C/N (w/w)	C (g/kg)	N (g/kg)	C/N (w/w)	E_C C(before) /C(after)	E_N N(before) /N(after)	
Clay	39.4	1.1	36	57.4	1.6	36	1.5	1.5	43
Fine Silt	26.4	0.6	46	53.6	1.3	40	2.0	2.4	34
Medium Silt	10.3	0.3	38	52.1	1.4	38	5.0	5.0	31

Enrichment of the samples in organic nitrogen was achieved by treatment with 10 % hydrofluoric acid as described by Schmidt *et al.* (1997). Solid-state CPMAS ^{15}N NMR spectra were obtained on a Bruker MSL 300 (7.05 T) applying the cross polarization magic angle spinning technique (Schaefer and Stejskal, 1976). A contact time of 1 ms, a pulse delay of 100 ms and a magic angle spinning speed of 4.3 kHz were used. A detailed description of the technique and the parameters used for acquisition of solid-state ^{15}N NMR spectra is given by Knicker and Lüdemann (1995).

The amount of total organic carbon and total nitrogen was determined in duplicate with a Leco CNS 2000 and an Elementar Vario EL (Table 1). The minimum detection levels were 0.1 ±0.3 g kg^{-1} for C and N. The content of amino acids was determined with a colorimetric procedure after hydrolysis with 6 N HCl (Table 1) as described in detail by Kögel-Knabner (1995).

3. RESULTS AND DISCUSSION

3.1. Alterations of Chemical Composition of the N-Fraction During HF-Treatment

Due to the low N-content of less than 15 g kg^{-1} (Table 1), solid-state ^{15}N NMR spectroscopy of the untreated fine particle size fractions of the Haplic Podzol resulted in spectra with extremely low signal-to-noise ratios, even after accumulation of more than one million scans. A qualitative interpretation of these spectra was not possible. Therefore, mineral matter was removed with 10% hydrofluoric acid (HF). Applying solid-state ^{13}C NMR spectroscopy, Schmidt *et al.* (1997) recently observed no major alterations of the organic carbon distribution by (10%) HF in soil samples. However, it cannot necessarily be assumed that this is also true for the behavior of the N-containing fraction, since solid-state ^{13}C NMR spectra of soil samples rarely allow conclusions about the chemical nature of their N-functional groups. In order to estimate possible HF-related alterations in the N-functionality of SON, the solid-state ^{15}N NMR spectra of ^{15}N-enriched degraded algal material obtained before and after treatment with 10% HF were compared. The solid-state ^{15}N NMR spectrum of the degraded algal material (Figure 1a) shows the typical signals expected for degraded biological material (Knicker and Lüdemann, 1995). A tentative assignment of peaks to chemical compounds is given in Table 2. The spectrum is dominated by a signal at -256 ppm in the chemical shift region of amide functional groups. Further signals can be observed at -294 ppm and -304 ppm, most probably assigned to NH_2-groups of basic amino acids or nucleosides (Knicker, 1993). Another resonance is clearly identified at -344 ppm and originates from free amino groups, most likely of amino acids and amino sugars. The heterocyclic-N in histidine, nuclei acid derivatives and substituted pyrroles, i.e. from porphyrin structures in chlorophyll are expected to contribute to the chemical shift region between -145 and -220 ppm (Martin *et al.*, 1981; Witanowski *et al.*, 1993). The signal at -355 ppm derives most probably from ammonium ions formed and accumulated during the mineralization process of N-containing algal compounds.

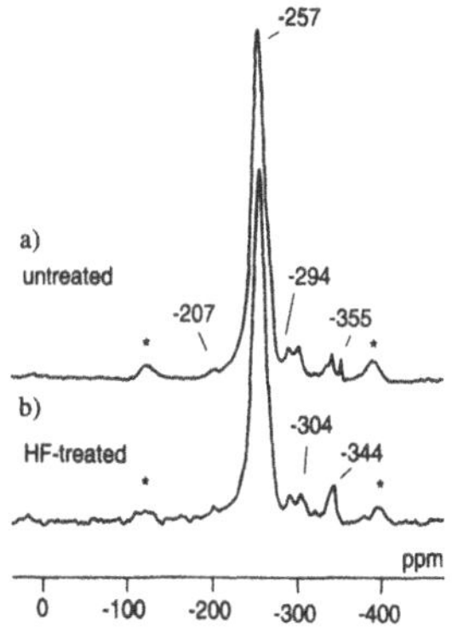

Fig. 1: Solid-state ^{15}N NMR spectra of ^{15}N-enriched algal material a) before and b) after treatment with 10% hydrofluoric acid (HF). Asteriks indicate spinning sidebands.

Table 2: Assignments for peaks in the solid-state ^{15}N NMR spectra (referenced to nitromethane = 0 ppm) (Martin *et al.*, 1981, Witanowski *et al.*, 1993).

Chemical shift range (ppm)	Assignment
25 to -25	Nitrate, nitrite, nitro groups
-25 to -90	Imine, phenazine, pyridine, Schiff-bases
-90 to -145	Purine, nitrile groups
-145 to -220	Chlorophyll-N, purine/pyrimidine, imidazole, substituted pyrroles
-220 to -285	Amide/peptide, N-acetylderivatives of amino sugars, tryptophane, proline, lactams, unsubstituted pyrroles, indoles and carbazoles
-285 to -325	NH in guanidine, NH_2- and NR_2-groups (N_d-arginine and N_a -citrulline, N_e-arginine, N_w-citrulline, urea, nucleic acids, aniline derivatives)
-325 to -350	free amino groups in amino acids and sugars
-350 to -375	NH_4^+

After HF-treatment only slight changes in the feature of the solid-state ^{15}N NMR spectrum of the degraded algal material can be observed. The most apparent one is the disappearance of the ammonium signal at -356 ppm (Figure 1b). This could be explained by ammonium removal during the HF treatment. Obviously ammonium was removed with the aqueous HF-solution. Integration of the two solid-state ^{15}N NMR spectra reveal a slight decrease of 5 % of the relative intensity in the amide region due to the HF-treatment (Table 3). This may result from hydrolysis of very labile amide structures.

Table 3: Signal intensity distribution of in the solid-state ^{15}N NMR spectra of ^{15}N-enriched degraded algal material before and after treatment with 10% hydrofluoric acid (HF).

Degraded algae	25/-25 ppm	-25/-90 ppm	-145/ -220 ppm	-220/-285 ppm	-285/-325 ppm	-325/-375 ppm
before HF-treatment	0	0	4	85	7	3
after Hf-treatment	0	1	6	80	8	5

These results suggest that treatment with 10% HF does not lead to major alteration of the organic nitrogen fraction. Thus, enrichment of organic nitrogen in mineral rich soil samples by removal of mineral matter with 10% HF, is a helpful tool for the improvement of the quality of the solid-state ^{15}N NMR spectra.

3.2. Soil Particle Size Fractions

The elemental composition of the clay, fine silt and medium silt fraction of the Haplic Podzol was analyzed before and after HF-treatment. Table 1 reveals that HF-treatment resulted in nitrogen enrichment factors of 1.5 for the clay fraction and 2.4 for the fine silt fraction, respectively. The medium silt fraction shows even higher nitrogen enrichment by a factor of 5. Comparable enrichment factors were achieved for carbon.

The solid-state ^{15}N NMR spectra of the fine silt, medium silt and clay fraction, treated with HF, show a similar pattern to those obtained from plants and their composts (Knicker and Lüdemann, 1995) (Figure 2). As revealed by the dominating peak at -256 most of the N is present in amide functional groups. Minor signals occur between -325 and -350 ppm and -285 and -325 ppm, originating from primary aliphatic amines and NH_2 or NR_2 derivatives, respectively (Table 2). Resonances, supporting the presence of higher amounts of N-containing heterocyclic compounds (pyridinic-N: -25 to -90 ppm; pyrrolic-N -145 to -240 ppm), cannot be observed in these spectra.

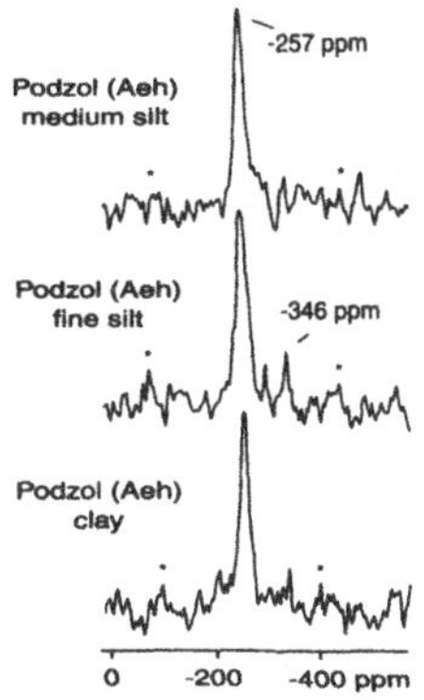

Fig. 2: Solid-state ^{15}N NMR spectra of medium silt, fine silt and clay fraction of a Haplic Podzol. Asteriks indicate spinning sidebands.

As samples were treated with 10% HF prior to NMR analysis, the missing of a signal in the chemical shift region between -350 and -375 ppm can be explained by the removal of NH_4^+ associated with the mineral fraction. NH_4^+ was most probably present in the original separates. The relative amount of NH_4^+, i.e. in the clay fraction, expressed as the percentage of the total nitrogen is generally < 5% (Christensen 1992).

The similarity of these spectra with those obtained from protein-rich biogenic material (algae, fungi and plants) (Knicker, 1993; Knicker and Lüdemann, 1995, Knicker *et al.*, 1995/1996) and the relative signal intensity distribution, typical for peptide-like material, suggest that the organic nitrogen in these fractions is most likely bound in proteinaceous compounds. Such proteinaceous material may derive from nitrogen immobilized in labile, newly synthesized microbial biomass. This nitrogen, however, cannot be considered to be part of the refractory N pool of the examined Podzol. On the other hand, nitrogen immobilized in such compounds should be easily hydrolyzed with 6N HCl, a method generally believed to degrade most, if not all, proteinaceous compounds in soils and sediments (Mayer *et al.*, 1988). Treating the samples, examined here, with 6N HCl, less than 43% of their total SON were hydrolyzed and identified as α-amino N (Table 1). Therefore labile amide functional groups incorporated in newly synthesized microbial biomass can only explain half of the amide signal intensity in the ^{15}N NMR spectra of the three particle size fractions. Thus, the remaining nitrogen must be related to refractory nitrogen, resistant to drastic chemical degradation.

4. CONCLUSION

The solid-state ^{15}N NMR spectroscopic investigation of the medium silt, fine silt and clay fraction of a Haplic Podzol treated with 10% HF, revealed that the major part of the organic nitrogen is bound in amide-N functional groups, most probably as part of proteinaceous material. Hydrolysis with 6 N HCl could only release less than 43 % of this total-N. Therefore at least some of the organic nitrogen in these samples, identified as amide-N, must be present in a form protected from microbial degradation and resistant to drastic chemical treatment. This resistance may explain to some degree the difficulties in identifying such structures with common wet analytical methods. The identification of refractory amide nitrogen in the fine particle size fraction of the Haplic Podzol, seeks for a explanation of the preservation mechanism. Further studies have to reveal how these amide functional groups are protected. They may be stabilized by chemical interactions of the substrate with the mineral matrix (Marshmann and Marshall, 1981, Mayer, 1994a/b; Christensen, 1992) or physically protected within micro- or macroaggregates (Christensen, 1992). The latter, however can be excluded, since during the pariticle size fractionation procedure such aggregates were completely destroyed by ultrasonic dispersion (unpublished results). Another possibility is the selective preservation of otherwise decomposable substrates in association with refractory biopolymers, recently proposed for the presents of amide-nitrogen in ancient organic-rich sediments (Knicker *et al.*, 1996) and only previously found in soils (Lichtfouse *et al.*, 1996). Trapped into aliphatic resistant biopolymers, amide-N could be unavailable to HF, HCl or bacteria.

5. ACKNOWLEDGMENTS:

Financial support was obtained from the Deutsche Forschungsgemeinschaft (Ko 1035/6-1). Experimental assistance was provided by W. Gosda and G. Wilde (University of Bochum. Prof. Dr. H.-D. Lüdemann (University of Regensburg, BRD) is gratefully acknowledged for the support in obtaining the solid-state ^{15}N NMR spectra.

6. REFERENCES:

Balesdent, J., A. Mariotti and B. Guillet, 1987. Natural ^{13}C abundance as a tracer for studies of soil organic matter dynamics. Soil Biol. Biochem., 19, 25-30.

Christensen B.T., 1992. Physical fractionation of soil and organic matter in primary particle size and density separates. Advances in Soil Science, 20, 1-90.

Christensen, B.T. and S. Bech-Andersen, 1989. Influence of straw disposal on distribution of amino acids in soil particle size fractions. Soil Biol. Biochem, 21, 35-40.

FAO, Food and Agriculture Organization of the United Nations (Ed.) 1990. FAO - Unesco. Soil Map of the world, Revised Legend. Rome.

Kelly, K.R. and F.J. Stevenson, 1996.Organic forms of nitrogen. In: A. Piccolo (Editor), Humic Substances in Terrestrial Ecosystems. Elsevier Science, pp. 407-427.

Knicke,r H. and H.-D. Lüdemann, 1995. N-15 and C-13 CPMAS and solution HR NMR studies on the chemical modifications of N-15 enriched plant material during 600 days of microbial degradation. Org. Geochem., 23, 329-341.

Knicker, H. 1993. Quanititative ^{15}N- und ^{13}C-CPMAS-Festkörper- und ^{15}N-Flüssigkeits-NMR-Spektroskopie an Pflanzenkomposten und natürlichen Böden. Dissertation, University of Regensburg, Germany.

Knicker, H., R. Fründ and H.-D. Lüdemann, 1993. The chemical nature of nitrogen in native soil organic matter. Naturwissenschaften, 80, 219-221.

Knicker, H., G. Almendros, F.J. González-Vila, H.-D. Lüdemann and F. Martín, 1995. ^{13}C and ^{15}N NMR analysis of some fungal melanins in comparison with soil organic matter. Org. Geochem., 23. 1023-1028.

Knicker, H., A.W. Scaroni and P.G. Hatcher, 1996. ^{13}C and ^{15}N NMR spectroscopic investigation on the formation of fossil algal residues. Org. Geochem., 24, 661-669.

Kögel-Knabner, I., 1995. Composition of soil organic matter. In: P. Nannipieri, K. Alef (Editors), Methods in Applied Soil Microbiology and Biochemistry. Academic Press, pp. 66-78.

Lichtfouse, É., C. Chenu an F. Baudin, 1996. Resistant ultralaminae in soils. Org. Geochem. 25, 263-265.

Magid, J., A. Gorissen and K.E. Giller, 1996. In search of the elusive „active" fraction of soil organic matter: three size-density fractionation methods for tracing the fate of homogeneously ^{14}C-labelled plant materials. Soil Biol. Biochem., 28, 89-99.

Marshmann, N.A. and K.C. Marshall, 1981. Bacterial growth on proteins in the presence of clay minerals. Soil Biol. Biochem., 13, 127-134.

Martin, G.J., M.L. Martin and J.-P. Gouesnard, J.-P., 1981. ^{15}N NMR Spectroscopy. In: P. Diehl, E. Fluck, R. Kosfeld (Editors). NMR Basic Principles and Progress 18, Springer-Verlag, Heidelberg.

Mayer, L.M, S.A. Macko and L. Cammen, 1988. Provenance, concentrations and nature of sedimentary organic nitrogen in the Gulf of Maine. Mar. Chem., 25, 291-304.

Mayer, L.M., 1994a. Surface area control of organic carbon accumulation in continental shelf sediments. Geochim. Cosmochim., 58, 1271-1284.

Mayer, L.M., 1994b. Relationship between mineral surfaces and organic carbon concentrations in soils and sediments. Chem. Geol., 114, 347-36.

Schaefer, J. and E.O. Stejskal, 1976. Carbon-13 nuclear magnetic resonance of polymers spinning at magic angle. J. Am. Chem. Soc., 98, 1031-1032.

Schmidt, M.W.I., H. Knicker, P.G. Hatcher and I. Kögel-Knabner, 1996. Impact of brown coal dust on a soil and its size fractions - chemical and spectroscopic studies. Org. Geochem., 25, 29-39.

Schmidt, M.W.I., H. Knicker, P.G. Hatcher and I. Kögel-Knabner, 1997. Improvement of ^{13}C and ^{15}N CPMAS NMR spectra of bulk soils, partice size fractions and organic material by treatment with 10 % hydrofluoric acid. Eur. J. Soil Sci., 48, 319-328.

Schulten, H.-R., P. Leinweber and C. Sorge, 1993. Composition of organic matter in particle-size fractions of an agricultural soil. J. Soil Sci., 44, 677-691.

Skjemstad, J.O., P. Clarke, J.A. Taylor, J.M. Oades and R.H. Newman, (1994). The removal of magnetic materials from surface soils. A solid-state ^{13}C CP/MAS n.m.r. study. Aust. J. Soil Res., 32, 1215-1229.

Witanowski, M., L. Stefaniak and G.A. Webb, 1993. Nitrogen NMR spectroscopy. In: G. Webb (Editor), Annual Reports on NMR Spectroscopy 25, Academic Press, London.

Zelibor, J.L., Jr., L. Romankiw, P.G. Hatcher and R.R. Colwell, 1988. Comparative analysis of the chemical composition of mixed and pure cultures of green algae and their decomposed residues by ^{13}C nuclear magnetic resonance spectroscopy. Appl. Environ. Microbiol., 54, 1051-1060.

POLYMERIZATION: A POSSIBLE CONSEQUENCE OF COPPER-PHENOLIC INTERACTIONS

Andrea Oess[1], Martin V. Cheshire[2], D.B. McPhail[2], and Jean-Claude Vedy[1]

[1]Laboratoire de Pedologie, IATE, Departement de Genie Rural,
Ecole Polytechnique Federale de Lausanne, 1015 Ecublens, Suisse.
TEL: (41) 21- 693 37 67 FAX: (41) 21-693 56 70
[2]Macaulay Land Use Research Institute, Craigiebuckler, Aberdeen
AB15 8QH, UK.

1. INTRODUCTION

Although many studies have been carried out on copper interactions with humic and fulvic acids, little is known about its interactions with phenolic compounds. In Alpine podzolic systems considerable amounts of phenols are released from the acidophilic vegetation cover (*Rh. ferrugineum, Vacc. myrtillus, Calluna vulgaris*), together with other primary and secondary metabolites present in leaves (Vaughan and Malcolm 1985). All year-round, and in particular during the period of senescence, these phenols are leached to the soil where they can participate in the humification processes. This study shows that under appropriate pH and redox conditions phenols interact with metallic cations and, therefore, may play a role in their intrapedic transfer. Moreover, the change in redox state as a result of Cu-phenolic interactions may modify the phenol stability itself and provoke the polymerisation of phenolic monomers.

The aim of this study was to apply electron spin resonance (ESR) spectroscopy to: 1) Identify which phenols are able form complexes with Cu(II) and 2) Determine the effect of copper on phenol polymerisation processes. Since the chemical structure and, in particular, the position of the hydroxyl groups on the benzene ring are important in these processes, a range of phenols which have previously been identified in the vegetation cover (Gallet 1992) were selected for this experiment: (Fig.1)

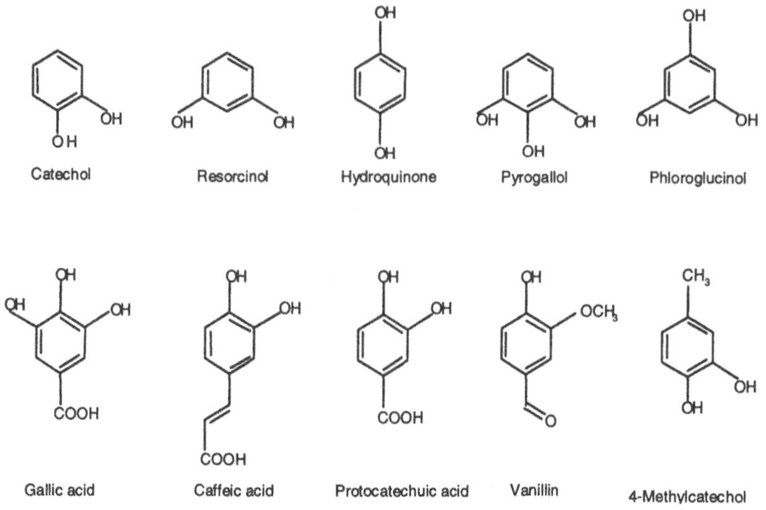

Figure 1. Phenols selected for interaction with Cu

1.1. Determination of copper complex formation by ESR

Cu(II) is a transition metal ion, which has a 9d-electron configuration. The presence of an unpaired electron results in paramagnetism and makes the ion and its complexes suitable for ESR analysis. Spectra may be obtained from a frozen solution or a fluid solution. To obtain maximum information on the complex under investigation it is often necessary to employ both approaches. Frozen solution spectra represent a summation of spectral contributions for all orientations of the principal axes of the complex with respect to the magnetic field of the spectrometer and are said to be anisotropic. Where the complexes approximate to axial symmetry, the anisotropic spectra display parallel and perpendicular features which can be characterised by $g_{//}$, $A_{//}$, $g\perp$ and A. The g-values are calculated from a knowledge of the microwave frequency at which the experiment was conducted and the magnetic field position of the parallel or perpendicular feature according to the equation:

$$g = h\nu/H$$

where h is Planck's constant, ν is the microwave frequency, is the Bohr magneton and **H** the magnetic field.

The A-values are calculated from the hyperfine structure which arises from the interaction of the unpaired electron with the spin (I = 3/2) on the copper nucleus. In many instances this structure only resolves in the parallel components. In fluid solution where the molecular tumbling rate is rapid in comparison to the timescale of the ESR transition, the anisotropic contribution is averaged out, and an isotropic spectrum results which may be characterised by g_{iso} and A_{iso} parameters. However, where complexes of higher molecular weight exist in solution, the tumbling rate may be insufficient to fully average out the anisotropic contribution which results in spectra intermediate between the isotropic and anisotropic case. Thus, information can be gained about the generation of polymeric copper species. The g- and A-values are sensitive to the coordination environment of the Cu(II) ion and can therefore be used to distinguish the presence of different complexes (Pilbrow, 1990).

ESR can also detect and characterise free radical species and thus may be of use in elucidating mechanisms of free radical-mediated polymerisation. Autooxidation of phenols in alkaline solutions in the presence of O_2, which produces free radical intermediates, is well documented and can lead to the formation of brown polymers. (Taylor and Battersby, 1967). Fe, Al and particularly Mn oxides have been shown to catalyse the oxidation of polyphenols (Wang et al., 1986). McBride et al. (1988) noted that several studies had shown that transition metals were able to catalyse polyphenol oxidation. They further showed that Al3+ catalyses catechol oxidation and suggested a mechanism whereby Al stabilises O semiquinone radicals at low pH by forming complexes and directs the type of polymerisation reactions (McBride and Sikora, 1990). O thought$_2$ by a similar mechanism.

2. MATERIALS AND METHODS

$CuSO_4$ solution (5 ml, 10 mM) was added to phenol solution (100 ml, 2.5 mM) in order to obtain a 1:5 Cu / phenol molar ratio. The pH was adjusted to 6 or 10 with NaOH (0.1 M). To prevent polymerisation solutions were frozen at -50 C until examination by ESR. Glycerine (10% by volume) was added to the samples to ensure the formation of a good aqueous glass, which needs to be formed to obtain resolved anisotropic (polycrystalline) ESR spectra from the frozen solution. ESR spectra were obtained at 100K, or ambient temperature, with a Bruker ECS 106 spectrometer operating at about 9.5 GHz and equipped with a TM_{110} cavity. Microwave power was 20 mW and the modulation amplitude 10 G.

3. RESULTS

3.1. Cu(II) in presence of catechol spectrum

At pH 6 the room temperature solution spectrum revealed four peaks with g_{iso} = 2.158 +/- 0.005 and A_{iso} = 64 +/-3 G indicating complexation with a low molecular weight ligand (Fig. 2). The frozen solution spectrum resolves parallel features with parameters different to those from the copper hexaaqua cation indicating complex formation (Fig. 3, Table 1). The isotropic parameters at pH 10 differ markedly from those at pH 6 with g_{iso} = 2.130 +/- 0.004 and A_{iso} = 81 +/- 2 G indicating a change in Cu coordination at higher pH (Fig. 4). The frozen solution spectrum (Fig. 5) shows that two new complexes have replaced the complex at pH 6.

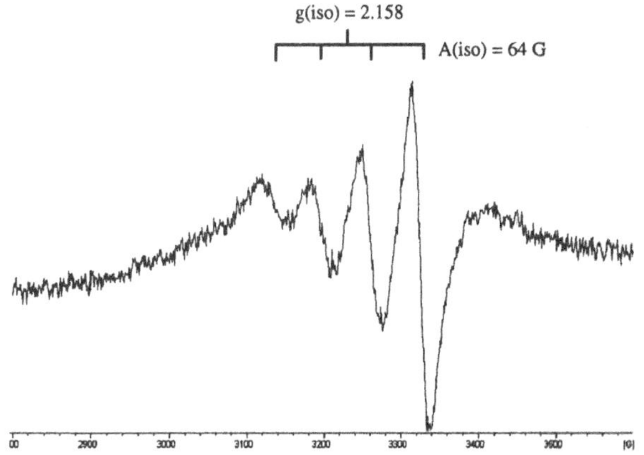

Figure 2. ESR fluid solution spectra of Cu^{2+} in the presence of catechol at pH 6

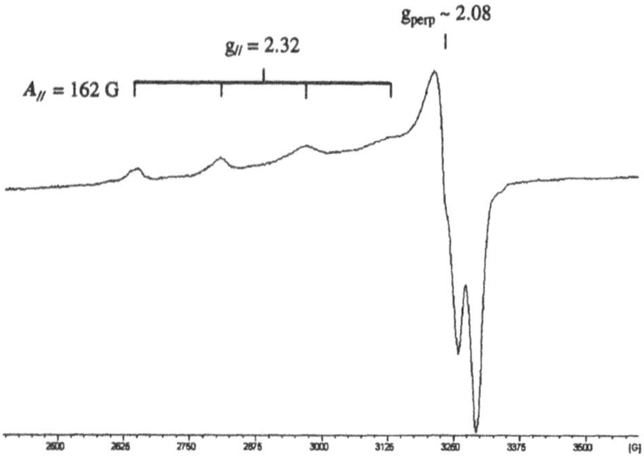

Figure 3. ESR frozen solution spectra of Cu^{2+} in the presence of catechol at pH 6

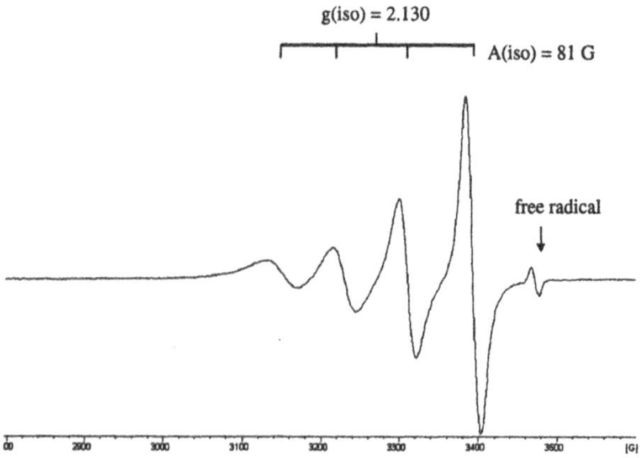

Figure 4. ESR fluid solution spectra of Cu^{2+} in the presence of catechol at pH 10.

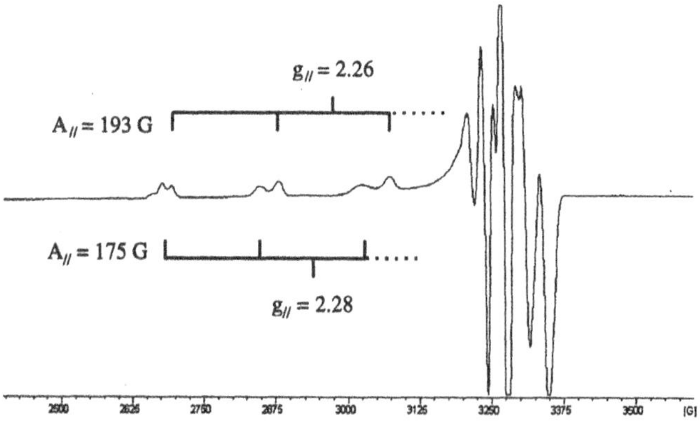

Figure 5. ESR frozen solution spectra of Cu^{2+} in the presence of catechol at pH 10. (The high-field (4th) parallel peaks overlap with perpendicular components of the spectrum).

3.2. ESR parameters of Cu(II) in presence of phenols at pH6 and pH10

Table 1 summarises the different g and A values obtained from the frozen solution spectra, which permitted us to attribute the number of species complexed to Cu(II). The A// and g// values for the uncomplexed Cu(II) were respectively 122 and 2.42. Although not all the phenols reacted with Cu(II), the changes in A and g values at pH 6 indicated that Cu(II) was mostly in the complexed form in a number of them. At pH10 additional complexes could sometimes appear which were superimposed on the first specie.

Sample	pH	number of species	A //	g //	g
$Cu(H2O)_6^{2+}$		1	122	2.42	~2.09
catechol	6	1	162	2.32	~2.08
	10	3	193	2.26	
			175	2.28	
resorcinol	6	2	152	2.36	~2.09
	10		166	2.3	~2.06
hydroquinone	6	1 or 2	122	2.42	~2.08
	10	2	152	2.32	~2.08
pyrogallol	6	1	140	2.36	~2.08
	10		not resolved	not resolved	not resolved
phloroglucinol	6		not resolved	not resolved	not resolved
	10	1	168	2.3	2.06
gallic acid	6	1	142	2.35	not resolved
	12	1	194	2.25	not resolved
caffeic acid	6	1 or 2	124	2.41	~2.08
	10	3	173	2.29	~2.08
			192	2.26	~2.08
protocatechuic acid	6	3	not resolved	not resolved	not resolved
	10	2	158	2.32	not resolved
			163	2.3	not resolved
vanillin	6	1	122	2.4	~2.08
	10		not resolved	not resolved	not resolved
methylcatechol	6		not resolved	not resolved	not resolved

Table 1. ESR spectra: frozen solution parameters for Cu^{2+} with different ligands

Comparing the A and g values for the whole range of molecules, 3 different groups of complexation behaviours were observed: Group 1, for phenols possessing a catechol function, able to form at least 1 complex at pH 6 and 2 or more complexes in presence of Cu(II) at pH 10 as represented by catechol, 4-methylcatechol, caffeic acid, and protocatechuic acid. These phenols all showed large changes in A- and g- values from that of the hydrated ion. Group 2, for phenols showing weaker interactions with Cu(II) which was deduced from the smaller shifts in A- and g- values from those of the hydrated ion, as for instance, resorcinol, hydroquinone, phloroglucinol, gallic acid and pyrogallol. Group 3, for phenols for which no complexes were observed as in vanillin, guaiacol, *trans*-cinnamic acid, *para*-hydroxybenzoic acid and *ortho*-, *para*-, *meta*-, coumaric acids. Generally the most reactive phenols had the greatest increase in A// values with pH and the greatest decrease in g// values.

3.3. Free radical signal and spectrum evolution with time

In the absence of Cu(II) most of the phenol solutions turned orange-brown after a few weeks due to oxidation and this behaviour was enhanced at high pH. In the presence of Cu(II) the reaction was considerably faster with the catechol solutions turning black in less than 5 hours. For catechol, gallic acid and protocatechuic acid, organic free radical signals were observed at pH 12 or lower pH, which may indicate the presence of free radical intermediates in the polymerisation of these monomers to polyhydroxy aromatic compounds. For hydroquinone the free radical signal was very stable and could be observed between pH 7 - pH 12. It was attributed to the formation of the stable semiquinone. It must be emphasised, however, that such radical formation can occur in the absence of Cu(II) as a consequence of autoxidation in which molecular oxygen acts as an electron acceptor from the phenoxyl anion. Consequently, further ESR studies need to be undertaken to provide direct evidence that Cu(II) enhances the formation of reactive free radical intermediates, which are involved in the polymerisation process.

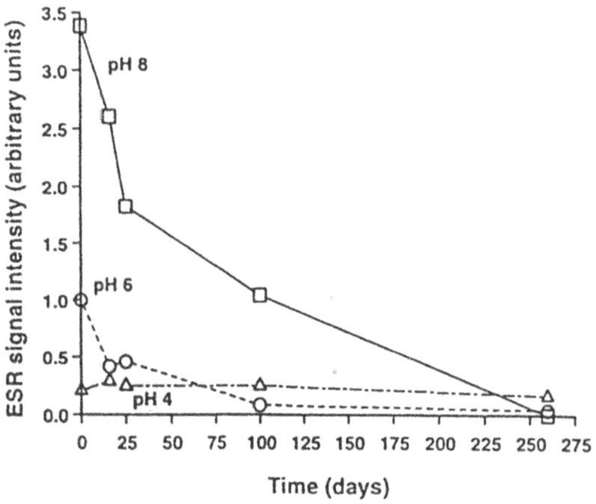

Fig 6. Change of Cu(II) ESR signal intensity with time

By following the loss of fluid solution signal intensity of Cu(II) in the presence of catechol with time (Fig. 6), the following deductions could be made:

At pH 4, intensity of the uncomplexed Cu(II) signal remains stable with respect to time. At pH 6 the spectral intensity of the dominant peak of the two species present gradually reduced with time which may indicate either that Cu(II) is bound to higher molecular weight species (which the ESR would be less sensitive in detecting) and/or that Cu(II) is being reduced to Cu(I), which is ESR silent. After 260 hours the Cu(II) signal has completely gone. The same behaviour can be observed at pH 8 although the loss in signal intensity is much more pronounced. The inability to detect an anisotropic signal in the frozen solution spectra and the lack of any precipitate occurring with respect to time is an indication that Cu(II) is being reduced to Cu(I) during the polymerisation process. It should be noted that signal intensity has been measured as a function of peak height and therefore accurate quantitative measurements cannot be made between signals derived from different species due to inherent differences in linewidths.

4. DISCUSSION

The A- and g-values of the fluid and frozen solution spectra of Cu(II) in the presence of different phenols indicates different interactions depending on the nature and position of the functional groups of the benzene ring. Compounds like catechol, caffeic acid, protocatechuic acid and 4-methylcatechol which have adjacent hydroxyl groups on the ring system were most likely to interact with Cu(II) forming up to 3 different species at pH 10. With catechol, two of the species detected may represent the mono- and bisbidentate chelates shown below in which coordination takes place within the equatorial plane of the copper. A third species occurring at higher pH may involve participation of OH^- in the coordination sphere.

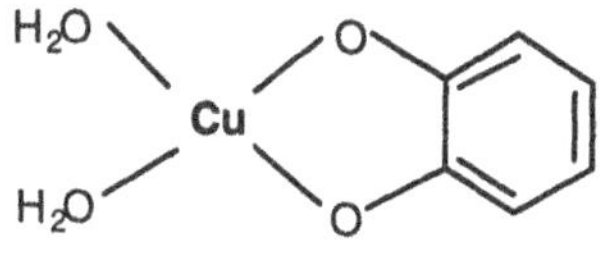

monobidentate catechol complex

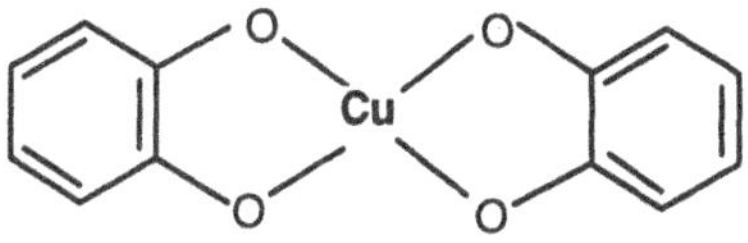

bisbidentate catechol complex

The presence of a carboxyl group on the phenol ring as occurs with protocatechuic acid, gallic acid and caffeic acid did not lead to the formation of any additional complexes which suggested that the carboxyl groups do not participate in the complexation reactions. The small changes in the ESR parameters as a result of methylation of catechol at the 4-position is thought to result from the weakly electron donating (positive induction) effect of the methyl group. For the 1,2,3-trihydroxyphenols a weaker interaction was observed compared to the 1,2-dihydroxyphenols. This may be a consequence of the third hydroxyl group destabilizing the enolate forms of the compounds.

With hydroquinone, where chelation is not possible, discrete changes in spectroscopic parameters from those of the Cu(II) hexaaqua cation indicates that weak interaction with the ring hydroxyls can occur. This interaction is also sufficient to avoid precipitation of the neutrally charged $Cu(OH)_2$ between pH6 and pH10.

The ESR data indicates a general trend for the $A_{//}$ values of the complexes to increase with pH and the $g_{//}$ values to decrease with pH which can be interpreted in terms of increased ligand - Cu(II) interaction as ligand hydroxyl groups sequentially deprotonate according to their pK_a values.

With catechol, HPLC analysis showed that the chemical structure of the original phenol had been modified. Furthermore, exclusion chromatography demonstrated an increase in molecular size (not presented here). The loss of Cu(II) ESR signal intensity with time strongly suggests that the polymerization process involve a redox interaction between the metal ion and the phenol.

From the colour change of the solutions caffeic acid, gallic acid and protocatechuic acid were less sensitive than catechol to oxidation and subsequent polymerisation. The presence of a substituent in position 4 of the benzene ring, may impede polymerisation through steric hindrance, directly blocking the site of reaction, or altering the unpaired electron spin density, and subsequent reactivity, at other sites.

Although concentrations of phenols used for this experiment were considerably greater compared with the concentrations observed in soil solutions, the differences in affinities observed for Cu(II) with the selected phenols could explain some selective Cu-solute interactions, which occur in natural samples.

The formation of higher molecular weight complexes a few hours after interaction between catechol and Cu(II), indicates that Cu(II) is a strong catalyst in the polymerisation process. This has direct environmental implications. In podzol soil solutions, the pH is usually around 4.5 but can sometimes reach 6.5 following rainfalls. The dynamics of polymerisation being very unstable between pH 4 and 8 (Fig.8) the mobility of the complexed metal could also change greatly.

5. ACKNOWLEDGEMENTS

Part of this work was funded by the Scottish Office Agriculture and Fisheries Department.

6. REFERENCES

Gallet,C.1992. Apports de la biochimie a la connaissance du fonctionement des écosystemes forestiers: rôle des composes phénoliques dans une pessière a myrtilles, thèse numero 278-92 de l'universite Claude Bernard-Lyon I.

McBride M.B. F.J. and Sikora 1990. Catalysed Oxidation Reactions of 1,2-Dihydroxybenzene (Catechol) in Aerated Aqueous Solutions of Al(3+). Journal of Inorganic Biochemistry, 39, 247-262.

McBride M.B., F.J. Sikora and L.G. Wesselink, 1988. Complexation and Catalysed Oxidative Polymerisation of Catechol by Aluminum in Acidic Solution. Soil Sci. Am. J. 52, 985-993.

Parish R.V.1990. NMR, NQR, EPR and Mossbauer Spectroscopy in Inorganic Chemistry, Ellis Horwood Series, 168-194.

Pilbrow J.R. 1990. Transition Ion Electron Paramagnetic Resonance, Clarendon Press, Oxford.

Taylor W. I. and A.R. Battersby 1967.Oxidative Coupling of Phenols, Edward Arnold Ltd., London.

Vaughan D. and R.E.Malcom 1985. Soil Organic Matter and Biological Activity, Developments in Plants and Soil Science, volume 16, Martinus Nijhoff / Dr W. Junk Publishers.

WangT.S.C., P.M. Huang, C-H. Chou and J-H. Chen, 1986. The role of soil minerals in the abiotic polymerization of phenolic compounds and formation of humic substances. In: Interactions of Soil Minerals with natural organics and Microbes. Eds P.M.Huang and M.Schnitzer. Chapter 8. Soil Sci. Soc. Amer. Spec. Pub. 17, 251-278.

EFFECT OF PH, EXCHANGE CATIONS AND HYDROLYTIC SPECIES OF AL AND FE ON FORMATION AND PROPERTIES OF MONTMORILLONITE-PROTEIN COMPLEXES

Annunziata De Cristofaro, Claudio M. Colombo,
Liliana Gianfreda and Antonio Violante

Dipartimento di Scienze Chimico-Agrarie,
Università di Napoli "Federico II",
Napoli, Italia.

1. INTRODUCTION

It is well established that the adsorption of proteins on clay minerals is influenced by many factors.

Certainly the pH of the system, the temperature and the physico-chemical properties either of the clay minerals (e.g. the surface area, the cation exchange capacity and the nature of the cation saturating the clay) or of the proteic molecules (e.g. molecular weight and isoelectric point) may determine changes in protein sorption on clays (Mc Laren et al., 1958; Armstrong and Chesters, 1964; Harter and Stotzky, 1971, 1973; Theng, 1979).
Proteins may be adsorbed on phyllosilicates on external surfaces or intercalated in the interlamellar spaces of expandible clays (Mc Laren et al., 1958; Harter and Stotzky, 1973; Fusi et al., 1989; Quiquampoix and Ratcliffe, 1992; Naidja et al., 1995).

Many authors gave different interpretations of the mechanisms for both adsorption and interlayering of proteins in the expansible clay minerals (Mc Laren et al., 1958; Harter and Stotzky, 1973; Larsson and Siffert, 1983; Norde, 1986; Fusi et al., 1989; Quiquampoix and Ratcliffe, 1992).

In acidic soil environments hydrolytic species of aluminium and iron are retained by expansible clay minerals, modifying the physicochemical properties and the mineralogy of clays (Huang and Violante, 1986; Barnhisel and Bertsch, 1989). However, the adsorption of proteins on clay minerals partially coated with hydrolytic products of Al and Fe has received poor attention (Gianfreda et al., 1991, 1992; Naidja et al., 1995; Violante et al., 1995)

In this work are reported some preliminary experiments, on the adsorption of various proteins, of different molecular weight and isoelectric points (urease, catalase, albumin, pepsin and lysozyme), on a montmorillonite differently saturated (Na, Ca or Mg

Effect of Mineral-Organic-Microorganism Interactions on Soil and Freshwater Environments
Edited by Berthelin, *et al*, Kluwer Academic / Plenum Publishers, New York, 1999

saturated) or coated by hydrolytic species of aluminum and iron, at pH values equal, below or above their isoelectric point.

2. MATERIALS AND METHODS

2.1. Clay Minerals and Chemicals

The < 2 μm fraction of a Crook montmorillonite (Wyoming, USA) was separated by sedimentation after dispersion in water. The Na-, Ca- and Mg-saturated clay minerals were prepared by washing suitable amounts of the clay suspension three times in 0.5 N NaCl, $CaCl_2$ or $MgCl_2$ respectively and then repeatedly in distilled water. The final suspensions were dialyzed until free of Cl^- ions and then freeze-dried.

The $Al(OH)_x$-montmorillonite and the $Fe(OH)_x$-montmorillonite (chlorite like) complexes (Barnhisel and Bertsch, 1989) were prepared by adding 0.1 M NaOH, dropwise (1 ml/min), to mixtures of $AlCl_3$ (or $FeCl_3$) and clay suspension, containing 3 mmol Al (Al-chlorite) or Fe (Fe-chlorite) per gram of clay, until pH 6.0 was reached. The final suspensions were immediately washed with water, dialyzed until free of Cl^- ions and freeze-dried. The presence of hydrolytic species of Al or Fe in the interlayers of the montmorillonite was ascertained by X-ray diffraction.

The proteins used were: urease (EC 3.5.1.5 from jack bean, Sigma Chemical Co., St Louis, MO, USA), catalase (EC 1.11.1.6 from bovine liver, Sigma Chemical Co., St Louis, MO, USA), albumin (from bovine serum, Boehringer Biochemia, Mannheim, Germany), pepsin (EC 3.4.23.1 from porcine stomach mucosa, Sigma Chemical Co., St Louis, MO, USA), lysozyme (EC 3.2.1.17 from hen egg white, Boehringer Biochemia, Mannheim, Germany).

Molecular weight (M.W.) and isoelectric points (ieps) of the proteins are reported in Table 1.

2.2. Adsorption of the Proteic Molecules on the Clay Minerals

The adsorption experiments of proteins on clay minerals were carried out, for 1 h at 208C, by adding to the clay suspensions in water the protein freshly prepared, keeping the pH of the complexes at a value equal to the iep of each protein, or 1.2 pH units below or above it. The pH of the suspensions (clay:protein ratio -w:w- of 1) was kept constant using an automatic Radiometer titrator. The final volume was 20 ml. In only some experiments carried out at pH 7.0 the final volume was 10 ml.

Adsorption isotherms of selected proteins on Na-saturated Mt were carried out at pH 7.0 as described by Violante et al.(1995).

The suspensions were centrifuged for 10 minutes at 10,000 rpm and washed twice (5 ml) with distilled water. The supernatants were collected and the amounts of proteins recovered were determined spectrophotometrically at 280 nm. The amount of protein adsorbed at equilibrium was calculated by the difference between the amount of protein recovered in the supernatant from that initially added.

The clay minerals and the washed protein-clay complexes were dispersed in 3 ml of distilled water and trasferred onto glass slides, dried at 20°C and X-rayed with a Rigaku diffractometer, using Co-Kα radiations generated at 20 kV and 30 mA.

3. RESULTS AND DISCUSSION

3.1. Adsorption Isotherms at pH 7.0

Figure 1 shows the adsorption isotherms at 20°C for lysozyme, catalase, albumin and pepsin on the Crook montmorillonite Na-saturated at pH 7.0. The isotherms showed typical Langmuir characteristics. According to the classification of Giles et al. (1974) lysozyme, catalase, albumin and pepsin showed a H-class, C-class, S-class and L-class adsorption isotherms, respectively. The adsorption isotherm for urease on Crook Mt was not studied. However, Gianfreda et al. (1992) showed that urease, at a final concentration ranging from 0 to 0.050 mg ml^{-1}, was adsorbed on a Uri Mt according to a H-class isotherm.

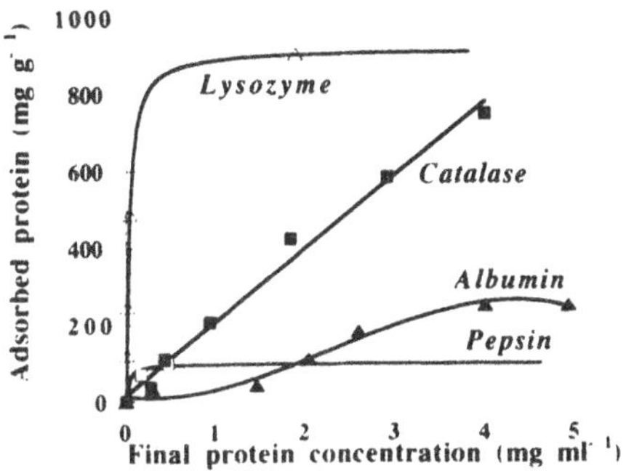

Figure 1. Adsorption isotherms, at pH 7.0, of catalase, albumin, pepsin and lysozyme on Na-saturated Crook montmorillonite.

The molecular weight and the isoelectric point had a great influence on the extent of adsorption of proteins on the Na-saturated montmorillonite (Table 1).

A large amount (909 mg/g, 65 µmol/g) of lysozyme was very easily adsorbed on the montmorillonite, clearly because of its low molecular weight and size and its high iep. In fact, at pH 7.0, lysozyme was in a cationic form and the sorption of this protein by montmorillonite was mainly due to electrostatic interactions between the opposite charges of clay surfaces and protein molecules. On the contrary, pepsin being negatively charged at pH 7.0, a pH very far from its iep, was poorly sorbed on clay surfaces (100 mg/g, 2.9 µmol/g).

Urease, albumin and catalase, characterized by quite similar ieps (4.9-5.6), showed a low net negative charge at pH 7.0, but were adsorbed in relatively high amounts. It was ascertained that the greater the M.W. of the protein, the higher the quantities (in mg/g) adsorbed. However, in spite of the fact that albumin had a M.W. 4 and 8 times smaller than that of catalase and urease, respectively, the µmoles of albumin adsorbed per gram of montmorillonite were only 1.6 and 1.7 greater than those of catalase and urease.

Table 1. Amounts* of lysozyme, pepsin, albumin, catalase and urease sorbed on Na-montmorillonite at pH 7.0.

Protein	*MW (kDa)*	*iep*	*Adsorbed amount*	
			(mg/g)	*(μmol/g)*
Lysozyme	14	11.1	909	65
Pepsin	34	2.5	100	2.9
Albumin	60	5.2	183	2.9
Catalase	238	5.6	425	1.8
Urease	480	4.9	872	1.8

*The final volume for the lysozyme-, pepsin-, albumin- and catalase-clay complexes was 10 ml, whereas for the urease-clay complexes was 20 ml. The clay-protein ratio for all the complexes was 1:1 (w:w).

These data seem to demonstrate that non-coulombic forces (hydrophobic forces, hydrogen bonding, Van der Waals forces) facilitate the sorption of large molecules at pH value near the iep.

All the proteins were intercalated, at pH 7.0, in the interlayer spaces of the Crook montmorillonite (Fig. 2), except pepsin. In fact, for the protein-clay complex obtained with pepsin any expansion of the basal distance was not observed, clearly because of the negative charge of the protein at pH >> 2.5 (its iep).

It appears interesting to note that in spite of the very high amounts of lysozyme adsorbed on the Na-montmorillonite the d-spacing of the lysozyme-clay complex (22.0 Å) was smaller than those obtained for the albumin- and catalase-clay complexes (29.7 Å and 34.1 Å respectively), evidently because of the greater size of albumin and catalase molecules.

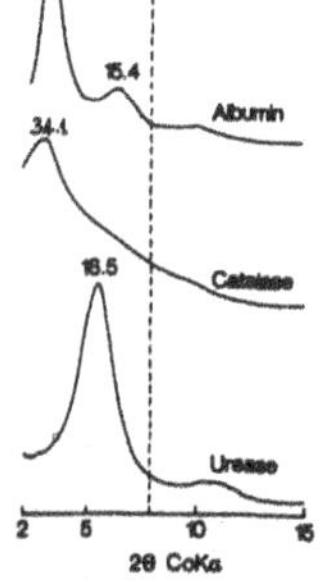

Figure 2. X-ray powder diffraction patterns at 20°C of lysozyme-, pepsin-, albumin-, catalase- and urease-montmorillonite complexes formed at pH 7.0. The dashed line indicates the d-spacing of Na-Mt.
The final volume for the lysozyme-, pepsin-, albumin- and catalase-clay complexes was 10 ml, whereas for the urease- clay complexes was 20 ml. The clay-protein ratio was 1:1 (w:w).

Furthermore, lysozyme and albumin were very well intercalated in the interlayer spaces of Mt, showing sharp and symmetric peaks.

Catalase and urease were intercalated in the interlamellar spaces of montmorillonite, in spite of the fact that they had large molecules negatively charged. However, catalase- and urease-complexes, often, showed very broad and/or asymmetric peaks. The basal distance of the urease-clay complex was smaller than those of the other protein-clay complexes. Probably the great size of urease was an obstacle for the protein intercalation in the interlayer of the clay.

3.2. Effect of pH, Exchange Cations and Hydrolytic Species of Al and Fe on Protein Adsorption

Figure 3 shows the amounts of albumin adsorbed on Na-, Ca- or Mg-saturated montmorillonite and on OH-Al and OH-Fe montmorillonite complexes at different pH values, and precisely at pH = 5.2 (pH = iep), 4.0 (pH < iep) and 6.4 (pH > iep).

It appears evident that the amounts of albumin fixed on the clay minerals changed depending on both the nature of the cation saturating the clay and the pH of the system.

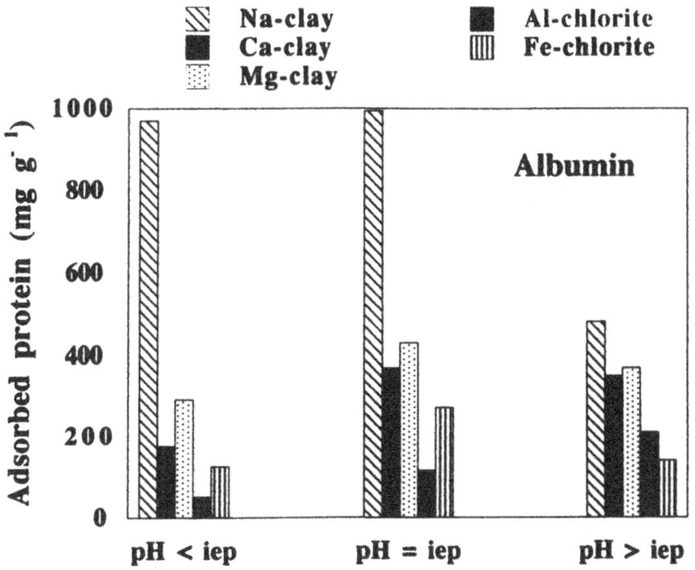

Figure 3. Amounts of albumin adsorbed on clay minerals at pH below (pH 4.0), equal (pH 5.2) or above (pH 6.4) its isoelectric point (iep).

The amounts of protein adsorbed on the Na-montmorillonite were very high at pH $\leq$ iep (970-995 mg/g) but deeply decreased at pH > iep to 480 mg/g, surely due to the electric charge repulsion between the protein and the surface of the clay both negatively charged.

Lower quantities of albumin were adsorbed on Mg- and Ca-montmorillonite and on the two chlorite-like complexes (Fig. 3), probably due to the lack of interlayering of the proteic molecules in these clays, as ascertained by X-ray analysis (Fig. 4).

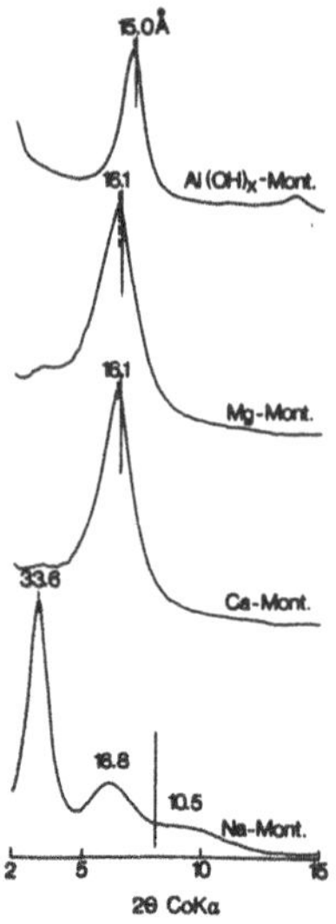

Figure 4. X-ray powder diffraction patterns at 20°C of albumin-clay minerals complexes formed at pH equal to the isoelectric point of the protein (pH 5.2). The lines indicate the d-spacings of $Al(OH)_x$-Mt complex and Mg-, Ca- and Na-saturated Mt, respectively.

It appears evident that the nature of the cation and of the hydrolytic species of Al and Fe saturating the clay had a great influence on the intercalation of the proteic molecules.

Albumin was very well intercalated in the interlayering of the Na-montmorillonite, as appears evident from the sharp and symmetric peak at 33.6 Å with some higher rational basal reflections, while no intercalation was observed for the Mg- and Ca-montmorillonite and for OH-Al- and OH-Fe-montmorillonite complexes (Fig. 4).

Indeed, it is well known that the intercalation of proteic molecules is easier to occur for Na-montmorillonite than for Ca- and Mg-montmorillonite (Theng, 1979; Larsson and Siffert, 1983), probably because of the easier "dispersion" of the Na-clay compared with the other clays. Larsson and Siffert (1983) and Violante et al. (1995) claimed that the process of radial diffusion into the interlayer space is questionable and demonstrated that the opening of the clay layers, when saturated with Na cations, is easier than a diffusional introduction of bulky molecules into the interlayer spacing.

Furthermore, the incorporation of proteins into the interlayers of the synthetic Al- and Fe-chlorite cannot take place easily also the interlayers are partially or completely filled by hydrolytic products of Al or Fe.

A maximum in adsorption was registered at pH equal to the iep of albumin, with a decrease in more or less symmetrical way on both sides of this pH value, either on Ca- and Mg-montmorillonite or on Fe-chlorite (Fig. 2), in accordance with many authors (Norde, 1986; Quiquampoix and Ratcliffe, 1992; Barroug et al., 1989).

Furthermore, for the Al-chlorite a continous increase in albumin sorption was registered by increasing the pH. The increase of protein adsorption at pH > iep may be due, at least in part, to the atractive forces between the biomolecules and the surface of the OH-Al-montmorillonite, respectively, negatively and positively charged.

4. CONCLUSIONS

This study on the adsorption of proteins on montmorillonite, either Na-, Ca- or Mg-saturated or coated by hydrolytic species of aluminium or iron, shows that:

- Urease, catalase, albumin, pepsin and lysozyme were differently adsorbed.
- More proteic molecules were adsorbed at pH equal to their iep than below or above it.
- More proteic molecules were sorbed on Na- than Ca- or Mg- montmorillonite.
- The presence of OH-Al- (mainly) or OH-Fe- species on the surfaces of montmorillonite reduced the adsorption of proteins.
- The order of adsorption of albumin molecules was Na-Mt > Ca-Mt ≈ Mg-Mt > OH-Fe-Mt > OH-Al-Mt.
- Low molecular weight proteins (lysozyme and albumin) appeared very well intercalated into the interlayer spaces of Na-montmorillonite, mainly at pH equal to their iep, showing sharp and symmetric peaks (up to 3.2 nm).
- High molecular weight proteins (catalase and urease) were intercalated into the interlayer spaces of Na-montmorillonite undergoing extensive unfolding.
- The intercalation of proteins into the Ca-, Mg- or OH-Al- and OH-Fe-montmorillonite was not observed probably because the opening of the interlayer spacings of clay became more difficult.

5. ACKNOWLEDGEMENTS

This study was partially supported by the International Scientific Cooperation (Contract No. CI1*-CT94-0048) of the European Community. Contribution No. 145 from the Dipartimento di Scienze Chimico-Agrarie (DISCA).

6. REFERENCES

Armstrong, D.E. and G. Chesters 1964. Properties of protein-bentonite complexes as influenced by equilibration conditions. Soil Sci., 98, 39-52.

Barnhisel, R.I. and P.M. Bertsch 1989. Smectites. In: J.B.Dixon & S.B. Weed (Editors). Minerals in Soil Environments, 2nd edn. Soil Science Society of America, Madison, Wisconsin, pp.675-727.

Barroug, A., J. Lemaitre and P.G. Rouxhet, 1989. Lysozyme on apatites: A model of protein adsorption controlled by electrostatic interactions. Colloids Surf., 37, 339-335.

Fusi, P., G.G. Ristori, L. Calamai and G.Stotzky, 1989. Adsorption and binding of protein on "clean" (homoionic) and "dirty" (coated with Fe oxyhydroxides) montmorillonite, illite and kaolinite. Soil Biol. Biochem., 21, 911-920.

Gianfreda, L., M.A. Rao and A. Violante, 1991. Invertase (β-fructosidase): Effect of montmorillonite, Al-hydroxide and $Al(OH)_X$-montmorillonite complexes. Soil Biol. Biochem., 23, 581-587.

Gianfreda, L., M.A. Rao and A. Violante, 1992. Adsorption, activity and kinetic properties of urease on montmorillonite, aluminium hydroxide and $Al(OH)_X$-montmorillonite complexes. Soil Biol. Biochem., 24, 51-58.

Giles, C.H., D. Smith and A. Huitson, 1974. A general treatment and classification of solute adsorption isotherm. I. Theoretical. J. Colloid Interface. Sci., 47, 755-765.

Harter, R.D. and G. Stotzky, 1971. Formation of clay-protein complexes. Soil Sci. Soc. Am. Proc., 35, 383-389.

Harter, R.D. and G. Stotzky 1973. X-ray diffraction, electron microscopy, electrophoretic mobility and pH of some stable smectite-protein complexes. Soil. Sci. Soc.Am. Proc., 37, 116-123.

Huang, P.M. and A. Violante, 1986. Influence of organic acids on the crystallization and surface properties of precipitation products of aluminum. In: P.M. Huang & M. Schnitzer (Editors). Interactions of

Soil Minerals with Natural Organics and Microbes. Soil Science Society of America, Spec. Pub. 17, Madison, Wisconsin, pp. 159-221.

Mc Laren, A.D., G.H. Peterson and I. Barshad, 1958. The adsorption and reactions of enzymes and proteins on clay minerals: IV. Kaolinite and montmorillonite. Soil Sci. Soc. Am. Proc., 22, 239-244.

Larsson, N. and B. Siffert, 1983. Formation of lysozyme containing crystals of montmorillonite. J. Colloid Interface Sci., 93, 424-431.

Naidja, A., A. Violante and P.M. Huang, 1995. Adsorption of tyrosinase on montmorillonite as influenced by hydroxyaluminum coatings. Clays Clay Miner., 43, 647-655.

Norde, W., 1986. Adsorption of proteins from solutions at solid-liquid interface. Adv. Colloid Interface Sci., 25, 267-340.

Quiquampoix, H. and R.G. Ratcliffe, 1992. A ^{31}P NMR study of the adsorption of bovine serum albumin on montmorillonite using phosphate and the paramagnetic cation Mn^{2+} : Modification of conformation with pH. J. Colloid Interface. Sci., 148, 343-352.

Theng, B.K.G., 1979. Formation and properties of clay-polymer complexes. Elsevier. New York.

Violante, A., A. De Cristofaro, M.A. Rao and L. Gianfreda, 1995. Physicochemical properties of protein-smectite and protein- $Al(OH)_x$-complexes. Clay Miner., 30, 325-336

ADSORPTION AND PROPERTIES OF UREASE IMMOBILIZED ON SEVERAL IRON AND ALUMINUM OXIDES (HYDROXIDES) AND KAOLINITE

Qiaoyun Huang, Minghua Jiang and Xueyuan Li

Department of Soil Science
Central China Agricultural University
Wuhan, Hubei 430070, China

INTRODUCTION

Soil enzymes and microorganisms promote the cycling of materials and energy in the soil ecosystem. The activity of soil enzymes can be modified by a variety of soil components but especially clay minerals, because extracellular enzymes in soil are usually physically or chemically bound to minerals. Many studies on the relationships between minerals and enzymes have been conducted by several soil scientists over the past few decades. Most of these works concern layer aluminum-silicate clays, including "clean" and "dirty" clays such as coated and interlayered aluminum silicates (McLaren *et al..*, 1958; Makboul and Ottow, 1979; Garwood *et al..*,1983; Dick and Tabatabai, 1987; Fusi *et al..*, 1989, Gianfreda *et al..*,1991, 1992, 1993; Huang *et al..*,1995). The advances and the state of knowledge in this field have been reviewed by Burns (1986), Boyd and Mortland (1990) and Huang and Li (1995).

Iron and aluminum oxides are very abundant and active in soils from tropical and subtropical regions. These oxides may exert profound influences on a series of soil characteristics, including biological properties such as enzyme activity and kinetics. Among soil enzymes, urease is one of the most important because it plays a significant role in the transformation of nitrogen compounds in soil. Early research works regarding soil urease were mainly devoted to the following aspects: 1) the estimation of urease activity in various soils (Paulson and Kurtz,1969; Tabatabai and Bremner,1972; Pettit *et al..*,1976; Dkhar and Mishra,1983; Baruah and Mishra,1984; Rao and Ghai,1985), 2) the examination of the stability and kinetics of soil urease (Burns *et al..*1972; Pettit *et al..*,1976), 3) the extraction of soil urease (Ceccanti *et al..*,1978; Nannipieri,1980) and 4) the influences on soil urease activity of some environmental factors such as trace elements (Tabatabai,1977), tillage (Klein and Koths,1980) and pesticides (Lethbridge and Burns,1976; Lethbridge *et al..*,1981; Gianfreda *et al..*, 1994). Studies concerning the interactions between clay minerals and urease have also been carried out by some investigators (Pinck and Allison,1961; Makboul and Ottow,1979; Boyd and Mortland,1990; Lai and Tabatabai,1992; Gianfreda *et al..*, 1992, 1995a, 1995b). However, very little information is available with regard to the interactions between oxides and urease (Huang, 1990; Huang and Li, 1995). The purpose of the present work was to examine the effects of some iron and aluminum oxides as well as kaolinite on the adsorption, activity and kinetics of urease.

Effect of Mineral-Organic-Microorganism Interactions on Soil and Freshwater Environments
Edited by Berthelin, *et al*, Kluwer Academic / Plenum Publishers, New York, 1999

MATERIAL AND METHODS

1. Minerals

Goethite was synthesized by neutralization of 0.15mol.L^{-1} $Fe(NO_3)_3$ solution with 2.5 mol.L^{-1} NaOH (5ml.min^{-1}) up to pH 12 under vigorous stirring. The colloid was aged for 48 h at 60°C, washed with water, dialyzed and freeze-dried. Hematite was obtained by heating goethite for 2 h at 300°C. The non-crystalline iron oxide (N-Fe) was prepared by adding 750 ml of 1mol.L^{-1} NaOH to 0.036 mol.L^{-1} $FeCl_3$ solution (20 ml.min^{-1}) to the final pH of 8.0. After constant stirring for 30 min, the suspension was allowed to stand and the precipitate was washed 3 times with NaOH solution at pH8.0. The residue was dialyzed and freeze-dried.

Bayerite was obtained by slowly adding 0.18 mol.L^{-1} NaOH to 0.06 mol.L^{-1} $Al(NO_3)_3$ solution (3 ml.min^{-1}) until the molar ratio of OH:Al reached 3. The suspension was aged for 15 days, washed with water, dialyzed and freeze-dried. The non-crystalline aluminum oxide (N-Al) was prepared by a rapid titration of 0.06mol.L^{-1} $Al(NO_3)_3$ solution with 0.18 mol.L^{-1} NaOH solution up to a final OH:Al molar ratio of 2.7. The suspension was washed with water and then dialyzed and freeze-dried.

The < 2μm fraction of kaolinite (chemical reagent) was separated by the sedimentation method. After saturation with Na^+ ion by 1mol.L^{-1} NaCl solution, the clay was washed with deionized water and then dialyzed until Cl^- free. The Na-kaolinite was freeze-dried and stored at room temperature.

The point of zero charge (PZC) of the minerals was determined by the titration method (Xiong, 1985). The surface area was evaluated by the retention of ethylene glycol monoethyl ether (Carter *et al.*., 1965). Some properties of the examined minerals are tabulated in Table 1.

Table 1. Selected properties of examined minerals

Minerals	pH	Surface area ($m^2.g^{-1}$)	Point of zero charge(PZC)
goethite	7.12	234.0	8.27
hematite	7.75	206.2	7.35
N-Fe	6.53	597.0	7.50
bayerite	6.55	122.9	8.23
N-Al	6.67	291.3	8.82
kaolinite	5.53	134.8	3.61

2. Enzyme

The urease (EC 3.5.1.5 from sword bean, 5U.mg^{-1}, optimal pH of 6.1) was purchased from Merck Company, Germany. The enzyme was stored in a refrigerator at 0–4°C until use.

3. Adsorption experiments

Fifty milligrams of each mineral were added to 5 ml of NaAc-HAc buffer solution (pH 5.5) containing 2, 4, 6, 8, 12, 16 or 20mg of urease respectively. The suspensions were shaken for 1 hr and then allowed to stand for 30 min. After centrifugation at 30,000 g and filtration through a 0.4 μm pore size membrane, the equilibrium supernatants were directly analyzed by spectrophotometry at 280 nm. Enzyme adsorption was also conducted in the pH range of 4.0 to 9.0 with a mineral quantity of 50 mg and urease concentration of 2.8 mg.ml^{-1}, using NaAc-HAc (pH 4.0~6.0) and Tris-HCl (pH 7.0~9.0) buffers.

4. Preparation of mineral-enzyme complexes

The ratio of enzyme to mineral (wt:wt) usually used in adsorption isotherm experiments is relatively higher than that in natural soils. Therefore, we employed lower enzyme/mineral ratio and different solid/liquid ratios (convenient for operation) in the study of mineral-enzyme complexes. In polypropylene tubes 625 mg of each mineral sample was added to 25 ml of 0.2 $mg.ml^{-1}$ urease solution. After 15 min of shaking, the suspension was centrifuged at 30,000 g. The residue was washed twice with deionized water and centrifuged. The residue was dispersed with 25 ml of water and stored at 4°C. The supernatant and two washings were also collected and used for enzyme assay so as to determine enzyme content. The amount of urease immobilized by each mineral was then calculated by subtracting the amount of enzyme in solution from the amount of enzyme added.

5. Enzyme assay

One milliliter of the colloidal mixture or supernatant and 0.2 ml of toluene (to suppress microbial activity) were added to each test tube and swirled. Three ml of 0.2 $mol.ml^{-1}$ NaAc-HAc buffer solution and 1ml urea were added to keep the final concentration of urea at 15 $mmol.L^{-1}$. The enzyme reaction was stopped after 2 h of incubation at 37°C by addition of 5ml of 2.5 $mol.L^{-1}$ KCl-0.01%Ag_2SO_4 solution. Enzyme activity was expressed as the concentration of ammonium in solution ($\mu mol\ NH_4^+.\ ml^{-1}.\ 2\ h^{-1}$) which was determined on a spectrophotometer at 420 nm. The specific activity was defined as the units measured per milligram of protein. The pH-activity profiles of both free and immobilized enzyme were obtained by determining the specific activity of enzyme from pH 4.5 to pH 6.5. The thermal stability of urease immobilized on different oxides and kaolinite was examined by incubating each enzyme-mineral colloidal complex for 1h at 20, 40, 60, 70 or 80°C. The activity of the complex treated at elevated temperature was then measured. The kinetic properties of the free and adsorbed enzyme were determined with five different substrate concentrations of 2.5, 5, 10, 20, 20 or 25 $mmol.L^{-1}$. The parameters K_m and V_{max} were calculated according to the Hanes-Wolf equation.

All of the experimental runs were performed in triplicate.

RESULTS AND DISCUSSIONS

1. Adsorption of urease on examined minerals

The adsorption of urease on iron, aluminum oxides and kaolinite showed different isotherms (Figure 1). Goethite, bayerite and N-Al isotherms showed typical Langmuir models that can be described by the following equation: $x = x_m kc/(1+kc)$, where x is the amount of enzyme adsorbed per unit mass of mineral (mg/mg); x_m is the maximum adsorption of enzyme on mineral; c is the equilibrium concentration of enzyme in solution and k is a constant related to the adsorption energy. The adsorption parameters were calculated and listed in Table 2. The best-fitting isotherms for urease adsorbed on hematite and N-Fe were cubic curves. The maximum adsorption of urease on these two minerals was present in the range of 2.0–2.2 enzyme concentrations. For kaolinite, the adsorption isotherm of urease showed an exponential pattern.

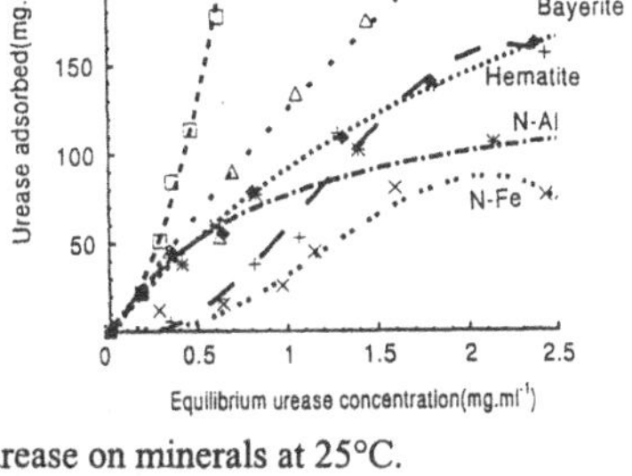

Figure 1. Adsorption isotherm of urease on minerals at 25°C.

Table 2 Langmuir and Freundlich parameters and correlation coefficients for the adsorption of urease on oxides (hydroxides) and kaolinite

Minerals	Langmuir equation			Freundlich equation		
	x_m	k	r	n	K	r
goethite	928	0.15	0.97	1.09	121.7	0.99
hematite				0.56	46.1	0.98
N-Fe				1.03	31.2	0.85
bayerite	361	0.34	0.94	1.36	85.6	0.99
N-Al	147	1.09	0.98	1.70	70.5	0.96
kaolinite				0.59	417.8	0.99

The adsorption isotherms also fitted the Freundlich equation: x=K$c^{1/n}$, where x and c are the same as in the Langmuir equation and K and n are constants that give estimates of the adsorptive capacity and intensity, respectively (Gianfreda *et al.*., 1992).

It can be observed from Figure 1 that the amount of enzyme held on examined minerals followed the sequence kaolinite> goethite > bayerite, hematite > N-Al > N-Fe, which had an enzyme concentration lower than 1.2 mg.ml^{-1}. Over the higher range of urease concentration (1.2-2.5 mg.ml^{-1}), the order for the amount of enzyme adsorption was as follows: kaolinite> goethite> bayeriteÅhematite> N-Al> N-Fe. A similar trend was found in the adsorption pattern of urease on minerals from pH 4.0 to 9.0; the trend can be described as kaolinite> hematiteÅgoethite> bayerite > N-Al> N-Fe (Fig.2). Within the pH range studied, the maximum adsorption of urease on all the oxides used was present around pH 5.0, which is close to the isoelectric point (p*I*) of the enzyme (Pinck and Allison,1961; Gianfreda *et al.*., 1992). This result was similar to those reported in studies on enzyme or protein adsorption on aluminum-silicate clays (Boyd and Mortland,1990; Huang *et al.*.,1995). With increasing or decreasing pH, there was a steady decrease in the amount of enzyme adsorbed on the examined oxides. For example, from pH 5.0 to 9.0, the ratio of enzyme adsorption decreased from 97.6% and 61.6% to 15.3% and 0 for goethite and N-Fe respectively. this suggested that urease is more easily adsorbed by iron and aluminum oxides near pH 5.0 in acidic soil environments. However, there was still a certain amount of urease adsorbed by examined oxides and kaolinite when the pH was above the p*I* of the enzyme and the PZC of minerals. This is attributed to the stronger effects of ligand exchange (Sepelyak *et al.*.,1984; Naidja *et al.*.,1997) and other factors such as saturating cations and steric hindrance (Naidja and Huang, 1996) on enzyme adsorption.

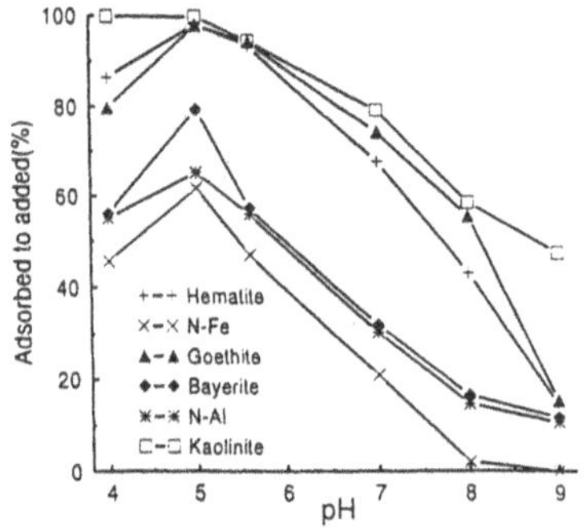

Figure 2. Percentage of adsorbed urease over a range of pH.

The different capabilities of minerals to adsorb urease is not easy to explain. There was no positive correlation between the surface area of minerals and the amount of enzyme adsorbed. For example, N-Fe and N-Al had surface area of 597 m^2.g^{-1} and 291 m^2.g^{-1}, respectively; both are higher than that of other oxides or kaolinite. The surface area of kaolinite (135 m^2.g^{-1}) was lower than those of goethite and hematite and higher than that of bayerite. However, the capacity for enzyme adsorption was in the order of kaolinite> crystalline oxides > non-crystalline oxides. It is known that the difference between kaolinite and oxide surfaces is that kaolinite has both hydrophobic and hydrophilic surfaces, while oxides have only a hydroxyl surface. The data obtained from our experiments imply that the siloxane hydrophobic surface of kaolinite may play a more important role in the adsorption of urease. A reasonable explanation for lower adsorption of enzyme on noncrystalline oxides is that the huge amounts of micropores on the surface of the oxides may not be available for the entry of high molecular weight organics, including protein molecules. This is different from the adsorption pattern of low weight molecules on clay minerals and oxides. The above-mentioned results suggest that kaolinite is probably the major determinant of the adsorption of some soil enzymes in acidic soil environments with the presence of very low content of humic substances. The crystalline oxides may be more important as the carriers of some enzymes compared with non-crystalline oxides.

2. Characteristics of immobilized enzyme

From Table 3 it can be observed that free and immobilized urease had different residual activities. The specific activity of adsorbed urease was in the following sequence hematite-U > N-Al-U > bayerite-U > goethite-U > N-Fe-U > kaolinite-U. The mean residual specific activities, as compared to that of free enzyme, were 95.3%, 62.0%, 57.3%, 54.9%, 47.6% and 37.7% for hematite-, N-Al-, bayerite-, goethite-, N-Fe- and kaolinite-urease complexes, respectively. The reduction of enzymatic activity for immobilized urease was in accordance with the findings of Boyd and Mortland (1990) and Gianfreda *et al.*. (1992). But a stronger inhibitory effect of non-crystalline aluminum hydroxide on urease activity (85%) than that of montmorillonite (29%) or Al(OH)x-montmorillonite complex (36%) was found by Gianfreda *et al.*.(1992).

Table 3 The specific activity of free enzyme and mineral-enzyme complexes

Complexes	Specific activity*	% of Free enzyme
free enzyme	28440	
hematite-U	27091	95.3
N-Al-U	17627	62.0
bayerite-U	16308	57.3
goethite-U	15626	54.9
N-Fe-U	13554	47.7
kaolinite-U	10726	37.7

*$\mu mol\ NH_4^+ .ml^{-1}.2h^{-1}.mg^{-1}$

The pH-activity profiles of adsorbed urease were not significantly different from those of free enzyme. Figure 3 shows the dependence of enzyme activity on pH. The activity of free urease reached its maximum at pH 5.5 and considerably decreased as pH decreased to 4.5 or increased to 6.5. The maximum activity of urease held on all the examined minerals was also at pH 5.5. But the changes in enzyme activity with pH variation for adsorbed urease were not so marked as that of free urease. These results were similar to those reported by Gianfreda *et al.*. (1992) who found that the urease immobilized by montmorillonite and Al(OH)x-montmorillonite complex was less sensitive to pH changes. Figure3 also indicates that throughout the pH range of 4.5 to 6.5, the specific activity of immobilized urease was in the following order: hematite-U, N-Al-U, bayerite-U, goetite-U, N-Fe-U and kaolinite-U. This is in agreement with the sequence of the specific activity of urease-mineral complexes in unbuffered systems.

The real reasons for the different specific activities displayed by the immobilized urease on examined minerals are not very clear. But based on our present data, it is certain that in acidic soil environments urease molecules showed remarkable denaturation when immobilized on these inorganic soil components. As the most important enzyme adsorbent and inhibitor, kaolinite may play a significant role in determining the overall activity of soil urease enzyme. In agricultural practices in acid soils, more attention should be paid to the adjustment of the activity of enzymes immobilized by the kaolinite fraction. Moreover, in view of the different specific activities of adsorbed urease on crystalline and noncrystalline oxides, it is important to notice the transformation of the forms of oxides influenced by environmental factors in natural soils.

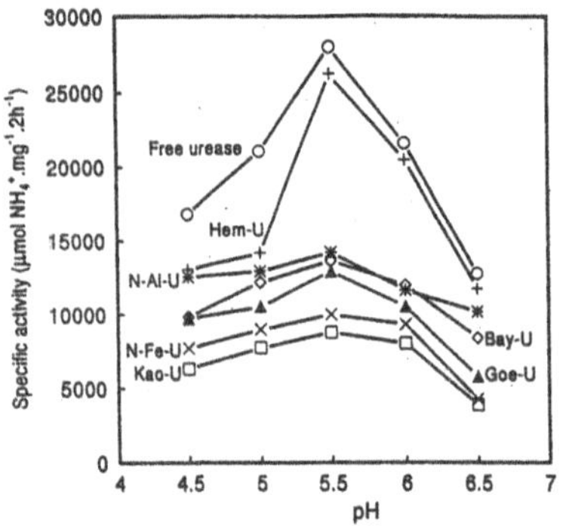

Figure 3. The pH-activity profiles of urease-mineral complexes.

The stability of urease-mineral complexes toward thermal denaturation was compared to that of free enzyme at 40~80°C (Figure 4). The immobilized urease showed different stabilities to elevated temperatures. For example, the ratios of the residual enzyme activities at 60°C to those at 20°C were 78.0%, 97.0%, 88.2%, 86.5%, 62.8%, 52.4% and 48.7% for free urease, hematite-U, bayerite-U, goethite-U, N-Al-U, N-Fe-U and kaolinite-U mixtures, respectively. This demonstrated that the urease immobilized on hematite, bayerite and goethite was more stable to temperature as compared with free enzyme. However, the urease held on N-Fe, N-Al and kaolinite was more sensitive than free enzyme, which agrees wiith the observations reported by Boyd and Mortland (1990) and Gianfreda *et al.*.(1992). It is especially worth mentioning that urease-hematite complex had the greatest specific activity and the highest thermal stability among all the enzyme-mineral mixtures. The hematite-enzyme complex was prepared in unbuffered system. The pH of hematite was 7.75, which is higher than the PZC of hematite. The binding of urease with hematite was no doubt by nonelectrostatic force. Our results indicate that this kind of adsorption of urease on hematite did not result in much enzyme deactivation and was also quite stable to higher temperature. Similar results were reported by Quiquampoix (1987) who observed very little deactivation of (β-D-glucosidase adsorbed on montmorillonite and kaolinite by nonelectrostatic force. If this is true for the complexes of hematite with other enzymes that have some significance in the degradation and bioremediation of environmental pollutants, hematite would be of great value for further study.

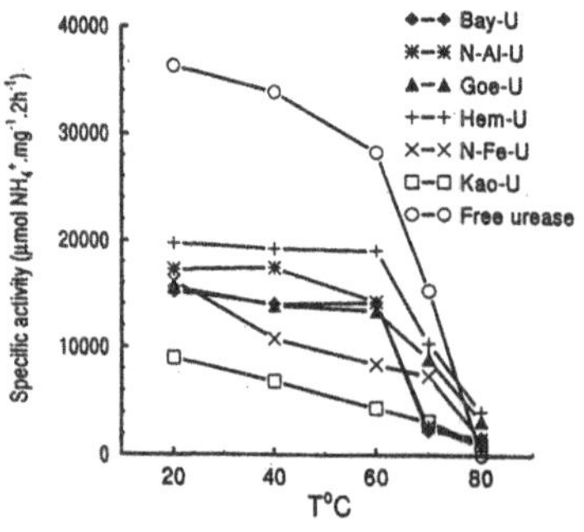

Figure 4. The thermal stability of urease-mineral complexes.

Urease, both free and immobilized on examined minerals, obeys Michaelis-Menten kinetics. The kinetic parameters calculated by the Hanes-Wolf equation are listed in Table 4. The K_m values of adsorbed enzyme were greater than that of free enzyme and the V_{max} values of adsorbed enzyme were always less than that of free enzyme. This indicates that lower affinity of urease to the substrate was obtained after immobilization on iron, aluminum oxides and kaolinite. However, a decrease of the K_m value of urease-clay complexes compared to free enzyme was observed by Gianfreda *et al.*.(1992). Table 4 also shows that the urease held on oxides or hydroxides displayed a greater catalytic efficiency than the enzyme immobilized on kaolinite, but less than free enzyme, as evaluated by the V_{max}/K_m ratio.

This study may give some useful information for the understanding of enzyme kinetics in soils with variable charges. However, the work regarding the interactions between oxides-humus mixtures and soil enzymes deserves close attention because it is that these minerals are ubiquitously associated with various kinds of organics especially humic substances.

Table 4 The kinetic parameters of urease in different mineral systems

Complexes	K_m	V_{max}	V_{max}/K_m
goethite-U	17.29	7.78	0.450
hematite-U	14.86	6.29	0.423
N-Fe-U	19.27	8.17	0.424
bayerite-U	9.81	5.72	0.583
N-Al-U	14.61	6.57	0.449
kaolinite-U	20.26	4.38	0.216
Free urease	9.34	9.82	1.051

K_m: $mmol.L^{-1}$, V_{max}: mmol $NH_4^+.ml^{-1}.2\ h^{-1}$

CONCLUSIONS

1. The amount of urease adsorbed on different minerals was in the following sequence: kaolinite > crystalline oxides (hydroxides) > non-crystalline oxides (hydroxides).

2. The specific activity of urease immobilized on iron and aluminum oxides (hydroxides) was greater than that on kaolinite. The highest specific activity and thermal stability of adsorbed urease were found in the hematite system.

3. The pH-activity profiles of adsorbed urease were similar to that of free enzyme. The thermal stability of urease-mineral complexes was in the following order: crystalline oxides (hydroxides) > free urease > non-crystalline oxides (hydroxides).

4. The urease immobilized on examined minerals obeyed Michaelis-Menten kinetics and showed decreased V_{max} and increased K_m values as compared with free enzyme.

5. The study of the influence of soil oxide (hydroxide)-humus complexes on enzyme activity and kinetics deserves further attention.

ACKNOWLEDGMENT

The research was carried out with the financial support of the International Foundation for Science (IFS). Project No. C/2527-1.

REFERENCES

Baruah M. and R.R.Mishra, 1984. Dehydrogenase and urease activities in rice-field soils. Soil Biol. Biochem., 16,423-424.

Boyd S.A. and M.M. Mortland, 1990. Enzyme interactions with clays and clay-organic matter complex. In: J. M. Bollag and G. Stotzky (Editors), Soil Biochemistry, Vol.6. Marcel Dekker, New York. pp.1-28.

Burns R.G., A.H.Pukite and A.D.McLaren, 1972. Concerning the location and persistence of soil urease. Soil Sci. Soc. Am. Proc., 36,308-311.

Burns R.G., 1986. Interaction of enzymes with soil mineral and organic colloids. In: P.M. Huang and M. Schnitzer (Editors), Interaction of Soil Minerals with Natural Organics and Microbes. Soil Sci. Soc. Amer., Madison, pp.429-451.

Carter D.L., M.D.Heilman and C.L.Gonzalez, 1965. Ethylene glycol monoethyl ether for determining surface area of silicate minerals. Soil Sci., 100, 356-360.

Ceccanti B., P.Nannipieri, S.Cervelli and P.Sequi, 1978. Fractionation of humus-urease complexes. Soil Biol. Biochem., 10, 39-45.

Dick W.A. and M.A. Tabatabai, 1987. Kinetics and activities of phosphatase-clay complexes. Soil Sci., 143, 5-14.

Dkhar M.S. and R.R.Mishra, 1983. Dehydrogenase and urease activities of maise (*Zea mays* L.) field soils. Plant & Soil, 70, 327-333.

Fusi P., G. G. Ristori, L. Calamai and G.Stotzky, 1989. Adsorption and binding of protein on "clean" (homoinic) and "dirty" (coated with oxyhydroxides) montmorillonite, illite and kaolinite. Soil Biol. Biochem., 21, 911-920.

Garwood G.A.; M.M. Mortland and T.J. Pinnavaia, 1983. Immobilization of glucose oxidase on montmorillonite clay: hydrophobic and ionic modes of binding. J. Mol. Catal., 22, 153-163.

Gianfreda L., M.A. Rao and A. Violante, 1991. Invertase (β-Fructosidase): effects of montmorillonite, Al-hydroxide and Al(OH)x-montmorillonite complex on activity and kinetic properties. Soil Biol. Biochem., 23, 581-587.

Gianfreda L., M.A.Rao and A.Violante, 1992. Adsorption, activity and kinetic properties of urease on montmorillonite, aluminum hydroxide and Al(OH)x-montmorillonite complexes. Soil Boil. Biochem., 24, 51-58.

Gianfreda L., M.A. Rao and A. Violante, 1993. Interactions of invertase with tannic acid, hydroxy-aluminium (OH-Al) species or montmorillonite. Soil Boil. Biochem., 25, 671-677.

Gianfreda L., F.Sannino, N.Ortega and P.Nannipieri, 1994. Activity of free and immobilized urease in soil: effects of pesticides. Soil Biol. Biochem., 26, 777-784.

Gianfreda L., A.D. Cristofaro; M.A. Rao and A.Violante, 1995a. Kinetic behavior of synthetic organo-and organo-mineral-urease complexes. Soil Sci. Soc.Am. J., 59, 811-815.

Gianfreda L., M.A.Rao and A.Violante, 1995b. Formation and activity of urease-tannate complexes affected by aluminum, iron, and manganese. Soil Sci. Soc.Am. J., 59, 805-810.

Huang P.M., 1990. Role of soil minerals in transformations of natural organics and xenobiotics in soils. In: J.-M.Bollag and G.Stotzky (Editors). Soil Biochemistry, Vol.6. Marcel Dekker, New York, pp.29-96.

Huang Q., H. Shindo and T.B.Goh, 1995. Adsorption, activities and kinetics of acid phosphatase as influenced by montmorillonite with different interlayer material. Soil Sci.,159, 271-278.

Huang Q. and X. Li, 1995. Influences of clay mineral and organic matter on the activities of soil enzymes, Progress in Soil Sci., 23, 12-18.

Klein T.M. and J.S.Koths, 1980. Urease, protease and acid phosphatase in soil continuously cropped to corn by conventional or no-tillage methods. Soil Biol. Biochem., 12, 293-294.

Lai C.M. and M.A.Tabatabai, 1992. Kinetic parameters of immobilized urease. Soil Biol. Biochem., 24, 225-228.

Lethbridge G. and R.G. Burns, 1976. Inhibition of soil urease by organo-phosphorus insecticides. Soil Biol. Biochem., 8, 99-102.

Lethbridge G., A.T. Bull and R.G. Burns, 1981. Effects of pesticides on 1,3-β-glucanase and urease activities in soil in the presence and absence of fertilisers, lime and organic materials. Pesticide Sci., 12, 147-155.

McLaren A.D., G.H. Peterson and I. Barshad, 1958. The adsorption and reactions of enzyme and proteins in clay materials. IV. Kaolinite and montmorillonite. Soil Sci. Soc. Am. Proc., 22, 239-244.

Makboul H.E. and J.C.G. Ottow, 1979. Clay minerals and the michaelis constant of urease. Soil Biol. Biochem., 11, 683-686.

Naidja A. and P.M. Huang, 1996. Deamination of aspartic acid by aspartase-Ca-montmorillonite complex. J. Mol. Cat., 106, 255-265.

Naidja A., P.M. Huang and J.-M. Bollag, 1997. Activity of tyrosinase immobilized on hydroxyaluminum-montmorillonite complexes. J. Mol. Cat., 115, 305-316.

Paulson K.N. and L.T. Kurtz, 1969. Locus of urease activity in soil. Soil Sci. Soc. Am. Proc., 33, 897-901.

Pettit N.M, A.R.J. Smith, R.B.Freedman and R.G.Burns, 1976. Soil urease: activity, stability and kinetic properties. Soil Biol.Biochem., 8, 479-484.

Pinck L.A. and F. E. Allison, 1961. Adsorption and release of urease by and from clay minerals. Soil Sci., 183-188.

Quiquampoix H., 1987. A stepwise approach to the understanding of extracellular enzyme activity in soil I: Effect of electrostatic interaction on the conformation of a β-D-glucosidase adsorbed on different mineral surfaces. Biochimie, 69, 753-763.

Rao D.L.N. and S.K. Ghai, 1985. Urease and dehydrogenase activity of alkali and reclaimed soils. Aust. J. Soil Res., 23, 661-665.

Sepelyak R.J., J.R. Feldkamp, T.E. Moody, J.L. White and S.L.Hem, 1984. Adsorption of pepsin by aluminum hydroxide I: Adsorption mechanism. J. Pharm. Sci., 73, 1514-1517.

Tabatabai M.A., 1977. Effects of trace elements on urease activity in soils. Soil Biol. Biochem., 9, 9-13.

Tabatabai M.A. and J.M.Bremner, 1972. Assay of urease activity in soils. Soil Biol.Biochem., 4, 479-487.

Xiong Y. (Editor), 1985. Soil Colloids,Vol. 2, Science Press, Beijing.

THE FATE OF ACID PHOSPHATASE IN THE PRESENCE OF PHENOLIC SUBSTANCES, BIOTIC AND ABIOTIC CATALYSTS

Maria Antonietta Rao, Antonio Violante and Liliana Gianfreda*

Dipartimento di Scienze Chimico-Agrarie,
Università di Napoli "Federico II"
Via Università 100, Portici, (Napoli), Italy

1. INTRODUCTION

Once secreted in soil, extracellular enzymes may undergo different fates, depending on reactive phenomena in which they are involved. Under the adverse conditions of soil environment, enzymes still exhibit a catalytic activity, if their conformation and catalytic sites are protected by a stabilization mechanism. The immobilization on different supports such as inorganic and organic soil components may improve the stability of enzymes and preserve their activity (Burns, 1986).

Phoshatases have a great importance in the metabolism of organic phospho-compounds in soil. They allow the organic phospho-fractions to produce inorganic phosphate, which is the only form available to plant roots and soil microorganisms.

One of the most common process occurring in soil is the formation of humus-like substances. Phenolic compounds, possibly present in soil as wastes, pesticide derivatives or degradation products, my undergo oxidation and polymerization processes and produce extended polymers or complexes with other inorganic and organic soil components. In particular, if active enzymatic proteins are involved, catalytically active aggregates may form.

The oxidative transformation of phenolic substances can be catalyzed by biotic catalysts such as extracellular phenoloxidases including laccases, peroxidases or tyrosinases (Bollag, 1992). Inorganic components such as clay minerals, Fe, Mn or Al oxides and short-range-ordered minerals may promote the polymerization of phenolic substances, as well (Huang, 1990).

As recently demonstrated by Pal *et al.* (1994), biotic and abiotic catalysts may differ in their catalytic behavior and function through different reaction mechanisms.

The purpose of this paper was to investigate the interaction of a phosphatase and two phenolic substances and the formation of active enzymatic phenolic copolymers in the presence of biotic and abiotic catalysts.

Effect of Mineral-Organic-Microorganism Interactions on Soil and Freshwater Environments
Edited by Berthelin, *et al*, Kluwer Academic / Plenum Publishers, New York, 1999

2. MATERIALS AND METHODS

2.1. Chemicals

Acid phosphatase (P) (EC. 3.1.3.2, from sweet potato, MW 100 kDal, Type I, 60 U mg^{-1}) was purchased from Boehringer Mannheim, Germany. Peroxidase (POD) (EC. 1.11.1.7 from horseradish, Type VI-A, 1100 U mg^{-1}) was from Sigma Chemical Co. St. Louis, MO. Pyrogallol (Pyr) and tannic acid (Tan) (MW 1701.23) were from Fluka AG, CH. A commercial MnO_2 preparation, showing x-ray diffraction peaks characteristic of pyrolusite, was used as abiotic catalyst. All other chemicals were reagent grade and were supplied by Analar, BDH, Ltd Poole, UK.

2.2. Activity assay

The activities of free and phenolic-phosphatase mixtures were measured by incubating at 10°C suitable volumes of free enzyme solutions and immobilized enzyme suspensions (generally 0.030 or 0.050 mL) with 1 mL of 6 mM *p*-nitrophenylphosphate (*p*NPP) in 0.1 M Na-acetate buffer at pH 5.0. After incubation, 1 mL of 1 M NaOH was added and the concentration of *p*-nitrophenol, product of the hydrolytic reaction, was determined directly with the spectrophotometer (Perkin Elmer Lambda 3B) by the absorbance at 405 nm (molar absorption coefficient 18.5 cm^{-1} mM^{-1}). The enzymatic activity unit was expressed in katal, i.e. the moles of *p*-nitrophenol produced by 1 mL of free and immobilized enzyme solution over 1 sec. at 10°C and at pH 5.0. The specific activity was expressed as the units measured per mg^{-1} of protein.

2.3. Phosphatase-phenol interaction studies

To assess the effect of phenols on phosphatase activity, mixtures containing 0.161 mg mL^{-1} of phosphatase and phenol concentrations ranging from 0 to 20 mM for pyrogallol and from 0 to 0.15 mM for tannic acid in 0.05 M Na-acetate buffer at pH 5.4 were incubated at 10°C. The mixing time was assumed as zero incubation time. At different time intervals, suitable aliquots of the mixtures were withdrawn and their residual activity was determined as stated above.

2.4. Preparation of phosphatase-phenolic copolymers

Enzyme-phenolic copolymers were prepared according to the procedure described in detail by Badalucco *et al.* (1993). Usually, 20 ml of 0.05 mM acetate buffer at pH 5.4, containing suitable amounts of phenols, and 20 ml of H_2O_2 (1% v/w) were simultaneously added to 2 ml of buffered peroxidase (0.1 mg mL^{-1}) solution (biotic catalysis) or to 20 mg of MnO_2 (abiotic catalysis). The addition lasted 24 min. After 3 min from the mixing of the solutions, 50 mL of acid phosphatase (3.5 mg mL^{-1}) were added every 3 min for a total of 2 mL in 2 h. The suspensions were kept at 10°C. After 4 h-incubation, the suspensions, previously tested for their residual activity, were centrifuged at 10,000 *g* for 10 min and the absorbances at 472 and 664 nm of the supernatants were measured. The E_4/E_6 ratio was calculated by dividing the absorbance values. The colour of the suspensions was also registered. Experiments without catalysts were carried out, as well.

Phosphatase-phenolic copolymers obtained by biotic and abiotic catalysis were subjected to an ultrafiltration procedure. Usually, a suitable volume of each suspension was injected into a stirred, plane membrane ultrafiltration (UF) cell, equipped with an ultrafiltration membrane whose molecular weight cut-off was 300 kDa. The apparatus was continuously fed with buffer in order to keep constant the volume within the cell. Permeate samples were collected by means of an automatic fraction collector and assayed for their residual enzymatic activity. Enzymatic activity tests were performed also on the solution accumulating within the cell, upstream of the membrane

All experiments were carried out at least in triplicate.

3. RESULTS

Figure 1 shows the residual activity of acid phosphatase in the presence of 10 mM pyrogallol and 0.7 mM tannic acid after 4 and 24 h-contact time. As typical enzymatic inhibition studies showed that tannic acid displayed a detectable inhibition on phosphatase activity (Rao *et al.*, 1996), a lower concentration of tannic acid was used in the experiments. However, considering the molecular structure of two compounds, comparable concentrations of reactive phenolic groups were utilized. The activity of the enzyme decreased by about 40 and 54% after 4 and 24 h of contact with pyrogallol. No detectable variation of the pH was measured. On the contrary, a much higher reduction (more than 80%) was observed with tannic acid ever after 4 h and then it remained almost constant with the time. Furthermore, pyrogallol-phosphatase (Pyr-P) mixtures showed no colour, whereas tannic acid-phosphatase (Tan-P) displayed a light pink colour and a E_4/E_6 value of 11 (Table 1).

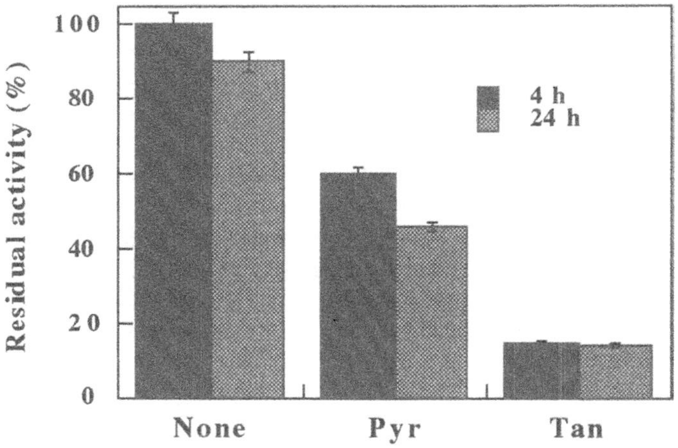

Figure 1. Effect of incubation time on the residual activity of pyrogallol- and tannic acid-acid phosphatase suspensions.

When the incubation was carried out in the presence of both the catalysts, a further decrease of activity and a browning of the mixtures occurred (Table 1).

With pyrogallol, the lowest activity as well as the strongest browning were observed in the presence of the biotic catalyst POD. The activities of copolymers obtained with pyrogallol (Pyr-POD-P) decreased to 15%, while the colour became red-brown. In addition, the E_4/E_6 ratio decreased as respect to that determined without catalyst or in the presence of MnO_2 (Table 1).

Table 1 : Residual activity and characteristics of phenolic-phosphatase copolymers obtained without and with biotic (POD) and abiotic (MnO_2) catalytsts.

	Residual activity (%)			E_4/E_6 ratio			Colour		
Compound	No catalyst	POD	MnO_2	No catalyst	POD	MnO2	No catalyst	POD	MnO2
Pyrogallol	60	15	51	–	7	16	colourless	red brown	yellow
Tannic acid	15	17	4	11	20	6	light pink	light brown	dark brown

An opposite behaviour was observed with tannic acid. The activity of tannic acid-phosphatase copolymers obtained in the presence of MnO_2 (Tan-MnO_2-P) decreased up to 4%, whereas no further reduction was observed for those with POD (Tan-POD-P mixtures). The preparation became brown and a lower E_4/E_6 ratio was measured (Table 1).

Activity assays performed on samples of Pyr-POD-P and Tan-POD-P mixtures after ultrafiltration showed that phosphatase activity was completely retained upstream of the membrane (i.e. it remained within the UF cell) for the Pyr-POD-P samples. On the contrary, positive activity tests were found in the permeate solution for the tannic acid-derivatives.

When similar experiments were carried out with the samples obtained in the presence of the inorganic catalyst MnO_2, a complete and opposite behaviour was observed. Indeed, enzyme activity was present within the UF cell for the Tan-MnO_2-P samples, whereas it was found in the permeate solution with Pyr-MnO_2-P.

4. DISCUSSION

The results shown in Figure 1 indicate that phosphatase interacted much more with tannic acid than with pyrogallol. In previous papers, Gianfreda *et al.* (1993, 1995) demonstrated that tannic acid gave rise to stable interactions with urease and invertase, as well. However, the association of phosphatase with tannic acid produced complexes with higher activity levels than with urease and invertase (Rao *et al.* 1996). Usually, phenolic substances and proteins may be involved in different phenomena. They imply the direct interaction of the two substances through their active sites as well as the polymerization of phenolic substances and the involvement of enzymatic molecules. Various phenolic-enzyme complexes with different enzymatic residual activities may be produced.

The data shown in Table 1 indicate that in the absence of catalysts pyrogallol was not able to form polymeric aggregates with the enzyme, whereas some polymeric aggregates probably formed with tannic acid. In fact, the pyrogallol-suspensions showed no colour nor a value for the E_4/E_6 ratio was measured. As reported by Chen *et al.* (1977), this latter parameter may provide indicative information on the size of humic-like substances. The higher/lower the value, the smaller/larger the size of humic substances.

When the incubation was, instead, performed in the presence of peroxidase, the polymerization of pyrogallol was strongly promoted: the suspensions became red brown and an E_4/E_6 ratio of 7 was measured. By contrast, a more efficient effect was demonstrated by the abiotic catalyst on tannic acid (Table 1). In both the cases, larger enzyme-phenolic copolymers formed, as demonstrated by the browning as well as the decrease of E_4/E_6 ratios of the mixtures.

These conclusions are partly confirmed by ultrafiltration results. They suggest that active enzymatic aggregates, characterized by a molecular weight higher than 300 kDa, were formed with pyrogallol and tannic acid in the presence of POD and MnO_2, respectively. Consequently, they were rejected by the UF membrane and remained within the cell. On the contrary, Pyr-MnO_2-P and Tan-POD-P associations were characterized by a size lower than 300 kDal and passed through the UF membrane in the permeate solution.

The decrease of the enzyme activity, measured for all the enzymatic aggregates, could be explained by the entrapment of enzyme molecules within phenolic polymers, or by their association through one or more sites on the protein surface to many phenolic molecules, or a direct inhibition of the bound enzyme by some of the aromatic constituents (Burns, 1986, a review). As a consequence, active enzymatic sites were masked to the substrate access and a reduction of the activity occurred. Moreover, a partial deactivation of protein conformation during the complexation process could not be excluded.

5. ACKNOWLEDGMENTS

DISCA Publication no. 146.

Research supported by National Research Council of Italy (CNR) and MURST - Research Programme 40%.

6. REFERENCES

Badalucco, L., A.M. Garzillo, F. De Cesare, M. De Bianchi., S. Grego and V. Buonocore, 1993. Sintesi e caratterizzazione di un copolimero umo-simile resorcinolo-fosfatasi acida e suo possibile impiego nel suolo. Proceedings of XI National Symposium of Società di Chimica Agraria, Cremona - Italy, September 21-24, 1993, pp. 387-392

Bollag, J.-M., 1992. Decontaminating soil with enzymes. Environ. Sci. Technol., 26, 1876-1881.

Burns, R. G., 1986. Interaction of enzymes with soil mineral and organic colloids. In: P.M. Huang and M. Schnitzer (Editors) Interactions of Soil Minerals with Natural Organic and Microbes Soil Science Society of America, Madison, pp. 429-451.

Chen, Y., N. Senesi and M. Schnitzer, 1977. Information provided on humic substances by E_4/E_6 ratios. Soil Sci. Soc. Am. J., 41, 351-358.

Huang, P.M., 1990. Role of soil minerals in transformations of natural organics and xenobiotics in soil. In: J.-M. Bollag and G. Stozky (Editors) Soil Biochemistry Vol. 6 Marcel Dekker, New York, pp. 29-115.

Gianfreda, L., M.A. Rao, and A. Violante, 1993. Interaction of invertase with tannic acid, hydroxy-aluminum (OH-Al) species or montmorillonite. Soil Biol. Biochem., 25, 671-677.

Gianfreda, L., M.A. Rao, and A. Violante, 1995. Formation and activity of urease-tannate complexes as affected by Al, Fe, and Mn. Soil. Sci. Soc. Am. J., 59, 805-810.

Pal, S., J.-M. Bollag, and P.M. Huang, 1994. Role of abiotic and biotic catalysts in the transformation of phenolic compounds through oxidative coupling reactions. Soil. Biol. Biochem., 26, 813-820.

Rao, M.A., L. Gianfreda, F. Palmiero, and A. Violante, 1996. Interactions of acid phosphatase with clays, organic molecules and organo-mineral complexes. Soil Sci., 161, 1-10.

KINETICS OF CATECHOL OXIDATION CATALYZED BY TYROSINASE OR δ-MnO_2

Abdallah Naidja[1], P.M. Huang[1], Jerzy Dec[2], and Jean-Marc Bollag[2]

[1]Department of Soil Science, University of Saskatchewan, 51 Campus Drive, Saskatoon, SK S7K 5A8 Canada

[2]Laboratory of Soil Biochemistry, Center for Bioremediation and Detoxification, The Pennsylvania State University, University Park, PA 16802 USA

1. INTRODUCTION

The enormous catalytic potential of enzymes was recognized in the beginning of this century. Ever since, extensive studies have been carried out to elucidate the principles of biological catalysis. The major research areas included the reaction kinetics, the structural and physico-chemical characteristics, and the specificity of enzymatic preparations (Dixon and Weeb, 1979). To facilitate the evaluation of complex enzymatic systems, chemical models have been developed to mimic enzyme activity. Such a biomimetic approach has been applied to model the oxidation reactions which are controlled by catalytic systems involving enzymes of high selectivity (Matsuura, 1977).

Recently, it has been shown that tyrosinase (EC 1. 14. 18. 1), a copper-containing polyphenol-oxidizing enzyme, and birnessite (δ-MnO_2) catalyze oxidative coupling of phenolic compounds resulting in the formation of similar polymeric products (Bollag *et al.*, 1995, Naidja *et al.*, 1998). Tyrosinase has an important role in natural synthesis of humic substances (Sjoblad and Bollag, 1981). It is widely distributed in plants and was found responsible for undesirable melanosis and blackening of agricultural products (Chen *et al.*, 1991). The tyrosinase-induced black melanin or eumelanin can be used as UV absorbers and cation exchange resins (Zawistowski *et al.*, 1991). Birnessite, on the other hand, is one of the most common forms of mineralized manganese in soils (McKenzie, 1971). Like tyrosinase, it has the ability to catalyze the polymerization of phenolic substrates (Naidja *et al.*, 1998), and is involved in humus formation (Shindo and Huang, 1982; McBride, 1989; Bollag *et al.*, 1995). In the presence of birnessite, toxic chlorophenols are polymerized with release of chloride ions (Pizzigallo *et al.*, 1995).

As shown by ESR, the oxidation of catechol mediated by δ-MnO_2 proceeds through the formation of semiquinone radicals (McBride, 1989). With tyrosinase, however, no semiquinone radicals are formed and the electron transfer is controlled by the binuclear copper active site (Himmelwright, 1980). The above difference in the kinetics and mechanisms of the oxidation has never been quantified. In order to fill this gap, the present study focused on comparing the reaction kinetics of the two pathways. The experimental objectives were to determine the effect of temperature on the initial rates of catechol oxidation catalyzed by

Effect of Mineral-Organic-Microorganism Interactions on Soil and Freshwater Environments
Edited by Berthelin, *et al*, Kluwer Academic / Flenum Publishers, New York, 1999

tyrosinase and δ-MnO_2 and to compare their activation energies.

2. MATERIAL AND METHODS

2.1. Substrate and enzyme

Catechol (1,2-dihydroxybenzene) and mushroom tyrosinase (EC 1. 14. 18. 1) with an activity of 2100 units/mg were purchased from Sigma Chemical Company (St. Louis, MO). One unit of tyrosinase activity is defined as the amount of enzyme that causes an increase in absorbance at 280 nm of 0.001/min at pH 6.5 and 25°C in 3 mL reaction mixture containing 0.33 mM L-tyrosine. The enzyme was stored in a desiccated state at - 5°C. Before use, it was dissolved in deionized distilled water.

2.2. Birnessite

Birnessite was prepared from potassium permanganate according to McKenzie (1971). The < 2 µm fraction of the mineral was separated by sedimentation and then freeze-dried. The specific surface area of this fraction, determined by BET method, was 128 m^2/g, representing the sum of mesopore (108 m^2/g) and micropore (20 m^2/g) surfaces; the total surface area determined by EGME (Eltantawy and Arnold, 1973) was 273 m^2/g. Before use, birnessite was dispersed in deionized distilled water by ultrasonication (3 min at 60 W), and the suspension pH was adjusted to 6.0 with 0.1 M NaOH, using a Brinkmann 672 titroprocessor.

2.3. Measurement of oxygen consumption as a function of time

The experiments were conducted at various temperatures (10-60°C) in 82-mL sterilized glass vessels which, in the case of enzyme-mediated reactions, were filled with 11.72 mL of 35.0 mM catechol solution, 69.04 mL of deionized distilled water, and 0.08 mL of toluene. The reaction was started by adding 1 mL of tyrosinase (0.148 mg) solution. During the reaction, a total volume of 0.16 mL of 0.02 M NaOH solution was added using the titroprocessor to adjust the pH to 6.0. In the case of birnessite-mediated reactions, the reaction vessels were filled with 11.72 mL of 35.0 mM catechol solution, 68.80 mL of deionized distilled water, and 0.08 mL of toluene; subsequently, 1 mL of δ-MnO_2 (2.0 mg) suspension was added to start the reaction. The pH of the reaction mixture was maintained at 6.0 by the addition of 0.02 M HCl at a total volume of 0.40 mL.

The oxygen consumption (µmol O_2) was monitored as a function of time by a membrane-covered Clark-type polarographic oxygen sensor (Naidja *et al.*, 1997). The measurements of oxygen consumption in both tyrosinase and δ-MnO_2 systems were the average of two replicates. There was no oxygen consumption in the enzyme solution or MnO_2 suspension in the absence of catechol during the reaction period studied. The small amount of the oxygen consumed in the catechol solution in the absence of catalysts (blank) was subtracted from that of the catechol-tyrosinase or the catechol-birnessite systems.

The activity of both catalysts at various temperatures (10-60°C) was determined by the initial reaction rates from the slope of the initial linear part of the curve (Allison and Purich, 1996) of the oxygen consumption as a function of time.

3. RESULTS AND DISCUSSION

The oxidation of catechol to melanins (dark color humic-like polymers) in the presence of tyrosinase (Himmelwright, 1980) and δ-MnO_2 (McBride, 1989) is controlled by different mechanisms :

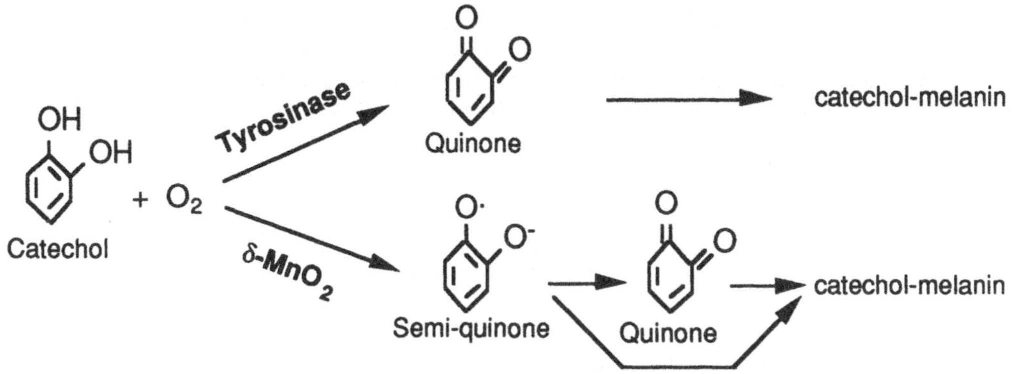

The purpose of this study was to determine how the initial rates of the oxygen consumption were influenced by each catalyst. The initial rates were obtained from : (1) the measurement of the slope of the initial linear part of the kinetic curve using the best fitting with a least square regression program (Cricket Graph III), and (2) the mathematical determination of the slope of the tangent to the kinetic curve of oxygen consumption as a function of time at the origin ($t = 0$). In the latter method, a least square regression program was also used to fit the experimental results in a mathematical equation. The most appropriate fitting was a polynomial equation which is continuous, differentiable, easy to evaluate for all values of its argument, and thus, in many respects the best-behaved of all mathematical functions (Goult, 1973) :

$$nO_2 = \sum_{i=0}^{m} a_i t^i \qquad (1)$$

where nO_2 is the number of μmoles of oxygen consumed, m is the degree of the polynomial, a_i are the polynomial coefficients, and t is the reaction time. The slope of the tangent to the curve at the origin is given by the first derivative at $t = 0$, and thus, can be written as :

$$\left(\frac{dnO_2}{dt}\right)_{t=0} = \left(\sum_{i=1}^{m} ia_i t^{i-1}\right)_{t=0} = \mathbf{a_1} \qquad (2)$$

For enzymatic or abiotic oxidations, the experimental data in this study were best fitted in a polynomial of degree 4 :

$$nO_2 = a_0 + \mathbf{a_1}t^1 + a_2t^2 + a_3t^3 + a_4t^4 \qquad (3)$$

with the r^2 values higher than 0.999 and the p values lower than 2×10^{-16} (Figure 1). The first derivative of equation (3) at $t = 0$ is :

$$\left(\frac{dnO_2}{dt}\right)_{t=0} = \left(\mathbf{a_1} + 2a_2t^1 + 3a_3t^2 + 4a_4t^3\right)_{t=0} = \mathbf{a_1} \qquad (4)$$

Thus, the initial rate (k) is given by the slope $\mathbf{a_1}$ of the tangent to the curves at $t = 0$ (Table 1).

The fitting in polynomials of the same fourth degree of the kinetic curves of oxygen consumption at different temperatures up to 30°C for both catalysts indicated a striking similarity in the kinetics behavior of tyrosinase and δ-MnO_2 (Figure 1). The initial rate of the oxygen consumption was about three to four times higher in the presence of tyrosinase than in the presence of δ-MnO_2 (Table 1). The results showed that initial rates calculated from the tangent to the curves $nO_2 = f(t)$ fitted in polynomials were slightly higher than those calculated from the initial linear part of the kinetic curves for both catalysts, indicating that the calculation of the initial rate from the linear part of the kinetic curves was somewhat underestimated.

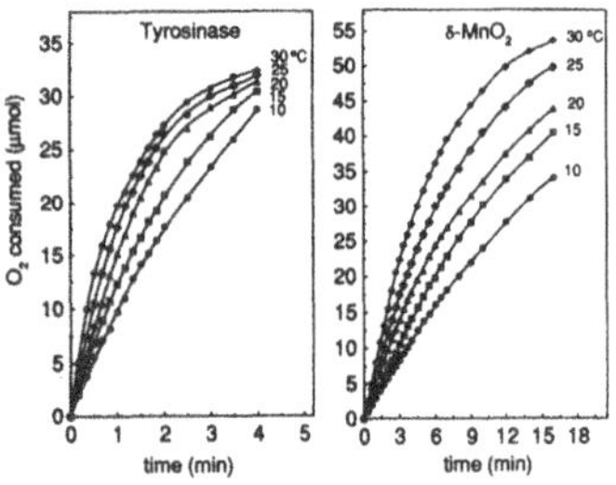

Figure 1. Time function of the oxygen consumption for the transformation of catechol catalyzed by tyrosinase (0.148 mg) or δ-MnO_2 (2.0 mg) at an initial pH of 6.0 and different temperatures. The concentration of catechol was 5.0 mM and the final volume of the solution was 82.0 mL. For all fitted curves, the r^2 values were higher than 0.999 and the p values were lower than 2.0×10^{-16}. Different scales were used for tyrosinase and δ-MnO_2 for clarity of presentation.

Table 1 : Initial rate (k) of the oxidation of catechol catalyzed by tyrosinase or δ-MnO_2.

Temperature (°C)	Tyrosinase		δ-MnO_2	
	a	b	a	b
	k (μmol O_2 consumed min^{-1})[c]			
10	9.60	10.92	3.36	3.06
20	16.74	17.46	4.68	5.34
30	26.64	31.50	7.92	9.42

[a] Calculated from the initial linear part of the kinetic curves.

[b] Calculated from the slope of the tangent at the origin (t = 0) to the kinetic curves as fitted in polynomials of degree 4.

[c] Experimental error ± 0.42.

The effect of the reaction temperature on the activity of both biotic and abiotic catalysts is shown in Figure 2. The determination of the activity by the initial reaction rate from the slope of the tangent at the origin to the kinetic curves was not conducted, because at temperatures higher than 30°C, the enzyme is inactivated and the consumption of oxygen as a function of time does not fit in the same mathematical equation. The activity of tyrosinase increased until an optimum temperature of 30°C, then sharply decreased with increasing temperature. In contrast, the activity of δ-MnO_2 steadily increased with increasing temperature up to 60°C. Apparently birnessite was more resistant to higher temperatures than tyrosinase which was denaturated upon heating. Therefore, in warm regions the abiotically catalyzed oxidation of phenols should prevail in soil environments.

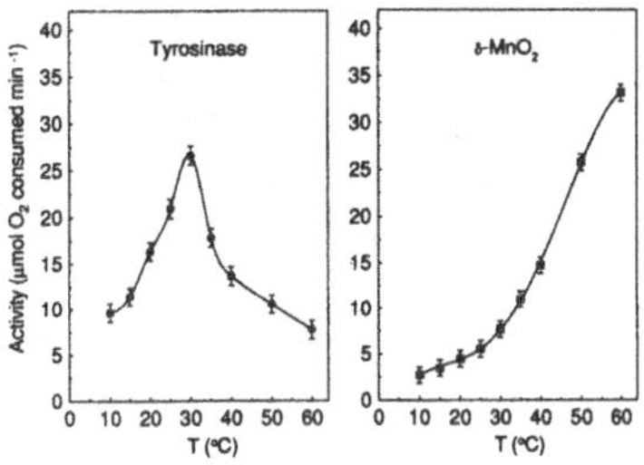

Figure 2. Effect of temperature on the activity of tyrosinase (0.148 mg) and δ-MnO_2 (2.0 mg) at an initial pH of 6.0. The concentration of catechol was 5.0 mM and the final volume of the solution was 82.0 mL.

The activation energy (E_a) of the abiotic oxidation by δ-MnO_2 was calculated from the initial rate and in the same temperature range (10-30°C) as for the tyrosinase system. E_a was calculated from the slope of the linear form of the Arrhenius equation :

$$\mathrm{Ln}\, k = \mathrm{Ln}\, A - \frac{E_a}{R}\frac{1}{T} \tag{5}$$

where k is the initial rate, A is the frequency or preexponential factor for the reaction and E_a is the activation energy (Laidler, 1984). The values of E_a and A, derived from the initial rate k determined at various temperatures in the range of 10 to 30°C, are shown in Figure 3.

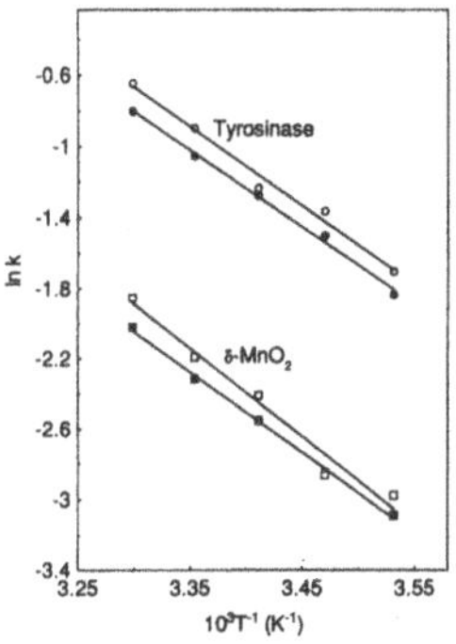

Figure 3. Arrhenius plots for the oxidation of catechol in the presence of tyrosinase and δ-MnO_2. Solid symbols stand for the initial rate calculated from the initial linear part of the curve, and open symbols stand for the initial rate calculated from the slope of the tangent to the polynomials fitted curve.

As indicated by the r^2 and p values (Table 2), the degree of data fitting to equation (5), using a least square regression program was satisfactory. The activation energy for the oxidation of catechol by tyrosinase was 35.8 kJ/mol when calculated from the initial part of the kinetic curves and 36.9 kJ/mol when calculated from the slope of the tangent to the kinetic curves as fitted in polynomials of degree 4 at the origin (t = 0). For the oxidation of catechol by δ-MnO_2, the activation energy values derived by the above described methods were 38.0 and 41.7 kJ/mol, respectively (Table 2). E_a values calculated from the slope of the tangent were slightly higher than those calculated from the slope of the initial linear part of the kinetic curves. The difference apparently arose from the fact that most of the initial linear parts of the kinetic curves did not intercept the origin at $t = 0$, which led to an error in calculating the rate at the beginning of the reaction. Thus, the E_a value calculated from the slope of the tangent at $t = 0$ to the kinetics curves as fitted in polynomials appears to better approximate the real E_a value.

Table 2 : Activation energy (E_a) and preexponential factor (A) of tyrosinase and δ-MnO_2 in the oxidation of catechol, calculated from the Arrhenius equation.

Catalyst	E_a	A	r^{2a}
	— $kJ\ mol^{-1}$ —	— min^{-1} —	
Tyrosinase	[b] 35.8 ± 0.9	$39.6x10^6$	0.997
	[c] 36.9 ± 0.9	$70.2x10^6$	0.981
δ-MnO_2	[b] 38.0 ± 1.0	$28.2x10^6$	0.996
	[c] 41.7 ± 1.0	$138.0x10^6$	0.964

[a] Least square regression coefficient of the linear form of the Arrhenius equation with $p < 2.0x10^{-3}$.

[b] Calculated from the initial linear part of the kinetic curves.

[c] Calculated from the slope of the tangent at the origin ($t = 0$) to the kinetic curves as fitted in a polynomials of degree 4.

Previous studies showed that the activation energy for the oxidation of catechol by tyrosinase was 16.7 kJ/mol if the enzyme was extracted from kiwi fruit (Park and Luh, 1985), and 21.8 kJ/mol if it was extracted from shrimps (Bailey *et al.*, 1959). The activation energies of the oxidation of DL-dihydroxyphenylalanine (dopa) by the shrimp tyrosinase were 48.1 and 58.2 kJ/mol for the pink and the white species (Simpson *et al.*, 1988), respectively, and 83.7 kJ/mol for Harding-Passey mouse enzyme (Martínez *et al.*, 1987). So far no tangible explanation has been proposed concerning this large variation in the trend of activation energies for oxidases of mammalian, crustacean or plant origin. Nevertheless, it was attributed to hypothetical differences in the protein structure of the enzyme molecules (Simpson *et al.*, 1987). The methods of enzyme purification may also have an effect on the activation energy (Simpson *et al.*, 1987).

A relatively small reduction in activation energy can greatly enhance the rate of reaction. The slightly lower activation energy of tyrosinase compared to that of δ-MnO_2 (Table 2) resulted in a more than three-fold enhancement of the initial rate of catechol oxidation (Table 1). The comparison of calculated frequency factors may be misleading. Originally the calculation of the preexponential factor was done using the kinetic theory of collisions, in which the molecules are treated as hard spheres (Laidler *et al.*, 1984). According to Ackerman (1962), the collision theory is always valid but is difficult to apply to enzyme reactions because it involves the product of two factors: $A = \alpha Z$, where α is the probability factor and Z is the collision rate, neither of which can be measured or computed directly. In addition, to date, no data were reported on the kinetic theory of collisions in heterogeneous colloidal phase during catalysis by mineral oxides.

4. CONCLUSIONS

The initial reaction rates calculated from the slope of the tangent to the fitted polynomial curve are more reliable than those calculated from the slope of the linear part of the kinetic curve which often does not intercept the origin. The fitting of the kinetic curves in polynomials indicated the striking similarity of the metalloenzyme, tyrosinase and the metal oxide, δ-MnO_2 in the kinetics of the catechol oxidation at temperatures up to 30°C. The slightly lower activation energy of tyrosinase compared to that of δ-MnO_2 caused the enzymatic reaction to occur more than three times faster than in the presence of birnessite. However, unlike birnessite, tyrosinase was deactivated at temperatures above 30°C because of the denaturation of the protein molecules. Therefore, in warmer regions metal oxides are expected to play an important role in the transformation of natural phenolic compounds and the detoxification of phenolic pollutants.

5. ACKNOWLEDGMENTS

This study was supported by Grant GP 2383-Huang of the Natural Sciences and Engineering Research Council of Canada, and by the Office of Research and Development, Environmental Protection Agency (USEPA; Grant No. R-823847).

6. REFERENCES

Ackerman, E., 1962. Biophysical Sciences. Prentice-Hall, Englewood Cliffs, NJ, pp. 401-418.

Allison, R D. and D.L. Purich, 1996. Contemporary enzyme kinetics and mechanisms. Academic Press, London, pp. 37-56.

Bailey, M.E., E.A. Fieger and A.F. Novak, 1959. Physico-chemical properties of the enzymes involved in shrimp melanogenesis. Food Res. 25, 557-564.

Bollag, J.-M., C. Meyers, S. Pal and P.M. Huang, 1995. The role of abiotic and biotic catalysts in the transformation of phenolic compounds. In : P.M. Huang, J. Berthelin, J.-M. Bollag, W.B. McGill and A.L. Page (Editors). Environmental Impact of Soil Component Interactions. CRC/Lewis Publishers, Boca Raton. Vol. I, pp. 299-310.

Chen, J.S., C. Wei and M.R. Marshall, 1991. Inhibitory effect of kojic acid on some plant and crustacean polyphenoloxidases. J. Agric. Food Chem. 39, 1896-1401.

Dixon, M. and E.C. Weeb, 1979. Enzymes. Academic Press, New York, pp. 138-164.

Eltantawy, I. M. and P.W. Arnold, 1973. Reappraisal of ethylene glycol mono-ethyl ether (EGME) method for surface area estimations of clays. J. Soil Sci. 24, 232-238.

Goult, R.J., R.F. Hoskins, J.A. Milner and M.M. Pratt, 1973. Applicable mathematics, a course for scientists and engineers, MacMillan, London pp. 241-291.

Himmelwright, R.S., N.C. Eickman, C.D. LuBien, K. Lerch and E.I. Solomon, 1980. Chemical and spectroscopic studies of the binuclear copper active site of Neurospora tyrosinase : Comparison to hemocyanins. J. Am. Chem. Soc. 102, 7339-7344.

Laidler, K.J., 1984. Chemical Kinetics, Tata McGraw-Hill, New Delhi, pp. 49-114.

Martínez, J.H., F. Solano, A. Arocas, J.C. García-Borrón, J.L. Iborra and J.A. Lozano, 1987. The existence of apotyrosinase in the cytosol of Harding-Passey mouse melanoma melanocytes and characteristics of enzyme reconstitution by Cu(II). Biochim. Biophys. Acta 923, 413-420.

Matsuura, T., 1977. Bio-mimetic oxygenation. Tetrahedron 33, 2869-2905.

McBride, M.B., 1989. Oxidation of dihydroxybenzenes in aerated aqueous suspensions of birnessite. Clays & Clay Miner. 37, 341-347.

McKenzie, R.M., 1971. The synthesis of birnessite, cryptomelane and some other oxides and hydroxides of manganese. Miner. Mag. 38, 493-502.

Naidja A., P.M. Huang and J.-M. Bollag, 1997. Activity of tyrosinase immobilized on hydroxyaluminum-montmorillonite complexes. J. Mol. Catal. A : Chemical 115, 305-316.

Naidja A., P.M. Huang and J.-M. Bollag, 1998. Comparison of reaction Products from the Transformation of Catechol Catalyzed by δ-MnO_2 or tyrosinase. Soil Sci Soc. Am. J. 62, 188-195.

Park, E.Y. and B.S. Luh, 1985. Polyphenol oxidase of kiwifruit. J. Food Sci. 50, 678-684.

Pizzigallo, M.D.R., P. Ruggiero, C. Crecchio and R. Mininni, 1995. Manganese and iron oxides as reactants for oxidation of chlorophenols. Soil Sci. Soc. Am. J. 59, 444-452.

Shindo, H. and P.M. Huang, 1982. Role of Mn(IV) oxide in abiotic formation of humic substances in the environment. Nature (London) 298, 363-365.

Simpson, B.K., M. R. Marshall and W.S. Otwell, 1987. Phenoloxidases from shrimp (*Penaeus setiferus*)

: Purification and some properties. J. Agric. Food Chem. 35, 918-921.

Simpson, B.K., M.R. Marshall and W.S. Otwell, 1988. Phenoloxidases from pink and white shrimp : Kinetic and other properties. J. Food Biochem. 12, 205-217.

Sjoblad, R.D. and J.-M. Bollag, 1981. Oxidative coupling of aromatic compounds by enzymes from soil microorganisms. In : E.A. Paul and J.N. Ladd (Editors). Soil Biochemistry. Marcel Dekker, New York, Vol. 5, pp. 113-152.

Zawistowski, J., C.G. Biliaderis and N.A. Michael, 1991. Polyphenol oxidase. In : D.S. Robinson and N.A.M. Eskin (Editors). Oxidative Enzymes in Foods, Elsevier Applied Science, London, pp. 217-273.

PLANT RESIDUE DECOMPOSITION: EFFECT OF SOIL POROSITY AND PARTICLE SIZE

Laetitia Fruit[1,2], Sylvie Recous[1*] and Guy Richard[1]

[1]INRA, Unité d'Agronomie, rue F. Christ,
02007 Laon Cedex, France
[2] present address: INRA Station de pathologie végétale,
Montfavet, France

1. INTRODUCTION

The decomposition of organic matter is a biological process that depends on climatic and edaphic factors. Numerous studies have been done to determine the effect of environmental factors such as temperature (Kirschbaum, 1995), N availability (Mary *et al.*, 1996), plant residue content (Reinertsen *et al.*, 1984), water potential (Sommers *et al.*, 1980) and oxygen concentration (Parr and Reuszer, 1959). Little information is available about the effects that different degrees of contact between soil and plant residues have on the decomposition of organic materials. This contact depends mainly on the physical and chemical characteristics of the residue and the physical properties (texture, structure, water content) of the soil. Previous studies have investigated the effect of residue particle size (Bremer *et al.*, 1991; Jensen, 1994; Angers and Recous, 1997) and soil texture (Hassink 1992; Scott *et al.*, 1996) on C decomposition, but further investigations of the interaction between these factors should improve the description of plant residue decomposition in C-N models. For instance, reducing the particle size of residues of high C:N ratio such as mature cereals, enhanced short-term C decomposition while it has nil or a reverse effect for residues of low C:N ratio (Bremer *et al.* 1991; Jensen, 1994, Ambus and Jensen, 1997; Angers and Recous, 1997). The main hypotheses invoked to explain these observations were that decreasing particle size of residues of high C:N ratio enhanced the availability of soil N to decomposing microorganisms, while for low C:N ratio residues, increasing contact enhanced the protection of C against biodegradation.

The contact between soil aggregates and residue has two components (figure 1). The first is the residue potential surface area for contact with soil aggregates and for colonisation by micro-organisms. This area mainly depends on the particle size of plant residues and their location in soil (Christensen 1985). The second is the inter-aggregate porosity of soil and the size of aggregates which define the actual area of contact between the soil and plant residues. In addition, soil characteristics also affect oxygen transfer between aggregates and the transfer of nutrients, such as nitrate, within these aggregates. Two types of nutrient transfer occur during residue decomposition : (i) transfer close to the residue, which depends greatly on the area of contact between the soil and the residue, and (ii) transfer within the soil aggregates, which mainly depends on soil characteristics. This work was undertaken to analyse the interaction between soil porosity and residue particle size on the decomposition of plant residues under controlled conditions, which allowed the modelling of gas and nitrate diffusion within the soil core.

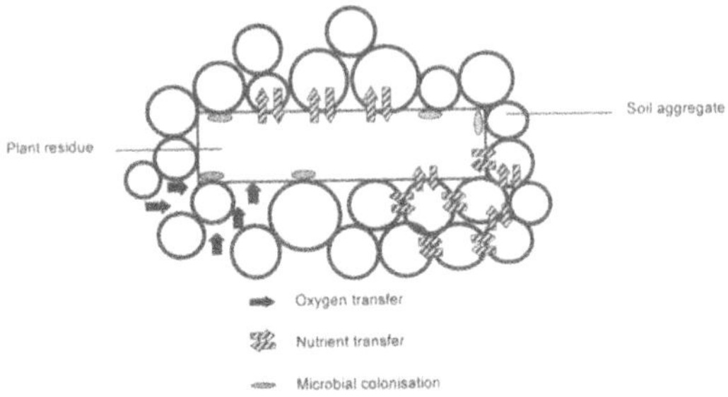

Figure 1. Simplified diagram of interactions between the soil and plant residue particle.

2. MATERIAL AND METHODS

2.1. Soil and experimental system

The soil used was a silty soil (clay 170 g kg^{-1}, silt 780 g kg^{-1}, sand 50 g kg^{-1}, pH_{H2O}= 7.2) and was obtained from the experimental site at Mons-en-Chaussées (North France), from a plot with a history of intensive cropping of corn, sugar beet and wheat. The C and N content of the soil were 10.0 and 1.0 g kg^{-1}, respectively. The soil was kept at its initial moisture (0.17 g g^{-1}) and at 5°C $\pm$ 0.5°C. It was sieved at its initial moisture so that the whole soil passed through a 3.15 mm sieve. Only aggreggates remaining over a 2-mm mesh were kept. Visible particles of organic matter were removed by hand. The experimental procedure used by Richard and Guérif (1988a) to study seed germination was used. It consists of a standard compaction of soil aggregates at a given water potential. A mass of aggregates was compacted in a cylindrical chamber of constant volume (7 cm diameter, 2.4 cm thickness) to obtain a soil core with a defined and homogeneous inter-aggregate porosity (figure 2). The volume of the structural pore space was calculated as the difference between the volume of the soil sample and the volume of the aggregates. This last volume was measured using the method of Monnier *et al.* (1973).

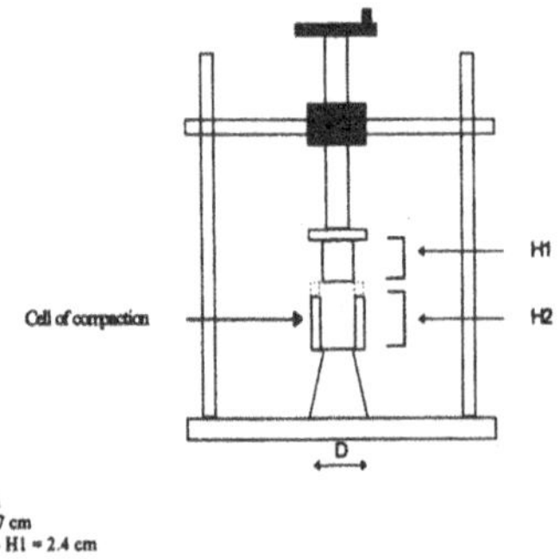

Figure 2. Experimental system used to prepare soil core.

2.2. Simulation of oxygen transfer

The incubation parameters were determined using a simulation of oxygen diffusion in the soil core in order to increase soil-residue contact and to reduce soil porosity without inducing anaerobic conditions. Several incubation conditions were tested as a function of temperature, amount of carbon added, and inter-aggregate porosity. This simulation was based on the physical equation of gas diffusion in porous media, i. e. Fick's laws, applied to a steady state. The oxygen concentration in a homogeneous, isotropic soil core with the two surfaces in contact with the atmosphere was calculated as follows (equation 1) for a uniform distribution of carbon in soil samples :

$$C = \frac{1}{2} \cdot \frac{P}{D} \cdot z^2 - \frac{1}{2} \cdot \frac{P}{D} \cdot h \cdot z + C_0$$

where C is the oxygen concentration (m^3 m^{-3}) with C_0 =0.21 m^3 m^{-3}, h is the soil core thickness (m) and z the depth into the soil sample (m) with $0 \leq z \leq h/2$; D is the oxygen diffusion coefficient in the soil sample (m^2 s^{-1}), and P (cm^3 O_2 h^{-1} cm^{-3}) is the rate of O_2 consumption by soil and plant residue. Data from Recous *et al.* (1995) were used to estimate P. This was the maximum daily oxygen demand for the decomposition of straw. P was considered to be independent of the oxygen concentration. Simulation was made for an inter-aggregate porosity greater than 0.10 m^3 m^{-3}, which is 1/1000 of the oxygen diffusion coefficient in air (Richard and Guérif, 1988b).

Table 1 : Minimal oxygen concentration (m^3 m^{-3}) in soil cores, during straw decomposition. Inter-aggregate porosity of 0.10 m^3 m^{-3}. Calculations were made with Ficks' law in steady state (equation 1)

Temperature	Carbon added (mg kg^{-1} soil)		
°C	1000	660	0
25	0.03	0.09	0.19
20	0.08	0.12	0.20
15	0.12	0.15	0.20
10	0.14	0.16	0.20

1a : Uniform distribution of the carbon in the soil core.

Temperature	Carbon added (mg kg^{-1} soil)		
°C	1000	660	0
25	0	0	0.19
20	0	0.04	0.20
15	0.04	0.09	0.20
10	0.07	0.12	0.20

1b : Carbon in a single layer in the centre of the soil core.

2.3. Incubation conditions and treatments

Soil samples were prepared with 163g aggregates at a final moisture content of 0.20g.g^{-1}, corresponding to a water potential of -80 kPa. This water potential should allow optimal conditions for microbial activity without inducing denitrification. Indeed at this water potential, the volume of the soil particles plus the volume of water are lower than the volume of aggregates, indicating that the aggregates contain air and that consequently they are not saturated with water. A solution of KNO_3 was added to the aggregates. The N concentration of the solution was calculated to obtain the appropriate final water and N contents for the incubation. N solution was added by spraying as labelled ^{15}N solution (atom ^{15}N excess 10 %) on thin layers of aggregates.

C was added at the rate of 800 mg C per kg soil as mature wheat straw (*Triticum aestivum,* L.) containing 46% C and 0.17% N (C:N=270). Only stems without nodes and leaves were used. Three particle sizes were prepared. The smallest size was prepared by grinding stems into particles of 0.5-1 mm diameter and sieved (>0.5 mm) in order to remove straw powder. The two other sizes were prepared by cutting straw with scissors to give particles 1 or 10 mm long. Controls (samples without added carbon) were incubated for each treatment. Three replicates were prepared per treatment.

The decomposition of the three sizes of straw particles was first compared using soil samples with an inter-aggregate porosity of 0.15 m^3 m^{-3}. The second experiment was done using the two extreme sizes (0.5 and 10 mm) which were incorporated into samples of compacted (0.15 m^3 m^{-3}) and uncompacted aggregates (inter-aggregate porosity of 0.35 m^3 m^{-3}). Lastly the decomposition of 10mm long straw particles with inter-aggregate porosity of 0.35 m^3 m^{-3} was compared at two initial nitrogen concentrations, 64 and 176 mg N kg^{-1} soil, corresponding to 80 and 220 mg of N per g of added C, respectively. These two concentrations were above the threshold for soil N availability allowing non limiting decomposition of straw (Recous *et al.*, 1995).

These conditions corresponded to a minimal oxygen concentration in the soil sample of 0.09 m^3 m^{-3} at 15°C temperature of incubation (table 1).

2.6. Incubation and analysis methods

Soil samples were incubated in 1L airtight jars for 35 days at 15±0.5°C. Two beakers were placed in each jar, the first one contained water to keep a water-saturated atmosphere, and the second contained 12 ml NaOH 0.25 *N* to trap CO_2 produced during decomposition. The CO_2 evolved was determined every 2-3 days at the beginning of incubation, then every week. At each sampling time, the jars were opened, aerated for few minutes, and the beakers of NaOH were changed.

Carbon mineralisation was monitored throughout the incubation (35 days) by the method of static trapping (Freijer and Bouten, 1991). The CO_2 trapped was measured after precipitating the carbonate with excess $BaCl_2$ (5 ml). The remaining NaOH was then titrated with HCl (0.25 *N*). The amount of CO_2 evolved (expressed as percentage of added carbon) was calculated as the difference between samples containing straw and the corresponding control sample.

A ^{15}N balance was made at the beginning and at the end of incubation on the whole soil without any fractionation. Soil was air-dried at 80°C for 48 hours, finely ground, and the N total and ^{15}N were measured with an CHN analyser (Carlo Erba, NA 1500) connected to a mass spectrometer (VG SIRA 9).

3. RESULTS

3.1. ^{15}N balance

Table 2: ^{15}N recovered in soil samples : wheat straw incorporated in soil samples with inter-aggregate porosity of 0.15 (compacted) or 0.35 m^3 m^{-3} (uncompacted) and corresponding controls, without straw.

Sample	Total ^{15}N % initial ^{15}N [a]
Compacted aggregates	
straw 0.5 mm	106.7 ± 7.0
straw 10 mm	102.8 ± 2.8
control	98.4 ± 4.1
Uncompacted aggregates	
straw 0.5 mm	98.2 ± 0.4
straw 10 mm	107.6 ± 9.8
control	96.3 ± 3.0

[a] initial ^{15}N = Total ^{15}N measured at the beginning of incubation.

The recovery of ^{15}N was determined for the two inter-aggregate porosities (0.15 and 0.35 $m^3\ m^{-3}$) and the 0.5 and 10 mm sizes (table 2). The ^{15}N balance was 96-98% and 106-107% in the two treatments, the coefficient of variation being < 9%. Consequently, no N losses were assumed to occur from the system during the experiment, particularly from denitrification.

3.2. Effect of residue particle size

The cumulative amounts of CO_2 produced for the three sizes of particles incorporated in soil samples having an inter-aggregate porosity of 0.15 $m^3\ m^{-3}$ (figure 3) showed a significant effect of residue particle size on C-mineralisation. Cumulative mineralisation was greater for the smaller particles than for the larger particles. There was a significant difference in rates of C mineralisation between days 5 and 15 after incorporation. The cumulative difference between 0.5 and 10 mm particles seemed to be constant after 15 days and was maintained until the end of incubation (35 days). The total amount of C-CO_2 evolved at the end of incubation was 34 ± 2% of added C for 10 mm particles and 41 ± 1% for the smallest particles.

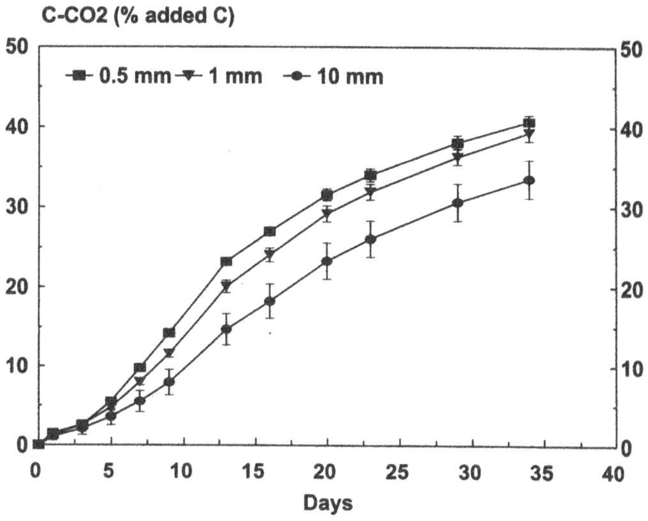

Figure 3. Cumulative amount of CO_2 evolved : wheat straw at three particle sizes (0.5, 1 and 10 mm length) incorporated into soil samples of 0.15 $m^3\ m^{-3}$ inter-aggregate porosity.

3.3. Effect of inter-aggregate porosity

The effect of inter-aggregate porosity on C mineralisation can be observed for the 0.5 and 10 mm straw particles (figure 4). Cumulative mineralisation was lower for residues incorporated into uncompacted aggregates (0.35 $m^3\ m^{-3}$ porosity) than into compacted aggregates. The greatest cumulative mineralisation occurred with the smallest particles, whatever the soil porosity. The difference between the amounts of C mineralised for the two residue sizes incorporated into uncompacted soil was similar to that for compacted soil samples. Again the difference occurred 5-20 days after the incorporation of straw and the maximal difference in the C-CO_2 evolved was 9% of added C. There was not effect of inter-aggregate porosity on the mineralisation of C from control samples (data not shown).

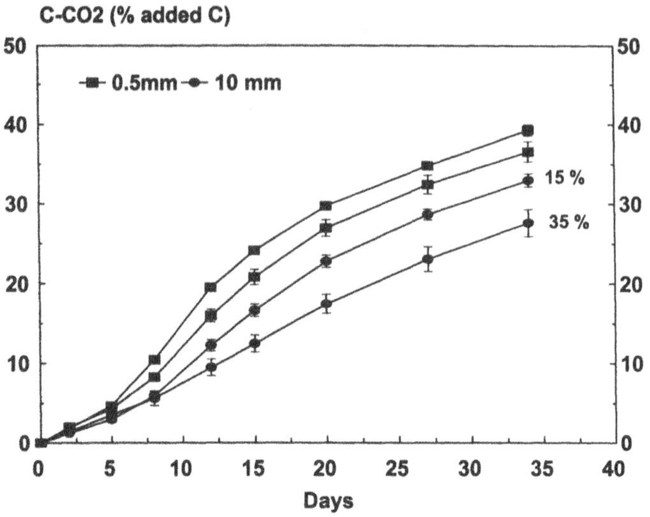

Figure 4. Cumulative amount of CO_2 evolved as a function of straw particle size and inter-aggregate porosities (——: 0.15 m^3 m^{-3} or ----: 0.35 m^3 m^{-3})

3.4. Effect of initial N concentration

The effect of high initial N concentrations was tested using the 10 mm size incorporated into soil samples of 0.35 m^3 m^{-3} inter-aggregate porosity (figure 5). The mineralisation of carbon did not vary with the soil N concentration.

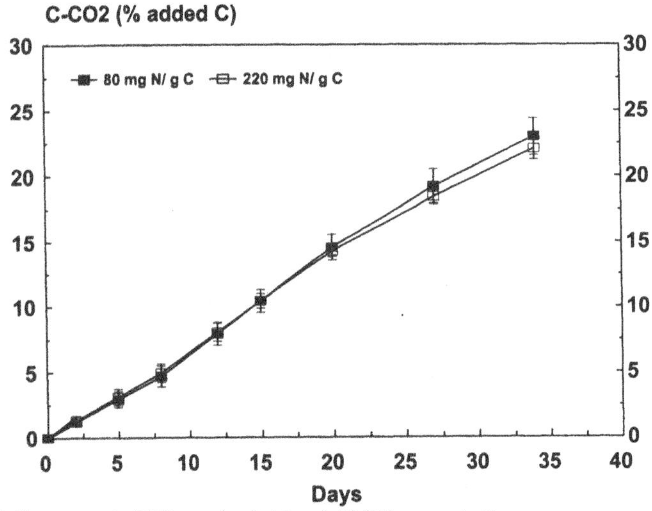

Figure 5. Cumulative amount of CO_2 evolved at two initial N concentrations (10 mm long wheat straw particles and 0.35 m^3 m^{-3} inter-aggregate porosity)

4. DISCUSSION AND CONCLUSION

We examined the effect of contact between soil aggregates and particles of plant residues in an experimental system that allowed regulation of inter-aggregate porosity and modelling of gas transfers. The objective was to create the greatest contact between soil and residue by using a low inter-aggregate porosity, but without inducing oxygen deficiency around residues. ^{15}N balance showed that there was no nitrogen loss during the experiment, indicating no denitrification and that aerobic conditions were maintained.

The area of contact between the soil and residue results from the interaction between soil characteristics, residues size and distribution. It could influence the microbial colonisation and the local transfer of oxygen and nitrogen between aggregates and residues. Soil properties also determine the transfer of oxygen and nitrogen within the soil towards the surface of the residues. We tested first the influence of residue characteristics by incorporating wheat straw of three particle sizes (0.5, 1 and 10 mm) into soil samples having a 0.15 m^3 m^{-3} of inter-aggregate porosity. The contact varied with the potential area of organic residue. C mineralisation decreased with increasing residue particle size. These results are similar to those obtained by Bremer *et al.*(1991) and Angers and Recous (1997). Their hypothesis was that availability and accessibility of N varied with particle size. The influence of soil porosity was then tested. Soil porosity alters the area of contact and also changes nutrient transfers. Decreasing the inter-aggregate porosity also increases the coefficient of diffusion of nitrate. Reducing inter-aggregates porosity enhanced C mineralisation. This effect seemed to be greater at the largest residue particle size. The mineralisation was increased by 9% of added C by reducing inter-aggregates porosity from 0.35 to 0.15 m^3 m^{-3}. We postulate that this positive effect of reducing the porosity is due to change in the N diffusion coefficient in soil. Previous studies (Bremer *et al.* 1991, Jensen, 1994) emphasised the interaction between soil-residue contact and N availability (in soil and residue). We evaluated the effect of two high initial soil N concentrations (64 and 176 mg N kg^{-1} soil). These concentrations had no effect on C mineralisation. Thus an overabundance of N in the soil does not overcome the effect of low soil-residue contact due to the residue size or soil characteristics. Hence, either the difference between the two N concentrations tested were not large enough to produce any difference in N transfer between the soil aggregates and the residues particles, or that N does not limit the response to soil-residue contact. If is so, the effect of particle size could be due to a difference in the microbial colonisation of straw particles.

5. ACKNOWLEDGMENTS

We thank J.Guerif for fruitful discussions, M. Boucher, C. Nice, and E. Venet for technical assistance. This work was supported by grants from INRA AIP Ecosol.

6. REFERENCES

Ambus P & Jensen E S 1997 Nitrogen mineralization and denitrification as influenced by crop residue particle size. Plant Soil, 197, 261-270.

Angers, D. and S. Recous, 1997. Decomposition of wheat straw and rye in soil as affected by particle size. Plant Soil, 189, 197-203.

Bremer, E., W. van Houtum and C. van Kessel, 1991. Carbon dioxide evolution from wheat and lentil residues as affected by grinding, added nitrogen, and the absence of soil. Biol. Fert. Soils, 11, 221-227.

Christensen, B. T., 1985. Wheat and barley straw decomposition under field conditions : effect of soil type and plant cover on weight loss, nitrogen and potassium content. Soil Biol. Biochem., 17, 691-697.

Freijer, J. I. and W. Bouten, 1991. A comparison of field methods for measuring soil carbon dioxide evolution : Experiments and simulation. Plant Soil, 135, 133-142.

Hassink, J. 1992. Effects of soil texture and structure on carbon and nitrogen mineralization in grassland soils. Biol. Fertil. Soils, 14, 126-134.

Jensen, E. G., 1994. Mineralization-Immobilization of nitrogen in soil amended with low C:N ratio plant residues with different particle sizes. Soil Biol. Bioch., 26, 519-521.

Kirschbaum, M. U. F., 1995. The temperature dependence of soil organic matter decomposition, and the effect of global warming on soil organic storage. Soil Biol. Bioch., 27, 753-760.

Mary, B., S. Recous, D. Darwis and D. Robin, 1996. Interactions between decomposition of plant residues and nitrogen cycling in soil. Plant Soil, 181, 71-82

Monnier, G., P. Stengel and J.C. Fies, 1973. Une méthode de mesure de la densité apparente de petits agglomérats terreux. Application à l'analyse des systèmes de porosité du sol. Ann. Agron., 24, 533-545.

Parr, J. F. and H. W. Reuszer, 1959. Organic matter decomposition as influenced by oxygen level and method of application to soil. Soil Sci. Soc. Proc., 214-216.

Recous, S., D. Robin, D. Darwis and B. Mary, 1995. Soil inorganic N availability : effect on maize residue decomposition. Soil Biol. Bioch., 27, 1529-1538.

Reinertsen, S. A., L. F. Elliot, V. Cochran and G., S. Campbell, 1984. The role of available C and N in determining the rate of wheat straw decomposition. Soil Biol. Bioch., 16, 459-464.

Richard, G. and J. Guérif, 1988. Influence of aeration conditions in the seedbed on sugar beet seed germination : experimental study and model. Proc. 11th international conference ISTRO, Edinburgh, U.K., 103-108.

Richard, G. and J. Guérif, 1988. Modélisation des transferts gazeux dans le lit de semences : application au diagnostic des conditions d'hypoxie des semences de betterave sucrière (*Beta vulgaris* L.) pendant la germination. I. Présentation du modèle. Agronomie, 8, 539-547.

Scott, N. A., C. V. Cole, E. T. Elliott and S. A. Huffman, 1996. Soil texture control on decomposition and soil organic matter dynamics. Soil Sci. Soc. Am. J., 60, 1102-1109.

Sommers, L. E., C. M. Gilmour, R. E. Wildung and S. M. Beck, 1980. The effect of water potential on decomposition processes in soil. In Water potential relation in soil Microbiology, SSSA Spec. publ. n° 9,. ASA, Madison, Wisconsin, pp. 97-117.

THE EFFECT OF HUMIC SUBSTANCES FROM OXYHUMOLITE ON PLANT DEVELOPMENT

Slawomir S. Gonet[1], Elzbieta Gonet[2] and Andrzej Dziamski[2]

[1]Department of Soil Chemistry
University of Technology and Agriculture
Bernardynska 6, 85-029 Bydgoszcz
Poland

[2]Department of Botany and Ecology,
University of Technology and Agriculture
Kaliskiego 7, 85-796, Bydgoszcz
Poland

1. INTRODUCTION

The results of many investigations have shown that soil organic matter (OM) significantly affects the growth and yield of plants. This effect can be indirect because OM in general influences soil fertility by determining soil physical and chemical properties (air and water conditions, sorption properties and structure) or direct - connected with the uptake of many other processes, for example, protein and nucleic acids synthesis, enzymatic activity, photosynthesis and respiration (Chen and Aviad 1990, Vaughan and Malcolm 1985). Mechanisms of direct interaction between humic substances and plants are not fully explained yet. Many other problems require profound studies. They are, among others, explanation in which phases of plant development humic substances are important and the description of connections between the structure and chemical properties of humic substances and their effectiveness.

From our earlier data (Gonet *et al.*, 1993, Gonet *et al.*, 1996) it can be concluded that humic substances from oxyhumolite can have a stimulating effect on the growth of seedlings of different plant species. Moreover, Ca and Mg humates stimulated the growth of leaves to a greater extend than that of roots.

The objective of our study was to determine the effects of humus preparation (humic acids fraction) obtained from the Czech oxyhumolite on the growth of barley and rye plants. The preparation was tested in germination, pot and field experiments.

Effect of Mineral-Organic-Microorganism Interactions on Soil and Freshwater Environments
Edited by Berthelin, *et al*, Kluwer Academic / Plenum Publishers, New York, 1999

2. MATERIAL AND METHODS

Humic acid (HA) fractions from dried oxyhumolite (Most, Czech Republic) sample was extracted with 0.1 M NaOH, precipitated at pH 2.0 with 5.0 M HCl, and then purified with the mixture HF-HCl, according to the Schnitzer and Skinner (1968) method. The extracted HAs were analysed for:

- elemental composition with 2400 Perkin-Elmer Analyser (C, H, N; O by difference. Internal HA oxidation index (ω) as defined by Zdanov (1965) was calculated from elemental composition as follows: $\omega = (2\,O + 3\,N - H) : C$;
- thermal properties with Derivatograph C, MOM, Hungary. Thermal analyses were performed on mixtures of 40 mg HAs in 360 mg KBr, within the range 20-700° C at a heating rate of 3° C per min.;
- IR absorption spectra with a Carl Zeiss Jena M80 spectrometer, on pellets of 3 mg HA in 800 mg KBr;
- UV-VIS absorption spectra and absorbances at 280, 465 and 665 nm, with a Hewlett Packard - UV-VIS spectrometer, on solutions of 10 mg HA/L in 0.05M $NaHCO_3$.

The preparation was tested in the case of three experiments with barley cv. Ars and rye cv. Dankowskie Zlote. Experiment *1* was performed on germination beds with 7 treatments: NPK - standard mineral fertilisers - and the HA preparation at amounts of 20, 50, 100, 200, 500 or 1000 mg HA/L, usually with NPK. Six replicates with 50 seeds each were done. Morphological parameters were determined on 30 seedlings taken from each replicate in each treatment. Average values from 180 determinations are presented in Tables 1 and 2. This experiment was carried out in an air-conditioned chamber Seed Germinator G-120 at 20° C and illumination intensity of 200 lux. Experiment *2* was done in pots. The rye and barley seeds were sowed in plastic pots containing 300 g of river sand, with 40 seeds per pot in six replicates. The sand material was enriched with the following concentrations of humus preparation: 2, 5, 10, 50, 100 mg HA/pot, usually with NPK. This experiment was made at constant temperature (14-16° C) and 60 % of moisture. Root and leaf dry matter was determined at the end of the experiment. All results are shown in Table 3. Experiment *3* was carried out in field conditions on a sandy soil by split-split-block method using three replicates. Plot area for harvest was 5 m^2. The humic preparation was top-dressed in doses of 0, 12, 20, 40, 100, 200 mg HA/m^2. The plants were sprinkled at the three-leafs phase and at the tillering phase, each time with half doses. The results are given in Table 4.

Table 1.Parameters of barley seedlings cv. Ars (Experiment *1*)

Treatment	Roots			Leaf
	number	mean length (cm)	total length (cm)	mean length (cm)
NPK-control	6.3	9.15	57.7	13.01
20 mgHA/L	5.9	10.24	60.4	13.62
50 mgHA/L	5.8	10.76	62.4	13.62
100 mgHA/L	5.9	9.89	58.3	13.87
200 mgHA/L	7.2	9.68	55.3	13.87
500 mgHA/L	5.7	5.91	33.5	12.38
1000 mgHA/L	5.8	3.95	22.9	11.18
LSD a=0.01	0.9	1.11	3.6	1.60

Table 2. Parameters of rye seedlings cv. Dankowskie Zlote (Experiment *1*)

Treatment	Roots			Leaf
	number	mean length (cm)	total length (cm)	mean length(cm)
NPK-control	5.1	7.21	37.7	12.16
20 mgHA/L	5.3	7.80	41.3	12.78
50 mgHA/L	4.9	7.90	38.4	12.77
100 mgHA/L	5.6	8.85	49.5	12.86
200 mgHA/L	5.3	7.24	38.6	12.03
500 mgHA/L	5.2	6.52	33.9	12.18
1000 mgHA/L	4.4	5.09	22.3	9.81
LSD a=0.01	0.8	0.60	2.8	0.43

Table 3. Parameters of barley and rye plants, in g of dry matter per plant. (Experiment *2*)

Treatment	Barley		Rye	
	roots	leaves	roots	leaves
NPK-control	0.983	0.617	0.528	0.622
2 mgHA	0.985	0.838	0.543	0.692
5 mgHA	1.495	0.813	0.602	0.738
10 mgHA	1.092	0.770	0.662	0.680
20 mgHA	1.340	0.847	0.650	0.722
50 mgHA	1.173	0.872	0.538	0.687
100 mg HA	0.965	0.757	0.437	0.553
LSD a=0.01	0.150	0.075	0.094	0.065

Table 4. Yields of grain and straw of barley cv. Ars, in g/plot (Experiment *3*)

Treatment	Grain	Straw
NPK-control	1.84	2.12
10 mgHA/m^2	1.94	2.20
20 mgHA/m^2	2.05	2.39
40 mgHA/m^2	2.34	2.43
100 mgHA/m^2	2.23	2.47
200 mgHA/m^2	2.20	2.40
LSD a=0.01	0.17	0.23

3. RESULTS

Elemental composition (in atomic percentage, on a ash-free basis) of humus preparation was as follows: 31.1 % C, 34.8 % H, 32.4 % O, 1.7 % N, H:C = 1.11, O:C = 1.00, N:C = 0.055. The HAs had a high internal oxidation degree (ω = 1.13) and a low carbon content. Their elemental composition was different for that of AHs from soils and brown coals. Specific character of the preparation was confirmed by spectral and differential thermal analysis (Fig. 1 and 2). IR-spectra showed that HA molecules are composed mainly of cyclic, possibly aromatic structures and oxygen containing functional groups. The spectra had absorption bands as follows: 2640 cm^{-1} (OH), 1730 cm^{-1} (C=O of different sources), 1620 cm^{-1} (aromatic C-C), 1400 cm^{-1} (-OH and C-C), and 1200 cm^{-1} (-OH). There were no absorption bands in the 2800-2900 cm^{-1} area, which excluded the existence of aliphatic structures.

The measurements of absorption in the UV-VIS region of the solution gave the following values: A_{280} - 5.92, A_{465} - 0.870, A_{665} - 0.141. High values of the ratios A_{280}/A_{665} (41.9) and A_{465}/A_{665} (6.2) suggested the existence of a large number of oxygen functional groups and lower humification degree than that of soil HAs.

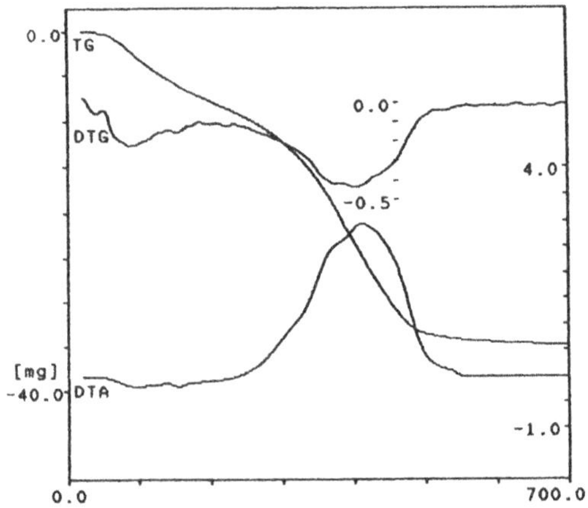

Figure 1. Thermogram of humic acid preparation (TG - thermogravimetric curve, DTG - derivative thermogravimetric curve, DTA - differential thermoanalytic curve).

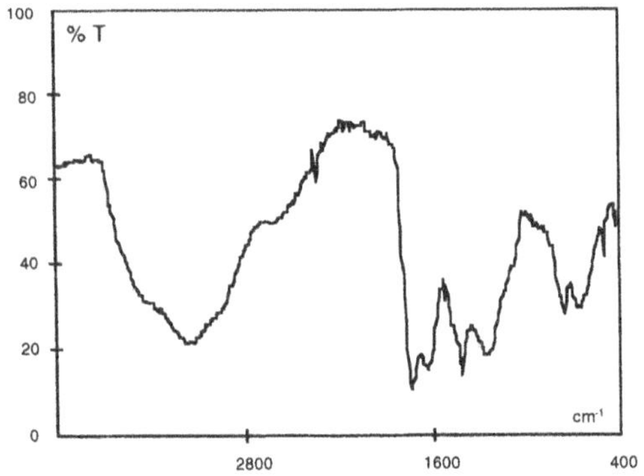

Figure 2. IR-spectrum of humic acid preparation.

Thermogram of this preparation is shown on Fig. 2. In comparison with thermal degradation of soil HAs, it was very specific. On the DTA curve one wide endothermic effect in the 66-158° C range and one exothermic with a wide maximum at 415° C were noted. During the endothermic reaction 20.4% of weight of the HAs preparation was decomposed. It was an unusual result for HAs from a soil source (Gonet and Wegner, 1994).

The endothermic effect can be the result of the hygroscopic water loss, volatilisation of low-macromolecular compounds and degradation of structures (mainly functional groups). During the exothermic reaction OM of fundamental skeleton of humic acids structure was disintegrated. In this effect reacting substance made up 79.6% of the total sample.

Results of Experiment *1* (Tables 1 and 2) demonstrated that the preparation had a positive influence on the first phase of plants development. This effect was also true for roots system and leaves. In the case of rye and barley plants this preparation had a strong influence on the growth of seedlings in the laboratory experiment. The doses of 50 mg/L preparation in the medium stimulated the length of barley roots, length of roots and leaves of rye seedlings. Rye seed material was of a better quality than that of barley, because a stimulating effect of the rye leaves was noticed. Generally, the doses of 50-100 mg/L were optimal to improve the parameters of seedlings. The preparation used at high doses of 500 mg/L had a negative influence on rye and barley roots development.

In the pot experiment (Table 3) rye and barley plants disclosed a positive reaction to the amendment of humus preparation when grown on a sandy substrate. An increase of dry matter of barley leaves (for all doses used) and rye leaves (in doses of 2-50 mg/pot) was observed. The preparation stimulated the development of plant roots, too. In comparison with the control assay (NPK) it was about 30% and 25% for barley and rye plants, respectively. Generally, it was noticed that optimal doses ranged between 5 and 20 mg of preparation for 300 mg of substrate.

The results given in Table 4 show that HAs enhanced the yields of barley grain and straw. However, in the field experiment, after application of this preparation to the soil no positive effect was obtained. Under the conditions of foliar spray application an increase of grain and straw yield of about 27% in comparison with the control treatment was observed, with relatively low dose (about 20 mg/m^2) of HAs preparation. The smallest dose of 10 mg/m^2 did not modify the grain yield.

4. DISCUSSION

As shown by the above results, the obtained HAs preparation was very specific and had a different chemical structure from than of HAs extracted from soils and brown coals (Gonet and Wegner, 1994). In comparison with soil HAs, they had a lower C and N content, a very high O content and a high O:C ratio. In comparison with HAs from brown coal they contained less C, much more O and had higher O:C and N:C ratios. It can be assumed that HAs from oxyhumolite had a cyclic ring structure, part of them having an aromatic nature (ring or cross-link carbon structure is not unlikely), and presented a large number of oxygen-containing functional groups, such as carboxyl, hydroxyl, and quinoic).

The obtained results show that there can be an interaction between humic substances and fertiliser components in the process of plant nutrition. Earlier data (Gonet et al. 1996) showed that humic substances may have a stimulating effect on the growth of barley seedlings, but that Ca and Mg-humates have a stronger influence than Na and K-humates. The results presented in this paper demonstrated that preparations of "pure" HAs have favourable biostimulating effects, too. In the case of barley such effect can be important, but fertilisation with suitable humic substances can lead to an increase of barley yields. It can be supposed that the effects of humic substances were very significant on sandy soils, which are poor in humus, in conditions of optimal mineral fertilisation. The results of the discussed studies do not permit to speculate about the mechanisms of processes undergoing in plants under the influence of humus substances. Full understanding of such phenomena needs more systematic physiological and biochemical analysis carried out along the whole vegetation period.

5. CONCLUSION

The obtained results demonstrated that humic substances extracted from oxyhumolite stimulate the growth of barley and rye plants in all phases of their development. It is very important to remind that all experiments were conducted under optimal conditions of mineral fertilisation. Moreover, humic preparation was applied at relatively low doses. The obtained results confirmed the hypothesis that humic substances can be a helpful agent in plant mineral nutrition, not without intrinsic fertilising character. Application of the preparation at doses of 200-500 g/ha can be beneficial for increasing cereal yields.

6. REFERENCES

Chen, Y. and T. Aviad, 1990. Effects of humic substances on plant growth. In: Humic Substances in Soils and Crop Sciences. Am.Soc.of Agronomy. Inc., Madison, 161-186.

Gonet, S.S., A. Dziamski and E. Gonet, 1993. Wplyw substancji humusowych na wzrost i rozwoj siewek salaty. Zesz.Probl.Post.Nauk Roln., 409, 182-189.

Gonet, S.S., A. Dziamski and E. Gonet, 1996. Application of humus preparations from oxyhumolite in crop production. Environ.Int., 22, 5, 559-562.

Gonet, S.S. and K. Wegner, 1994. The effect of mineral and organic fertilization on properties of soil humic acids. In: N.Senesi and T.Miano (Editors). Humic substances in the global environment and implications on human health. Elsevier, Amsterdam, 607-612.

Schnitzer, M. and S.I.M. Skinner, 1968. Gel filtration of fulvic acid, a soil humic compound. In: Isotopes and radiation in Soil Organic Matter Studies. IAEA Vienna, 41-55.

Vaughan, D. and R.E. Malcolm, 1985. Influence of humic substances on growth and physiological processes. In: D.Vaughan and R.E. Malcolm (Editors). Soil organic matter and biological activity. M.Nijhoff Publ./Dr.W.Junk Publ., Dortrecht, 37-75.

Zdanov, J.A., 1965. Srednaja stepen okislenija ugleroda i nezamenimost aminokislot. Biochimija, 30, 1257-1259.

CHANGES IN SOME PROPERTIES OF HUMIC SUBSTANCES FROM MELANUDANDS INDUCED BY VEGETATIONAL SUCCESSION FROM GRASS TO DECIDUOUS TREES

Teruo HIGASHI, Teiji SAKAMOTO and Kenji TAMURA

Laboratory of Soil Science
Institute of Applied Biochemistry,
University of Tsukuba
Tsukuba, 305-8572, Japan

1. INTRODUCTION

Melanudands are widely distributed in Japan and have been considered to be formed from volcanic ashes under the prevailing influence of grass vegetation such as Miscanthus sinensis (Yamane,1973; Kato, 1978). Recent studies on gramineous phytolith separated from the soil samples collected at different parts of Japan also support the intensive effect of grass vege-tation on the genesis of Melanudands in Japan (Kato, 1960; Sase and Kondo, 1974; Kondo and Sase, 1986; Sase *et al.*, 1985; Kawamuro and Torii, 1986; Hosono *et al.*, 1992). Melanudands in Soil Taxonomy (Soil Survey Staff, 1992) are characterized and classified by andic soil properties and melanic epipedon under udic moisture regime and are thought to be corre-lated with the soils within the different subgroups of Non-allophanic Andosols and Andosols groups in Japanese Soil Classification System (Classification Committee of Cultivated Soils, 1996).

High aromaticity in the chemical structure of accumulated humic substances in Ah horizon of Melanudands is probably one of the reason for such a blackish color of melanic epipedon, frequently corresponding to the existence of A-type humic acids (Kumada, 1987; Tate *et al.*, 1990). Moreover, a large portion of the humic substances in Melaudands are likely to form complexes with positively charged hydroxy-ions of Al and Fe, that is probably responsible for their elevated accumulation of humic substances, up to 25% as carbon in Ah horizons, and also for their high resistance to microbial decomposition (Kobo and Oba, 1974; Wada and Higashi, 1976; Higashi, 1983; Inoue and Higashi, 1988).

However, it has been pointed out that the blackish color of Ah horizon of Melanudands becomes somewhat lighter in color, so-called "decolorized process", under a certain vege-tational condition (Kawada, 1975). A similar phenomenon was observed for Melanudands under the secondary vegetational succession from grass to deciduous trees (Tamura *et al*, 1993a, b). These studies suggest that the change in the vegetation seems to be highly influ-ential on the "decolorized process" of blackish humic substances in Melanudands.

The objective of the present study is to examine the effects of vegetational succession from grass to deciduous trees on some chemical and optical properties of humic substances in Ah horizons of Melanudands, and to elucidate the plausible mechanism for their "decolorized process" observed at Sugadaira Montane Research Center, University of Tsukuba.

Effect of Mineral-Organic-Microorganism Interactions on Soil and Freshwater Environments
Edited by Berthelin, *et al*, Kluwer Academic / Plenum Publishers, New York, 1999

2. STUDY AREA AND SAMPLING SITES INFORMATIONS

As shown in Figure 1, Sugadaira Montane Research Center is located at the footslope of Mt. Neko-dake in Sugadaira Highland, Nagano prefecture, central Japan (36°36'N, 138°27'E). The elevation of this area is 1320 m a.s.l.. Mean annual temperature and annual precipitation is 6.5° and 1200mm, respectively. Although *Fagus crenata* forest is thought to be the climax forest in the natural vegetational succession in this area, the interstages of several secondary forests such as *Pinus densiflora, Betula platyphylla, Quercus crispula* and the afforested *Larix leptolepis* are frequently encountered, probably due to the strong human impacts as well as natural incidents of fire. In addition to these forests, there are several kinds of artificially maintained and managed grassland from older times which are used for grazing, acquisition of roofing materials, agricultual composts etc.. However, once the management of the grass vegetation by human beings is weakend or ceased natural vegetational succession is advanced gradually to the climax forest of *Fagus crenata* through *Pinus densiflora* and/or *Betula platyphylla* and *Quercus crispula* as the interstages at Sugadaira Highland (Hayashi, 1967; 1981).

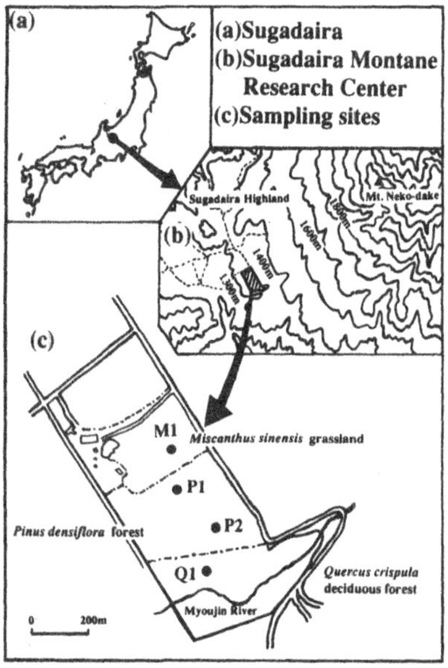

Figure 1. Location of sample sites.

Detailed map of sampling sites are shown in Figure 1. This sampling sites are now covered with three dominant species, *Miscanthus sinensis*, *Pinus densiflora* and *Quercus crispula*. This allocation of each species has been realized by the use of time-lag in the abandonment of human impacts on the original grassland of *Miscanthus sinensis*. That is, for the experimental purpose on the secondary succession, the management for the maintenance of grassland was ceased at *Quercus crispula* and *Pinus densiflora* site at about 50 and 25 years before, respectively. Four pedons were selected and observed under these vegetation, considering the previous studies by Tamura *et al.* (1993a, b). Soils are classified as Melanudands under *Miscanthus sinensis* and *Pinus densiflora*, while those under *Quercus crispula* as Hapludands. Especially, the soils under the latter two vegetation show the clear increase in the value and chroma of soil color, resulting in the "decolorized process" of blackish Ah horizon of Melanudands. Brief descriptions for the depth, soil color and structure of the subhorizons of Ah from four pedons are shown in Table 1.

Table 1. Some morphological characteristics, and physical and chemical properties of soil samples

Site Hor.	Depth (cm)	Color (moist)	Structure[1)]	pH H_2O	pH KCl	CEC[2)]	BS (%)	Tex-ture	BD[3)]	Poro-sity(%)
M1 site (*Miscanthus sinensis*)										
Ah1	0-14	7.5YR2/1	3fcr+sbk	5.27	4.26	43.1	21	LiC	0.45	79.4
Ah2	14-35	7.5YR2/1.5	2msbk	5.16	4.44	38.0	5	LiC	0.50	78.3
Ah3	35-48	7.5YR2/1.5	2f-mcr+sbk	5.26	4.49	37.5	3	LiC	0.53	78.2
P1 site (*Pinus densiflora*)										
Ah1	0-12	7.5YR2/2	2fcr+sbk	4.80	4.13	38.4	9	LiC	0.41	81.9
Ah2	12-25	7.5YR2/2	2f-mcr+sbk	5.00	4.43	34.5	2	LiC	0.42	83.0
Ah3	25-39	7.5YR2/1.5	2f-msbk	5.06	4.62	31.5	2	LiC	0.49	80.1
P2 site (*Pinus densiflora*)										
Ah1	0-10	7.5YR2/2.5	3f-mcr	5.38	4.37	32.8	21	LiC	nd	nd
Ah2	10-25	7.5YR2/3	2f-mcr+sbk	5.16	4.47	27.1	12	LiC	nd	nd
Ah3	25-45	10YR3/3	2f-msbk	5.23	4.74	16.5	7	LiC	nd	nd
Q1 site (*Quercus crispula*)										
Ah1	0-11	7.5YR2/3	2fcr+1fsbk	5.18	4.33	33.8	14	LiC	0.39	83.2
Ah2	11-23	7.5YR2.5/3	1fcr+2msbk	5.38	4.61	26.8	5	LiC	0.51	79.0
Ah3	23-41	10YR3/3.5	2f-msbk	5.45	4.87	6.8	6	HC	0.67	74.2

1)after Soil Survey Manual (USDA, 1951), 2) expressed as cmole kg^{-1}, 3)expressed as Mgm^{-3}.

3. MATERIAL AND METHODS

Soil samples were taken from Ah horizons of four pedons, one from *Miscanthus sinensis* site, two from *Pinus densiflora* site and one from *Quercus crispula* site. These sampling sites are named as M1, P1and P2, and Q1, respectively. P2 site is in more advanced stage in vegetational succession than P1, judging from the average DBH (Diameter of Breast Height) of *Pinus densiflora* and the predominance of *Miscanthus sinensis* undergrowth. Each Ah horizon of four pedons were further divided into three subhorizons, namely Ah1, Ah2 and Ah3. Total of these twelve samples were air-dried and passed through 0.5mm sieve and used for the following experiments.

Besides some routine chemical and physical analyses of soil samples, including Melanic Index (Honna *et al.*., 1988), extraction of humic substances with 0.1M sodium pyrophosphate (Blakemore *et al.*, 1987) was done to determine the amounts and FA/HA ratio of extracted humic substances, and E4/E6 ratio and RF values (calculated as the absorbance at 600nm per unit of carbon; Kumada, 1987) of humic acids. Solid state CP/MAS 13C-NMR analysis for whole soil samples were done after Golchin *et al.* (1994). Amounts of ten kinds of low-molecular-weight aliphatic carboxylic acids (LACAs) were extracted with distilled water and determined by the method of Tani *et al.* (1993), where HPLC analysis was done on the purified water extracts with polyvinyl pyrrolidone(PVP) and liquid-liquid continuous extraction with diethyl ether. Determination of Al, Fe and Si in acid ammonium oxalate (0.2M, pH3) extracts as well as those in sodium pyrophosphate (0.lM, pH10) extracts were done by ICP and AAS analysis following after Blakemore *et al.*(1987).

4. RESULTS AND DISCUSSIONS

4.1. Some morphological characteristics of soils, and some physical and chemical properties of soil samples

As shown in Table 1, when proceeded from M1 site to Q1 site, the thickness of Ah horizon tends to be thinner, and the color becomes lighter, although the structure showed no distinct change among the four sites. Data of bulk density (BD) and porosity(%) showed the develop-ment of pores with vegetational succession, especially in Ah1 horizons, as pointed out by Tamura *et al.* (1993a). No clear tendency was observed for base saturation degree (BS) and texture. CEC was clearly decreased from the samples in M1 site to those in Q1 site, being probably reflected by the decrease in total carbon contents (Table 2). pH value seems to be the lowest in Ah1 horizon sample from P1 site.

Table 2. Total carbon and nitrogen contents, and characteristics of extracted humic substances

Site Hor.	Total-C (g/kg)	Total-N (g/kg)	C/N	EC/TC[1] (%)	FA/HA[2]	E4/E6	RF[3]	HA[4] type	Melanic Index
M1 site (*Miscanthus sinensis*)									
Ah1	193	11.9	16.2	37	0.30	3.33	108	A	1.566
Ah2	131	7.1	18.5	40	0.39	3.41	117	A	1.641
Ah3	134	6.4	20.9	42	0.22	3.44	132	A	1.549
P1 site (*Pinus densiflora*)									
Ah1	166	9.8	16.9	37	0.77	3.58	85	A	1.641
Ah2	124	6.8	18.2	39	0.81	3.40	89	A	1.633
Ah3	99	5.1	19.4	34	0.93	3.37	125	A	1.579
P2 site (*Pinus densiflora*)									
Ah1	147	8.9	16.5	35	0.72	3.61	71	B	1.677
Ah2	115	6.9	16.6	34	0.81	3.78	85	A	1.618
Ah3	66	3.9	16.9	34	0.93	3.45	106	A	1.585
Q1 site (*Quercus crispula*)									
Ah1	133	8.2	16.2	33	0.92	3.58	71	B	1.683
Ah2	100	6.4	15.6	27	1.49	3.22	78	B	1.663
Ah3	50	3.1	16.1	25	1.67	3.33	93	A	1.606

[1]ratio of extracted carbon to total carbon, [2]ratio of the amount of fulvic acid to humic acid, determined by $KMnO_4$ wet oxidation, [3]absorbance at 600nm divided by unit of carbon comsumed by $KMnO_4$ wet oxidation (Kumada, 1987), [4]humic acid type according to Kumada(1987), [5] determined after Honna *et al.* (1988).

4.2. Total carbon and nitrogen contents, and some characteristics of extracted humic substances

As shown in Table 2, it was clear that the total carbon and nitrogen contents of all sub-horizons samples decreased regularly from M1 site to Q1 site. C/N ratio was somewhat fairly constant at Ah1 horizons, while in Ah2 and Ah3 horizons it tended to decrease with vege-tational succession. Higher values of C/N ratio for Ah2 and Ah3 compared with Ah1 horizon is frequently encountered for Melanudands in Japan. Extractability of humic substances with sodium pyrophosphate, as expressed by the ratio of extracted carbon to total carbon (EC/TC), is comparable to those values in the previous studies (Wada and Higashi, 1976; Inoue and Higashi, 1988). EC/TC ratio and RF values of humic acids decreased, while FA/HA ratio (fulvic to humic acid ratio) increased with vegetational succession. However, no distinct change was observed in the values of E4/E6 of humic acids. These data suggest that humic substances in original grassland soils are likely to be subjected to the decomposition and to the changes in their structure under advanced stages of forest vegetation. This phenomena is well reflected to the type of humic acids (Kumada, 1987) and also to the value of Melanic Index, which changes from A-type to B-type and increases to the values of almost 1.7, respectively.

4.3. Chemical composition of carbon revealed by solid state CP/MAS ^{13}C-NMR for whole soil samples

For a semi-quantitative comparison, as shown in Figure 2, the spectra were divided into four chemical shift ranges, 0-46, 46-110, 110-164, 164-190 ppm, which were used to estimate alkyl, O-alkyl, aromatic and carboxyl (carbonyl) carbon, respectively (Baldock *et al.*, 1992; Golchin *et al.*, 1994).

As it is clear from Table 3, aromatic carbon apparently decreased from 21.5 to 13.7% of total signal intensity, while both alkyl and carboxyl (carbonyl) carbon increased from about 25 to 30% and from 9 to 13%, respectively, with the vegetational succession from M1 to Q1 site. Such a tendency was not evident for O-alkyl carbon. Decomposition processes of original soil organic matter under forest vegetation , shown by the decrease in the total carbon content in P1, P2 and Q1 sites (Table 2), may proceed with the increase of alkyl and carboxyl (carbonyl) carbon as well as alkyl/O-alkyl ratio, as proposed by Baldock *et al.* (1992).

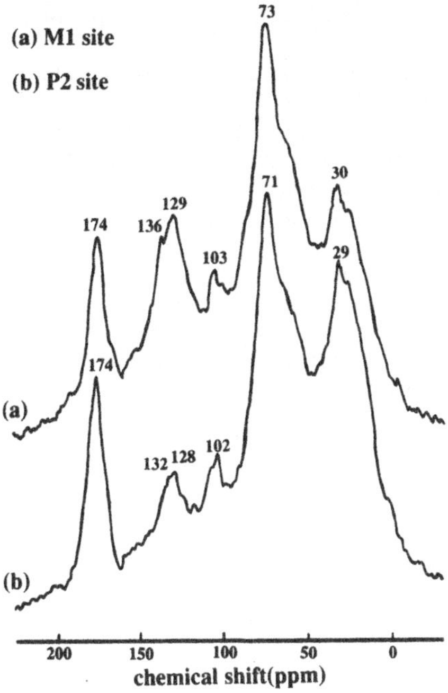

Figure 2. Solid state of CP/MAS ^{13}C NMR spectra of whole soil samples.

Table 3. Distribution of the respective signal intensities collected in carboxyl(carbonyl), aromatic, O-alkyl and alkyl carbon spectral regions of solidCP/MAS ^{13}C-NMR spectra.

Site Horizon	Alkyl (0-46ppm)	O-alkyl (46-110ppm)	Aromatic (110-164ppm)	Carboxyl (164-190ppm)	Alkyl/O-alkyl ratio
M1 site(*Miscanthus sinensis*)					
Ah1	24.8	44.7	21.5	9.0	0.55
P1 site (*Pinus densiflora*)					
Ah1	27.6	48.1	15.2	9.1	0.57
P2 site (*Pinus densiflora)*					
Ah1	30.9	42.3	14.4	12.4	0.73
Q1 site (*Quercus crispula*)					
Ah1	31.4	42.1	13.7	12.8	0.75

Figures, except for Alkyl/O-alkyl ratio, are the relative percentage of signal intensities in respective spectral regions to the total signal intensities.

Although the spectra of whole soil samples are not strictly identical to the extracted humic substances, the decrease in aromatic carbon with vegetational succession is conclusive. Thus, the changes in chemical composition of carbon revealed by solid state CP/MAS 13C-NMR spectra are quite consistent with those observed in some chemical and optical properties of extracted humic substances, such as FA/HA ratio and RF values of humic acids (Table 2). It is well expected that the vegetational succession from grass to deciduous trees brings the marked changes in the microorganisms induced by the change in substrate, which results in the changes in the turnover rate of carbon (Oades, 1988), and further leads to the "decolorized process" of humic substances with higher aromaticity in Melanudands by the assimilation/decomposition processes.

4.4. Amounts of low-molecular-weight aliphatic carboxylic acids(LACAs)

Table 4 shows the amounts of five major LACAs and the sum of LACAs including other five minor carboxylic acids, extracted by phosphate buffer solution . As it is clear, sum of LACAs increased regularly with the vegetational succession from about 211É mol/kg in Ah1 horizon of M1 site to 316É mol/kg in Q1 site. These data are comparable with those observed for some forest soils in Japan (Tani *et al.*, 1993). Since the production of these LACAs is attributed to root acivities and the intermediate decomposition products of organic debris, changes in vegetation are expected to have great influence on LACAs. Although these LACAs are present in small quantity in soils as the equilibrium concentration between their production and decomposition rate, the total quantity produced in a certain long period, e.g. 50 or 25 years of duration in this study, would be quite sufficient to their complexing and dissolution reaction with polyvalent metals/hydroxides, such as Al and Fe (Stevenson, 1967; Pohlman and McColl, 1986, 1988; Tam and McColl, 1991).

Table 4. Amounts of five major LACAs and the sum LACAs in Ah1 horizon samples

LACA's name and sum	M1 site (*Miscanthus sinensis*)	P1 site (*Pinus densiflora*)	P2 site (*Pinus densiflora*)	Q1 site (*Qurercus crispula*)
formic acid	60.4	68.8	84.8	79.4
acetic acid	37.5	46.6	64.8	66.2
oxalic acid	34.8	30.6	55.0	54.1
succinic acid	25.5	27.6	28.0	27.1
citric acid	39.1	44.3	50.6	53.3
Sum of LACAs	210.6	242.1	309.0	315.8

Figures are shown as mmol per 1kg of oven-dried soils. Sum of LACAs includes other five minor LACAs of propionic acid, lactic acid, butyric acid, fumaric acid and malic acid.

Out of the five major acids, the amount of formic acid was the highest, followed by acetic acid, citric acid, oxalic acid and succinic acid, where oxalic acid and citric acid are supposed to be highly responsible for the complexing and dissolution reaction with Al and Fe ions in soil solution as well as at the interface between solid and soil solution (Stevenson, 1967, Huang and Violante, 1986). So, we believe that, in a long span of time as required for the vegetational succession, the increased amounts of LACAs in addition to fulvic acids (Table 2) under forest vegetation would cause the "decomplexing process" of Al and Fe ions from biochemically-stable original humus complexes in grassland soil.

4.5. Amounts of pyrophosphate and acid oxalate extractable Al, Fe and Si with respect to the vegetational succession

As shown in Table 5, amounts of sodium pyrophosphate extractable Al and Fe apparently decreased with the vegetational succession. This tendency was in accordance with the regular decrease in the amounts of extracted humic substances (Table 2). In contrast, the amounts of acid ammonium oxalate extractable Al, Fe and Si seem to increase regularly with the vegetational succession. The difference in those amounts between sodium pyrophosphate and acid ammonium oxalate extraction, designated as Alox-pyr and Feox-pyr in Table 5, also increased. This tendency is more evident for Al than Fe. The increase in Alox-pyr, referred to amorphous Al in this study, could be interpreted as follows: the originally complexed Al as stable humus complexes in M1 site is gradually removed by the increased amounts of LACAs and fulvic acids (Table 4), and is migrated downward the profile followed by the subsequent microbial decomposition. Liberated Al in this manner will form amorphous Al. This postulated transformation process of Al is to be more pronounced in forested sites, since the more Al is removed from original humus complexes in M1 site under the increased influences of organic ligands, the more the amounts of amorphous Al will increase after the decompos-ition of organic ligands.

Table 5. Amounts of pyrophosphate and acid oxalate extractable Al, Fe and Si(g/kg)

Site Horizon	Alpyr[1)]	Fepyr	Alox[2)]	Feox	Siox	Alox-pyr[3)]	Feox-pyr	Alox-pyr/Sio[4)] atomic ratio
M1 site (*Miscanthus sinensis*)								
Ah1	18.2	6.6	22.7	14.9	2.3	4.5	8.3	2.04
Ah2	17.8	6.5	35.	19.0	6.8	17.3	12.5	2.65
Ah3	20.7	5.3	39.7	18.7	7.2	19.0	13.4	2.75
P1 (*Pinus densiflora*)								
Ah	17.1	6.2	23.5	12.5	3.2	6.4	6.3	2.08
Ah2	15.5	5.5	39.0	15.7	9.5	23.5	10.3	2.58
Ah3	13.8	4.1	46.1	16.7	13.2	32.4	12.6	2.56
P2 (*Pinus densiflora*)								
Ah1	14.0	5.0	26.1	12.7	6.3	12.1	7.7	2.00
Ah2	13.3	4.5	34.0	14.7	9.6	20.7	10.2	2.24
Ah3	12.4	3.3	39.3	15.0	12.2	26.9	11.7	2.30
Q1 (*Quercus crispula*)								
Ah1	15.6	4.4	33.2	14.9	9.0	17.6	10.5	2.04
Ah2	11.7	2.9	45.8	17.0	14.2	30.1	14.1	2.21
Ah3	8.6	1.0	53.9	17.9	9.9	45.3	17.0	2.47

[1)]denotes pyrophosphate extraction, [2)]denotes acid oxalate extraction, [3)]difference in the amounts between both extractions, [4)]roughly corresponding to Al/Si atomic ratio of allophane/imogolite (Parfitt, 1988; 1989).

Since the amounts of Si extracted with acid ammonium oxalate, roughly corresponding to allophane/imogolite contents (Parfitt and Childs, 1988), also increased with the vegetaional succession (Table 5), liberated Al from humus complexes explained above may combine with Si to form allophane/imogolite in addition to amorphous Al. Al/Si atomic ratio of allophanes calculated after Parfitt and Childs (1988) in Ah1 horizons in all sites showed the values of about 2.0, which are frequently encountered for Andisols from different countries under udic moisture regime (Parfitt and Kimble, 1989). Allophanes in Ah2 and Ah3 horizons were aluminum-rich in composition and showed the decrease in their Al/Si atomic ratio from about 2.7 in M1 site to about 2.3 in Q1 and P2 sites. However, this speculation for the forma-tion of allophane/imogolite derived from the removed Al from humus complexes shoud be studied in detail in future.

5. CONCLUSION

Effects of vegetational succession from grass to deciduous trees on some optical and chemical properties of humic substances as well as on Al transformation will be summarized and concluded as follows. The processes of changes involved in some properties of humic substances together with Al are shown in Figure 3 as a schematic representation.

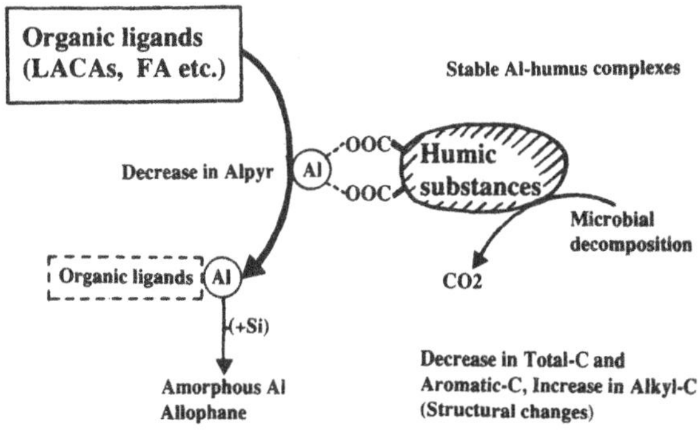

Figure 3. Schematic representation of plausible mechanism for "decolorized processes".

With the vegetational succession, complexed Al and Fe ions with humic substances having high aromaticity in their strucutre, originally present as biochemically-stable humus com-plexes in grassland site, are to be removed or decomplexed by the increased amounts of active organic ligands, such as LACAs and fulvic acids in forested sites. Thus removed Al, possibly Fe, in the forms of organic ligands complexes are moved down the Ah horizons with the movement of water where the porosity of soils is increased under forest vegetation. However, such complexes are somewhat unstable and subjected to microbial decomposition. Meanwhile, humic substances without complexed Al and Fe ions thus formed may be easily assimilated as useful sbstrate by microorganisms. And, original aromatic character of humic substances in grassland is weakened, where the simultaneous enrichment of alkyl and carboxyl (carbonyl) carbon in humic substances occurs. In this course of "decolorized process" of Ah horizon of Melanudands, Al liberated from the migrated organic ligands complexes will then form amorphous Al or even combine with Si to form allophane/imogolite *in situ.*

6. ACKNOWLEDGEMENT

We express the sincere thanks to Dr P. Clarke and Prof. J. M. Oades for providing the oppor-tunity to study on solid state CP/MAS 13C-NMR analysis of humic substances, while one of the authors, Dr. T. Higashi, stayed at Waite Campus, University of Adelaide, South Australia, as Oversea Research Fellow of the Ministry of Education, Science and Culture, Japan. Thanks are also extended to Misses N. Inoue and Y. Yamakaji for their technical assis-tance in some analyses of soil samples.

7. REFERENCES

Baldock, J. A., J. M. Oades, A. G. Waters, X. Ping, A.M. Vassallo and M. A. Wilson, 1992. Aspects of chemical structure of soil organic materials as revealed by solid state ^{13}C NMR spectroscopy. Biogeochemistry, 16, 1-42.

Blakemore, L.C., P.L.Searle and B.K.Daly, 1987. Extractable iron, aluminum and silicon, In: Methods for chemical analysis of soils. NZ Soil Bureau Scientific Report No. 80, pp. 71-76, DSIR, Lower Hutt.

Classification Committee of Cultivated Soils, 1996. Classification of cultivated soils in Japan, third Approximation. National Institute of Agroenvironmental Sciences, Tsukuba, 62pp.

Golchin A., J. M. Oades, J. O. Skjemstad, and P. Clarke, 1994. Study of free and occluded particulate organic matter in soils by solid state 13C CP/MAS NMR spectroscopy and scan-ning electron microscopy. Aus. J. Soil Res., 32, 285-309.

Hayashi, I., 1967. Study on vegetational succession in Sugadaira Highland(Part I). Bull. Sugadaira Montane Research Center, 1, 1-18.

Hayashi, I., Y. Nishimura, T. Yanasawa, 1981. Structure and functioning of *Miscanthus sinensis* grassland in Sugadaira, central Japan. Vegetatio, 48, 17-25.

Higashi, T. 1983. Characterization of Al/Fe-humus complexes in Dystrandepts through comparison with synthetic forms. Geoderma, 31, 277-288.

Inoue, K. and T. Higashi, 1988. Al- and Fe-humus complexes in Andisols, In: Proc. 9th International Soil Classification Workshop(August 1987, Japan), pp.81-96.

Honna, T., S. Yamamoto, and K. Matsui, 1988. A simple procedure to determine Melanic Index that is useful for differentiating Melanic from Fulvic Andisols. Pedologist, 32, 69-78 and also in" ICOMAND Circular Letter", No.10, pp. 76-77.

Hosono, M.T. Sase, and K. Aoki, 1992. Estimation of the basal age of Kuroboku soils by using time maker tephras: an example at tephra profile in Yunodai, the area of Towada volcanic ashes. The Quaternary Research, 46, 121-132.

Huang, P.M. and A. Violante, 1986. Influence of organic acids on crystallization and surface properties of precipitation products of aluminum. In: P.M. Huang and M. Schnitzer (Editors), Interaction of Soil Minerals with natural Organics and Microbes, SSSA Special Publication, N° 17, Madison, p. 159-221.

Kato, Y., 1960. Opal phytolith in Kuroboku soils. Jpn J. Soil Sci. Plant Nutr., 30, 549-552.

Kato, Y., 1978. Genesis and classification of humus-rich soils(Kuroboku soils).

Jour. JSIDRE, 46, 869-876.

Kawada, H., 1975. Study on humus in forest soils(Part 3), Forms of humus in Black soils, Degraded Black soils, Red soils, Yellow soils and Rendzina-like soils. Bull. Gov. For. Exp. Sta., No. 278, pp. 51-74.

Kawamuro, K. and A. Torii, 1986. Difference in past vegetation between Black soils and Brown Forest soils derived from volcanic ash at Mt. Kurohime, Nagano Prefecture, Japan. The Quaternary research, 25, 81-98.

Kobo K. and Y. Oba, 1974. Factors for humus accumulation in volcanic ash soils and its effects on some properties of soils. Jpn J. Soil Sci. Plant Nutr., 45, 293-297.

Kondo, R. and T. Sase, 1986. Opal phytolith, their nature and application. The Quaternary research, 25, 31-63.

Kumada, K., 1987. Chemistry of Soil organic matter. Japan Scientific Societies Press (Tokyo)-Elsevier(Amsterdam), 241 pp.
Oades, J. M., 1988. The retention of organic matter in soils. Biogeochemistry, 5, 35-70.
Parfitt, R.L. and C.W.Childs, 1988. Estimation of forms of Fe and Al: A review and analy-sis of contrasting soils using dissolution and Moessbauer methods. Aus. J. Soil Res., 26, 121-144.
Parfitt, R. L. and J. M. Kimble, 1989. Conditions for formation of allophane in soils. Soil Sci. Soc. Am. J., 53, 971-977.
Pohlman, A. A., and J. G. McColl, 1986. Kinetics of metal dissolution from forest soils by soluble organic acids. J. Environ. Qual., 15, 86-92.
Pohlman, A. A. and J. G. McColl, 1988. Soluble organics from forest litter and their role in metal dissolution. Soil Sci. Soc. Am. J., 52, 265-271.
Sase, T. and R. Kondo, 1974. The study of opal phytolith in the humus horizon of burried volcanic ash soils in Hokkaido. Res. Bull. Obihiro Univ., 8, 465-483.
Sase, T. , Y. Kato and S. Makino, 1985. Plant opal analysis of volcanic ash soils at the footslopes of Mt. Fuji and Mt. Amagi. Pedologist, 29, 44-59.
Soil Survey Staff, 1992. Key to Soil Taxonomy, 5th edition, SMSS technical monograph No. 19. Blacksburg, Virginia, Pocahontas Press, 556 pp.
Stevenson, F. J., 1967. Organic acids in soil. In: Soil Biochemistry, vol. 6(ed. by McLaren, A.D. and G.H.Peterson), 119-146, Marcel Dekker, NY.
Tam, S. C. and J.G. McColl, 1991. Aluminum-binding ability of soluble organics in doug-las fir litter and soil. Soil Sci. Soc. Am. J., 55, 1421-1427.
Tate, K.R., K. Yamamoto, G.J.Churchman, R.Meinhold and R.H. Newman, 1990. Re-lationship between the type and carbon chemistry of humic acids from some New Zealand and Japanese soils. Soil Sci. Plant Nutr, 36, 611-621.
Tamura K., S. Nagatsuka, and Y. Oba, 1993a. Effects of secondary succession on physical and chemical properties of Ando soils in central Japan. Jpn. J. Soil Sci. Plant Nutr., 64, 166-176.
Tamura, K. S. Nagatsuka, and Y. Oba, 1993b. Effects of secondary succession on humus characteristics of Ando soils in central Japan. Jpn. J. Soil Sci. Plant Nutr., 64, 177-182.
Tani, M., T. Higashi and S. Nagatsuka, 1993. Dynamics of low-molecular-weight aliphatic carboxylic acids(LACAs) in some forest soils. Part I: Amounts and composition of LACAs in different types of forest soils in Japan. Soil Sci. Plant Nutr., 39, 485-495.
Wada, K. and T. Higashi, 1976. The categories of aluminum- and iron-humus complexes in Ando soils determined by selective dissolution. J. Soil Sci., 27, 357-368.
Yamane, I., 1973. Significance of Miscanthus sinensis for the genesis of Kuro-boku soils. Pedologist, 17, 84-94.

CHARACTERIZATION OF THE ORGANIC SUBSTANCES IN RECLAIMED SOILS

L.Petrova[1], M.G. Sokolovska[2], M.Gaiffe[3]

[1]"N. Poushkarov" Research Institute, Sofia, Bulgaria
[2] Forest Research Institute, Bulg. Acad. of Sci., Sofia, Bulgaria
[3] University "Franche - Comte", Besançon, France

1. INTRODUCTION

The formation of recent soils during the performance of recultivation activities in ecosystems of technogenic landscapes originating from coal-mining and coal-processing, is related to the type of zonal soils (the main soil unit of every ecological zone) (Trofimov et al., 1979). This is because the processes of humus accumulation and stabilization of its fractions in technogenic ecosystems, are very close to the processes in zonal soils with nearly the same factors of plants decomposition and microbial transformations.

2. MATERIALS AND METHODS

The present investigation has been carried out in the region of the Pernik Coal Basin (Bulgaria) with the aim of elucidating the more substantial specific features of the group and fractional composition of soil organic matter in ecosystems after coal production. The experimental sample areas are situated on one of the old dump fills near the T. Nenkov mine. The site is in immediate proximity to the residential quarters of the town of Pernik. Reclamation activities for its planting took place in 1957.

An afforested park with a mixed tree-shrub vegetation and naturally pollinated grass, was formed. The dump fill that had drastically disturbed the ecological balance has since become one of the most attractive places for recreation.

The composition of the fill materials making up the refuse dump has exerted a direct impact on the soil formation processes. These materials consisted mainly of clayey schists and marls with admixtures of coaly shale and coal. The filling started as early as 1891 and was finished in 1954. Humus soils were also introduced from other places during this period.

The soils of 3 sites under tree stands of different composition (acacia, oak, ash-tree) and one under mixed grass verdure were studied on various exposures and slopes. The samples were collected according to either morphological horizons, if the latter were differentiated or at two depths for a profile. The organic substance from the reclaimed soils was fractionated according to the method of Kononova-Belchikova (Kononova, 1961). Total nitrogen was

Effect of Mineral-Organic-Microorganism Interactions on Soil and Freshwater Environments
Edited by Berthelin, *et al*, Kluwer Academic / Plenum Publishers, New York, 1999

determined by the Kjeldhal procedure (Bremner, 1965), organic carbon by the modified method of Tyurin (Kononova, 1961), Humus status criteria were determined by the system of Grishina and Orlov (1978).

3. RESULTS AND DISCUSSION

The results showed that the investigated soil substrates contained organic substances with characteristic components. The ratios between the individual components were rather different in quantity and according to the distribution in the profile. The studies were carried out for each of the experimental areas (see Table 1). The differences in the humus content were considerable, since in the case of two-layer profiles a top surface black layer and a lower cindered orange-red one were morphologically established. The organic carbon content in the dark horizons varied between 6% and 16% — a high to very high level, but organic carbon was low to very low (0.30-0.70%) in the brightly colored underlying layers. A great part of the organic carbon was in fact the coal carbon admixtures in the substrate. The real humus carbon in these soils was mainly the carbon obtained from the introduced soil layers used to prevent self inflammation of the fill materials. The newly formed organic substances represented only a minor part of the total carbon content. Both types of humus — the introduced and the newly formed — had the characteristics of the zonal ecologically balanced humus which is of humate type, rich in nitrogen and has humic acids molecules with average molecular weight.

Regardless of the origin of the humus horizon in the recultivated soils (either during the new natural soil formation or from the dump layer), it is very important for the humus substances to have sufficient resistance against oxidation-hydrolization actions. It has been proved (Schnitzer and Skinner, 1969) that the more unstable forms of humic acids that are easily subjected to chemical transformations prevail at the early stages of humus substance formation. A general feature of the newly formed humus profile is its small thickness connected to the surface. This fact is related to the initial character of the settling organisms and to the metabolism products and residues, which are fundamental to the formation of the soil organic matter.

The assessment of the possibilities for the formation of different organic-mineral complexes in the soils is of principal importance for the creation of a new humus horizon. The penetration of humic acids into the clayey minerals is quite possible and is one of the ways for the formation of organic-mineral compounds in soils (Kobo and Fujisawa, 1964). The formation of stable adsorption complexes of humic acids and the clayey minerals during their surface interaction is another way this mechanism works. The common feature of the organic substances in newly formed soils is the high percent C in the insoluble or non-extractable fraction. According to our data it varies slightly for the different areas and represents more than 66% of the total organic carbon. These values are extremely high under the canopy of the ash-tree stands and in some cases are higher in the lower layers.

According to Schnitzer and Skinner (1969) data the character of the interaction between the humic acids and the clayey minerals is influenced to a great extent by the pH of the soil solution. It is well known that under conditions of more alkaline reaction and in the presence of carbonates, the additional carbon admixtures accelerate the processes of humus formation and humus accumulation in the soil (for example, under the ash-tree stands and grass verdure). The degree of base saturation (V%) here is rather high. The total acidity is negligible and is represented by slightly acidic exchangeable hydrogen. All of the examined soil substrates possess an irregular colloid structure and demonstrate low reactivity. These soils are developing into illite-kaolinite structures, according to the contemporary clay evolution.

The data obtained for the organic substance in the studied sites exhibited a broader ratio between total C and total N (C:N - exceeding 20). Analogous data have been reported by other authors (Taranov et al., 1979). This ratio is lower (8-19) in the organic substance from the acacia sites, this is explained by the presence of mineral, inorganic and unexchangeably fixed ammonium nitrogen (Black, 1973). The recultivated areas were especially rich in total nitrogen. Its values changed insignificantly throughout the studied areas. For the acacia cultures alone this quantity was sharply changed from the highest value at the surface to the lowest one in depth (0.636 % - 0.021 %). The chemical composition of the plant remains is of major importance in this case, followed by the mineralogical composition of the brightly colored substrates layers.

The influence of the forest vegetation on the humus characteristics is expressed only as a trend over the 40 year period. Humic acids complexed with calcium prevailed in the whole profile under the ash-tree cultures and the grass formations. Free or R_2O_3 bound humic acids prevailed under the oak and acacia stands. The greater degree of humic acids bonded with Ca in the first case is explained by the greater silt content under the ash-tree canopy (20-40 %). The coal presence in the lower layers hampered the normal humic acid distribution and they even were completely absent in some cases. The higher amount of humic acids in the surface layers at this early stage of the soil formation process resulted from the more intensive deposition and transformation of the fresh plant remains. Moreover, the humic acids from oxidized coal have a structure which is similar to that of the chernozem-type humic acids (Komissarov et al., 1979). In this respect, humic acids from oxidized coal will favor the soil formation processes taking place under the different cultures. Hence, the profile distribution of carbon and humic acids can indicate whether the basic source of the organic substance is of phytogenic origin or not. Accumulation in the surface soil layer and sharp decrease at depth is observed in the first case. The example with acacia cultures is a very typical one; these differences are on the order 10-14 times (see Table 1).

A substantial peculiarity of the organic substances in the studied areas is the small amount of fulvic acids, as well as, the absence of the fraction 1a ("aggressive" fulvic acids, extractable with 0.1 N H_2SO_4 solution, the most dynamic fulvic acids with low molecular weight fulvic acids fraction) in their composition. Another peculiarity is the ratio of the humic acids alkaline solutions optical densities at 465nm and 665nm, E4/E6. This ratio varied from 4.4 to 5.7 independently on the vegetation type. Under normal conditions of humus formation, the recorded ratio E4/E6 should indicate mature condensed humic acids with higher amounts of carboxil groups (Chen et al., 1977). The studied organic substances have a humate type of humus, i.e., the ratio Cha/Cfa (carbon of humic to carbon of fulvic acids ratio) has values higher than 1. In the case with oak-tree vegetation it reaches the values of 20-22. All these facts can be explained by keeping in mind that one of the factors of humus formation in the experimental areas was the coal admixtures. They positively affected formation of extractable humic acids, and consequently the low amount of fulvic acids, E4/E6 ratio and its variations as well as the widening of the Cha/Cfa ratio.

Table 1: Organic carbon amount and characteristics, total nitrogen

Depth cm	Total organic C (%)	Total N (%)	C/N	Extractable carbon (%) total	humic acid	fulvic acid	Cha/ Cfa	Humic acid carbon (%) Free or R_2O_3 complexed	Ca complexed	Unextrac-table C (%)	E4/ E6	pH
1	2	3	4	5	6	7	8	9	10	11	12	13
P1	ash tree											
Ah	8.01	0.34	23.4	1.09	0.82	0.27	3.04	0.06	0.76	6.92	5.6	7.6
0- 8				13.6	10.24	3.37		7.32	92.68	86.4		
ABI	8.15	0.30	26.9	1.03	0.71	0.32	2.22	0.06	0.65	7.12	5.7	7.5
8-20				12.6	8.71	3.94		8.45	91.55	87.4		
II	7.69	0.25	30.0	0.62	0.42	0.20	2.10	0	100	7.07	5.2	7.7
20-48				8.06	5.46	2.60				91.94		
III	7.34	0.26	31.1	0.43	0.43	0	0	0	100	6.91	5.0	7.9
48-70				5.86	5.86					94.14		
P3	ash tree											
A	8.27	0.26	31.0	1.10	0.74	0.36	2.05	0.04	0.70	7.17	5.3	7.7
0-10				13.3	8.95	4.35		5 .41	94.59	86.70		
I	9.55	0.29	32.1	1.31	1.07	0.24	4.46	0	100	8.24	5.0	7.3
10-31				13.7	11.20	2.51				86.3		
P2	acacia											
Ah	5.71	0.63	8.98	1.31	1.02	0.29	3.92	0.42	0.60	4.40	5.1	6.0
0-15				22.9	17.86	5.08		41.18	58.82	77.1		
I	0.41	0.04	8.72	0.08	0	0 08	0	0	0	0.33	_	5.5
15-31				19.5		19.5				80.5		

II	0.31	0.02	14.7	0.05	0	0.05	0	0	0	0.26	–	6.4
31-60				16.1		16.1				83.9		
P46	acacia											
A	7.63	0.40	19.0	2.38	2.14	0.24	8.92	1.82	0.32	5.25	5.4	5.4
0-13				31.2	28.05	3.15		85.05	14.95	68.8		
AI	0.70	0.04	18.4	0.19	0.12	0.07	1.71	100	0	0.51	5.5	6.4
13-18				27.1	17.14	10.0				72.9		
P4	meadow											
A	9.14	0.34	26.6	1.38	0.79	0.59	1.34	0.09	0.70	7.76	5.4	7.5
0-21				15.1	8.84	6.46		11.39	88.61	84.90		
I	7.28	0.23	27.6	0.53	0.44	0.09	4.90	0.07	0.37	6.75	4.4	7.8
21-65				7.28	6.04	1.24		15.91	84.09	92.72		
P5	oak											
A	14.91	0.53	27.7	3.03	2.89	0.14	20.6	1.40	1.49	11.88	5.4	5.8
0-19				20.3	19.38	0.94		48.44	51.56	79.7		
I	16.14	0.44	36.3	3.00	2.42	0.58	4.17	1.49	0.93	13.14	5.1	5.2
19-39				18.6	14.99	3.59		61.57	38.43	81.4		
P8	oak											
Ah	15.9	0.58	27.1	2.63	2.52	0.11	22.91	2.33	0.19	13.27	5.4	5.0
0-14				16.5	15.85	0.69		92.46	7.54	83.5		
AI	10.81	0.57	18.7	1.76	1.68	0.08	21.00	100	0	9.05	5.3	3.8
14-26				16.3	15.54	0.74				83.7		

characters over the line represent percent of dry matter.

characters under the line represent percent of organic carbon.

4. CONCLUSION

The performed investigations show that the humification of plant residue in forest and grass biocenoses on recultivated soils proceeds to formation of ecologically balanced humus. All of the parameters for the humus condition of soils prove positive effects of the phytocenoses. The formed ecosystems gradually reach the state of dynamic equilibrium with the environmental factors. At the same time a constant, although irregular, increase of the productivity of plant associations is observed and hence the income of plant materials to soils is increased. The humus resources increase until, as a result of the processes of mineralization and washing away, humus losses become equal to the amount of plant organic substances entering the soil. Hence, humus formation and humus accumulation at the early stages of soil formation are the basic elementary soil formation processes.

5. ACKNOWLEDGEMENTS

The investigations have been performed with the financial support of the National Fund "Scientific Investigations" of the Bulgarian Ministry of Education, Science and Technologies.

6. REFERENCES

Black, C. A., 1973. Plant and Soil. M., p.503 (in Russian).

Bremner, J. M., 1965. Total nitrogen. In: Methods of soil analysis, Part 2, p.p. 1149-1178, Ed Black, C. A., American Society of Agronomy, Medison, Wisconsin, USA.

Chen, Y., N. Senesi and M. Shnitzer, 1977. Information Provided on Humic Substances by E4/E6 Ratios. Soil Science Society of America Journal, v.41, N 2, p.p. 352-358.

Grishina, L. A., and D. S. Orlov, 1978. Soil Organic Matter Status System of Criteria. Soil Science Problem, M.

Kobo, K, and T. Fujisawa. 1964. Preferential Absorption of Humic Acids by Clay. J. Sci. Soil Manure, v.35, N 2.

Komissarov, I. D., I. N. Streltsova and N. V. Kouznetsova. 1979. Chemical Nature of Humus Substances of Recent Soils, Technogenic Alluvial and Oxidized Coal in the Kouzbas and Their Interaction with Minerals. Soil Formation in Technogenic Landscapes. Nauka, Novosibirsk, 212-258 (in Russian).

Kononova, M. M. 1961. Soil Organic Matter, its Nature , its Role in Soil Formation and in Soil Fertility. Pergamon Press. Oxford-London- New York- Paris, 450p.

Schnitzer, M., and S. I. M. Skinner. 1969. Free Radicals in Soil Humic Compounds. Soil Sci., v.108, N 6, 383-390.

Taranov S. A., F. A. Fatkulin and I. S. Rodinyuk. 1979. Conditions of Natural Regeneration of Soils in Disturbed by Industry Landscapes along the Upper Section of the Tomi River. In: Soil Formation in Technogenic Landscapes. Nauka, Novosibirsk, 156-163 (in Russian).

Trofimov S. S., A. A. Titlyanova and I. L. Klevenskaya. 1979. A System Approach to the Study of Soil Formation Processes in Technogenic Landscapes. In:Soil Formation in Technogenic Landscapes. Nauka, Novosibirsk, 3-19 (in Russian).

Trofimov S. S., and N. N. Naplekova. 1986. Humus Formation in Technogenic Ecosystems. Humus Formation in Technogenic Landscapes. Nauka, Novosibirsk, 160p. (in Russian)

INTERACTIONS BETWEEN POLYCHLORINATED BIPHENYLS AND YEAST CELLS IN LIQUID MEDIUM

Oudin P. (1,2), J.A. Toth (2), R. Bonaly (1), M.D. Toth (2), M. Nagy (4)

(1) Lab. Biochimie microbienne Faculté de Pharmacie de NANCY
2 r. A. Lebrun F-54000 NANCY
Tel: 33/383 17 88 42 Fax: 33/383 32 30 58
Email: oudin@scbiol.u-nancy.fr

(2) Ecological departement Kossuth Lajos Tudományiegyetem Pf. 71 H-4010 DEBRECEN
Tel: 36/52 316 666 (26 19) Fax: 36/52 431 148
Email: oudin@quant.ecol.klte.hu

(4) Central Laboratory Agricultural University of Debrecen
H-4032 DEBRECEN (Hungary)

INTRODUCTION

PCBs represent a large group of toxic environmental chemicals. In the world, the global estimation of water and soil polluted by PCBs is about 400 000 tons, with 60% in sea water and 40% in soils and sediments (TANABE 1988). The main characteristic of these compounds is their chemical stability and non-reactivity (GERVASON 1987).

The first direct problem with PCBs is that the World Health Organization classified this group of substances as probably carcinogen (HOOPER 1990). Another problem is the production of highly toxic derivatives such as dioxins at high temperature (above 650 °C) (MILLERET 1987).

The complete elimination by incineration with post combustion is not available for wastes with low concentration of toxins, but for these materials it appears that a biological treatment may be an efficient solution. Two biological methodes are possible: biodegradation and bioaccumulation.

Biodegradation was the first method investigated. All the studies were completed with selected bacteria (BAXTER *et al.* 1975, FURUKAWA *et al.* 1978). Results were efficient only for the less chlorinated congeners (BOYLE *et al.* 1992 and BURKHARD *et al.* 1996), although the most important kinds of PCBs in the environment are from electric transformer fluid and are the most chlorinated mixes (HOOPER *et al.* 1990). Microorganisms can assimilate and accumulate toxics as well as xenobiotics without any degradation. This phenomenon needs a structure which favors the interactions between microorganisms and PCBs. This structure depends on the cell envelope's chemical composition.

In the present study two yeast species were used on account of their ubiquity in nature: *Saccharomyces cerevisiae* is the most produced yeast in the world; *Rhodotorula glutinis* produces extracellular lipids able to enhance PCBs solubility in liquid medium. This yeast species is also widespread in sea water and in soils.

We have studied the interactions between PCBs and the two yeasts in order to investigate a possible way to eliminate these toxic substances from the environment.

Effect of Mineral-Organic-Microorganism Interactions on Soil and Freshwater Environments
Edited by Berthelin, *et al*, Kluwer Academic / Plenum Publishers, New York, 1999

MATERIAL and METHODS

1. Microorganisms and culture conditions.

The two strains used were: *Saccharomyces cerevisiae uvarum* 009 and *Rhodotorula glutinis* CBS 3044.

Three culture media were used:

Medium D: Glucose 40g; Yeast extract 1.5g; KH_2PO_4 1.0g; $MgSO_4,7H_2O$ 0.2g; H_2O 1l.

Medium D+actidione : containing variable concentrations of actidione (cycloheximide).

Medium V: Sucrose 20g; Yeast extract 1.5g; KH_2PO_4 2.0g; $(NH_4)_2$ SO_4 3.0g; NaCl 0.1g; Sol A. 1ml ($CaCl_2$ 0.25g; $MnCl_2,4H_2O$ 0.001g; $FeCl_3,6H_2O$ 0.0005g, H_2O 1l); Sol B 1ml ($MgSO_4$, $7H_2O$ 0.25g; $ZnSO_4,7H_2O$ 0.001g; H_3BO_3 0.001g; $CuSO_4,5H_2O$ 0.0001g; KI 0.0001g; H_2O 1l).

Yeast growth was achieved in liquid medium and followed spectrophotometrically at 620 nm. We used 100 ml culture flasks containing 20 ml of culture medium and aerated by magnetic stirring at 24°C. The PCBs used were a commercial mix : « PYRALENE » containing 60% of chlorine with an average of 8 chlorine atoms per molecule (BALLSCHMITER and ZELL 1980). The PCBs concentrations in the media are expressed in percentage (v/v). A 0.5% concentration represents 8100 ppm. Microemulsions of PCBs in media were obtained by sonication with ultrasounds of autoclaved media at 50 W for 20 min at room temperature.

2. Effects of PCBs on yeast cells.

The cultures in D medium were achieved in the presence of different PCBs concentrations. The toxic effect of the derivatives was estimated by comparison of the growth of the cells and by morphological observations using electron microscopy.

3. Electron microscopy observations

3.1 Scanning electron microscopy

Samples were fixed in a 2.5% glutaraldehyde solution pH 7.2 at 4°C for 24h, then fixed with 1% osmic acid at 4° C for 2h. Dehydration of the samples was obtained by several washings with acetone solutions (concentration of acetone in order after fixation: 20, 50, 70, 90, 100%). Acetone was then extracted from the sample by CO_2 supercritical fluid extraction. The samples were then glued on to a special board and metalized with gold. Observations were performed on a scanning electron microscope "CAMBRIDGE Stereoscan S 240" with 20 KV electron acceleration.

3.2. Transmission Electron Microscopy

We used the same fixation as in scanning microscopy, then the following treatments were completed : Dehydration was obtained by several washings in ethanol solutions (concentration of ethanol in order after fixation: 30, 50, 70, 90, 100%). After inclusion in standard inclusion polymer, samples were cut into fine and ultrafine slides with a diamond knife. Slides were laid on a copper grid for Reynolds staining (REYNOLDS 1963) and on a gold grid for Thiery staining (THIERY 1967). Observations were performed on a transmission electron microscope "SIEMENS MPE 101" with 80KV electron acceleration.

4. General analytical procedures

4.1 PCBs quantification

The process used to identify concentrations of PCBs was GC/ECD (Gas Chromatography with Electron Capture Detector). We compared 18 peaks with a standard PCBs mix. In liquid medium, two hexane extractions 1/5 (v/v) were completed and the two extracts were diluted to obtain a concentration between 1 ppm and 50 ppb. For quantification in the biomass, cells were harvested by membrane filtration and PCBs were extracted with hexane for 8 hours using the standard Soxhlet extraction process (BURKHARD 1996).

RESULTS .

Effects of PCBs on yeast growth and morphology in relation to culture conditions

As shown in Fig 1 and Fig 2, the presence of PCBs at concentrations up to 1% did not alter the growth of yeast strains. A concentration of 5% decreased slightly the development of *R. glutinis*. It can be postulated that *R. glutinis* growth is perturbed by the presence of high amounts of insoluble oil in the culture medium.

For *S. cerevisiae*, after culturing for 24h and even 48h on glucose medium, the growth curves in medium suplemented with PCBs were always higher than in the control without PCBs. We postulated that this could be explained by a degradation of the less chlorinated derivatives by the yeast strain, and a stimulation of growth by the products.

Figure 1.a

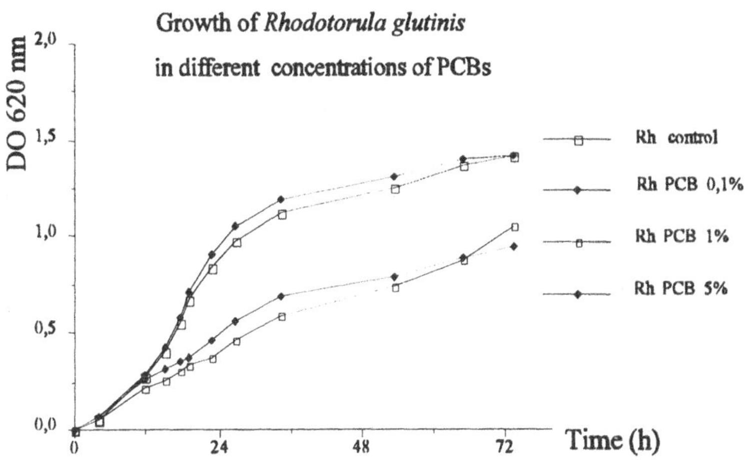

Figure 1.b

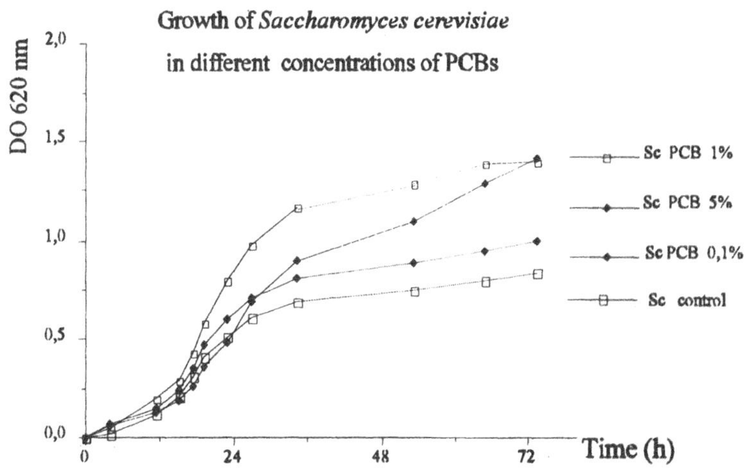

Fig 1.a: Growth of *R. glutinis* in the presence of different PCBs concentrations
Fig 1.b: Growth of *S. cerevisiae* in the presence of different PCBs concentrations

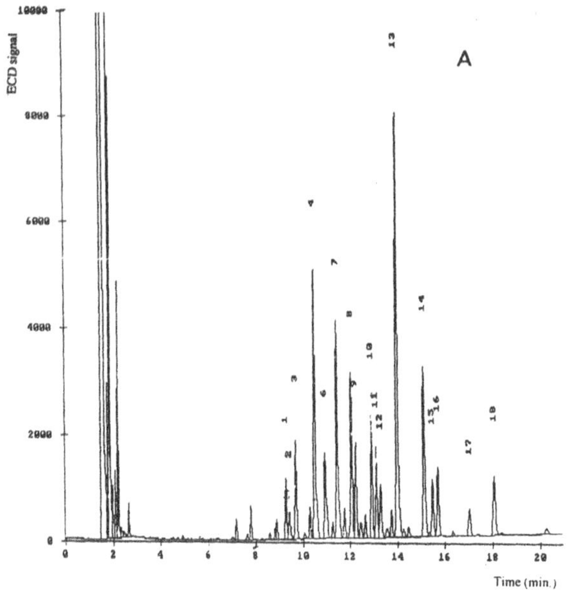

Fig. 2.a: GC/ECD PCBs stock solution

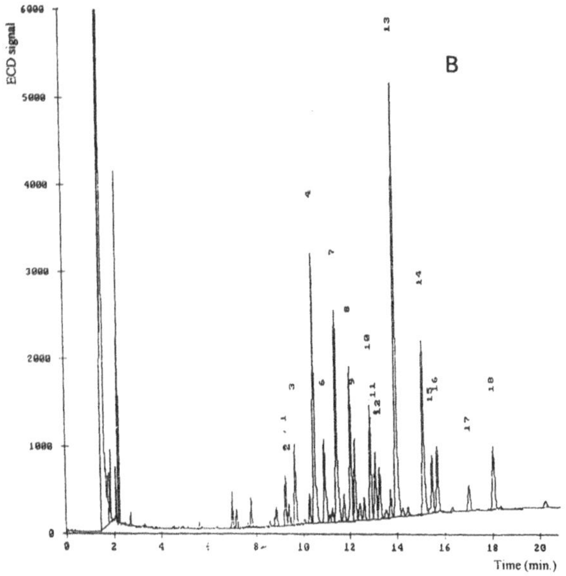

Fig. 2.b: GC/ECD PCBs extracted from the biomass of *S. cerevisiae*

Scanning electron microscopy

Observations performed on both strains after growth in the presence of PCBs revealed invaginations and roughness of the cell surface (Fig 3). This also demonstrates an interference by the derivatives on cell metabolism.

Transmission electron microscopy

After growth in the presence of PCBs, the cells were stained following the technique of Reynolds and observed by transmission electron microscopy. Characteristic dark areas were detected in the cytosol. These dark areas were not detected after Thiery staining using thiocarbosemihydrazide. This result revealed an elimination of PCBs incorporated in the cells. Such an observation has already been reported by BULLY and REISINGER (1994). These authors suggested that Thiery staining provokes an elimination of included PCBs in electron microscopy slides.

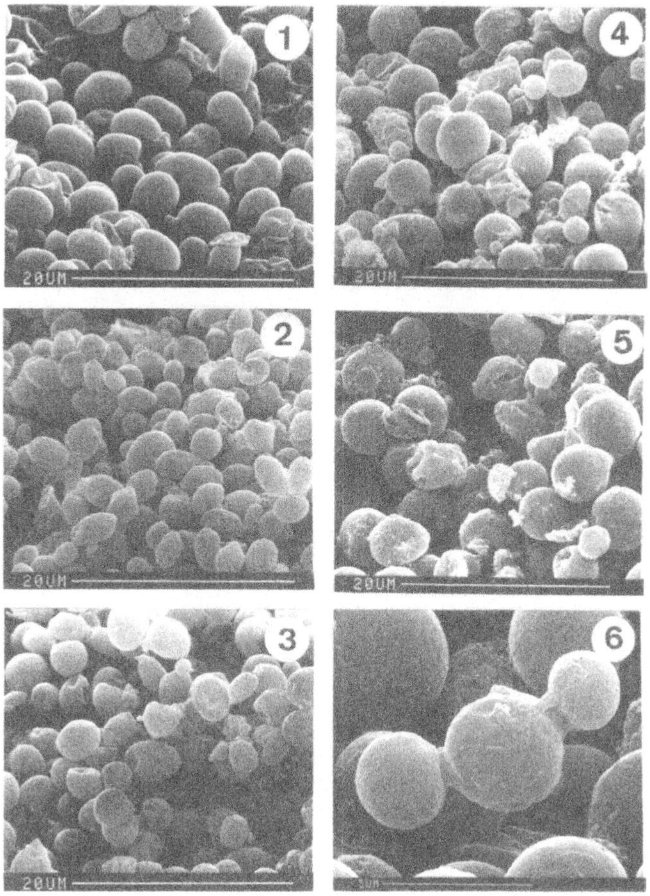

Fig 3: Scanning electron microscopy
3.1: *R. glutinis* culture D medium without PCBs
3.2: *R. glutinis* culture D medium 1% PCBs
3.3: *R. glutinis* culture D medium 5% PCBs
3.4: *S. cerevisiae* culture D medium without PCBs
3.5: *S. cerevisiae* culture D medium 1% PCBs
3.6: *S. cerevisiae* culture D medium 5% PCBs

PCBs quantification in biomass

The PCBs concentration were measured in the whole yeast biomass. Results in Table 1 show high concentrations after ultrasounds treatment. It is highly probable, upon comparison with transmission electron microscopy observations (Fig. 4) that the values are not representative of a real bioaccumulation. Indeed, with such PCBs concentrations, cells should appear much more dense to electron beam.

Intracellular PCBs analysis in *R. glutinis* by gas chromatography did not show any differences in the congener's distribution. This may be due to an incomplete extraction of the derivatives from this yeast strain. More study of this problem is in progress.

Concerning *S. cerevisiae*, intracellular PCBs analysis (Fig. 2b) showed a signal decrease of several peaks in the extract. Indeed, by comparison with the distribution of the congeners in the initial mix used (Fig. 2a), it appears that there is a characteristic modification of this distribution. For PCBs extracted from the biomass, the peak 1 disappeared and peaks 2 to 6 represent respectively 2, 5, 2, 8 and 44% of peak 14. For PCBs mix used peak 1 represent 6% and peaks 2 to 6 respectively : 9%, 12%, 6%, 25%, 71% of peak 14. For Peak 17 (more than 8 Chlorine atoms) for both chromatograms represent 9.4% of peak 14. These results show that this *S. cerevisiae* strain was able to degrade the less chlorinated PCBs.

Table 1 : PCBs titration in *S. cerevisiae* and *R. glutinis*

	PCB 1% in D medium	PCB 1% in D medium + ultrasounds
Rhodotorula	44 000	335 000
Saccharomyces	N.D.	295 000

Results are expressed in ppm
N.D.: not determined

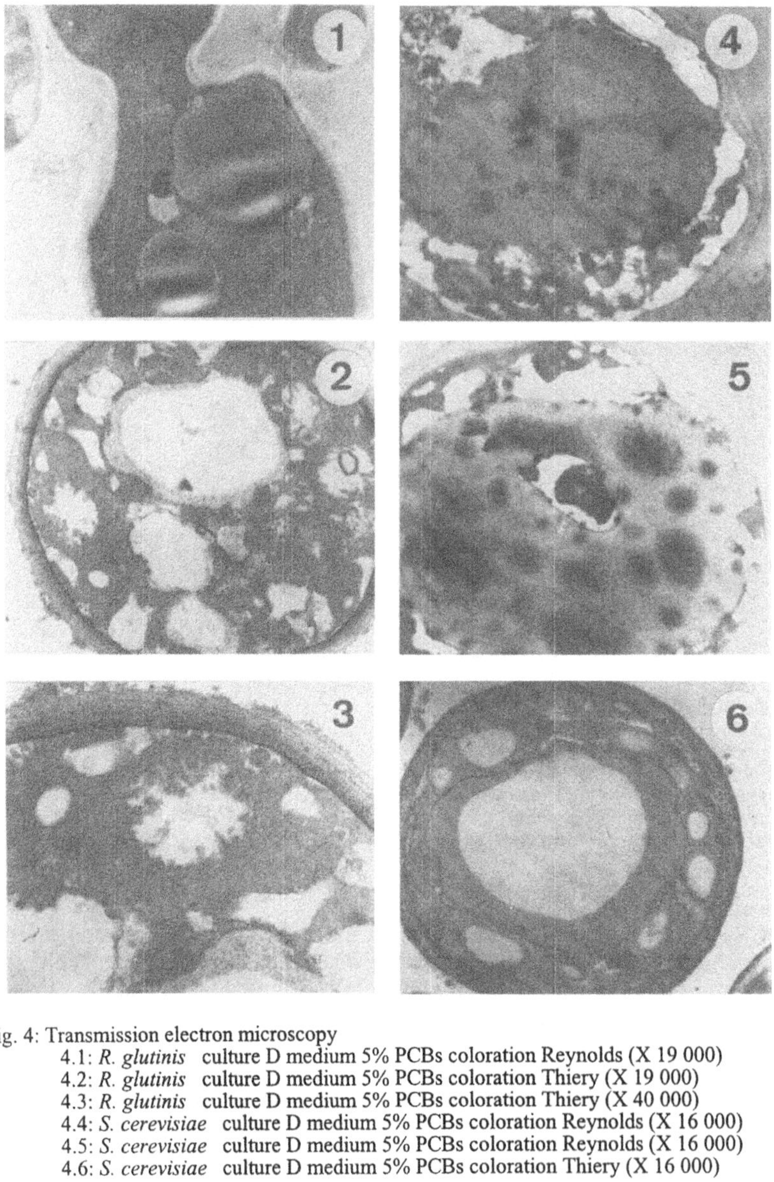

Fig. 4: Transmission electron microscopy
4.1: *R. glutinis* culture D medium 5% PCBs coloration Reynolds (X 19 000)
4.2: *R. glutinis* culture D medium 5% PCBs coloration Thiery (X 19 000)
4.3: *R. glutinis* culture D medium 5% PCBs coloration Thiery (X 40 000)
4.4: *S. cerevisiae* culture D medium 5% PCBs coloration Reynolds (X 16 000)
4.5: *S. cerevisiae* culture D medium 5% PCBs coloration Reynolds (X 16 000)
4.6: *S. cerevisiae* culture D medium 5% PCBs coloration Thiery (X 16 000)

DISCUSSION

Elimination of PCBs from the environment remains a problem since these toxins are spread at low concentrations over large areas and since they are chemically stable and non-reactive. For this reason, bioremediation was considered as a possible economical method of removal. Microorganisms especially bacteria belonging to the genera Pseudomonas and Acinetobacter (FURUKAWA 1978) and mixed inoculum (NATARAJAN 1996) have been assayed for their potential to bioeliminate PCBs. The results were a significant biodegradation of all the less chlorinated PCBs congeners. For the most chlorinated congeners, the elimination was not efficient and the authors observed a bioimmobilisation of those derivatives. For yeast, no potential degradation of PCBs was describe, but their structure could allow a bioaccumulation phenomenon.

In the present study, we demonstrate that yeasts are also an appropriate biological material for bioaccumulation of PCBs. Low concentration of these derivatives in the liquid culture medium did not provoke any significant inhibition of growth of *R. glutinis*, the growth of *S. cerevisiae* was enhanced. The size of the tainted cells was similar to control cells, but the morphology was affected when the yeasts developed in the presence of 1 to 5% of toxics. As observed by scanning electron microscopy, invaginations and roughness appeared on the cell surface.

PCBs, which are electron-dense substances, accumulate inside the yeast cells. Indeed, after Reynolds staining, transmission electron microscopy observations revealed in the cytosol the presence of many dark areas which were not detectable after Thiery staining. Such a result has already been obtained by BULLY and REISINGER (1994), suggested that in microorganisms those dark areas were included PCBs.

Our results demonstrate that yeasts can be a useful material for PCBs bioremediation of low concentrations of PCBs. After bioaccumulation by yeasts of PCBs extracted from slightly polluted wastes (BURKHARD *et al.* 1996), it is possible to eliminate this biologically concentrated pollution by classical incineration at extremely low costs. However as shown by preliminary assays with *S. cerevisiae* it also can not be excluded to take advantage of biodegradation of these congeners. It remains to search for yeast strains that most efficiently only accumulate these toxics and effect in their biodegradation.

REFERENCES

Ballschmiter K., Zell M., 1980. General nomenclature for Polychlorinated Biphenyl's. Fresenius . Anal of Chemistry , 302 20.

Baxter R.A., P.E. Gilbert, R.A. Lindgett, J.H. Mainprize, H.A. Vodden, 1975. The degradation of polychlorinated biphenyls by microorganisms. The science of total environment. 4, 53-61.

Boyle A.W., C.J. Silvin, J.P. Hassett, J.P. Nakas, S.W. Tanenbaum, 1992. Bacterial PCB biodegradation. Biodegradation 3 285-298.

Bully F., O. Reisinger, 1994. Relation des PCBs avec des microorganismes. Environmental microbiology. 1, 117 - 123.

Burkhard J., B. Polakova, K. Demnerova, J. Pazlarova, 1996. PCBs biodegradation in water suspension of contaminated soil - the evaluation of the course. Fresenius Environmental Bulletin. 5 392-396

Furukawa K., K. Tonomura, A. Kamibayashi, 1978. Effect of chlorine substitution on the biodegradability of polychlorinated biphenyls. Applied and environnemental microbiology. 35, 223-227.

Gervarson P., 1987. Les PCB : leurs propriétés et leurs applications dans l'électochimie. Revue générale de l'électricité . 8, 5-11.

Hooper S.W., C.A. Pettigrew, G.S. Sayler, 1990. Ecological fate, effects and prospects for the elimination of environnemental polychlorobiphenyls (PCBs). *Environnemental toxicology and chemistry.* 9, 655-667

Milleret G., 1987. Destruction des PCB : le procédé par incinération. Revue générale de l'électricité. 8, 151-155.

Nataranjan MR, W.M. Wu, J. Nye, H. Wang, (1996) Dechlorination of polychlorinated biphenyl congeners by an anaerobical microbial consortium. Applied Microbiology Biotechnology 46 : 673-677

Reynolds E. S., 1963. The use of lead citrate at high pH as an electron opaque stain in electron microscopy. Journal of cellular biology 17, 208-212.

Tanabe S., 1988. Problems in the future : foresight from current knowledge. Environmental pollution 50, 439-443.

Thiery J.P., 1967. Mise en évidence des polysaccharides sur coupes fines en microscopie électronique. Journal microscopie. 6, 987-1018.

EFFECTS OF PH, ELECTROLYTES AND MICROBIAL ACTIVITY ON THE MOBILIZATION OF PCB AND PAH IN A SANDY SOIL

Bernd Marschner, Christiane Baschien, Maik Sarnes and Ulrike Döring

TU Berlin, FG Bodenkunde, Salzufer 11-12, D-10587 Berlin,
Tel.: +49-30-314 73527 , Fax: +49-30-314 73548,
e-mail: marsnghg@mailszrz.zrz.TU-Berlin.de.

1. INTRODUCTION

In the terrestrial environment, soils are the most important sinks for many anthropogenic xenobiotics. Airborne or otherwise introduced hydrophobic organic compounds such as polycyclic aromatic hydrocarbons (PAH) or polychlorinated biphenyls (PCB) accumulate mainly in the organic topsoils due to their high affinity for organic matter and their low water solubility. The further fate of these substances is dependent on their chemical structure and may include volatilization as the major pathway for the loss of lower chlorinated biphenyls from sewage sludge-treated soils (Alcock *et al.*, 1993). Well documented is the biochemical degradation of low molecular weight PAHs in soils by autochthonous microflora (Wild *et al.*, 1990) or by introduced specialists such as certain genera of white rot fungi (Barr and Aust, 1994, Kästner and Mahro, 1996). However, 5- and 6-ring PAHs and PCBs seem to be largely resistant against biochemical attacks in the soil environment and are therefore regarded as highly persistent (Wild *et al.*, 1990, Alcock *et al.*, 1993). Furthermore, the bioavailability of organic chemicals, which is a prerequisite for degradation, can be greatly reduced due to incorporation into the macromolecular structures of humus molecules and the formation of so-called bound residues.

While these degradation and immobilization processes contribute to the elimination of the compounds from the soil solution, their apparent solubility and their mobility may, on the other hand, be enhanced in the presence of surfactants, organic cosolvents or by sorption to colloids such as particulate organic matter (POM) or to dissolved organic matter (DOM). In soil solutions and aquatic systems, the fraction sorbed to such organic colloids may approach or even exceed the truly dissolved phase of highly hydrophobic compounds (Chiou, 1989, Magee *et al.,* 1991, Murphy and Zachara, 1995). As a consequence, this third phase has to be considered in the determination of partition coefficients (K_d, K_{OC}) and in the modeling of transport processes (Chiou, 1989, Magee *et al.*, 1991). Apart from DOM concentration, this mobilizing effect is mainly dependent on DOM composition since this

Effect of Mineral-Organic-Microorganism Interactions on Soil and Freshwater Environments
Edited by Berthelin, *et al*, Kluwer Academic / Plenum Publishers, New York, 1999

determines (a) the affinity for the organic xenobiotic and (b) the sorption of DOM to the solid soil phase.

Generally, DOM sorption coefficients for hydrophobic organic compounds (K_{DOM} or K_{DOC}) increase with increasing molecular weight of DOM and its degree of aromaticity, which is attributed to the increased presence of hydrophobic regions or cavities (Gauthier *et al.*, 1987, Chiou *et al.*, 1987). At low pH, sorption coefficients are higher because the DOM molecules are more hydrophobic and more condensed due to the protonation of functional groups (Schlautmann and Morgan, 1993). High electrolyte concentrations also cause organic colloids to assume a more condensed configuration (Engbretson and Wandruszka, 1994), but effects on K_{DOC} are cation-dependent. At the same time, low pH and high salt concentrations reduce DOM solubility and especially favor the precipitation and solid-phase sorption of the more hydrophobic compounds (Murphy and Zachara, 1995) so that the overall effect of the chemical environment on DOM properties is difficult to predict.

This is even more complicated by the fact that DOM formation or release in soils is largely a biologically mediated process (Evans *et al.*, 1988) which therefore is influenced by chemical factors as well as by temperature, moisture content and substrate availability. The formation and release of DOM is reflected in the distinct seasonality of DOM concentrations in soil solutions (e.g., Campbell, 1989). However, little is known about the effects of different biological activities or microbial population compositions on the properties of DOM, especially with regard to its sorption coefficients for hydrophobic organic compounds. Brusseau et al. (1994) and Dohse and Lion (1994) showed that microbially produced polymers such as cyclodextrin and other unidentified compounds can increase PAH and PCB solubility and mobility in laboratory columns. But these experiments were conducted with very high concentrations of the microbial exudates under sterile conditions and therefore give little clue to the relevance of such processes under field conditions.

In the present study, several sorption and short-term incubation experiments were conducted with soils artificially contaminated with benzo(a)pyrene or 2,2′,5,5′-tetrachlorobiphenyl (PCB 52) in order to delineate purely physico-chemical from biological effects of pH, electrolyte and substrate variations on the mobilization of DOM and the hydrophobic compounds.

2. MATERIALS AND METHODS

All experiments were conducted with a sandy topsoil from a field that had received untreated sewage waters for over 100 years. The untreated soil material had a pH (0.01 m $CaCl_2$) of 5.3, C_{org} of 1.7% and showed moderate total heavy metal concentrations (Cd: 5.9 mg/kg, Cu: 60 mg/kg) and extractable PAH and PCB contents (Σ16 PAH: 1.33 mg/kg, Σ6 PCB: 0.13 mg/kg). Subsamples of the collected material were additionally contaminated with benzo(a)pyrene (Sigma, No. B-1760) and 2,2′,5,5′-tetrachlorobiphenyl (PCB 52, synthesized by A. Schuphan, RWTH Aachen, Germany) by mixing with quartz sand (at 10 g/kg) that had been enriched with the acetone-dissolved compounds and then left to dry. The target soil concentrations were 100 mg BaP/kg and 20 mg PCB 52/kg were reached within ± 5% and designated as "BaP" and "P52" respectively. All samples were stored covered in an open shed prior to use in the experiments.

After 10 months aging, subsamples of the spiked materials were acidified according to a previously determined buffering curve with 0.5 M HCl at 0.08 mol/kg to give pH 3.3 (designated as "P52a" and "BaPa"). The acidified and untreated soils were then used in a

14-month column experiment with periodic leaching with deionized water (Marschner, 1998). Material from the beginning and end of this experiment (Table 1) was used for the extraction and incubation experiments.

2.1 Incubations with different substrates

This experiment consisted of different substrate and biocide additions to the column-derived moistened soil samples (50 g DW) giving the following treatments: control, 0.3 g glucose, glucose + penicillin/streptomycin to suppress bacterial activity (PCB 52 only), glucose + actidion to suppress fungal activity, 1 g cellulose, 1 g ground pine heartwood (BaP only) and 1 g ground leaves of *Agropyron repens*, which is the dominant grass on the field plot. The samples were incubated for 17 days (PCB 52) or 14 days (BaP) with three replicates per treatment.

2.2 Extraction and analysis

Soil extracts were prepared with 50 g soil and 150 mL aqueous solutions containing no additives (millipore-H_2O), $CaCl_2$ (0.01 M), NaCl (0.02 M) or Soerensen-buffer (adjusted to pH 3 or pH 6.5). In the incubation experiment, only millipore-H_2O was used to characterize BaP and PCB 52 extractability before and after the incubation. The slurry was shaken for 24 h and followed by centrifugation (15 min at 3000 rpm) and analysis of the supernatant. PCB 52 and BaP were determined in the unfiltered solutions because preliminary tests with pure aqueous solutions had shown strong sorption of the compounds by filter materials. In this way, no differentiation between freely dissolved or colloid-sorbed fractions is possible in the solutions. Depending on the expected concentration range, 20 to 50 mL of the aqueous soil extracts were passed through 500 mg C18 reverse phase columns at approximately 1 mL/min. BaP was eluted with 20 mL of toluene and transferred into acetonitrile for measurement on a HPLC with fluorescence detector. PCB 52 was eluted with 20 mL of hexane which was evaporated to 1 mL prior to injection into a GC with ECD-detector.

Total dissolved and colloidal organic matter in the solutions was measured with an elemental C-N-analyzer and characterized by UV/VIS-spectra (272, 465 and 665 nm) of the solutions. Other parameters, including pH, electrical conductivity, cations and anions, were analyzed with standard procedures.

3. RESULTS

The data in Table 1 show that the 10 month aging period hardly reduced extractable PCB 52 and BaP concentrations from the originally applied amounts. After incubation in the columns, however, BaP retrieval was greatly reduced in both treatments and PCB 52 concentrations also were significantly lower than at the beginning of the experiment. Apparently, the 10 month storage period was inadequate to promote aging effects which then occurred in the column experiment. This may be due to the low moisture content during storage, thus reducing dissolution, diffusion and biological processes. In the column experiment, the pH of the untreated materials declined by about one unit due to strong nitrification and subsequent leaching of base cations (Marschner, 1998) and approached the values of the acidified samples P52a and BaPa. Differences in organic carbon contents between the treatments or between sampling time were not significant.

Table 1: Chemical properties of the original uncontaminated soil material (Ref), the 10-months aged and the column derived contaminated soil materials used in the sorption and incubation assays (P52s and BaPs denote the acid treatment).
Concentrations of BaP and PCB 52 determined after extraction with a mixture of petrolether/acetone/water (1:2:1) according to Reese-Stähler *et al.*, (1995).

treatment		Ref	P52 aged	P52 column	P52a column	BaP aged	BaP column	BaPa column
pH ($CaCl_2$)		5.7	5.3	4.1	3.6	5.3	4.4	3.7
C_{org}	[%]	1.66	1.66	1.59	1.69	1.66	1.59	1.69
PCB 52	[mg/kg]	0.02	18.6	17.0	17.2	nd	nd	nd
BaP	[mg/kg]	0.12	nd	nd	nd	98.5	52.5	62.3

The extractability of the two tested organic compounds with aqueous solutions was strongly influenced by pH and electrolyte concentrations (Figure 1). At low pH, solution concentrations of BaP and PCB 52 were only 30 to 40% of the values obtained in the near-neutral extracts (pH 6.5). While PCB 52 concentrations were all within the range of water solubility of this compound (120 to 140 µg/L), BaP concentrations were far above its water solubility of 2 to 3 µg/L. Since no organic co-solvents or surfactants were present in the systems, most of the BaP in the extracts must have been sorbed to a colloidal phase or to dissolved organic matter to account for the high concentrations. Unfortunately, quantification of TOC was not possible in these extracts due to the use of a citrate buffer, but higher concentrations in the pH 6.5 extract were indicated by higher absorbance at 272 nm.

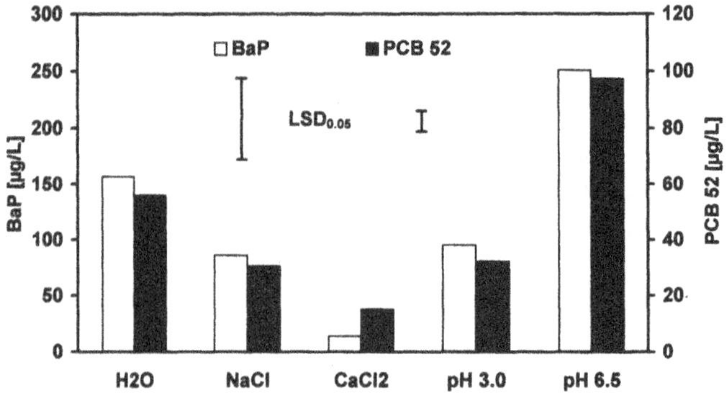

Figure 1: Mean BaP and PCB 52 concentrations in aqueous extracts with different pH and electrolytes from the aged spiked sewage farm soil samples. LSD for $p < 0.05$ derived from ANOVA with t-test.

To a certain degree, this is also true for the electrolyte effects (Figure 1). While BaP; PCB 52 and TOC concentrations (140 mg/L) are all highest in the pure water extracts, the different xenobiotic concentrations in the salt solutions are not associated with similar TOC variations. At the same ionic strength as the $CaCl_2$ extract, the NaCl extract contains almost 2-fold higher PCB 52 and 2 to 5-fold higher BaP concentrations, although TOC concentrations are lower (55 vs. 66 mg/L). Again, the BaP concentrations in all extracts surpass water solubility by far and therefore must be attributed to solubility enhancement by suspended or dissolved sorbents, namely POM or DOM. In contrast, PCB 52 concentrations are consistently below water solubility, making the participation of sorbents in the solutions not necessary.

From the spectroscopic properties of the water and salt extracts, certain DOM quality properties were estimated. The E_{465}:E_{665} ratios are indicators for DOM molecular size (Chen *et al.*, 1977) and the specific absorptivity at 272 nm (E_{272}:DOC) is an estimator for DOM aromaticity (Chin *et al.*, 1994). The data indicate that DOM in the aqueous extracts was characterized by the highest aromaticity and the largest molecules (E_{465}:E_{665} = 2.8). In the NaCl extract, the high E_{465}:E_{665} ratio of 8.3 indicates much smaller molecules which are also less aromatic. The lowest DOM aromaticity was calculated for the $CaCl_2$ extracts, while the E_{465}:E_{665} ratio is intermediate between the other two extracts. Although no direct molecular size classes or distributions nor percentage of aromatic structures can be calculated from this data, they indicates that apart from TOC concentrations, DOM quality is affected by the chemical composition of the extraction solution.

In the substrate incubation experiments, substrate additions initially had no effect on the extractability of PCB 52, as is evidenced by the narrow concentration ranges of 85 to 113 μg/L in the extracts from the non-acidified samples and 78 to 102 μg/L in the extracts from the acidified soil. However, after 17 days of incubation at room temperature, extractability of PCB 52 had changed considerably in most treatments apart from the control (Figure 2). In the non-acidified samples, all substrate treatments produced an increase in PCB 52 extractability. The highest PCB 52 concentrations of 170 to 195 μg/L were found in the glucose and *Agropyron* treatments. Interestingly, the mobilizing effect of the glucose treatment was slightly reduced in the samples containing the procaryote-toxins penicillium/streptomycin and even more in the fungicide (actidion) treated samples.

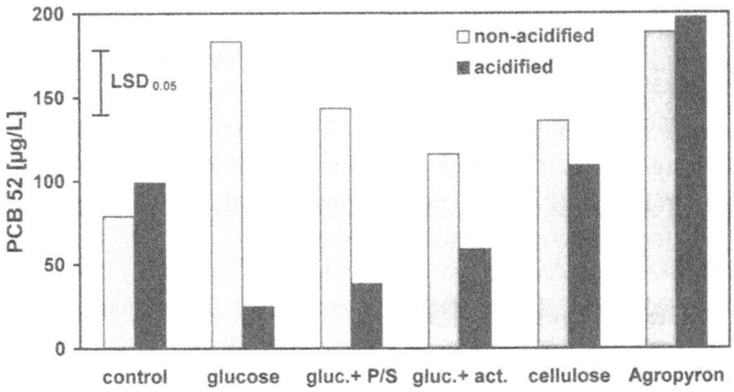

Figure 2: Mean PCB 52 concentrations in aqueous extracts from the non-acidified and acidified spiked sewage farm soil samples after incubation with different substrate additions. LSD for $p < 0.05$ derived from ANOVA with t-test.

The acidified soil samples showed completely different reactions to the glucose treatments (Figure 2). Incubation with pure glucose reduced PCB 52 extractability to about 30% of the mean control value. These reductions were less pronounced in the two biocide treatments (glucose+P/S and glucose+act), indicating that similar to the other soil, the glucose effects were depressed by the toxins. Maximum concentrations of PCB 52 were found in the extracts from the *Agropyron* supplemented samples which corresponds with the effects observed in the non-acidified soil. Cellulose had no significant effect on PCB 52 extractability in either soil.

These different substrate effects showed no consistent relation to any of the other analyzed solution parameters. TOC concentrations were similar in both soils and showed maximum values in the glucose+actidion treatment and little differentiation among the other treatments. Electrical conductivity as a measure for the salt concentrations in the extracts was elevated in the glucose and *Agropyron* treatments and extremely high in the glucose+penicillium/streptomycin (P/S) soil extracts, since these antibiotics were added in "physiological" 0.9% NaCl solution.

The only factor that was related to PCB 52 extractability was pH (Figure 3). The lowest pHvalues (around 3.5) were recorded in the glucose treatment of the acidified soil, while maximum values occurred in the *Agropyron* treatments of both soils. However, the high PCB 52 concentrations in the glucose treatment of the non-acidified samples occurred at pH values between 4.4 and 5.2 and thus do not follow this trend. No improvement of regression parameters was achieved by introducing other variables into the equation.

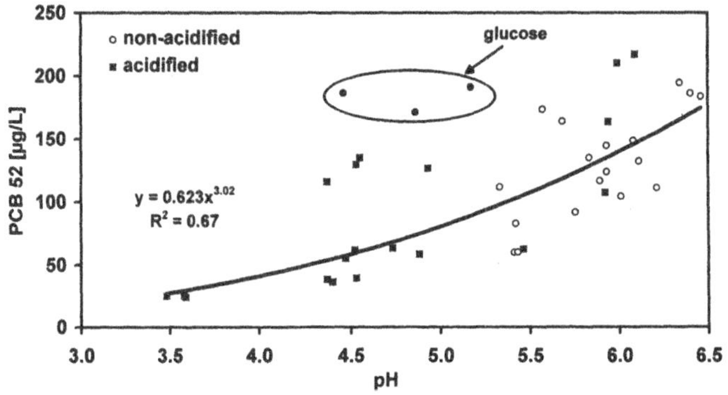

Figure 3: Relationship between pH and PCB 52 concentrations in the aqueous extracts from the spiked sewage farm soils after incubation with different substrates. The fitted regression curve was calculated with all data points except for the circled glucose treatment of the non-acidified samples.

The reaction of the BaP-spiked soils to the substrates showed a completely different pattern than the PCB 52-contaminated soils. In the non-acidified soils, no pre- or post-incubation effects of the substrates on BaP extractability were observed except in the pine-wood treatment, where before incubation BaP concentrations were lower than control values (31 vs. 40 µg/L). At the end of the incubation, BaP concentrations were also lowest in the pine wood treatment (22 µg/L), but not significantly different from the other extracts where BaP concentrations ranged between 28 and 34 µg/L.

The acidified soils reacted very differently (Figure 4). The glucose + actidion, cellulose and pine-wood treatments had immediate diminishing effects on BaP extractability, but BaP concentrations in the *Agropyron* and glucose treatments were not different from the control. After incubation, control and glucose treatments had the lowest BaP concentrations in the extracts (9 to 14 µg/L), while all other treatments were in the range of 15 to 25 µg/L, which generally was lower than before incubation.

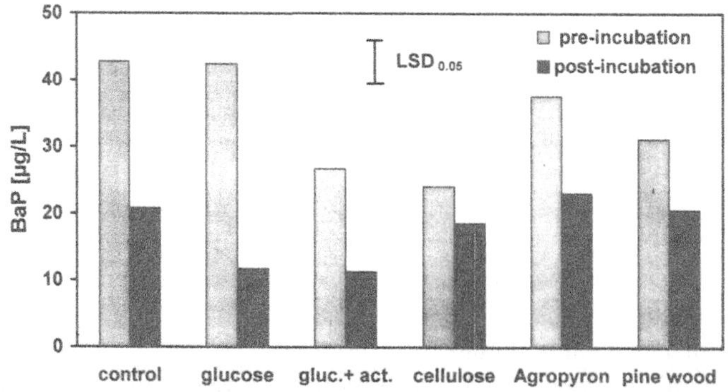

Figure 4: Mean BaP concentrations in aqueous extracts from acidified spiked sewage farm soil samples before and after incubation with different substrate additions. LSD for $p < 0.05$ derived from ANOVA with t-test

Similar to PCB 52, the differences in BaP concentrations fail to show any relationship to TOC or most other solution parameters. But the strongly reduced BaP extractability in the control and glucose treatments corresponds with a pH decline from 4.6 before to 3.7 to 4.1 after incubation. Only slight pH changes occurred in the other treatments (Figure 5). The resulting close relationship between BaP concentrations and pH is expressed in the linear correlation coefficient of 0.86. But the regression equation based on pH as the only independent variable does not fit to the pre-incubation data of the acidified soil or the non-acidified soil data, which showed virtually no pH variation. However, by introducing electrical conductivity (EC) into the regression analysis, the whole data set of the acidified samples can be described by the following equation, indicating that an increase in EC during incubation due to mineralization may be responsible for the decrease in BaP-concentrations.

$$\log [BaP] = 0.33\ pH - 0.20 \log [EC] \qquad R^2 = 0.73 \qquad \text{Eq. 1}$$

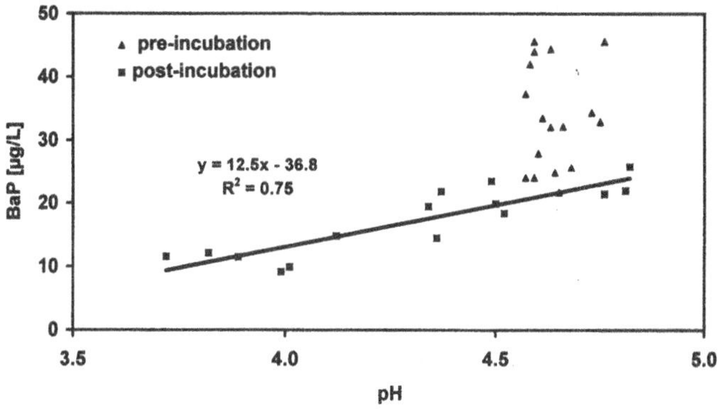

Figure 5: Relationship between pH and BaP concentrations in the aqueous extracts from the acidified sewage farm soils before and after incubation. The fitted regression curve only applies to the data from the post-incubation samples.

3. DISCUSSION

The observed effects of inorganic solution parameters on BaP and PCB 52 mobilization are unlikely caused by direct interactions, such as "salting out", at the pH ranges and salt concentrations used in these experiments (Schwarzenbach *et al.*, 1993). Therefore, indirect effects on structure and reactivity of sorbents in the solid or solution phase are made responsible for the observed phenomena.

The low extractability of BaP and PCB 52 at low pH can be explained by several processes. First, the negative charges of mineral surfaces and especially of organic matter are reduced through protonation of functional groups such as -COOH and -OH (Stevenson, 1982, Ryan and Elimelech, 1996). As a result, the diffuse double layer is less developed (Ryan and Elimelech 1996) so that hydrophobic interactions are more likely to occur between dissolved compounds and the surfaces. In organic matter, the protonation of functional groups can enhance the accessibility of hydrophobic sorption sites such as aromatic structures on the surfaces or even within the macromolecules (Schwarzenbach *et al.*, 1993, Engebretson and Wandruszka, 1994, Murphy *et al.*, 1994). While such hydrophobic interactions are probably the main sorption mechanism for BaP, the electronegative Cl-groups of PCB 52 may also participate in dipole-dipole interactions (Schwarzenbach *et al.*, 1993), which should be favored at less negative surface charges.

A second explanation for the reduced extractability of the two compounds at low pH may be found in the reduced solubility and colloid release of organic matter. This is caused by the reduced polarity of the protonated functional groups and the subsequent increased adsorption or coagulation of potentially mobile organic macromolecules (Ghosh and Schnitzer, 1980, Ryan and Elimelech, 1996). As a result, dissolved organic matter under acidic conditions consists mainly of relatively hydrophilic small molecules with low sorption coefficients (K_{DOC}) for hydrophobic compounds (Chiou *et al.*, 1987, Murphy *et al.*, 1994). But BaP concentrations of almost 100 μg/L in the pH 3 extract show that even under these conditions large amounts of suspended and soluble sorbents must be released from the soil solid phase.

At higher pH, the dissociation of the functional groups causes the organic molecules to become more hydrophilic and less condensed, resulting in a lower affinity for hydrophobic compounds (Ghosh and Schnitzer, 1980, Schlautmann and Morgan, 1993). For PCB 52, this may be the dominating process, since solution concentrations do not exceed water solubility in the extracts. For BaP, mobilization must be additionally enhanced by soluble or colloidal sorbents, since solution concentrations are about 100-fold higher than water solubility. This may simply be the result of co-mobilization of BaP-containing organic colloids or molecules. In addition, larger DOM molecules are released at higher pH (Marschner, 1998) which can have higher K_{DOC} values than smaller size fractions (Herbert *et al.*, 1993, Murphy *et al.*, 1994).

The effect of elevated electrolyte concentrations on BaP and PCB 52 extractability may be the result of similar mechanisms. First, with increasing ionic strength of the solution, the double layer of charged surfaces is reduced in thickness (Ryan and Elimelech, 1996). This would enable a closer approach of hydrophobic substances and an increased interaction with the solid phase by van der Waals forces. Different effects of Na^+ and Ca^{2+} were also observed by Murphy and Zachara (1995) and attributed to some form of "Ca-facilitated exposure of hydrophobic binding sites" of the organic matter. A resultant increased sorption of BaP and PCB 52 to the solid phase organic matter would therfore partly explain the differences in extractability.

But in the case of the largely sorbent-associated BaP in the extracts, BaP solution concentrations are probably further reduced by a second salt effect. At elevated electrolyte concentrations, the release of colloids and soluble macromolecules is reduced due to the diminished double layers (Ryan and Elimelech, 1996). Apart from this quantity effect, the sorptive properties of the released sorbents appear to be affected by the cationic species in solution, since BaP concentrations are much lower in the $CaCl_2$ extract than in the NaCl extract despite similar TOC concentrations.

In the incubation experiments, the substrate effects on PCB 52 extractability appear to be largely due to changes in the sorptive properties of the organic matter due to pH changes as described above. The high PCB 52-concentrations in the extracts from the soils amended with *Agropyron* powder were associated with a pH increase which probably was caused by either the buffering action of base cations from the introduced grass or by decarboxylation of organic anions as proposed by Yan *et al.*, (1996). Solubility enhancement by mobilized organic matter is most likely in this treatment since PCB 52 concentrations were far above water solubility in both soil extracts. However, with our experimental setup, it is not possible to determine whether these effects were caused by microbial exudates as demonstrated by Dohse and Lion (1994), by soluble degradation products of the plant biomass or soil organic matter, or by suspended microbial biomass itself.

The pH reductions associated with the glucose treatments are most likely caused by the formation of carboxylic groups during glycolysis (Yan *et al.*, 1996). The more pronounced pH decrease in the acidified soil samples is attributed to their decreased acid buffering capacity due to a low base saturation. Consequently, PCB 52 concentrations are lowest in this treatment. But the high PCB 52 concentrations in the glucose-amended non-acidified soil samples show that pH alone is inadequate to account for the different PCB 52 extractability. Obviously, the expected increased solid-phase PCB 52 sorption in that treatment is counterbalanced by other mechanisms. Solution enhancement by glucose degradation products seems unlikely since these are mainly low molecular hydrophilic organic anions. Instead, microbial exudates or organic colloids from the degradation of soil organic matter could account for the observed phenomena. This mechanism may only be active in the non-acidified samples because of the higher pH (> 4.5) and associated differences in DOM solubilty or composition of the microbial community.

The pH-associated effect on BaP extractability was also only observed in those acidified samples where incubation reduced pH below 4.5 (glucose treatments). This may indicate that the release or production of organic molecules with a high affinity for hydrophobic compounds is greatly reduced below this threshold pH. Interestingly, this value lies within the same range as the pKa of humic and fulvic acids below which humic acids become insoluble (Stevenson, 1982). A decrease of humic acid-like DOM could account for the reduced K_{DOC} values, since Gauthier *et al.* (1987) and Chen *et al.* (1992) report 2 to 5 times lower K_{OC} values for fulvic acids than for humic acids. This indicates that structural characteristics that determine the solubility of organic macromolecules in acidic solutions may also be responsible for their affinity towards hydrophobic compounds.

5. ACKNOWLEDGMENTS

This study was part of an interdisciplinary research project on the behavior and effects of PAHs and PCBs in urban soils, financed by the German Ministry of Education, Research, Science and Technology (BMBF-grant No. 07 OTX 08 A).

6. REFERENCES

Alcock, R. E., A. E. Johnston, S. P. McGrath, M. L. Berrow and K. C. Jones. 1993. Long-term changes in the polychlorinated biphenyl content of United Kingdom soils. Environ. Sci. Technol., 27, 1918-1923.

Barr, D. P. and S. D. Aust. 1994. Mechanisms white rot fungi use to degrade pollutants. Environ. Sci. Technol., 28, 78A-87A.

Brusseau, M. L., X. Wang and Q. Hu. 1994. Enhanced transport of low-polarity organic compounds through soil by cyclodextrin. Environ. Sci. Technol., 28, 952-956.

Campbell, D. J. 1989. The soil solution chemistry of some Oxfordshire soils: temporal and spatial variability. J. Soil Sci., 40, 321.339.

Chen, S., Inskeep, W. P., Williams, S. A. & Callis, P. R. 1992. Complexation of 1-naphtol by humic and fulvic acids. Soil Sci. Soc. Am. J., 56, 67-73.

Chen, Y., Senesi, N. & Schnitzer, M. 1977. Information provided on humic substances by E4/E6 ratios. Soil Sci. Soc. Am. J., 41, 352-358.

Chin, Y.-P., Aiken, G. & O'Loughlin, E. 1994. Molecular weight, polydispersity, and spectroscopic properties of aquatic humic substances. Environ. Sci. Technol., 28, 1853-1858.

Chiou, C. T. 1989. Theoretical considerations of the partition uptake of nonionic organic compounds by soil organic matter. In: Sawhney, B. L. and K. Brown (Eds.) Reactions and Movement of Organic Chemicals in Soils. Soil Sci. Soc. Am. Spec. Publ. No. 22, pp. 1-30.

Chiou, C. T., D. E. Kile, T. I. Brinton, R. E. Malcolm, J. A. Leenheer and P. MacCarthy. 1987. A comparison of water solubility enhancement of organic solutes by aquatic humic materials and commercial humic acids. Environ. Sci. Technol., 21, 1231-1234.

Dohse, D. M. and L. W. Lion. 1994. Effect of microbial polymers on the sorption and transport of phenanthrene in a low carbon sand. Environ. Sci. Technol., 28, 541-548.

Döring, U.; Marschner, B. 1998. Water solubility enhancement of benzo(a)pyrene and 2,2′,5,5′-tetrachlorobiphenyl. Proceedings of the EGS-conference, Vienna April 1997 (in press).

Engebretson, R. R. and R. v. Wandruszka. 1994. Mikroorganisation in dissolved organic acids. Environ. Sci. Technol., 28, 1934-1941.

Evans, A., L. W. Zelazny and C. E. Zipper. 1988. Solution parameters influencing dissolved organic carbon levels in three forest soils. Soil Sci. Soc. Am. J., 52, 1789-1792.

Gauthier, T. D., W. R. Seitz and C. L. Grant. 1987. Effects of structural and compositional variations of dissolved humic materials on pyrene Koc values. Environ. Sci. Technol., 21, 243-248.

Ghosh, K. and M. Schnitzer. 1980. Macromolecular structures of humic substances. Soil Sci., 129, 266-276.

Herbert, B. E., Bertsch, P. M. & Novak, J. M. 1993. Pyrene sorption by water-soluble organic carbon. Environ. Sci. Technol., 27, 398-403.

Kästner, M. & Mahro, B. 1996. Microbial degradation of polycyclic aromatic hydrocarbons in soils affected by the organic matrix of composts. Appl. Microbiol. Biotechnol., 44, 668-675.

Magee, B. R., L. W. Lion and A. T. Lemley. 1991. Transport of dissolved organic macromolecules and their effect on the transport of phenanthrene in porous media. Environ. Sci. Technol., 25, 323-331.

Marschner, B. 1998. DOM-enhanced PAH- and PCB-mobilization in contaminated soils under different chemical conditions. Proceedings of the EGS-conference, Vienna April 1997 (in press).

Murphy, E. M. and J. M. Zachara. 1995. The role of sorbed humic substances on the distribution of organic and inorganic contaminants in groundwater. Geoderma, 67, 103-124.

Murphy, E. M., Zachara, J. M., Smith, S. C., Phillips, J. L. & Wietsma, T. W. 1994. Interaction of hydrophobic organic compounds with mineral-bound humic acid. Environ. Sci. Technol., 28, 1291-1299.

Reese-Stähler, G., Klementz, D., Volk, C. & Pestemer, W. 1995. Gezielte Herstellung eines mit PCB 52 und BaP dotierten Bodens und rückstandsanalytische Bestimmung von PCB und PAK in Rieselfeldböden und Streu. VDLUFA-Schriftenreihe-Kongressband, 40, 909-912.

Ryan, J. N. & Elimelech, M. 1996. Colloid mobilization and transport in groundwater. Colloids Surfaces A, 107, 1-56.

Schlautmann, M. A. and J. J. Morgan. 1993. Effects of aqueous chemistry on the binding of polycyclic aromatic hydrocarbons by dissolved humic materials. Environ. Sci. Technol., 27, 961-969.

Schwarzenbach, R. P., Gschwend, P. M. & Imboden, D. M. 1993. Environmental Organic Chemistry. John Wiley & Sons, New York.

Stevenson, F. J. 1982. Humus Chemistry. John Wiley & Sons, New York.

Wild, S. R., K. S. Waterhouse, S. P. McGrath and K. C. Jones. 1990. Organic contaminants in an agricultural soil with a known history of sewage sludge amendments: Polynuclear aromatic hydrocarbons. Environ. Sci. Technol., 24, 1706-1711.

Yan, F., S. Schubert and K. Mengel. 1996. Soil pH increase due to biological decarboxylation of organic anions. Soil Biol. Biochem., 28, 617-624.

MODIFICATION OF HERBICIDE MINERALIZATION AND EXTRACTABILITY IN SOIL BY ADDITION OF ORGANIC MATTER IN MODEL EXPERIMENTS

Sabine Houot, Enrique Barriuso and Valérie Bergheaud

I.N.R.A., Unité de Science du Sol, B. P. 01,
78850 Thiverval-Grignon, FRANCE

1. INTRODUCTION

Atrazine is the most commonly used herbicide for maize in France. Although atrazine and other s-triazines have been termed as recalcitrant (Kaufman and Kearney, 1970), biodegradation remains the principal process of atrazine dissipation in soils, as for the other pesticides. A large variety of soil microorganisms are able to degrade atrazine partially by N-dealkylation or dehalogenation reactions (Kaufman and Kearney, 1970; Behki and Khan, 1986; Mougin et al., 1994). Recently, complete mineralization of the triazine ring has been reported (Gschwind, 1992; Mandelbaum et al., 1993a; Mandelbaum et al., 1995; Radosevich et al., 1995). Ring cleavage apparently occurs only after hydroxylation (Kaufman and Kearney, 1970). The formation of hydroxyatrazine has been thought to be of chemical origin, occurring in acidic conditions and involving acid functions of humic substances (Khan, 1978). Recently its microbial formation has been demonstrated also (Mandelbaum et al., 1993b).

Despite potentially rapid atrazine degradation, this molecule has been frequently detected in waters. Because of this, a new herbicide has been proposed for maize treatment by ICI Agrochemicals, the sulcotrione, a more easily degraded molecule (Compagnon and Béraud, 1992).

The addition of organic amendments to soils increases their organic matter content and generally stimulates the soil microbial activity. The consequent modification of pesticide behavior varies with the nature of the organic amendments and with their effect on the microbial activity (Benoit, 1994; Alvey and Crowley, 1995). The most important effect of organic amendment addition to soil is the increase of pesticide sorption (Bellin et al., 1990; Martinez-Iñigo and Almendros, 1992; Guo et al., 1993). Pesticide degradation is often modified also. A reduction in degradation is usually explained by the decrease of the pesticide availability after sorption increases (Doyle et al., 1978). In contrast, an increase in degradation may be explained by the soil microbial activation after the organic amendment application which favors pesticide degradation by co-metabolism (Hance, 1973). Concerning atrazine, its degradation varies with carbon and nitrogen availability in soils (Alvey and Crowley, 1995; Topp et al., 1996).

The purpose of the work presented here was to compare the modification of the behavior of atrazine and sulcotrione in a loamy soil after addition of two exogenous organic matters (EOM) representative of organic amendments frequently applied to soils: a wheat straw and a municipal solid waste compost. Today, many soils are likely to receive these kinds of organic amendments since composting represents a developing alternative for municipal solid waste management. Atrazine and sulcotrione dissipation in the soil via degradation and stabilization through formation of non extractable, so-called "bound residues" was followed during laboratory incubations in controlled conditions.

Effect of Mineral-Organic-Microorganism Interactions on Soil and Freshwater Environments
Edited by Berthelin, *et al*, Kluwer Academic / Plenum Publishers, New York, 1999

2. MATERIAL AND METHODS

2.1. Soil and Exogenous Organic Matters

The soil (typic Eutrochrept) was sampled from the surface layer (0-20 cm) of a bare experimental plot located at Grignon (France). It had a pH of 7.3 (in water) with 220 g kg^{-1} of clay, 730 g kg^{-1} of silt and 50 g kg^{-1} of sand. The organic carbon and nitrogen contents were 14.7 and 1.2 g kg^{-1} respectively.

The two EOM used were a municipal solid waste compost and a composted straw. The waste compost was formed from the organic fraction of the municipal solid household wastes of Bapaume (France). The windrow fermentation lasted 10 weeks and was followed by 3 months of maturation. The waste compost pH was 8.6 (in water) and its organic C and N contents were 154.2 and 13.8 g kg^{-1}, respectively. Wheat straw was composted for 6 months as explained in Morel et al. (1984). The composted straw had a pH of 6.5 (in KCl) and its organic C and N contents were 424.3 and 33.1 g kg^{-1}, respectively. The compost was passed through a 2 mm sieve and the straw was grounded in a plant-blender into particles smaller than 2 mm.

2.2. Chemicals

Two herbicides were used: atrazine [6-chloro-$\underline{N}^2$-ethyl-$\underline{N}^4$-isopropyl-1,3,5-triazine-2,4-diamine], and sulcotrione [2-(2-chloro-4-mesyl-benzoyl)-cyclohexane-1,3-dione]. Analytical standards of atrazine and its metabolites were purchased from ChemService (West Chester, PA, USA). The [U-ring-^{14}C]atrazine (specific activity 659 MBq $mmol^{-1}$; radiopurity > 97 %) was purchased from Amersham (Buckinghamshire, UK). Two solutions of ^{14}C-atrazine were prepared in water: solution A at 10.87 mg l^{-1} and 21.16 10^3 kBq ml^{-1}, and solution B at 5.06 mg l^{-1} and 9.85 10^3 kBq ml^{-1}. The [U-benzoyl ring-^{14}C]-sulcotrione (specific activity 849 10^3 kBq $mmol^{-1}$; radiopurity > 97 %) and analytical standards of sulcotrione and its metabolites were supplied by ICI Agrochemicals (Richmond, California, USA). A solution of ^{14}C-sulcotrione was prepared in water at 7.24 mg l^{-1} and 49.12 10^3 kBq ml^{-1} .

2.3. Incubation Experiment

The herbicides'degradation was followed during laboratory incubations with 9.7 g of soil added or not to 0.92 g of compost or 0.38 g of straw in hermetically stoppered jars at 28 ± 1 °C. The amount of EOM was calculated in order to double the organic C content of the soil in the mixtures. The incubations were realized in two steps. Preliminary incubations were done with soil + herbicide, soil + EOM, and EOM + herbicide. Then EOM, herbicide or soil were respectively added, and the ternary mixtures put back to incubate. Both the preliminary and the second incubations lasted 1 and 3 months each, respectively, for sulcotrione and atrazine. The behavior of sulcotrione and atrazine was also followed with only soil or EOM (same quantities as previously). The water content was adjusted to -10 kPa in the different treatments. The herbicide water solutions were complemented with water.

In the incubations with atrazine, 0.5 ml of solution A was used for the incubation with only EOM and 1 ml of solution B for the soil. When soil or EOM was added at the end of the pre-incubations, the mixture humidity was adjusted as necessary. Because of the addition of atrazine at the beginning of the second incubation, the mixture soil + EOM was incubated at 60% of the water holding capacity during the pre-incubation, then 0.5 ml of solution A of atrazine was added. In the incubations with sulcotrione, 0.5 ml of solution was added to all the jars. The water content of the different mixtures was kept constant by monthly weighing and adjusting with water. The incubations were done in triplicate. The ^{14}C-CO_2 evolved was trapped in 10 ml of 0.5 M NaOH and periodically measured by scintillation counting. The traps were replaced at each measurement date which allowed the jar atmosphere to be renewed and prevented the development of anaerobic conditions.

2.4. Herbicide Analysis

At the end of the incubations, three samples of soil, EOM and soil-EOM mixtures were extracted first with 50 ml of 0.01M $CaCl_2$, then three successive times with 50 ml of methanol by an end-over-end shaking for 24 hours each time. Every extract was recovered by centrifugation and its radioactivity was measured by liquid scintillation counting with a Kontron Betamatic V counter (Kontron Ins., Montigny le Bretonneux, France). The extractable radioactivity was calculated by addition of water and methanol extracts. The non-extractable radioactivity, corresponding to the "bound residues" was measured by scintillation counting of the $^{14}CO_2$ evolved after combustion of the solid residues (Sample Oxidizer 307, Packard, Meriden, CT, USA). The pooled methanol extracts were concentrated until dryness by evaporation with a TurboVap II concentrator (Zymark, Hopkinton, MA, USA) at 45°C on an helical flow of air with an operating pressure of 800 kPa. The residue was then dissolved in 3-ml of the solvent used for the HPLC analysis and then filtered through a Cameo 13N syringe nylon filter of 0.45 µm (MSI, Westboro, MA, USA). ^{14}C-atrazine, ^{14}C-sulcotrione and ^{14}C-metabolites were analyzed in all samples using a Waters HPLC appliance (600E Multisolvent Delivery System, 717 Autosampler and a Novapak C18 column of 250 x 4.6 mm and 5 µm - Waters, Milford, MA, USA) equipped with a radioactive flow detector (Packard-Radiomatic Flo-one A550, Packard, Meriden, CT, USA). For atrazine, the mobile phase was methanol/water buffered with 50 mM ammonium acetate with pH adjusted to 7.4. The chromatography started in water, then reached 40/60 methanol/water (vol./vol.) after 2 min with a linear gradient, then reached 80/20 methanol/water after 15 min with a concave gradient (gradient 7 in Waters Software) until the end of the chromatography. For sulcotrione the mobile phase was methanol/water with 10 mM tetrabutylammonium chloride. The chromatography started in methanol/water 30/70 then reached 65/35 methanol/water after 10 min with a concave gradient (gradient 8 in Waters Software) until the end of the chromatography. For both molecules, the mobile phase flow was 1.0 ml min^{-1}, and the injected sample volume was 600 µl.

3. RESULTS

3.1. Mineralization of Atrazine

Atrazine mineralization reached 19 to 24% of the initial ^{14}C-atrazine at the end of the control incubations of soil alone (Fig. 1). In both cases, after a lag phase of 30 days, the mineralization rate increased then reached a plateau after 110 to 150 days of incubation. Both compost and straw addition to soil at the end of the pre-incubation tended to decrease the atrazine ring mineralization to 15 to 16% of the initial radioactivity. Nevertheless, the differences were not significant.

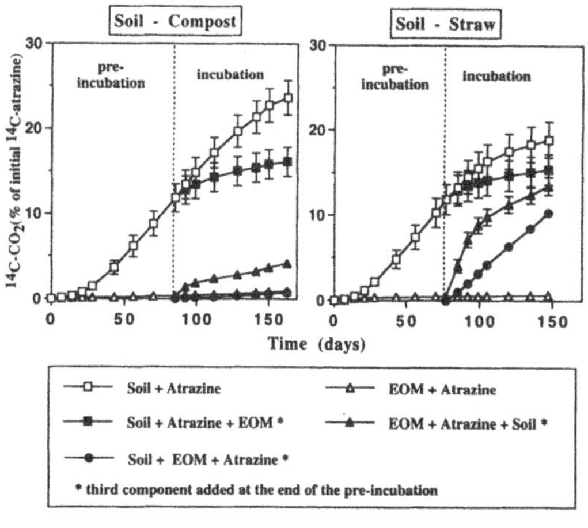

Figure 1: Kinetics of ^{14}C-atrazine mineralization during the pre-incubations and the incubations with soil and/or different exogenous organic matters (EOM-compost or straw). Results are expressed as a percent of the initial radioactivity. Standard deviations are indicated only when larger than symbols.

Less than 1 % of the ^{14}C-atrazine was mineralized at the end of the incubations of compost and straw. Soil addition at the end of the pre-incubation with compost significantly increased the atrazine mineralization to 4%. Soil addition to straw at the end of the pre-incubation strongly stimulated the atrazine mineralization. After 150 days, 14% of the initial atrazine was mineralized.

Only 0.7% of the ^{14}C-atrazine was mineralized when added to pre-incubated soil-compost mixture and atrazine mineralization was very slow. On the contrary, when atrazine was added to pre-incubated soil-straw mixture, the atrazine mineralization started rapidly without the lag phase observed in soil alone. The rate of mineralization increased linearly with time until the end of the incubation when it reached 10% of the initial radioactivity.

3.2. Extractability and Chemical Nature of the Residual Radioactivity at the End of the Incubations with Atrazine

At the end of the incubations of soil alone with ^{14}C-atrazine, 34 to 37% of the initial radioactivity remained extractable and 35 to 40% formed bound residues (Table 1). Atrazine was the main component and hydroxyatrazine the main metabolite in the extracts; they represented 73 to 82% and 17 to 25% of the extracted radioactivity, respectively (Table 2).

Table 1: Distribution of the radioactivity from the ^{14}C-atrazine between the mineralized fraction, the extractable (water plus methanol) and non extractable (bound residues) fractions at the end of the incubations.

Pre-incubation	Incubation	Mineralized CO_2	Extractable Residues	Bound Residues
		(—— % of applied ^{14}C-atrazine	——)	
Soil + Atrazine	Soil + Atrazine	23.6 ± 2.0 *(a*)*	33.7 ± 2.3 *(a)*	39.9 ± 2.8 *(a, d)*
	Soil + Atrazine + Compost	16.1 ± 1.7 *(b, e)*	37.9 ± 2.9 *(a, c)*	44.5 ± 3.9 *(a, c, e)*
Compost + Atrazine	Compost + Atrazine	0.9 ± 0.1 *(c)*	47.3 ± 0.3 *(b)*	52.6 ± 0.2 *(b)*
	Compost + Atrazine + Soil	4.1 ± 0.3 *(d)*	41.4 ± 0.9 *(c)*	52.1 ± 2.2 *(b, c)*
Soil + Atrazine	Soil + Atrazine	18.9 ± 2.1 *(a, b)*	36.7 ± 2.0 *(a)*	35.4 ± 2.4 *(d, g)*
	Soil + Atrazine + Straw	15.4 ± 1.8 *(b, e)*	34.8 ± 1.8 *(a)*	42.7 ± 5.9 *(a, b, g)*
Straw + Atrazine	Straw + Atrazine	0.6 ± 0.0 *(c)*	45.7 ± 1.0 *(b)*	56.5 ± 0.4 *(f)*
	Straw + Atrazine + Soil	13.5 ± 1.0 *(e)*	29.2 ± 0.7 *(c)*	55.7 ± 4.5 *(b, e, f)*

* In a column, numbers noted with the same letter are not significantly different at the level of α=5%.

Both compost and straw addition to soil at the end of the pre-incubations tended to increase the proportion of non extractable residues, but only compost addition allowed the maintenance of a larger proportion of extractable radioactivity (Table 1). Again, the differences were not significant. The compost did not modify the nature of the extracted radioactivity with 73% of atrazine and 24% of hydroxyatrazine (Table 2). On the other hand, straw addition to soil enhanced the partial degradation of atrazine and the formation of hydroxyatrazine up to 30% of the extracted radioactivity.

At the end of the incubations of EOM alone, 53 to 56% of the initial radioactivity was stabilized as bound residues and 46 to 47% remained extractable (Table 1). The main extracted component was still atrazine in the incubation with compost (74% of the extracted radioactivity). Hydroxyatrazine represented 21% of the extracted radioactivity in the case of compost. At the end of the incubation of straw, hydroxyatrazine represented 84% of the extracted radioactivity and atrazine only 5% (Table 2).

Soil addition to compost or straw at the end of the pre-incubation significantly enhanced the mineralization of atrazine as compared to the incubation of EOM alone; this enhancement came at the expense of the extractable radioactivity, which decreased (Table 1). On the other hand, soil addition to EOM did not modify the formation of bound residues, or the nature of the extractable radioactivity. Similar proportions of bound residues were formed as during the incubations of EOM alone.

Table 2: Characterization of the extracted radioactivity from the ^{14}C-atrazine at the end of the incubations. The other metabolites detected were deethylatrazine, deisopropylatrazine and deethyldeisopropylatrazine. Results of analyses of water and methanol extracts were cumulated.

Pre-incubation	Incubation	Atrazine	Hydroxy-atrazine	Others
		(———	% of extractable [^{14}C]	———)
Soil + Atrazine	Soil + Atrazine	72.7 ± 8.7	25.4 ± 0.4	1.5 ± 0.7
	Soil + Atrazine + Compost	73.4 ± 2.3	23.6 ± 0.5	2.9 ± 0.3
Compost + Atrazine	Compost + Atrazine	74.1 ± 2.1	20.8 ± 1.1	3.0 ± 0.5
	Compost + Atrazine + Soil	75.2 ± 2.5	20.1 ± 0.8	4.0 ± 0.9
Soil + Atrazine	Soil + Atrazine	81.5 ± 6.0	16.8 ± 4.4	1.6 ± 0.7
	Soil + Atrazine + Straw	68.4 ± 5.6	29.3 ± 1.8	2.1 ± 0.3
Straw + Atrazine	Straw + Atrazine	4.8 ± 1.1	84.1 ± 4.7	2.3 ± 0.7
	Straw + Atrazine + Soil	6.9 ± 1.8	84.1 ± 12.6	0.4 ± 0.2

3.3. Mineralization of Sulcotrione

Sulcotrione mineralization reached 45% of the ^{14}C-sulcotrione applied at the end of the incubation of soil alone (Fig. 2). Almost no lag phase was observed, and after a week the mineralization increased linearly at a rate of 1% of initial ^{14}C per day during the entire incubation. Compost addition to soil at the end of the pre-incubation significantly decreased the sulcotrione mineralization to 31% of the initial radioactivity. On the other hand, straw addition did not affect the mineralization as much; it reached 40% at the end of the incubation.

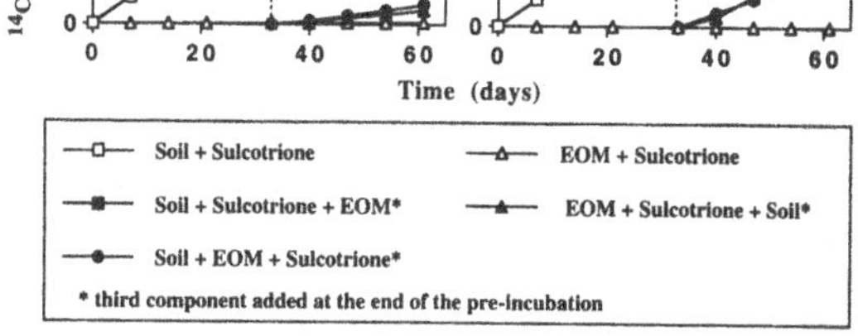

Figure 2: Kinetics of ^{14}C-sulcotrione mineralization during the pre-incubation and the incubation with soil and/or different exogenous organic matters (EOM-compost or straw). Results are expressed as a percent of the initial radioactivity. Standard deviations are indicated only when larger than symbols.

Less than 0.5% of the initial sulcotrione was mineralized at the end of the incubations of compost and straw (Fig. 2). Soil addition at the end of the pre-incubation with straw increased the sulcotrione mineralization to 11%. Soil addition to compost at the end of the pre-incubation significantly increased the mineralization of sulcotrione; it reached 2% of the initial amount at the end of the incubation.

When sulcotrione was added to pre-incubated soil-EOM mixture the mineralization was similar to the treatments EOM + sulcotrione, then addition of soil.

3.4. Extractability and Chemical Nature of the Residual Radioactivity at the End of the Incubations with Sulcotrione

At the end of the incubations of soil alone with ^{14}C-sulcotrione, 25% of the initial radioactivity remained extractable and 26% formed bound residues (Table 3). Only traces of sulcotrione remained in the extracts and the only detected metabolite was CMBA (2-chloro-4-(methyl-sulphonyl)benzoic acid), which represented 99.7% of the extracted radioactivity (Table 4).

Table 3: Distribution of the radioactivity from the ^{14}C-sulcotrione between the mineralized fraction, the extractable (water plus methanol) and non-extractable (bound residues) fractions at the end of the incubations.

Pre-incubation	Incubation	Mineralized CO_2	Extractable Residues	Bound Residues
		(—— % of applied ^{14}C-sulcotrione		——)
Soil + Sulcotrione	Soil + Sulcotrione	44.9 ± 2.3 *(a*)*	25.3 ± 1.7 *(a)*	26.3 ± 1.2 *(a)*
	Soil + Sulcotrione + Compost	31.1 ± 0.6 *(b)*	47.8 ± 0.9 *(b)*	21.9 ± 0.8 *(b)*
Compost + Sulcotrione	Compost + Sulcotrione	0.3 ± 0.0 *(c)*	79.4 ± 1.0 *(c)*	20.9 ± 1.0 *(b, c)*
	Compost + Sulcotrione + Soil	1.9 ± 0.1 *(d)*	79.9 ± 0.7 *(c)*	19.2 ± 0.6 *(c)*
Soil + Sulcotrione	Soil + Sulcotrione	44.9 ± 2.3 *(a)*	25.3 ± 1.7 *(a)*	26.3 ± 1.2 *(d)*
	Soil + Sulcotrione + Straw	40.3 ± 1.4 *(a)*	9.0 ± 3.8 *(d)*	52.4 ± 4.9 *(e)*
Straw + Sulcotrione	Straw + Sulcotrione	0.0 ± 0.0 *(c)*	82.8 ± 3.4 *(c)*	18.7 ± 0.5 *(c)*
	Straw + Sulcotrione + Soil	11.4 ± 1.1 *(e)*	27.4 ± 3.3 *(a)*	59.2 ± 1.9 *(e)*

* In a column, numbers affected of the same letter are not significantly different at the level of α=5%

Compost addition to soil at the end of the pre-incubations strongly increased the proportion of extractable residues to 48% of the initial radioactivity (Table 3). Simultaneously, the proportion of bound residues significantly decreased to 22%. The compost did not modify the nature of the extracted radioactivity, with 99% of CMBA in the extract (Table 4). On the other hand, straw addition to soil enhanced the formation of bound residues; their proportion reached 52% at the end of the incubations, while the extractable fraction decreased to 9% of the initial radioactivity. Only CMBA was detected in the extracts.

At the end of the incubations of EOM alone, 21 and 26% of the ^{14}C-sulcotrione was stabilized as bound residues, 79 and 83% remained extractable with compost and straw respectively (Table 3). With both EOM incubation the main extracted component was sulcotrione which represented 95 to 96% of the extracted radioactivity (Table 4).

Soil addition to compost at the end of the pre-incubation had a very modest effect on the behavior of sulcotrione. The distribution of the initial radioactivity at the end of the incubation was similar to that with the compost alone (Table 3). On the other hand, soil addition to straw significantly increased the formation of bound residues to 59% of the applied ^{14}C-sulcotrione. Simultaneously, the extractable fraction decreased to 27% of the initial radioactivity. Both sulcotrione (60%) and CMBA (40%) were detected in the extractable fraction (Table 4).

Table 4: Characterization of the extracted radioactivity from the ^{14}C-sulcotrione at the end of the incubations. Results of analyses of water and methanol extracts were added together.

Pre-incubation	Incubation	Sulcotrione	CMBA
		(% of extractable [^{14}C])	
Soil + Sulcotrione	Soil + Sulcotrione	0.3 ± 0.1	99.7 ± 0.4
	Soil + Sulcotrione + Compost	1.4 ± 1.2	98.6 ± 1.3
Compost + Sulcotrione	Compost + Sulcotrione	96.3 ± 0.4	1.8 ± 0.1
	Compost + Sulcotrione + Soil	28.1 ± 0.6	72.3 ± 0.6
Soil + Sulcotrione	Soil + Sulcotrione	0.3 ± 0.1	99.7 ± 0.4
	Soil + Sulcotrione + Straw	0.0 ± 0.0	100.0 ± 0.0
Straw + Sulcotrione	Straw + Sulcotrione	94.8 ± 0.5	5.2 ± 0.8
	Straw + Sulcotrione + Soil	60.0 ± 1.8	40.0 ± 1.8

CMBA: 2-chloro-4-(methyl-sulphonyl)benzoic acid

4. DISCUSSION

A larger fraction of sulcotrione was mineralized in a shorter time than atrazine, confirming the suspected high biodegradability of the molecule. Sulcotrione's half life has been found to be less than one week under field conditions (Compagnon and Beraud, 1992). Atrazine retention was more important than for sulcotrione on soil (Barriuso et al., 1996). The retention of atrazine was less reversible than for sulcotrione: (1) at the end of the incubation 99 and 59% of the extractable radioactivity was water extractable for sulcotrione and atrazine respectively; (2) 35 to 40% of the initial radioactivity formed bound residues for atrazine and only 26% for sulcotrione.

Hydroxyatrazine and CMBA appeared as intermediate metabolites in the pathways of degradation of atrazine and sulcotrione, respectively. Again sulcotrione appeared as the more easily degraded; 75% of the extractable fraction remained as atrazine but all the sulcotrione was degraded as CMBA. Atrazine remained the main extractable component during the incubation but hydroxyatrazine represented 30%. This may confirm that hydroxylation is a necessary step before ring cleavage and mineralization (Kaufman and Kearney, 1970). Hydroxyatrazine accumulation has been observed also by Assaf and Turco (1994) and Mandelbaum et al. (1993a) who demonstrated its possible microbial origin (Mandelbaum et al., 1993b).

Almost no atrazine or sulcotrione was mineralized during the incubations with compost and straw. For both molecules, the microflora responsible for their mineralization were present in the soil but not in the EOM. No degradation at all was observed for sulcotrione; all the extractable radioactivity remained as sulcotrione. A partial degradation of atrazine was observed with straw; hydroxyatrazine represented 84 % and atrazine only 5% of the extractable radioactivity. During the incubation of compost, atrazine remained the principal component of the extractable fraction, with 74 and 21% of atrazine and hydroxyatrazine, respectively. Both biotic and abiotic activity could contribute to the hydrolysis of atrazine in the EOM. The hydrolysis of atrazine in acidic solutions of humic substances decreased with the increase of pH from 2 to 7 (Khan, 1978). The humic fraction of the composted straw and the compost represented 50 and 20% of total organic C, respectively (Benoit et al., 1995; Serra-Wittling et al., 1995). The lower pH of the composted straw than of the waste compost, respectively 6.5 and 8.6, could partly explain the different proportion of hydroxyatrazine in the extractable radioactivity. On the other hand, important enzymatic activity has been measured in the two EOM, particularly peroxidase activity (Benoit et al., 1995; Serra-Wittling et al., 1995) that contribute to the formation of hydroxyatrazine. Many other hydrolase enzymes have been detected in degrading straw (Cao and Crawford, 1993; Trigo and Ball, 1994) that could also be involved in atrazine hydrolysis.

Addition of compost at the end of the pre-incubation slowed down the mineralization of both atrazine and sulcotrione. A decrease in the rate of mineralization of atrazine has been observed in a previous experiment with a similar compost in the same soil (Barriuso et al., 1997). This was related to the larger sorption of atrazine and sulcotrione on compost than on soil (Barriuso et al., 1996). For both herbicides, compost addition increased the available fraction of the molecules without any effect on their degradation since the composition of the extractable fraction was identical during the incubation of soil with or without addition of compost.

Straw addition also decreased the mineralization of both molecules, with a more minor effect than compost. It increased the formation of bound residues mainly in the case of sulcotrione for which the bound residue fraction doubled when straw was added. Simultaneously the extractable fraction decreased. The hydrolytic capacity of straw detailed previously increased the partial degradation of atrazine into hydroxyatrazine during the incubation of soil with straw. Straw addition did not modify the composition of the extractable fraction in the case of the sulcotrione.

Atrazine hydroxylation during the pre-incubation with straw stimulated the ring mineralization when soil was added. No latency time was observed, as occured at the beginning of the incubation of soil alone. After 20 days of rapid mineralization, the rate decreased and the kinetics seemed to reach the same asymptote as when straw was added to soil, corresponding to the potential non-reversible retention on the soil-straw mixture.
No degradation of sulcotrione occurred with compost or straw and 80% of the initial radioactivity remained extractable as sulcotrione. When soil was added to compost, little mineralization was observed except the important degradation of the sulcotrione into CMBA. With straw, soil addition increased the mineralization, although CMBA represented only 40% of the extractable radioactivity. Simultaneously, the proportion of bound residues more than doubled. A possible explanation could be the important biological activity in the soil-straw mixtures responsible for sulcotrione mineralization and incorporation into the microbial biomass, forming bound residues of biological origin.

5. CONCLUSION

The degradation of the two herbicides in soil was related to their structural characteristics. The triazine ring seemed more recalcitrant to biodegradation and mineralization than the benzoyl ring. Molecular structure governs herbicide retention on soil or organic matter. The higher the herbicide retention is, the lower is its biodegradation. The stabilization of the herbicide residues as non-extractable residues inversely depends on their transformations, mainly their mineralization. Microflora able to mineralize both herbicides were present in the soil, but not in the EOM used. Practically no mineralization was observed when both molecules were directly added to the EOM. However, their inoculation with soil allowed the herbicide mineralization to start. The two EOM components modified the behavior of both herbicides in the soil by different processes. Compost addition increased the retention and allowed the maintenance of a larger extractable fraction without modifying the pathway of degradation. Straw increased the formation of bound residues in soil. Partial degradation of atrazine into hydroxyatrazine occurred with straw, which favored the following complete mineralization of the triazine ring in the soil. Biotic or abiotic processes could be involved, including the important enzymatic activity or the acidity of the humic components present in the straw. Concerning sulcotrione, the important biological activity in the soil-straw mixtures could produce the large proportion of biologically formed bound residues.

6. ACKNOWLEDGMENTS

This work has been financially supported by the INRA program "Ecodynamique des substances à caractère polluant". The authors would like to thank the ICI company for supplying the labeled molecule and analytical standards of sulcotrione, and J.N. Rampon and V. Etievant for their technical assistance.

7. REFERENCES

Alvey, S. and D.E. Crowley, 1995. Influence of organic amendments on biodegradation of atrazine as a nitrogen source. J. Environ. Qual., 24, 1156-1162.

Assaf, A.A. and R.F. Turco, 1994. Accelerated biodegradation of atrazine by a microbial consortium is possible in culture and soil. Biodeg., 5, 29-35.

Barriuso, E., O.M. Eklo, E.J. Iglesias and S. Houot, 1996. Modification de la mobilité de pesticides dans les sols après addition de matières organiques exogènes. In : C. Walter and C. Cheverry (Editors). Proceedings of the "5èmes Journées Nationales de l'Etude des Sols, Rennes 96: Sols et transferts des polluants dans les paysages", AFES, pp. 99-101.

Barriuso, E., S. Houot and C. Serra-Wittling, 1997. Influence of compost addition to soil on the behaviour of herbicides. Pestic. Sci., 49, 65-75.

Behki, R.M. and S.U. Khan, 1986. Degradation of atrazine by *Pseudomonas*: N-dealkylation and dehalogenation of atrazine and its metabolites. J. Agric. Food Chem., 34, 746-749.

Bellin, C.A., G.A. O'Connor and Y. Jin, 1990. Sorption and degradation of pentachlorophenol in sludge-amended soil. J. Environ. Qual., 19, 603-608.

Benoit, P., 1994. Rôle de la nature des matières organiques dans la stabilisation des résidus de polluants organiques dans les sols. Thesis Institut National Agronomique Paris-Grignon, France.

Benoit, P., S. Houot, V. Bergheaud and E. Barriuso, 1995. Modification du comportement d'herbicides dans le sol par l'addition de matières organiques exogènes. ANPP-XVI Conférence COLUMA, Reims, pp. 117-124.

Cao, W. and D.L. Crawford, 1993. Carbon nutrition and hydrolytic and cellulolytic activities in the ectomycorrhizal fungus *Pisolithus tinctorius*. Can. J. Microbiol., 39, 529-535.

Compagnon, J.M. and J.M. Béraud, 1992. ICIA0051, un nouvel herbicide polyvalent de post-levée sélectif du maïs. ANPP - XV Conférence COLUMA, Versailles, pp. 349-356.

Doyle, R.C., D.D. Kaufman and G.W. Burt, 1978. Effect of dairy manure and sewage sludge on ^{14}C-pesticide degradation in soil. J. Agric. Food Chem., 26, 987-989.

Gschwind, N., 1992. Rapid mineralization of the herbicide atrazine by a mixed microbial community. Proceedings of the International Symposium on Environmental Aspects of Pesticide Microbiology, Department of Microbiology, Swedish University of Agricultural Sciences, Uppsala, pp. 204-206.

Guo, L., T.J. Bicki, A.S. Felsot and T.D. Hinesly, 1993. Sorption and movement of alachlor in soil modified by carbon-rich wastes. J. Environ. Qual., 22, 186-194.

Hance, R.J., 1973. The effect of nutrients on the decomposition of the herbicides atrazine and linuron incubated with soil. Pestic. Sci., 4, 817-822.

Kaufman, D.D. and P.C. Kearney, 1970. Microbial degradation of *s*-triazine herbicides. Residue Rev., 32, 235-265.

Khan, S.U., 1978. Kinetics of hydrolysis of atrazine in aqueous fulvic acid solution. Pestic. Sci., 9, 39-43.

Mandelbaum, R.T., L.P. Wackett and D.L. Allan, 1993a. Mineralization of the *s*-triazine ring of atrazine by stable bacterial mixed cultures. Appl. Environ. Microbiol., 59, 1695-1701.

Mandelbaum, R.T., L.P. Wackett and D.L. Allan, 1993b. Rapid hydrolysis of atrazine to hydroxyatrazine by soil bacteria. Environ. Sci. Technol., 27, 1943-1946.

Mandelbaum, R.T., D.L. Allan and L.P. Wackett, 1995. Isolation and characterization of a *Pseudomonas* sp. that mineralizes the *s*-triazine herbicide atrazine. Appl. Environ. Microbiol., 61, 1451-1457.

Martinez-Iñigo, M.J. and G. Almendros, 1992. Pesticide sorption on soils treated with evergreen oak biomass at different humification stages. Commun. Soil Sci. Plant Anal., 23, 1717-1729.

Morel, R., T. Lasnier and S. Bourgeois, 1984. Les essais de fertilisation de longue durée de la station agronomique de Grignon: Dispositif Dehérain et des 36 parcelles. INRA, Paris.

Mougin, C., C. Laugero, M. Asther, J. Dubroca, P. Frasse and M. Asther, 1994. Biotransformation of the herbicide atrazine by the white rot fungus *Phanerochaete chrysosporium*. Appl. Environ. Microbiol., 60, 705-708.

Radosevich, M., S.J. Traina, Y.L. Hao and O.H. Tuovinen, 1995. Degradation and mineralization of atrazine by a soil bacterial isolate. Appl. Environ. Microbiol., 61, 297-302.

Serra-Wittling, C., S. Houot and E. Barriuso, 1995. Soil enzymatic response to municipal solid waste compost addition. Biol. Fertil. Soils, 20, 226-236.

Topp, E. , L. Tessier and E.G. Gregorich, 1996. Dairy manure incorporation stimulates rapid atrazine mineralization in an agricultural soil. Can. J. Soil Sci., 76, 403-409.

Trigo, C., and A.S. Ball, 1994. Production of extracellular enzymes during the solubilization of straw by *Thermomonospora fusca* BD25. Appl. Microbiol. Biotechnol., 41, 366-372.

SOLUBILIZATION OF PHOSPHORUS FROM APATITE BY SULFURIC ACID PRODUCED FROM THE MICROBIOLOGICAL OXIDATION OF SULFUR

P. CANTIN[1], A. KARAM[2] and R. GUAY[1]

[1]Département de Biologie Médicale, Faculté de Médecine, et
[2]ERSAM, Faculté des Sciences de l'Agriculture et de l'Alimentation, Université Laval, Ste-Foy (Québec), CANADA, G1K 7P4

INTRODUCTION

The phosphorus (P) present in apatitic materials is largely in forms that are not readily available for most arable crops. Many attempts have been made to increase the dissolution of phosphate rock. Hence, soluble P or superphosphate used as fertilizer has been produced by the industrial acidification of apatite (Becker, 1980). Alternatives to high cost superphosphate have been studied. These are direct application of either apatite (Bolland *et al.*, 1988), or biological superphosphate or biosuper (Swaby, 1975). Biosuper has been produced by mixing apatite rock phosphate with elemental sulfur (Kittams, 1963; Rajan and Edge, 1980). Pathiratna *et al.* (1989) conducted a series of laboratory incubation studies and found that incubating mixtures of apatite and sulfur in powder or pellet form with a soil inoculum of *Thiobacillus* spp. increased the water-soluble P extracted and decreased the pH of the soil. They indicated that sulfuric acid produced from biological oxidation of elemental sulfur reacts with the apatite rock phosphate to produce *in situ* superphosphate in the soil. An acid pH is a necessary prerequisite for the dissolution of an essentially water-insoluble phosphate rock in soil (Robinson and Syers, 1990).

The biological dissolution of phosphate rock materials has usually been studied using soil as bacterial inoculum (Pathiratna *et al.*, 1989). No published work, however, appears to have examined the effect of sewage sludge applied to soils as organic amendment on the dissolution of phosphate rock materials in soils. Recent studies have suggested that heterotrophic bacterial oxidation of elemental sulfur is stimulated by additions of organic matters (Cifuentes and Lindemann, 1993). Cowell and Shoenau (1995) reported that sulfur oxidation was enhanced by sewage sludges. Sewage sludges could contain sulfur-oxidizing bacteria such as *Thiobacillus thioparus* and *T. thiooxidans* (Dufresne *et al.*, 1993).

Effect of Mineral-Organic-Microorganism Interactions on Soil and Freshwater Environments
Edited by Berthelin, *et al*, Kluwer Academic / Plenum Publishers, New York, 1999

This paper reports a series of experiments that were performed to determine the effects of combination of commercial elemental sulfur with either a sewage sludge containing *thiobacilli* bacteria, or a mixture of *thiobacilli* bacteria strains, *T. thioparus* and *T. thiooxidans*, on solubilizing P from a Florida apatite.

MATERIAL AND METHODS

Two apatite rock phosphate samples from the state of Florida, USA, were used. The two apatite samples, namely apatite 1 and apatite 2, contained respectively 26.7 and 29.8% P_2O_5 and had pH_w of 6.9 and 7.9.

Municipal sewage (waste water) sludge containing *thiobacilli* (Dufresne *et al.*, 1993) were used (Blais *et al.*, 1993).

Strains of sulfur oxidizing bacteria *T. thioparus* ATCC 23645, *T. thiooxidans* ATCC 8085 and ATCC 55128 came from the American Type Culture Collection, Rockville (USA). *T. thioparus* C5 was given by Dr. Jean-François Blais (Université du Québec). Both *T. thioparus* strains were grown in the Barton and Shively (1968) culture medium adjusted to pH 7.0 (BS7) with elemental sulfur as an energy source, and *T. thiooxidans* was grown in the same culture medium adjusted to pH 4.0 (BS4). Subculturing was carried out at 30°C in Erlenmeyer flasks on a gyratory action shaker for 7 days.

The effectiveness of the waste water sludge and the strains of *thiobacilli* to solubilize P from the Florida apatite was evaluated in an apatite-sulfur-culture medium (ASM) containing 1, 10 or 20% (P/V) of apatite, elemental sulfur (precipitated sulfur, Fisher Scientific) at a constant ratio with apatite of 1:5 (sulfur:apatite), and BS7-m culture medium (0.6 g KH_2PO_4, 0.4 g Na_2HPO_4, 0.4 g $(NH_4)_2SO_4$, 0.5 g $MgSO_4{\cdot}7H_2O$, 0.20 g $CaCl_2$ and 0.01 g $FeSO_4{\cdot}7H_2O$). The pH of the culture medium was adjusted to 7.0 before autoclaving.

In experiment 1, three replicates of ASM were inoculated with 1% (V/V) of a seven day old culture of *T. thioparus* ATCC 23645 and incubated for 15 days.

In experiment 2, two replicates of ASM were inoculated with both bacteria, *T. thioparus* C5 and *T. thioparus* ATCC 8085 (1% V/V of 7 days old culture of each culture), and incubated for 21 days.

In experiment 3, three replicates of ASM were inoculated with *T. thioparus* ATCC 23645 and *T. thiooxidans* ATCC 55128 (1% V/V of 7 day old culture of each culture) and incubated for 33 days. Culture medium was then replaced with fresh BS7-m culture medium and the ASM was incubated for 24 days. The total duration of this experiment was 57 days.

In experiment 4, two replicates of the ASM were inoculated with 2% (V/V) of a municipal waste water sludge and incubated for 33 days.

All incubation experiments were conducted in Erlenmeyer flasks and incubated at 30°C on a gyratory action shaker. Sterile controls with and without sulfur and/or apatite were used in the same conditions. The pH of the apatite suspension was measured at increasing time intervals in a sample withdrawn aseptically from apatite suspension. The phosphorus in the apatite suspension was determined colorimetrically as ascorbic acid reduced phosphomolybdate (Watanabe and Olsen, 1965). The presence of active sulfur oxidizing bacteria was assessed by plating dilutions of ASM samples on BS7 and BS4 solid media.

RESULTS

In experiment 1, the pH value of apatite suspension containing only *T. thioparus* (Fig. 1) decreased from 6.8 (apatite 1) or 7.8 (apatite 2) to 3.9. This observation confirms that *T. thioparus* oxidized S^0 into sulfuric acid, but only traces of P was solubilized from apatite (data not shown). However, viable cells of *T. thioparus* were detected in the culture medium from day 0 to 8 but not at day 15 (data not shown).

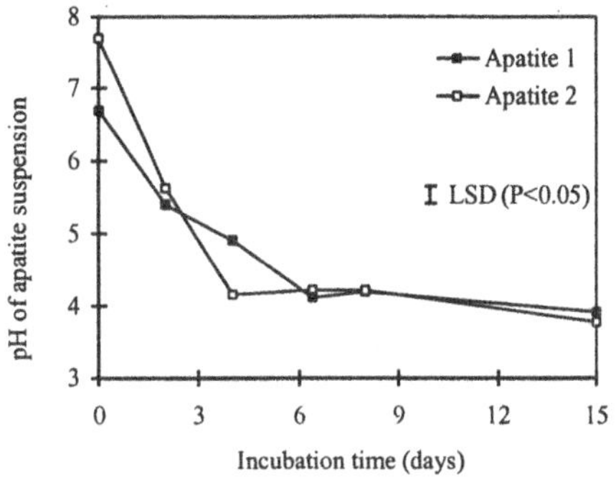

Figure 1. Experiment 1 - Changes in pH over incubation time following addition of sulfur and *T. thioparus* ATCC 23645 to apatite (10% W/V).

Combination of *T. thioparus* C5 and *T. thiooxidans* ATCC 8085 in experiment 2 decreased the pH of the apatite suspension to a lower level (Fig. 2a) than *T. thioparus* alone and resulted in a greater accumulation in soluble P (Fig. 2b). As illustrated in Fig. 2, soluble P concentration increased as pH decreased at all levels of apatite applied. In contrast, the pH and solubilized P curves for the sterile control showed no change over 21 days, thus indicating that these two species of *thiobacilli* bacteria play a significant role in solubilizing P from apatite. The pH values tended to decrease for all level of apatite applied with time, and then stayed constant up to the 21st day.

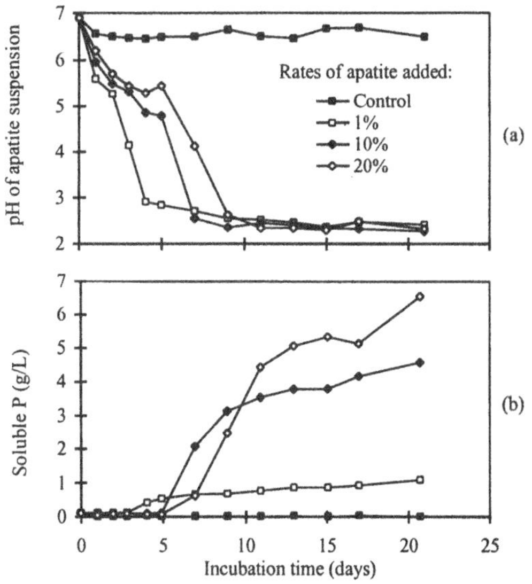

Figure 2. Experiment 2 - Changes in pH (a) and in soluble P (b) over incubation time following addition of sulfur, *T. thioparus* C5 and *T. thiooxidans* ATCC 8085 to apatite 1.

In the same way, *T. thioparus* ATCC 23645 combined with *T. thiooxidans* ATCC 55128 used in experiment 3 also decreased the pH of the apatite suspension and solubilized P from apatite (Fig. 3). Replacement of the spent culture medium at day 33 enhanced P solubilization although pH remained fairly constant. P solubilization kinetic increased after the replacement of the high soluble P content culture medium with fresh culture medium. Viable *T. thioparus* cells were detected in the culture medium up to day 5 whereas *T. thiooxidans* was present up to day 57 (data not shown). Over a 57-day period, 57% of the initial total P was solubilized.

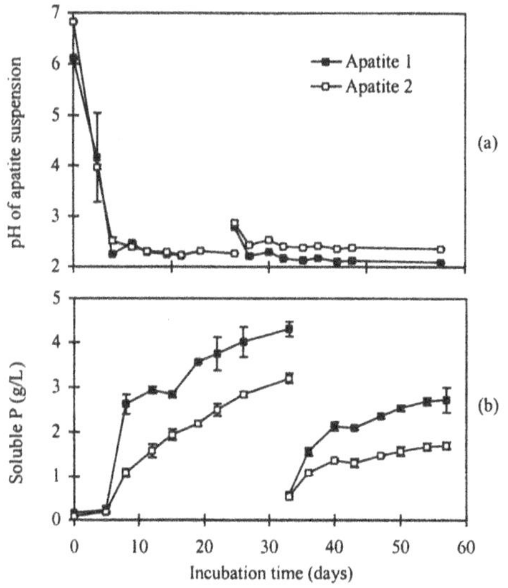

Figure 3. Experiment 3 - Changes in pH (a) and in soluble P (b) over incubation time following addition of sulfur, *T. thioparus* ATCC 23645 and *T. thiooxidans* ATCC 55128 to apatite (10% W/V) samples. Bars represent SD of the observations.

Finally, the results of the experiment 4 showing the effect of waste water sludge application on the acidification of apatite amended with sulfur is illustrated in Fig. 4. Changes in pH and soluble P patterns tended to follow the same trend as the other bacterial P solubilization curves (Fig. 2 and Fig. 3). After a relatively short period (3 days), the pH dropped rapidly and maintained at about 2.5. Over a 33-day period, only 28% of the initial total P was solubilized when the apatite-sulfur-sewage-sludge consisted of 20% apatite, but this proportion increased to 86% when the mixture was 1% apatite (Fig.4). The average release of P from all combinations of apatite with sewage sludge and sulfur was 53%.

DISCUSSION

The sequence of events that occured in the apatite-sulfur mixtures might be the following. The first part of the pH drop was most probably caused by the growth of *T. thioparus* for which the growth range is between pH 7.8 to 4.4 (Kelly and Harrison, 1984). There was very little P solubilization at this pH range. During this first drop in pH, *T. thiooxidans* remained viable. As the pH decreased to about 2.5, growth conditions became suitable for *T. thiooxidans*, which then took over and produced sulfuric acid from S^0. According to Kelly and Harrison (1984), the optimal pH for *T. thiooxidans* is between 2.0 and 3.0. At this pH level, P solubilization from apatite occured and *T. thiooxidans* remained active as long as S^0 was available to oxidation in the apatite suspension.

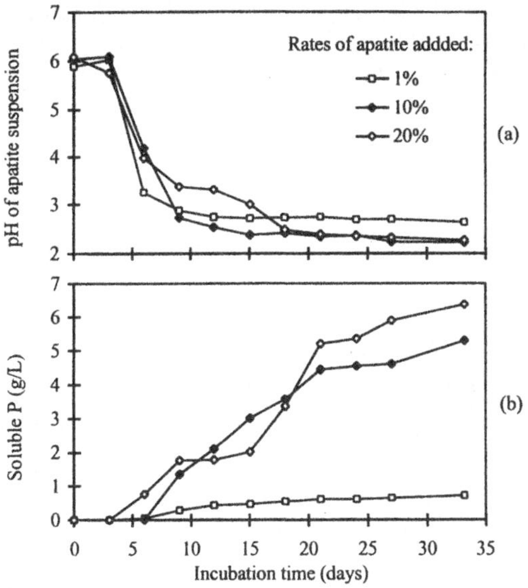

Figure 4. Experiment 4 - Changes in pH (a) and in soluble P (b) over incubation time following addition of sulfur and waste water sludge to apatite 1.

Our study indicated that soluble P increased with the amount of apatite added to the ASM and that the percent P solubilized from apatite decreased with the increasing rate of apatite added to the ASM. This observation is consistent with the results of Robinson and Syers (1991), who showed that increasing Ca (a product of apatite dissolution) concentration in soil solution decreased the dissolution of a phosphate rock. They suggested that in an open system, such as soil, the extent of the percentage of P solubilized from apatite could be higher.

Municipal sewage sludge was effective to solubilize P in a way similar to mixed bacterial cultures. The literature indicates that naturally occurring bacterial species in the sewage sludge are capable of oxidizing elemental sulfur into sulfuric acid (Blais *et al.*, 1992a, b). Consequently, sulfur oxidizing microflora such as *T. thioparus* and *T. thiooxidans* present in sewage sludge (Dufresne *et al.*, 1993) produces sulfuric acid from the elemental S, and sulfuric acid in turn reacts with the apatite rock phosphate to release further P in the solution.

CONCLUSION

A mixture of two *Thiobacillus* species, *T. thioparus* and *T. thiooxidans*, solubilized much more P from apatite amended with elemental sulfur than the treatment with *T. thioparus* alone.

The inoculation of phosphate rock with both bacterial strains would potentially improve the efficacy of apatite as a source of fertilizer P. The less acidophilic bacteria *T. thioparus* would help to establish suitable pH conditions in the apatite-sulfur pellet micro-environment for the growth of the acidophilic *T. thiooxidans*, even if the former does not produce enough sulfuric acid to solubilize P. Then, *T. thiooxidans* would produce enough sulfuric acid to solubilize P from apatite.

An indigenous population of sulfur oxidizing bacteria in the sewage sludge may explain the dissolution of apatite amended with S^0. In Canada, the interest to sewage sludge agriculture is well established. Combinations as pellet form of sulfur, apatite and stabilized sewage sludge as a source of *thiobacilli* for agricultural use, would provide an effective P fertilizer source for farmers in the future.

REFERENCES

Barton, L.L. and J.M. Shively, 1968. Thiosulfate utilization by *Thiobacillus thiooxidans* ATCC 8085. J. Bacteriol., 95, 720.

Becker, P., 1980. Phosphates and phosphoric acid : raw materials, technology, and economics of the wet process. Dekker, New-York.

Blais, J. F., J.C. Auclair and R.D. Tyagi, 1992a. Cooperation between two thiobacillus strains for heavy metal removal from municipal sludge. Can. J. Microbiol., 38,181-187.

Blais, J. F., R.D. Tyagi, J.C. Auclair and M.C. Lavoie, 1992b. Indicator bacteria reduction in sewage sludge by a metal bioleaching process. Wat. Res., 26, 487-495.

Bolland, M.D.A., R.J. Gilkes and M.F. D'Antuono, 1988. The effectiveness of rock phosphate fertilisers in Australian agriculture: a review. Aust. J. Exp. Agric., 28, 655-668.

Cifuentes, F.R. and W.C. Lindemann, 1993. Organic matter stimulation of elemental sulfur oxidation in calcareous soil. Soil Sci. Soc. Am. J., 57, 727-731.

Cowell, L.E. and J.J. Schoenau, 1995. Stimulation of elemental sulphur oxidation by sewage sludge. Can. J. Soil Sci., 75, 247-249.

Dufresne, S., J.-F. Blais, C. Roy and R. Guay, 1993. Municipal waste water treatement plant sludges: a source of organic carbon-tolerant, sulfur-oxidising *Thiobacillus* and *Sulfobacillus* strains. In: A.E. Torma, M.L. Apel and C.L. Brierley (Editors). Biohydrometallurgical Technologies. The Minerals, Metals & Materials society.

Kelly, D.P. and P. Harrison, 1984. Genus *Thiobacillus*. In: Bergey's manual of systematic bacteriology (vol. 3). William and Wilkins Co., Baltimore, USA.

Kittams, H.A., 1963. Use of sulfur for increasing the availability of phosphorus in rock phosphates. Dissertation Abstract, 24, 1323.

Pathiratna, L.S.S., U.P. De S. Waidyanatha and O.S. Peries, 1989. The effect of apatite and elemental sulphur mixtures on growth and P content of *Centrocema pubescens*. Fert. Res., 21, 37-43.

Rajan, S.S.S and E.A. Edge, 1980. Dissolution of granulated low grade phosphate rock, phosphate rock sulphur (Biosuper), and superphosphate in soil. New Zealand J. Agric. Res., 23, 451-456.

Robinson, J. S. and J. K. Syers, 1990. A critical evaluation of the factors influencing the dissolution of Gafsa phosphate rock. J. Soil Sci., 41, 597-605.

Robinson, J. S. and J. K. Syers, 1991. Effects of solution calcium concentration and calcium sink size on the dissolution of Gafsa phosphate rock in soil. J. Soil Sci., 42, 389-397.

Swaby, R.J., 1975. Biosuper-Biological superphosphate. In: McLachlan (Editor). Sulphur in Australasian Agriculture. Sydney University Press, pp. 213-220.

Watanabe, F.S. and S.R. Olsen, 1965. Test of an ascorbic acid method for determining phosphorus in water and $NaHCO_3$ extracts from soil. Soil Sci. Soc. Am. Proc., 29, 677-678.

AVAILABILITY OF SOIL PHOSPHORUS TO THE GREEN ALGAE *SELENASTRUM CAPRICORNUTUM*

Kirsti Krogerus[1] and Petri Ekholm[2]

1Regional Environment Agency of Häme,
P.O. Box 297, FIN-33101 Tampere, Finland,
Internet: kirsti.krogerus@vyh.fi
2Finnish Environment Institute,
P.O. Box 140, FIN-00251 Helsinki, Finland,
Internet: petri.ekholm@vyh.fi

1. INTRODUCTION

In freshwater environments, phosphorus (P) is usually the limiting nutrient for algal growth. Erosion of agricultural land forms a major source of P loading in many lakes. The contribution of soil P to the eutrophication of recipient waters depends on the rate at which it is transformed into dissolved orthophosphate, the only P compound that can be utilized directly by all planktonic algae (Cembella *et al.*, 1984; Boström *et al.*, 1988). The mobilization of orthophosphate from soil particles is largely controlled by desorption reactions (Yli-Halla *et al.*, 1995; Yli-Halla and Hartikainen, 1996). Owing partly to a gradually increasing water-to-soil ratio, the desorption of P starts as soon as rain or snow-melt water is mixed with surface soil and continues during the subsequent transport of eroded soil particles via ditches and rivers to lakes. In addition to this physically and chemically induced P-release, a biological mechanism has been proposed: when in direct contact with soil particles, algae may convert soil P into an available form with their surface-bound enzymes (see Williams *et al.*, 1980; Hegemann and Keenan, 1985; Boström *et al.*, 1988).

The availability of P to algae has commonly been determined with assays in which the sample under study provides the only source of P for algal growth. A batch assay allowing direct contact between algae and particles has been considered the most reliable technique (Lee *et al.*, 1980; Boström *et al.*, 1988). However, when the P source and the P sink are mixed, measuring the exchange of P is problematic. Usually the amount of available P is determined by counting algal cells at stationary growth phase and by comparing the yield with that obtained in a set of standard cultures using various amounts of dissolved orthophosphate.

When algae and P-carrying particles are separated from each other, for example by a membrane filter (DePinto *et al.* ,1981), the increase in algal P can be determined directly. In addition, the algal culture can be replaced during the assays to ensure P-starved conditions.

Effect of Mineral-Organic-Microorganism Interactions on Soil and Freshwater Environments
Edited by Berthelin, *et al*, Kluwer Academic / Plenum Publishers, New York, 1999

When analyzing agricultural P sources, the reported values for algal-available P obtained with batch assays are usually higher than those obtained with dual culture assays (Rekolainen *et al.*, 1997). However, the difference cannot be attributed solely to the effect of surface-bound enzymes; the results may also reflect differences in sample material and its pretreatment and other variance in assay conditions, such as pH (Rekolainen *et al.*, 1997). As far as we know, no direct comparison between these two assay techniques has been made using the same sample material.

In this paper we report the results of a study in which the availability of soil P to algae was estimated using both the batch assay and dual culture assay technique. Our aim was to determine whether a commonly used test organism, the green alga *Selenastrum capricornutum*, can, when in direct contact with soil particles, utilize more P than would be released by effective desorption only. The chemical and physical conditions during the assays were set to be as similar as possible. In addition to algal assays, we estimated availability of P by analyzing reactive P from the soils. Krogstad and Løvstad (1991) found that the amount of reactive P, a chemical fraction of soil P, was similar to that of P available to algae in Norwegian soils. We tested surface soil instead of eroded material in runoff or river waters because eroded material may already have lost a substantial amount of its labile P reserves. In this case, the testing of eroded material, which is the most common procedure, may underestimate the potential of soil as an algal P source.

2. MATERIALS AND METHODS

2.1. Soils and Their Analysis

We tested altogether 7 surface soil (0–10 cm) samples from experimental fields at Aurajoki and Jokioinen (Table 1). Both fields are situated in southern Finland and have slightly acidic (pH ~6) clay soils. The soils of both fields were classified as (very) fine, mixed Typic Cryaquepts (Soil Survey Staff, 1996). The soil at Aurajoki was richer in labile P, as determined by an ammonium acetate extraction, than the soil at Jokioinen (Table 1). This difference reflects the higher degree of cumulative P fertilization at the Aurajoki field. Each Aurajoki sample represented a plot in which different management practices were applied. Two of the Jokioinen samples were taken from buffer strips and one from a cultivated plot (Table 1). Further data on these soils are detailed by Puustinen (1994), Yli-Halla *et al.* (1995) and Uusi-Kämppä and Yläranta (1996).

Table 1. Characteristics of the soil samples. The clay percentage is based on sedigraph analyses; labile P was extracted by acid ammonium acetate.

Field/plot	Clay %	Total P $mg\ g^{-1}$	Labile P[1] $mg\ dm^{-3}$	Reactive P % of total P	TN $mg\ g^{-1}$
Aurajoki (all plots fertilized):					
A3 Winter wheat	55	1.6	22	15	1.9
A4 Wheat, stubble	55	1.8	28	15	1.9
A6 Wheat, normal ploughing	56	1.6	27	18	1.7
A7 Wheat, stubble tillage	53	1.2	24	14	2.0
Jokioinen:					
J1 Vegetated buffer strip	64	1.1	5	11	1.8
J2 Barley, normal ploughing + fertilization	67	1.1	7	15	1.8
J3 Natural buffer strip	73	1.3	9	11	2.3

1=Data for the Aurajoki and Jokioinen soils were provided by M. Yli-Halla and J. Uusi-Kämppä, respectively.

The Aurajoki soils were air-dried, homogenized, rewetted to a moisture content of 20% and stored at 5°C for almost 2 years before the algal assays. The Jokioinen soils were well ground but otherwise untreated before being used in the assays one week after sampling.

Before conducting the assays, we analyzed the particle size distribution of the soils with a Micromeritics SediGraph 5100 and their total P and total N content by means of H_2SO_4+$Na_2S_2O_8$ digestion, followed by P analysis using the ammonium molybdate method with ascorbic acid as the reducing agent, and N analysis using the Kjeldahl method (Ahl and Lindeval, 1974). Reactive P was analyzed according to the procedure applied by Krogstad and Løvstad (1989).

2.2 Algal Assays

For both assay techniques, the soil was first suspended in P-free algal nutrient medium (5% Z8, Kotai 1972) to obtain an initial P concentration of 140 to 300 $\mu g\ l^{-1}$. Part of each suspension was immediately poured into an assay vessel and the remainder was analyzed for P, total suspended solids and pH.

Dissolved reactive P was analyzed from a filtrate using the ammonium molybdate method. In total P and dissolved total P analyses, the original sample or the filtrate, respectively, was digested by $K_2S_2O_8$ before being analyzed with ammonium molybdate. To improve the digestion, the samples were diluted 1:4 (algal suspensions) or 1:10–1:20 (soil suspensions) with distilled water prior to analysis of total P (see Turtola, 1996). The concentration of dissolved unreactive P was calculated by subtracting dissolved reactive P from dissolved total P. Total suspended solids were analyzed gravimetrically. In all filtrations, Nuclepore polycarbonate filters (0.4 µm pore size) were used. The analyses have been described in greater detail by the National Board of Waters (1981).

In some samples, there was marked variation in the replicate determinations of soil suspensions, especially in the case of total P. For the suspensions made from the Aurajoki soils, the coefficient of variation for 4 replicate total P analyses ranged from 0.08 to 0.21. These variations were probably due to uneven distribution of easily settling soil material in the sample aliquots. There were no clear differences between total P values obtained from samples diluted to 1:10 and 1:20 with distilled water, suggesting that degradation of P was sufficient.

A total of 3 to 8% of soil P was immediately released as dissolved reactive P when soil was suspended in the P-free medium. The suspensions made from the Jokioinen soils also contained 2 to 4 $\mu g\ l^{-1}$ of dissolved unreactive P.

The batch assays were started by pipetting 1 ml of 7-day-old algal culture of *Selenastrum capricornutum* Printz (2×10^5 cells ml^{-1}) into Erlenmeyer flasks containing 100 ml of soil suspension or nutrient medium with 30 and 100 $\mu g\ l^{-1}$ P as K_2HPO_4 (the controls). The flasks were incubated on a shaking table for 2 weeks (Jokioinen soils) or 3 weeks (Aurajoki soils) at 20±1 °C and 4300±200 lux. After each week, the number of algal cells was counted using blood cell counting chambers. At least 200 cells or the entire area of the chamber were counted. Although the soils were not sterilized, no growth of indigenous algae was observed in the batch assays. Algal-available P (AAP_{batch}) was calculated from a regression equation determined earlier for obtained cell numbers in a set of solutions containing 0 to 100 $\mu g\ l^{-1}$ of K_2HPO_4 (Fig. 1). As in the case of total P, the variation in AAP_{batch} between the replicates was high, particularly for the Aurajoki soils: the coefficients of variation were 0.26 to 0.49 and 0.05 to 0.16 for the Aurajoki and Jokioinen soils, respectively. The results are presented as the mean values of three replicates with a 95% confidence limit of interpolation.

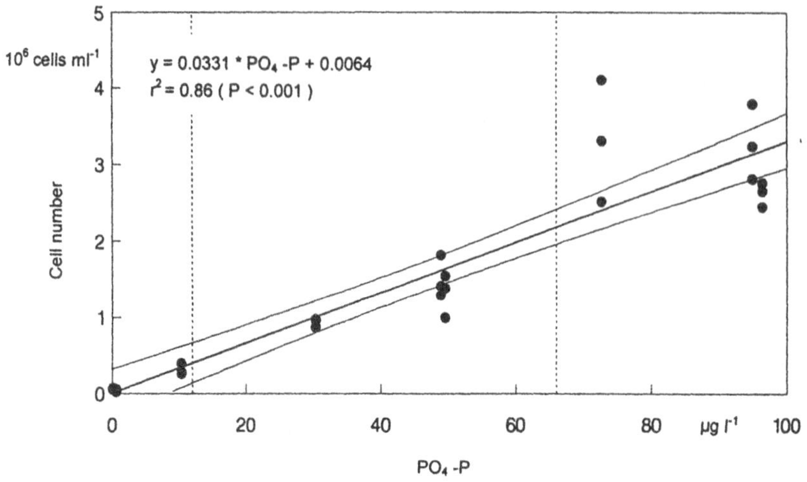

Figure 1. Algal cell number as a function of the concentration of dissolved orthophosphate. The broken lines show the calculated range of phosphorus released by the tested soil samples.

In the dual culture assays (DePinto *et al.*, 1981; Ekholm, 1994; Ekholm and Krogerus, 1998), a P-starved algal culture and soil suspension were incubated in a two-chambered vessel, separated from each other by a Nuclepore polycarbonate membrane (0.4 µm pore size). The vessels were incubated on a shaking table for 3 weeks at 20±1 °C and 4300±200 lux. After each week the algal culture was replaced by a fresh P-starved culture. The fresh and harvested cultures were analyzed for P and pH. After the experiments were completed, the sample chambers were emptied, rinsed with 10 ml of distilled water and the contents were analysed. In each test one vessel was treated as a control by adding a P-free nutrient medium with 73 µg P l^{-1} as K_2HPO_4 into the sample chamber. Algal-available P (AAP_{dual}) equaled the cumulative amount of P taken up by the algal cultures during the assay. The algal P uptake was determined from the increase in particulate P (total P - dissolved total P) in the algal suspension during the incubation. The tests, except the controls, were performed in duplicate and the results are presented as the mean values of the duplicates. The mean difference between the duplicates in the total increase in algal P content was 14 µg l^{-1}. This means that duplicate results deviated 18% on average from the presented mean value.

We estimated the long-term availability of P by assuming that ultimately available P (P_0) was released in the dual culture assays according to first-order kinetics (DePinto, 1982; Young *et al.*, 1982):

$$P_a(w) = P_0(1 - e^{-kw}) \qquad \text{Eq. 1.}$$

where P_a is the cumulative amount of P taken up by the algae and k the rate at which P becomes available during w weeks. P_0 and k were determined from the assay data by the method of least squares.

We tested the differences in the results between the two techniques by one-way analysis of variance (ANOVA). The ANOVA procedure was based on the raw data, i.e., the variation in the replicates was included in the analysis.

In both assays, the pH of all liquids used either as algal nutrient medium or in the dilution of soils was buffered to pH 8 by Tris (see Ekholm, 1994). During the assays the pH was 7.6 to 8.2.

3. RESULTS AND DISCUSSION

3.1. Dual Culture Assays

In the dual culture assays, the algae utilized 17 to 24% of soil P (see AAP_{dual} in Table 2). Although a higher proportion of P was labile in the Aurajoki soils than in the Jokioinen soils (Table 1), no clear differences between the two soils existed in the proportion of algal-available P. A relatively low amount of P in soil A6 was utilized by algae. However, when available P was expressed per one gram of dry soil, the values for all the Aurajoki soils were similar. As the Aurajoki soils were richer in P than the Jokioinen soils, AAP_{dual} expressed per one gram of dry soil was on average 19% higher in the Aurajoki soils (Table 2).

Available P was released more rapidly from the Jokioinen soils than from the Aurajoki soils (see k values in Table 2). The reserves of available P seemed to be depleted, however, from both soils during the 3-week incubation. This was concluded from the fact that P_0 was at the same level as AAP_{dual} (Table 2).

The experiment did not allow precise measurement of the availability of different P fractions. However, the availability of dissolved reactive P could be estimated tentatively from the observed changes in its concentration during the assay. Dissolved reactive P appeared to be available in marked quantities; the concentration in the algal cultures had mostly decreased below the limit of detection (2 $\mu g\ l^{-1}$) during the first week of incubation. The availability of the small amount of dissolved unreactive P released from the Jokioinen soils could not be estimated due to the difficulty of analysing such low concentrations of P.

Table 2. Algal-available phosphorus (AAP_{dual}), ultimately available phosphorus (P_0) and the release rate of algal-available phosphorus (k) in soil as tested with dual culture assays. Total P concentrations refer to soil suspensions used in the tests.

Field	Plot	Total P		AAP_{dual}			P_0	k
		$\mu g\ l^{-1}$	$mg\ g^{-1}$	$\mu g\ l^{-1}$	% of total P	$mg\ g^{-1}$	% of total P	
Aurajoki	A3	230	1.6	49	21	0.33	24	0.11
	A4	170	1.5	41	24	0.36	26	0.14
	A6	300	2.0	50	17	0.33	18	0.13
	A7	180	1.5	38	22	0.33	25	0.09
	Mean	220	1.6	45	21	0.34	23	0.12
Jokioinen	J1	140	1.3	32	23	0.30	22	0.35
	J2	160	1.3	30	19	0.24	18	0.40
	J3	150	1.3	34	23	0.30	22	0.29
	Mean	150	1.3	32	22	0.28	21	0.35

3.2. Batch Assays

In tests performed on the Aurajoki soils, the algal cell numbers increased during the first and second week and declined slightly during the third week. The Jokioinen soils were assayed for only 2 weeks and the results presented here are based on the cell numbers after 2 weeks' incubation for both soils. In batch assays 10 to 25% of soil P was utilized by the algae (see AAP_{batch} in Table 3). In the Jokioinen soils, the availability of P was 40% lower than in the Aurajoki soils. This result may partly be an artifact, however. In the test with the Jokioinen soils only 73% of P was utilized in the control flasks with 30 $\mu g\ P\ l^{-1}$, whereas in the test with the Aurajoki soils, a total of 91% of P was utilized in the corresponding controls. However, when corrected by the P utilization percentage in the controls, the AAP_{batch} values for the Jokioinen soils were still 27% lower than those for the Aurajoki soils. In contrast to the results obtained with the dual culture assays, those from batch assays correlated with labile P ($P<0.1$).

Our results are markedly lower than those of Krogstad and Løvstad (1991), whose batch assays indicated that natural populations of blue-green algae often utilized 20 to 70% of P in surface soil samples; the availability of P is lower with higher proportions of organic P in the soils. We did not determine the quantity of organic P in the soils. Its amount was probably rather low, since the soils contained only 2 to 4% organic carbon (Puustinen, 1994; Uusi-Kämppä and Yläranta, 1996).

Table 3. Algal-available phosphorus (AAP_{batch}) in soil as tested with batch assays. The values for AAP_{batch} correspond to the means of three replicates presented with a 95% confidence limit of interpolation. Total P concentrations refer to soil suspensions used in the tests.

Field	Plot	Total P	AAP_{batch}	
		$\mu g\ l^{-1}$	$\mu g\ l^{-1}$	% of total P
Aurajoki	A3	210	34 ± 6	16 ± 4
	A4	250	37 ± 6	16 ± 4
	A6	200	48 ± 6	25 ± 4
	A7	180	39 ± 6	23 ± 4
	Mean	210	40	20
Jokioinen J1	150	14 ± 8	10 ± 5	
	J2	130	14 ± 8	11 ± 6
	J3	140	20 ± 7	15 ± 5
	Mean	140	16	12

3.3. Comparison of the Assay Techniques

We assumed that in the dual culture assays, available P was released from soil mainly by way of desorption reaction. However, in the batch assays, the activity of algal surface-bound enzymes may also have affected P mobilization. Nevertheless, in 5 out of 7 samples, the dual culture assays gave higher values for the algal-available P than the batch assays. Only for sample A6 did the batch assays give a clearly higher result. Although the variation in the replicates was high in both assays, the difference between the results obtained by the two techniques was confirmed by ANOVA ($P<0.05$).

It is tempting to attribute the higher utilization of P in the dual culture assays to more efficient desorption brought about by the replacement of algal culture during the assay. From the Jokioinen soils with 44% lower AAP_{batch} than AAP_{dual}, negligible P was, however, released after the replacement of algal culture; most of the available P was already released during the first week of incubation. Nevertheless, the utilization of P may have been more efficient in the dual culture assays, since the algal cultures were grown in low P concentration before being used in the tests (see Ekholm, 1994). In contrast, the algae used in the batch assays were not P-starved and thus may have been unable to adapt rapidly enough to the low P concentration prevailing in the assays. Alternatively, the direct contact with soil may have adversely affected the algae in batch assays; microscopic inspections showed that part of the algae growing in the soil were somewhat deformed.

Our results show that the direct contact between algae and soil is not a prerequisite for the release of available P from mineral soil. Although the enzymes capable of degrading extracellular P sources are largely located along the cell wall, some of them are also released into solution (Cembella *et al.*, 1984). Thus, these enzymes may also have acted in the dual culture assays, since any dissolved compounds are rapidly exchanged between the chambers of the test vessels. In this case desorption was not the only factor controlling the release of available P.

Reactive P was approximately at the same level as AAP_{batch} (Tables 1 and 3). In contrast to the results of Krogstad and Løvstad (1991), no correlation was found either between reactive P and AAP_{batch} or between reactive P and AAP_{dual}. Thus, the measurement of reactive P seems not to serve as a surrogate for algal assays.

The estimation of the availability of P from agricultural sources is generally based on tests with eroded soil particles collected from rivers and streams (e.g., Williams *et al.*, 1980; DePinto *et al.*, 1981; Ekholm 1994). Using dual culture assays Ekholm (1994) found that 0 to 13% of P (or 0 to 0.18 mg g^{-1}) was available in suspended solids in agriculturally loaded Finnish rivers. The fact that for both soils tested in this study P availability was much higher (17 to 24% or 0.24 to 0.36 mg g^{-1} as tested with dual culture assays) indicates that the eroded particles had already lost most of their desorbable P. This P loss may have occurred due to the gradually increasing water-to-soil ratio during the transport from field to river. Alternatively, erosion may have selectively removed the finest particles with the best ability to bind P and, consequently, a low tendency to release P. In addition, drying of the soils may have increased P availability (Klotz, 1988). In tests with riverine particles, the samples were not dried (Ekholm, 1994).

4. ACKNOWLEDGMENTS

We are grateful to Jaana Uusi-Kämppä for providing the Jokioinen soil samples and to Markku Yli-Halla for providing the Aurajoki samples and for classifying the soils. Hertta Ilola performed the bioassays and Terttu Finni counted most of the algal samples. Michael Bailey and Silja Kudell revised the English.

5. REFERENCES

Ahl, T. and L. Lindeval, 1974. Intercalibration of sediment analyses. (Interkalibrering av sedimentkemiska analysemetoder.) Nordforsk, miljövårdssekretariatet, Publ. 5. (In Swedish).

Boström, B., G. Persson and B. Broberg, 1988. Bioavailability of different phosphorus forms in freshwater systems. Hydrobiologia, 170, 133–155.

Cembella, A.D., J.A. Naval and P.J. Harrison, 1984. The utilization of inorganic and organic phosphorus compounds as nutrients by eukaryotic microalgae: A multidisciplinary perspective: Part 1. Crit. Rev. Microbiol., 10, 317–391.

DePinto, J.V., 1982. An experimental apparatus for evaluating kinetics of available phosphorus release from aquatic particulates. Wat. Res. 16, 1065–1070.

DePinto, J.V., T.C. Young and S.C. Martin, 1981. Algal-available phosphorus in suspended sediments from lower Great Lakes tributaries. J. Great Lakes Res., 7, 311–325.

Ekholm, P., 1994. Bioavailability of phosphorus in agriculturally loaded rivers in southern Finland. Hydrobiologia 287, 179–194.

Ekholm, P. and K. Krogerus, 1998. Bioavailability of phosphorus in purified municipal wastewaters. Wat. Res. 32, 343–351.

Hegemann, D.A. and J.D. Keenan, 1985. Measurement of watershed phosphorus: A review. Toxicol. Envir. Chem., 9, 265–289.

Klotz, R.L., 1988. Sediment control of soluble reactive phosphorus in Hoxie Gorge Creek, New York. Can. J. Fish. Aquat. Sci., 45, 2026–2034.

Kotai, J., 1972. Instructions for preparation of modified nutrient solution Z8 for algae. Norwegian Institute for Water Research (NIVA), Publ. B-11/69.

Krogstad, T. and Ø. Løvstad, 1989. Erosion, phosphorus and phytoplankton response in rivers of South-Eastern Norway. Hydrobiologia, 183, 33–41.

Krogstad, T. and Ø. Løvstad, 1991. Available soil phosphorus for planktonic blue-green algae in eutrophic lake water samples. Arch. Hydrobiol., 122, 117–128.

Lee, G.F., R.A. Jones and W. Rast, 1980. Availability of phosphorus to phytoplankton and its implications for phosphorus management strategies. In: R.C. Loehr, C.S. Martin and W. Rast (Editors). Phosphorus management strategies for lakes. Ann Arbor Sci. Ann Arbor, pp. 259–308.

National Board of Waters, 1981. Methods of water analyses employed by the water administration. (Vesihallinnon analyysimenetelmät.) Helsinki. Report 213. (In Finnish).

Puustinen, M., 1994. Effect of soil tillage on erosion and nutrient transport in plough layer runoff. National Board of Waters and the Environment, Finland. Publ. Water Environ. Res. Inst., 17, 71–90.

Rekolainen, S., P. Ekholm, B. Ulén and A. Gustafson, 1997. Phosphorus losses from agriculture to surface waters in the Nordic countries. In: H. Tunney, O.T. Carton, P.C. Brookes and A.E. Johnston (Editors). Phosphorus loss from soil to water. CAB International, pp. 77–93.

Soil Survey Staff, 1996. Keys to Soil Taxonomy. USDA. Natural Resources Conservation Service. 7. Ed.

Turtola, E., 1996. Peroxodisulphate digestion and filtration as sources of inaccuracy in determination of total phosphorus and dissolved orthophosphate phosphorus in water samples containing suspended soil particles. Boreal Envir. Res., 1, 17–26.

Uusi-Kämppä, J. and T. Yläranta, 1996. Effect of buffer strip on controlling soil erosion and nutrient losses in southern Finland. In: G. Mulamootil, B.G. Warner and E.A. McBean (Editors). Wetlands: Environmental gradients, boundaries, and buffers. CRC Press, pp. 219–233.

Williams, J.D.H., H. Shear and R.L.Thomas, 1980. Availability to *Scenedesmus quadricauda* of different forms of phosphorus in sedimentary materials from the Great Lakes. Limnol. Oceanogr., 25, 1–11.

Yli-Halla, M. and H. Hartikainen, 1996. Release of soil phosphorus during runoff as affected by ionic strength and temperature. Agric. Food Sci. Finl., 5, 193–202.

Yli-Halla, M., H. Hartikainen, P. Ekholm, E. Turtola, M. Puustinen and K. Kallio, 1995. Assessment of soluble phosphorus load in surface runoff by soil analyses. Agric. Ecosyst. & Environ., 56, 53–62.

Young T.C., J.V. DePinto, S.E. Flint, M.S. Switzenbaum and J.K. Edzwald, 1982. Algal availability of phosphorus in municipal wastewater. J. Water Pollut. Control Fed., 54, 1505–1516.

ROLE OF PROTEINS IN THE ADHESION OF AZOSPIRILLUM BRASILENSE TO MODEL SUBSTRATA

Y. F. Dufrêne, C. J. -P. Boonart and P. G. Rouxhet

Unité de Chimie des interfaces, Université catholique de Louvain, Place Croix du Sud 2/18, 1348 Louvain-la-Neuve, Belgium Tel.: (32) 10473589 - Fax: (32) 10472005 - E-mail: rouxhet@cifa.ucl.ac.be

INTRODUCTION

In the natural environment, microbial cells adhere to a large variety of solid surfaces, from inanimate materials to living tissues (Savage and Fletcher, 1985; Characklis and Marshall, 1990; Marshall, 1991). In aquatic habitats (e.g. streams, lakes, oceans), microorganisms accumulate on living organisms, suspended particles, rocks and sediments. In many instances, these surfaces are rapidly colonized by bacteria and biofilms are formed as a result of cell multiplication and production of extracellular substances. Soil is undoubtedly the most complex microbial habitat (Stotzky, 1985) because of the high variability of the composition and size of solid constituents, the amount of water, nutrients and gases, and the physicochemical characteristics (e.g. pH, E_h, ionic strength). Soil particles that are colonized by bacteria are often coated by clay minerals, hydrous metal oxides, and organic matter. Clay minerals, and surfaces in general, may affect microbial activity in different ways, e.g., by modifying the physicochemical characteristics of the microbial habitat (Filip and Hattori, 1984; Rouxhet and Mozes, 1990; Mozes and Rouxhet, 1991, 1992). In the environment of plant roots (rhizosphere), bacteria benefit from root exudates as carbon and energy sources and, in turn, may promote plant growth by nitrogen fixation and production of various substances (Pueppke and Kluepfel, 1985).

Different models are available to describe microbial adhesion. The DLVO (Derjaguin, Landau, Verwey and Overbeek) theory accounts for long-range van der Waals and electrostatic forces (Rutter and Vincent, 1980; Rijnaarts *et al.*, 1995). Another approach is based on the balance of interfacial free energies involved in the creation of a cell-substratum interface and in the disappearance of cell-water and substratum-water interfaces (Busscher *et al.*, 1984). The hydrophobicity of cells has indeed been found to be an important property affecting adhesion (van Loosdrecht *et al.*, 1987, 1990). Both approaches assume that the surfaces are smooth and chemically homogeneous which is undoubtedly a rough approximation: bacterial cells bear appendages and/or solvated

Effect of Mineral-Organic-Microorganism Interactions on Soil and Freshwater Environments
Edited by Berthelin, *et al*, Kluwer Academic / Plenum Publishers, New York, 1999

macromolecules on their surfaces. These surface structures give rise to additional features that must be incorporated in the models describing microbial adhesion. Although the crucial role of interactions involving solvated "macromolecules-steric interactions" has been frequently mentioned in the literature (van Loosdrecht et *al*., 1990), there are practically no direct experimental data on these interactions.

Azospirillum is a Gram-negative diazotrophic bacterium that colonizes the rhizosphere of plants and promotes plant growth under certain conditions. This genus has become the focus of intense research efforts in view of its potential as a biofertilization agent (Okon, 1994). An important property that is thought to increase competitiveness of *Azospirillum* in the rhizosphere is the capacity to attach to plant roots and soil particles. Inasmuch as most experiments have been conducted using complex and undefined solid surfaces (e.g., living tissues, heterogeneous and porous systems) (Bashan and Levanony, 1988; Michiels *et al*., 1991), the physicochemical mechanisms governing adhesion are poorly understood. Moreover, the role played by cell surface components has often been speculated on the basis of indirect observations.
This contribution presents an overview of recent work aiming at a better understanding of the adhesion mechanism of *Azospirillum brasilense* to inert surfaces, special attention being devoted to the influence of extracellular compounds (Dufrêne and Rouxhet, 1996; Dufrêne *et al*., 1996a, 1996b, 1996c). Therefore flat model supports were used and the surface of cells and supports was analysed directly.

MATERIAL AND METHODS

Details can be found in papers cited above. The strain *A. brasilense* Sp7 (ATCC 29145) used in this study has been kindly supplied by Professor J. Vanderleyden, F.A. Janssens Laboratory of Genetics (Katholieke Universiteit Leuven, Belgium). The cells were grown in Luria-Bertani complex medium supplemented with 2.5 mM $CaCl_2$ and 2.5 mM $MgSO_4$ (LB*). They were resuspended in demineralized water, in phosphate buffer saline (PBS, 8.5 g l^{-1} NaCl, 3 g l^{-1} K_2HPO_4 and 1.15 g l^{-1} $NaH_2PO_4.H_2O$) or in a solution of Pluronic F68 (Sigma), a poly(ethylene oxide)-poly(propylene oxide)-poly(ethylene oxide) copolymer surfactant (0.1% w/v).

Glass and polystyrene were selected as model substrata. They were taken as models for oxide surfaces with a low isoelectric point (e.g. quartz, mica) and for surfaces coated by hydrophobic organic compounds, respectively, and were choosen for the sake of convenience and reproducibility. Glass samples were microscope slides (Menzel Glazer, Germany); they were cleaned by immersion overnight in sulfochromic mixture and rinsed with demineralized water prior to the adhesion tests. Polystyrene substrata were either Petri dishes (Merck) or substrata (5 x 7.6 cm) cut from a large polystyrene plate protected by an adhesive film (Vink, Belgium). The latter were cleaned by ultrasonication in isopropanol and rubbed dry with absorbing paper.

RESULTS AND DISCUSSION

Influence of cell surface properties and cell physiology

Adhesion of *Azospirillum brasilense* Sp7 to model substrata, glass and polystyrene, was characterized in static conditions by a test that brings the cells into contact with the substratum via sedimentation (Dufrêne and Rouxhet, 1996). Following harvesting and washing, the cells were resuspended in demineralized water and the suspension was poured into a Petri dish, containing a glass plate if required. The suspension was left undisturbed for 24 h at 30°C. The glass plate and the polystyrene dish were rinsed (vertical immersion and gentle agitation for 15 s in three baths of demineralized water for plate; filling three times with demineralized water and gentle stirring for dish) to remove the non-adhering or loosely bound cells. They were then observed under the microscope. Cell adhesion to glass and polystyrene substrata was denser for early stationary phase cells compared to exponential phase cells (Dufrêne and Rouxhet, 1996).

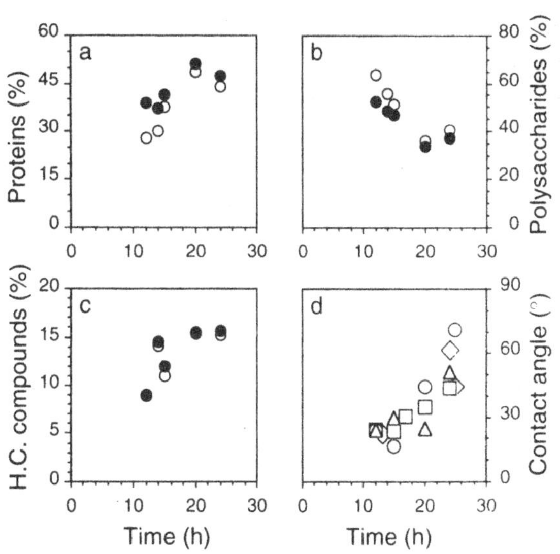

Figure 1. Evolution of the surface molecular composition (a-c) - estimated on the basis of the elemental composition (○) or the intensity of the components of the carbon peak (●) - or the water contact angle (d) *A. brasilense* during growth; for water contact angles, the different symbols correspond to independent sets of determinations carried out on separate bacterial cultures (Dufrêne and Rouxhet, 1996).

The surface chemical composition of freeze-dried cells (Dufrêne and Rouxhet, 1996) was determined by X-ray photoelectron spectroscopy (XPS), a technique that provides a direct chemical analysis of the outermost molecular layers (2-5 nm) of surfaces (Ratner and McElroy, 1986; Rouxhet and Genet, 1991). Figure 1 presents a typical set of data showing the evolution of the molecular composition of the cell surface as a function of culture time. The molecular composition was roughly modeled by considering three classes of basic constituents - proteins, polysaccharides and hydrocarbonlike compounds - and estimated on the basis of the elemental composition or the intensity of the components of the carbon peak. Clearly, variations of the surface molecular composition were observed during growth: the concentrations of proteins and hydrocarbonlike compounds increased, concomitantly with a decrease of the polysaccharide concentration. Figure 1d shows that the increase of the surface concentration of proteins and, to a certain extent, hydrocarbonlike compounds, was directly correlated with the cell surface hydrophobicity, as determined by water contact angle.

Accordingly, it appears that the increase of cell adhesiveness during growth is correlated with an increase of cell surface protein concentration and cell surface hydrophobicity.

The influence of contact time, temperature, and tetracycline addition on the adhesion of *A. brasilense* in water was further investigated (Dufrêne *et al.*, 1996a). Adhesion to glass and polystyrene substrata depended on time and temperature, lower adhesion densities were observed when the contact time was only 2 h or 6 h, compared to 24 h, or when the test was performed at 4°C, compared to 30°C. This pointed to the influence of cell physiology, which was further demonstrated by the inhibition of adhesion in the presence of tetracycline, an antibiotic that inhibits protein synthesis. Observations suggesting that active bacterial metabolism is required for attachment of *A. brasilense* have already been reported (Bashan *et al.*, 1986; Gafny *et al.*, 1986; Bashan and Levanony, 1988; Eyers *et al.*, 1988). However, most of these observations are somewhat doubtful since cells were submitted to treatments like γ-irradiation or heat which may also alter cell surface properties.

Involvement of proteins at the cell/substratum interface

The nature of the macromolecules produced at the cell-substratum interface was further investigated. To this end, surface analysis was performed by XPS on cell sediments and on polystyrene substrata after cell adhesion and subsequent detachment (Dufrêne *et al.*, 1996a). Cell sediments were obtained as follows: exponential phase cells resuspended in demineralized water were allowed to settle down onto polystyrene substrata and to age for 24 h at 30°C and 4°C; the unrinsed substrata were then immediately immersed in liquid nitrogen and freeze-dried. Polystyrene substrata were analysed after adhesion of exponential phase cells and cell detachment (repeated ultrasonication for 15 s in demineralized water; control by scanning electron microscopy).

Figure 2 presents O_{1s} peaks of a cell sediment obtained at 30°C and of polystyrene substratum after 24 h contact time at 30°C and detachment of the adhering cells. After adhesion and detachment of the adhering cells, the surface composition of polystyrene substratum was rich in oxygen, nitrogen and oxidized carbon, reflecting the presence of compounds of biological origin. The shape of the O_{1s} peak was very different from that of cell sediments: for the former, the prevailing component was due to $\underline{O}$=C, probably reflecting proteins, whereas for the latter the major component was due to C-$\underline{O}$H +C-$\underline{O}$-C, reflecting essentially polysaccharides.

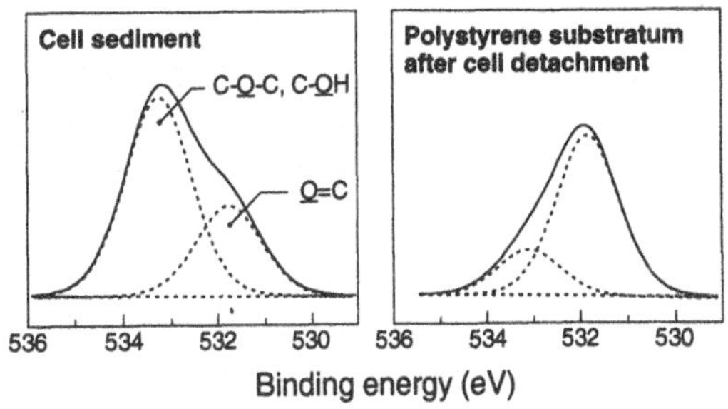

Figure 2. Representative O_{1s} peaks of a cell sediment (exponential phase cells, 24 h contact time at 30°C) and polystyrene substratum after 24 h contact time with exponential phase cells at 30°C and detachment of the adhering cells (Dufrêne et al., 1996a).

The surface elemental composition was converted into molecular composition as explained above. The protein to polysaccharide ratio of the cell sediment prepared at 4°C was similar to that observed with free cells freshly harvested from the exponential phase, the major constituent being polysaccharide. However, when metabolism was favored (30°C), the protein surface concentration of cell sediments increased and became similar to that of polysaccharides, as observed for free cells during growth. The surface composition of the substrata examined after adhesion in standard conditions (24 h contact time at 30°C in water) and subsequent cell detachment was very different from that of free cells and cell sediments, the protein to polysaccharide ratio being clearly higher in the former case. Lowering the contact time or performing adhesion under unfavorable metabolic conditions (4°C) or in the presence of tetracycline resulted in a decrease of protein concentration at the substratum surface, in the same way as the adhesion density.
The following conclusions may thus be drawn: i) for both cell sediments prepared at 30°C in water and free cells in culture, time leads to an enrichment of the cell surface in proteins; ii) proteins are the major constituent at the substratum surface after adhesion in standard conditions and subsequent cell detachment; iii) changes of experimental conditions affect the protein concentration at the cell surface (e.g. growth phase, aging temperature) or at the substratum surface (e.g. time, temperature, addition of tetracycline) and influence the density of adhering cells in the same direction.

These observations, based on direct surface analysis, provide an unambiguous indication of the involvement of extracellular proteins in the adhesion of *A. brasilense* to inert surfaces. In the literature, the involvement of surface proteins in the attachment of *Azospirillum* had been suggested only on the basis of indirect observations. Bashan and Levanony (1988) proposed that attachment of *A. brasilense* Cd to sand particles is actively mediated by a network of proteinaceous material. Michiels *et al.* (1991) showed, by the use of protease treatments, that initial adhesion of *A. brasilense* Sp7 to wheat roots was possibly mediated by proteinic compounds; furthermore, a mutant deficient in both polar and lateral flagella expression failed to adhere to wheat roots. Croes *et al.* (1993) showed that adhesion of *A. brasilense* Sp7 to wheat roots is mediated by the polar flagellum.

The results of the present study are of ecological significance. In the bulk soil, microorganisms rapidly utilize available energy substrates and create a situation in which most of the organisms are probably starving (Marshall, 1991). The accumulation of extracellular proteins and the increase of cell adhesiveness resulting from starvation conditions may therefore improve the immobilization of *Azospirillum* in the upper layers of the soil profile, thereby keeping the cells in the vicinity of the rhizosphere.
At this stage, two questions may be raised: i) how do proteins reach the cell-substratum interface, and ii) what are the mechanisms by which proteins influence adhesion? To elucidate these questions, cell adhesion to polystyrene substrata was examined in a parallel-plate chamber in absence of flow, thus allowing cell transport to the substratum by sedimentation. After 2 h or 24 h contact time, the chamber (7.8 x 3.8 x 0.06 cm) was turned upside down for 10 min in order to separate the non-adhering cells from the adhering cells by sedimentation. Water was then circulated (flow rate 0.11 ml/s; 15 min) to remove all non-adhering cells. Compared to Petri dish tests, the parallel-plate chamber provides more refined information by allowing a better control of hydrodynamic conditions during rinsing, as well as in situ observation of adhesion.

Protein adsorption onto substrata

The liquid phase of aged cell suspensions was characterized by protein assay (Bradford method) and UV-visible spectrophotometry (Dufrêne *et al.*, 1996b). The protein concentration of the supernatants of cell suspensions aged for 24 h at 30°C increased with the concentration of the cell suspension and was close to zero when the suspension was aged for only 2 h. The absorption spectra of the supernatants after cell ageing for 24 h at 30°C revealed an absorption band with a maximum at 260 nm, which indicates also the presence of nucleic acids. These results show that proteins and nucleic acids are released into the solution when *A. brasilense* cells are allowed to age in water at 30°C.
Cell adhesion on the bottom plate and top plate of the chamber was investigated after 24 h in water at 30°C,for increasing concentrations of the cell suspension, and the plates were analysed by XPS after detachment of adhering cells (Dufrêne *et al.*, 1996b).

Figure 3 presents the variation of the adhesion density on polystyrene plates and of the N/C atomic concentration ratio determined on the plates after cell detachment, as a function of the cell concentration. The relationship between the adhesion density observed after 24 h and the N/C ratio determined after detachment of adhering cells was clearly different when comparing the bottom plate and the top plate. For the bottom plate, the increase of N/C as a function of the cell concentration was correlated with an increase of adhesion density; for the top plate, the N/C ratio increased with the suspension concentration as for the bottom plate, but the adhesion density remained very low. This means that the direct contact between the cells and the substratum is not required for accumulation of proteins at the substratum surface, and thus suggests that proteins adsorb from the solution. The N/C ratio of 0.04 is consistent with a continuous film (0.85 nm thick) of adsorbed proteins, a thick proteinaceous film covering 30% of the surface or any combination of these extremes.

Taken together, the above results show that, during the adhesion test, proteins are released progressively by the cells into the solution and adsorb onto the substratum.

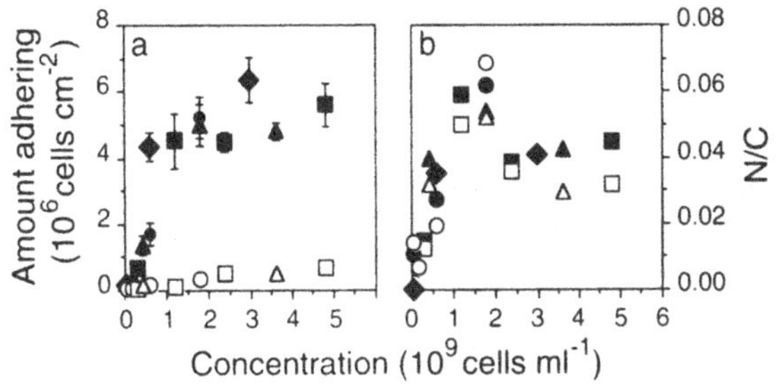

Figure 3. Variation of the density of exponential phase cells adhering on polystyrene plates after 24 h in water at 30°C (a) and of the N/C atomic concentration ratio determined by XPS on the plates after detachment of the adhering cells (b), as a function of the cell concentration. Data obtained on the bottom plate (closed symbols) and on the top plate (open symbols). The different symbols correspond to four independent sets of experiments (Dufrêne *et al.*, 1996b).

Role of proteins in the adhesion process

To assess the role of extracellular proteins in adhesion, the influence of cell aging and substratum preconditioning was examined (Dufrêne *et al.*, 1996b). Aged cells were obtained following a procedure similar to that used for preparing the cell sediments: cells harvested in the exponential phase were resuspended in demineralized water and the suspension was allowed to age for 24 h at 30°C, without agitation. The cells were then separated from their supernatant by centrifugation and resuspended in water. Polystyrene plates were preconditioned by contact with the liquid phase of an aged suspension.

Figure 4 presents the adhesion densities obtained with fresh exponential phase cells and with aged cells, after 2 h and 24 h contact time with bare polystyrene and with preconditioned polystyrene. Mean values from four independent experiments are presented. After short contact time (2 h), cell aging and substratum preconditioning, two treatments which were shown to increase the protein concentration at the cell-substratum interface, significantly favored adhesion. This may be related to the correlation observed above between the increase of cell adhesiveness and the increase of cell surface protein concentration during growth.

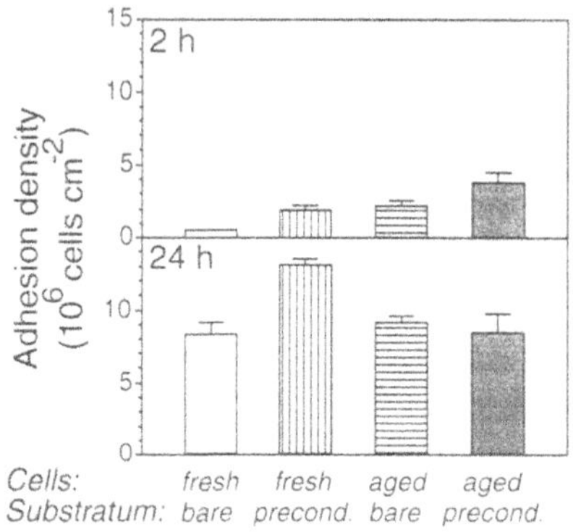

Figure 4. Density of *A. brasilense* cells, either freshly harvested in the exponential phase or aged 24 h in water, adhering after 2 h (top) and 24 h (bottom) contact time, on bare or preconditioned polystyrene substrata. Mean values from four independent experiments are presented; bars: standard deviations (Dufrêne *et al.*, 1996b).

The role played by adsorbed proteins and cell surface proteins was further demonstrated by the influence of the surfactant Pluronic F68: performing the adhesion test in the presence of Pluronic F68 resulted in a concomitant inhibition of protein adsorption and cell adhesion. A similar effect was reported in the field of mammalian cell adhesion: reconditioning polystyrene by a solution of collagen supplemented with Pluronic F68 impaired the adhesion of human epithelial cells due to prevention of collagen adsorption or to its desorption by rinsing (Dewez *et al.*, 1996).

Thus, both proteins that adsorb onto the substratum and proteins that accumulate at the cell surface play a role in adhesion. However, Figure 4 shows that when the cell-substratum contact time was prolonged (24 h), substratum preconditioning and cell aging had no significant effect any more and adhesion was always denser compared to 2 h. The observation that the prolonged contact between the cells and the substratum promotes adhesion may be attributed either to physicochemical processes characterized by slow kinetics or to the period required to induce the active production of adhesive polymers by the cells. The second hypothesis is supported by the dependence of adhesion on active cell metabolism and by scanning electron microscope observations showing that adhering cells produce extracellular material when left in contact with a substratum for 24 h. From these data, it is concluded that the prolonged contact between cells and substratum affects adhesion through the in situ secretion of anchoring proteins.

The role of proteins in the adhesion of *A. brasilense* to inert surfaces is thus twofold, as illustrated in Figure 5. In the first stage, proteins accumulate at the cell surface, are liberated into the solution and adsorb onto the substratum; the increase of the protein concentration at the cell-substratum interface promotes initial adhesion. In the next stage, the prolonged contact between adhering cells and the substratum leads to the in situ secretion of proteins which ensures cell anchorage.

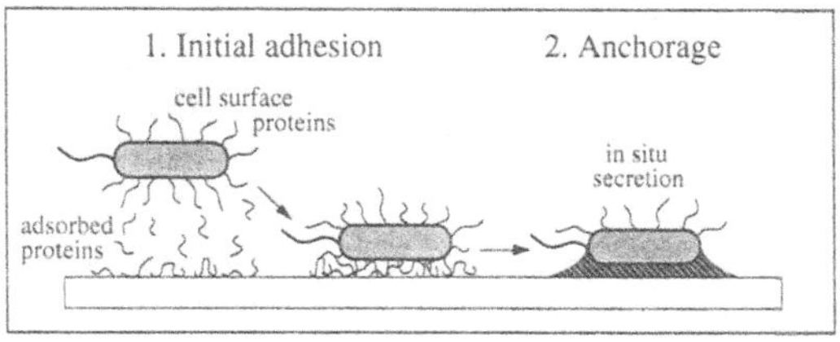

Figure 5. Two-step adhesion process of *A. brasilense* to inert surfaces: 1) initial adhesion, through protein accumulation at the cell-substratum interface; 2) anchorage, through the in situ secretion of proteins.

Influence of ionic strength

The influence of ionic strength on cell adhesion has been examined by comparing water and phosphate buffer saline (PBS) as suspending media (Dufrêne *et al.*, 1996c). Figure 6 shows that polystyrene substrata analysed by XPS after adhesion in PBS and detachment of the adhering cells clearly presented a higher protein surface concentration compared to substrata analysed after adhesion in water. To elucidate this effect, substrata were analysed after contact with three cell supernatants obtained respectively by separation of cells resuspended in water, separation of cells resuspended in water and addition of the constituents of PBS, separation of cells resuspended in PBS. The results showed that PBS favors both the release of proteins by the cells into the solution and the tendency of proteins to adsorb onto the substratum.

After 2 h contact time, a denser adhesion was observed in PBS, as compared with water (Fig. 6c and d). This may be explained either in the light of the DLVO theory, namely by a decrease of electrostatic repulsion between the cells and the substratum due to the increase of ionic strength, or by a more efficient bridging due to the increase of the concentration of adsorbed proteins.

The adhesion density in PBS after 24 h was the same as after 2 h and was lower than in water after 24 h. This is in contradiction with double layer interactions and does not fit with the effect of adsorbed proteins discussed above. It appears that in the presence of PBS, the bacterial cells are embedded in a thick layer of proteinaceous material, which leads to a heterogeneous distribution of adhering cells and to a relatively low adhesion density due to detachment of pellicles upon rinsing.

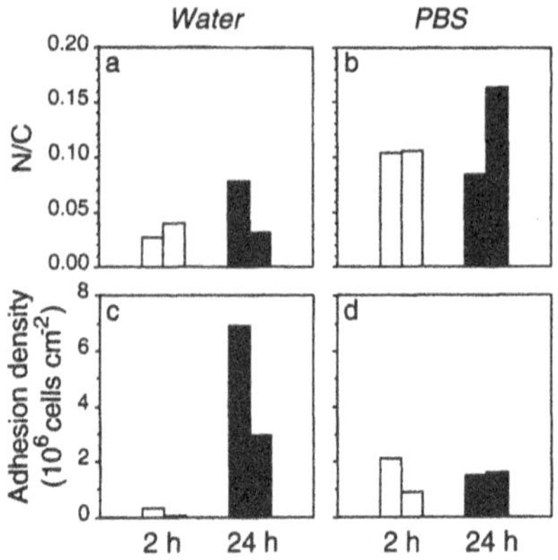

Figure 6. Variation of the N/C atomic concentration ratio determined by XPS on polystyrene substrata after cell detachment (a, b) and variations of the density of adhering cells (c, d) as a function of contact time (2 h (□); 24 h (■)) and the liquid phase (water (a, c); PBS (b, d)). Two sets of independent experiments are presented (Dufrêne *et al.*, 1996c).

CONCLUSION: A COMPREHENSIVE VIEW OF THE ADHESION MECHANISM

This work represents the first attempt to apply a modern surface chemical analysis technique (XPS) to the characterization of bacterial macromolecules at interfaces, in relation to cell adhesion. It is demonstrated that extracellular proteins play a key role in the adhesion of *A. brasilense* to model substrata. In a first stage, proteins accumulate at the cell surface, are liberated into the solution and adsorb at the substratum surface; the increase of the protein concentration at the cell-substratum interface promotes initial adhesion. Then, in situ secretion of proteins during the prolonged contact between the cells and the substratum strengthens adhesion and leads to cell anchorage.

The adhesion mechanism of *A. brasilense* can be discussed further in terms of molecular interactions. At physiological pH and low ionic strength, both cells and substrata possess a negative zeta potential. In such conditions, the DLVO theory predicts that the energy barrier encountered by the cells as they approach the substratum is hundreds of kT (Rijnaarts *et al.*, 1995). This is too high to be overcome by whole cells and thus adhesion in the primary minimum is prevented. Accordingly, the DLVO theory alone does not account for the considerable adhesion densities found in this study.

The DLVO theory considers the surfaces of both cells and substratum to be molecularly smooth. This study shows that this is clearly a rough approximation: proteinaceous structures located at the cell-substratum interface play a central role in adhesion. The effect of both cell surface appendages and macromolecules must be considered. Surface appendages, by virtue of their small radius, may cross the potential barrier between the cells and the substratum. Indeed, fibrils (Bashan *et al.*, 1986; Bashan *et al.*, 1991) and flagella (Croes *et al.*, 1993) have been reported to play a role in the adhesion of *Azospirillum*. The presence of macromolecules at the cell-substratum interface may affect adhesion in different manners, depending on their conformation and on their affinity for the solvent. Compact layers of macromolecules may influence adhesion by creating new surfaces and modifying van der Waals and electrostatic interactions, or hydrophobic interactions. Highly hydrated polymers with loops and tails protruding into the solution can bridge the cells to the substratum through attractive steric interactions. Ionic strength has often been reported to influence microbial adhesion by affecting electrostatic repulsion between the cells and the substratum (van Loosdrecht *et al.*, 1989; Rijnaarts *et al.*, 1995). For the system investigated here, the influence of ionic strength is far beyond affecting electrostatic interactions between surfaces. An increase of ionic strength causes an increase of the concentration of proteins at the substratum surface by both favoring protein release by the cells and enhancing the tendency of proteins to adsorb. The formation of a thick layer of proteinaceous material in which the bacterial cells are embedded leads to a heterogeneous distribution of adhering cells and to a relatively low adhesion density due to detachment of pellicles upon rinsing.

The correlation between cell adhesiveness and surface hydrophobicity observed during growth points to the involvement of hydrophobic interactions in the adhesion process. *A. brasilense* was shown to possess one of the highest cell surface hydrophobicity amongst several rhizobacteria, as evaluated by adhesion to hydrocarbon and to polystyrene (Achouak *et al.*, 1994). Although the influence of surface hydrophobicity on cell adhesion has often been interpreted according to a balance of interfacial free energies (Busscher *et al.*, 1984), caution is urged in using such an approach in view of its severe limitations: consideration of smooth surfaces in molecular contact, assumption of reversibility, approximations involved in computing surface free energy and lack of consideration of the role of double layer interactions as two surfaces approach each other. As pointed out in the present work, the effect of surface hydrophobicity may be related to the presence of surface proteinaceous structures: cell surface proteins may indeed possess hydrophobic moieties, thus allowing hydrophobic interactions to take place with the substratum through the repulsive barrier. This is supported by literature data. Bacterial fimbriae are proteinaceous appendages possessing hydrophobic properties; they have been reported to play a central role in the adhesion to solid surfaces (Irvin, 1990).

Finally, proteins located at the cell-substratum interface may also give rise to specific interactions involving molecular recognition. Adsorption of fibronectin onto polymethylmethacrylate was shown to promote the adhesion of *Staphylococcus aureus*, apparently due to specific interactions between the adsorbed fibronectin and cell surface receptors (Vaudaux *et al.*, 1984). Literature data suggest the involvement of specific interactions in the attachment of *A. brasilense* to plant roots and in cell aggregation (Madi *et al.*, 1988; Yagoda-Shagam et al., 1988; Del Gallo *et al.*, 1989). Further research is required to assess whether, in the system studied here, extracellular proteins adsorbed onto the substratum are specifically recognized by cell surface receptors.

Acknowledgements

The authors thank V.B. Wiertz for her help, J. Vanderleyden, N. Mozes and M.J. Genet for fruitful discussions. P.G.R. is a member of the Research Center for Advanced Materials. The support of the National Fund for Scientific Research (to Y.F.D.), of the Department of Scientific Policy (PAI-Supramolecular Chemistry and Catalysis) and of the Department of Education and Scientific Research (Concerted Action Physical Chemistry of Interfaces and Biotechnology) is gratefully acknowledged.

References

Achouak, W., F. Thomas and T. Heulin, 1994. Physico-chemical surface properties of rhizobacteria and their adhesion to rice roots. Colloids Surfaces B: Biointerfaces, 3, 131-137.

Bashan, Y. and H. Levanony, 1988. Active attachment of *Azospirillum brasilense* Cd to quartz sand and to a light-textured soil by protein bridging. J. Gen. Microbiol., 134, 2269-2279.

Bashan, Y., H. Levanony and E. Klein, 1986. Evidence for a weak active external adsorption of *Azospirillum brasilense* Cd to wheat roots. J. Gen. Microbiol., 132, 3069-3073.

Bashan, Y., G. Mitiku, R.E. Whitmoyer and H. Levanony, 1991. Evidence that fibrillar anchoring is essential for *Azospirillum brasilense* Cd attachment to sand. Plant and Soil, 132, 73-83.

Busscher, H.J., A.H. Weerkamp, H.C. van der Mei, A.W.J. van Pelt, H.P. de Jong and J. Arends, 1984. Measurement of the surface free energy of bacterial cell surfaces and its relevance for adhesion. Appl. Environ. Microbiol., 48, 980-983.

Characklis, W.G. and K.C. Marshall (Editors), 1990. Biofilms. Wiley-Interscience, New York.

Croes, C.L., S. Moens, E. van Bastelaere, J. Vanderleyden and K.W. Michiels, 1993. The polar flagellum mediates *Azospirillum brasilense* adsorption to wheat roots. J. Gen. Microbiol., 139, 2261-2269.

Del Gallo, M., M. Negi and C.A. Neyra, 1989. Calcofluor- and lectin-binding exocellular polysaccharides of *Azospirillum brasilense* and *Azospirillum lipoferum*. J. Bacteriol., 171, 3504-3510.

Dewez, J.-L., Y.-J. Schneider and P.G. Rouxhet, 1996. Coupled influence of substratum hydrophilicity and surfactant on epithelial cell adhesion. J. Biomed. Mater. Res., 30, 373-383.

Dufrêne, Y.F. and P.G. Rouxhet, 1996. Surface composition, surface properties and adhesiveness of *Azospirillum brasilense* - variation during growth. Can. J. Microbiol., 42, 548-556.

Dufrêne, Y.F., H. Vermeiren, J. Vanderleyden and P.G. Rouxhet, 1996a. Direct evidence for the involvement of extracellular proteins in the adhesion of *Azospirillum brasilense*. Microbiology, 142, 855-865.

Dufrêne, Y.F., C.J.-P. Boonaert and P.G. Rouxhet, 1996b. Adhesion of *Azospirillum brasilense*: role of proteins at the cell-support interface. Colloids Surfaces B: Biointerfaces, 7, 113-128.

Dufrêne, Y.F., V.B. Wiertz and P.G. Rouxhet, 1996c. Adhesion of *Azospirillum brasilense* to polystyrene: influence of ionic strength through protein adsorption. Biofouling, 9, 307-315.

Eyers, M., J. Vanderleyden and A. Van Gool, 1988. Attachment of *Azospirillum* to isolated plant cells. FEMS Microbiol. Lett., 49, 435-439.
Filip, Z. and T. Hattori, 1984. Utilization of substrates and transformation of substrata. In: K.C. Marshall (Editor). Microbial Adhesion and Aggregation. Springer Verlag, Heidelberg, Berlin, pp. 251-288.
Gafny, R., Y. Okon and Y. Kapulnik, 1986. Adsorption of *Azospirillum brasilense* to corn roots. Soil Biol. Biochem., 18, 69-75.
Irvin, R.T., 1990. Hydrophobicity of proteins and bacterial fimbriae. In: R.J. Doyle and M. Rosenberg (Editors). Microbial cell surface hydrophobicity. American Society for Microbiology, Washington, pp. 137-177.
Madi, L., M. Kessel, E. Sadovnik and Y. Henis, 1988. Electron microscopic studies of aggregation and pellicle formation in *Azospirillum* spp. Plant and Soil, 109, 115-121.
Marshall, K.C., 1991. The importance of studying microbial cell surfaces. In: N. Mozes, P.S. Handley, H.J. Busscher and P.G. Rouxhet (Editors). Microbial Cell Surface Analysis: Structural and Physicochemical Methods. VCH Publishers, New York, pp. 3-19.
Michiels, K.W., C.L. Croes and J. Vanderleyden, 1991. Two different modes of attachment of *Azospirillum brasilense* Sp7 to wheat roots. J. Gen. Microbiol., 137, 2241-2246.
Michiels, K., J. Vanderleyden and C. Elmerich, 1994. Genetics and molecular biology of *Azospirillum*. In: Y. Okon (Editor). *Azospirillum* /Plant Associations. CRC Press Inc., Boca Raton, Fl., pp. 41-56.
Mozes, N. and P.G. Rouxhet, 1991. Effects of the microenvironment on metabolic activity of immobilized yeast cells. Med. Fac. Landbouww. Rijksuniv. Gent, 56, 1761-1768.
Mozes, N. and P.G. Rouxhet, 1992. Influence of surfaces on microbial activity. In: L.F. Melo, T.R. Bott, M. Fletcher and B. Capdeville (Editors). Biofilms - Science and technology. Kluwer Academic Publishers, Dordrecht, pp. 125-136.
Okon, Y. (Editor), 1994. *Azospirillum* /Plant Associations. CRC Press Inc., Boca Raton, Fl.
Pueppke, S.G. and D.A. Kluepfel, 1985. Responses of plant cells to adsorbed bacteria. In: D.C. Savage and M. Fletcher (Editors). Bacterial Adhesion. Plenum Press, London, pp. 401-435.
Ratner, B.D. and B.J. McElroy, 1986. Electron spectroscopy for chemical analysis: applications in the biomedical sciences. In: R.M. Gendreau (Editor). Spectroscopy in the Biomedical Sciences. CRC Press, Boca Raton, Fl., pp. 107-140.
Rijnaarts, H.H.M., W. Norde, E.J. Bouwer, J. Lyklema and A.J.B. Zehnder, 1995. Reversibility and mechanism of bacterial adhesion. Colloids Surfaces B: Biointerfaces, 4, 5-22.
Rouxhet, P.G. and N. Mozes, 1990. The micro-environment of immobilized cells : critical assessment of the influence of surfaces and local concentrations. In: J.A.M. de Bont, J. Visser, B. Mattiasson and J. Tramper (Editors). Proc. Int. Symp. on Physiology and Immobilized Cells. Elsevier Science Publishers, Amsterdam, pp. 343-354.
Rouxhet, P.G. and M.J. Genet, 1991. Chemical composition of the microbial cell surface by X-ray photoelectron spectroscopy. In: N. Mozes, P.S. Handley, H.J. Busscher and P.G. Rouxhet (Editors). Microbial Cell Surface Analysis: Structural and Physicochemical Methods. VCH Publishers, New York, pp. 173-220.

Rutter, P.R. and B. Vincent, 1980. The adhesion of micro-organisms to surfaces: physico-chemical aspects. In: R.C.W. Berkeley, J.M. Lynch, J. Melling, P.R. Rutter and B. Vincent (Editors). Microbial Adhesion to Surfaces. Ellis Horwood Ltd, Chichester, pp. 79-92.

Savage, D.C. and M. Fletcher (Editors), 1985. Bacterial adhesion. Plenum Press, London.

Stotzky, G., 1985. Mechanisms of adhesion to clays, with reference to soil systems. In: D.C. Savage and M. Fletcher (Editors). Bacterial adhesion. Plenum Press, London, pp. 195-253.

van Loosdrecht, M.C.M., J. Lyklema, W. Norde, G. Schraa and A.J.B. Zehnder, 1987. Electrophoretic mobility and hydrophobicity as a measure to predict initial steps of bacterial adhesion. Appl. Environ. Microbiol., 53, 1898-1901.

van Loosdrecht, M.C.M., W. Norde and A.J.B. Zehnder, 1989. Bacterial adhesion: a physicochemical approach. Microb. Ecol., 17, 1-15.

van Loosdrecht, M.C.M., J. Lyklema, W. Norde, J. Lyklema and A.J.B. Zehnder, 1990. Hydrophobic and electrostatic parameters in bacterial adhesion. Aquat. Sci., 52, 103-114.

Vaudaux, P.E., F.A. Waldvogel, J.J. Morgenthaler and U.E. Nydegger, 1984. Adsorption of fibronectin onto polymethylmethacrylate and promotion of *Staphylococcus aureus* adherence. Infect. Immun., 45, 768-774.

Yagoda-Shagam, J., L.L. Barton, W.P. Reed and R. Chiovetti, 1988. Fluorescein isothiocyanate-labeled lectin analysis of the surface of the nitrogen-fixing bacterium *Azospirillum brasilense* by flow cytometry. Appl. Environ. Microbiol., 54, 1831-1837.

XYLANASE, INVERTASE AND UREASE ACTIVITY IN PARTICLE - SIZE FRACTIONS OF SOILS

Ellen Kandeler*, Michael Stemmer** , Sabine Palli *, Martin H.Gerzabek **

* Federal Agency and Research Centre for Agriculture,
Spargelfeldstr. 191, A-1220 Vienna, Austria
** Austrian Research Centre Seibersdorf, A-2444 Seibersdorf, Austria
Corresponding author: Ellen Kandeler, present address: University of Hohenheim,
Institute of Soil Science, D-70599 Stuttgart, Germany (kandeler@uni-hohenheim.de)

INTRODUCTION

Sources of intracellular and extracellular enzymes in soils are primarily the microbial biomass and to a lesser extent plant and animal residues. Free enzymes in soil are adsorbed on organic and mineral constituents or complexed with humic substance, or both (Kiss *et al.* 1975, Skujins 1978, Burns 1982). The amount of free enzymes in the soil solution is much lower than that in adsorbed state (Kandeler 1990). Microbial cells or cell fragments may also exist be adsorbed or in suspension (Tabatabai and Fu 1992).

Several attempts have been made to describe the location of soil enzymes, the microenvironment in which they function, and how they are bound or stabilized in that environment. The interaction between enzymes and humic substances were examined by chemical extraction of natural complexes from soil or by preparation of model humic-enzyme copolymers (Tabatabai and Fu 1992, Ruggiero *et al.* 1996). The recovery of enzyme activity was generally less than 20 % (Ruggiero *et al.* 1996). In addition, humic-enzyme complexes may be modified during the extraction procedure. The interaction between enzymes and clay minerals has been investigated by modeling natural clay-enzyme complexes or by immobilizing enzymes on various matrices (for reviews see Burns 1986, Boyd and Mortland 1990, Stotzky and Burns 1982, Robert and Chenu 1992, Ruggiero *et al.* 1996).

Because the classic granulometric methods use chemical reagents to remove organic matter and amorphous compounds and disperse clay minerals, the information on the distribution of soil microbial biomass and enzymes in soil particles after physical fractionation is scarce (Ladd *et al.* 1996). If chemicals are avoided, the particle size separation of mineral and organic constituents can only be achieved by applying mechanical or ultrasonical energy. None of the fractionation methods based on mechanical disruption completely disperses the constituents. The physical fractionation procedures widely differ in moisture content of samples (air-dried versus field-moist samples), type of dispersion, separation and size of particles, and recovery of enzyme activity (Tab.1 and Tab.2). This makes it difficult to compare the results of the various investigations. In general, cell numbers and microbial biomass are higher in the smaller fractions (fine silt and clay size), whereas the enzyme activity of fractions largely depends on the enzyme investigated and the fractionation

Effect of Mineral-Organic-Microorganism Interactions on Soil and Freshwater Environments
Edited by Berthelin, *et al*, Kluwer Academic / Plenum Publishers, New York, 1999

procedure (Lensi *et al.* 1995, Ladd *et al.* 1996). By separating organic and mineral particles with a wet sieving and sedimentation procedure, Kanazawa and Filip (1986) found that ß-glucosidase, ß-acetylglucosaminidase and proteinase decreased with the size of the organic soil particles, but increased with decreasing size of the mineral soil particles. Mateos and Carcedo (1985) described the predominance of catalase, dehydrogenase and urease in soil structural microunits with a diameter less than 50 μm. In contrast, phosphatase activity was concentrated in the larger soil fractions (2000-100 μm) (Rojo *et al.* 1990). The authors attributed this result to the association of this enzyme with plant debris and less humified organic matter.

The aim of our study was to test whether microbial biomass (FE), xylanase, invertase and urease are mainly associated with the clay fraction, which provides most of the surface available for interaction with microorganisms, or with larger plant debris not associated with the mineral fractions. In order to separate fractions containing most of the particulate organic matter and to preserve microaggregates, soil samples were fractionated after ultrasonical destruction of the unstable macroaggregates without dispersion by chemicals or removal of organic matter and calcium carbonate cement (Stemmer *et al.* 1998). In a second experiment we hypothesized that tillage may change the distribution of soil organic matter and enzyme activities not only in the soil profile, but also in the particle-size fraction. It is well known that tillage increases disaggregation processes by disrupting macroaggregates and exposing aggregates to rain drops (Marshall and Holmes 1988). In addition, organic matter is more evenly distributed in conventionally cultivated soils than in reduced or minimum tillage plots, where crop residues are concentrated at the soil surface (Arshad *et al.* 1990). Therefore, microsites of soil microbial biomass and soil enzymes involved in carbon mineralization (xylanase, invertase) were investigated on particle-size fractions from a Chernozem under conventional, reduced and minimum tillage at depths of 0-10 cm and 20-30 cm.

Table1: Methods for the determination of soil enzyme activities in particle-size fractions

soil enzyme	fractionation procedure		size of fractions (μm)	ref.
	dispersion	separation of particles		
protease, phosphatase dehydrogenase, oxidase	air-dried soil, mechanical dispersion	centrifugation at 60 *g*, 1000 *g* and 12,000 *g*, ultrafiltration	>2.0, 2.0-0.2, 0.2-0.04	(1)
urease	air dried soil, ultrasonical dispersion	not specified	2000-50, 50-2, <2	(2)
catalase, dehydrogenase urease, protease	air-dried soil, no or mechanical dispersion (16 h with 5 agate marbles)	wet sieving	2000-200, 200-100,100-50,<50	(3)
ß-glucosidase, proteinase ß-acetylglucosaminidase	field-moist soil, mechanical dispersion (gently brushing of soil crumbs)	wet sieving, sedimentation	mineral particles: 500-200, 200-100, 100-50, <50 organic particles: >5000, 5000-2000, 2000-1000, 1000-500, 500-200, 200-100, 100-50	(4, 5)
catalase, dehydrogenase urease, protease	air-dried soil, no or mechanical dispersion (16 h with 5 agate marbles)	wet sieving	2000-100, 100-50, <50	(6)
acid phosphatase, sulfatase	field-moist soils, no dispersion	sieving	>1000, 1000-500, 500-250, 250-100	(7)
alkaline, acid phosphatase	air-dried soil, no or mechanical dispersion (16 h with 5 agate marbles)	wet sieving	2000-200, 200-100, 100-50, <50	(8)
denitrification enzyme activity (DEA)	field-moist soil, mechanical dispersion (16 h with 5 agate marbles)	wet sieving, sedimentation at 1 *g*, centrifugation at 90 *g*	2000-250, 250-50, 50-20, 20-2, <2	(9)
invertase, xylanase	field-moist soil, ultrasonical dispersion (50 Js^{-1}, 2 min)	wet sieving, centrifugation at 150 *g* and 3900 *g*	2000-200, 200-63, 63-2, 2-0.1	(10)

ref.1: Ladd and Paul (1973), ref.2: Tabatabai (1973), ref.3: Mateos and Carcedo (1985), ref.4: Kanazawa (1979), ref.5: Kanazawa and Filip (1986), ref.6: Mateos and Carcedo (1987), ref.7: Gupta and Germida (1988), ref.8: Rojo et al. (1990), ref.9: Lensi et al. (1995), ref.10: Stemmer et al. (1997)

Table 2: Methods for the determination of soil microbial biomass in particle-size fractions

soil microbial biomass	fractionation procedure		size of fractions (μm)	ref.
	dispersion	separation of particles		
ATP	field-moist soil, ultrasonical dispersion	wet sieving, sedimentation, centrifugation at 320 *g*	>53, 53-20, 20-15, 15-10, 10-5, 5-2, 2-1, 1.0-0.5, 0.5-0.1, <0.1	(1)
microbial counts ATP	field-moist soil, mechanical dispersion (gently brushing of soil crumbs)	wet sieving, sedimentation	mineral particles: 500-200, 200-100, 100-50, <50 organic particles: >5000, 5000-2000, 2000-1000, 1000-500, 500-200, 200-100, 100-50	(2)
diaminopimelic acid	air-dried soil, ultrasonic dispersion	gravity sedimentation	2000-20, 20-2, <2	(3)
microbial biomass C	field-moist soil, mechanical dispersion (16 h 5 agate marbles)	wet sieving, sedimentation, centrifugation at 90 *g*, flocculation with $CaCl_2$	>250, 250-50, 50-20, 20-2, 2-0.1	(4)
microbial biomass C, N, P	field-moist soils, no dispersion	wet sieving through a series of sieves	>4750, 4750-2000, 2000-500, 500-300, 300-53, <53 macroaggregates: >300, microaggregates: <300	(5)
^{14}C microbial biomass ^{14}C microbial products	air-dried soil, no dispersion	dry sieving	>2000, 2000-1000, 1000-500, 500-250, <250	(6)
microbial biomass C	moist and dried-remoistened soils, mechanical dispersion (16 h 5 agate marbles)	wet sieving, sedimentation, centrifugation at 90 *g*, flocculation with $CaCl_2$	>200, 200-50, 50-20, 20-2, <2	(7)

ref.1: Ahmed and Oades (1984), ref.2: Kanazawa and Filip (1986), ref.3: Christensen and Bech-Anderson (1989), ref.4: Monrozier et al. (1991), ref.5: Singh and Singh (1995), ref.6: Degens and Sparling (1996), ref.7: van Gestel et al. (1996)

MATERIAL AND METHODS

Sites and treatments

For the preliminary experiment, we used a Haplic Chernozem from Jedenspeigen/Lower Austria, which was sampled from the depth of 0-20cm in March 1990 and stored at -20°C prior to analysis. In the second experiment, we studied the effect of agricultural tillage practices on soil microbial processes in an experimental field with a randomized block design, which was laid out in Fuchsenbigl (Lower Austria, Austria) in 1988. The soil is classified as a fine-sandy loamy Haplic Chernozem. Chemical and physical properties of the soils used are shown in Tab.3. The mean long-term annual temperature is 9.3°C, the mean long-term annual precipitation 519 mm. Sugar beet (Axel) was cultivated on all plots in 1994, barley (Viva) was sown on 30.3.1995. The preparation of the soil was the same for all plots. Microbial biomass N as well as xylanase and invertase activities were monitored in three of the treatments, namely:

Table 3: Chemical and physical properties of the Chernozems used (0-10cm)

soil type/site	tillage	C_{org} (mg g^{-1})	N_t (mg g^{-1})	C/N	pH ($CaCl_2$)	fraction size (%) 2000-200 μm	200-63 μm	63-2 μm	2-0.1 μm	<0.1 μm	recovery (%)
Calcic Chernozem Jedenspeigen	conv.	15.0	1.61	9.32	7.50	14.6	18.7	57.1	9.2	ND*	99.6
Haplic Chernozem Fuchsenbigl	conv.	15.6	1.60	9.75	7.65	8.3	27.4	42.2	19.2	2.8	99.5
Haplic Chernozem Fuchsenbigl	red.	18.2	1.60	11.38	7.65	8.9	29.7	39.3	18.8	3.2	99.9
Haplic Chernozem Fuchsenbigl	min.	18.7	1.60	11.69	7.65	8.8	26.9	42.2	19.5	2.7	98.1

(Particle size distribution was determined according to the combined ultrasonication and particle size separation by centrifugation)
*not determined, con. (conventional), red. (reduced), min. (minimum)

I. **Conventional tillage:** soils were plowed in autumn to a depth of 25-30 cm with a reversible plow. The plow mixed the soil and turned it over to incorporate crop residues. In spring the seedbed was prepared with a zig-zag drag harrow and a crumbling roller. At the same date the seed was sowed.

II. **Reduced tillage**: soils were treated with a cultivator in autumn to a depth of 20 cm. The soil was mixed but not turned over. In spring the seedbed was prepared with a zig-zag drag harrow and crumbling roller. The seed was sowed at the same date.

III. **Minimum tillage**: plots were treated with a rotary-driller on 3.4.1995. No primary treatment occurred on plots in autumn. The rotavator loosened the soil till to a depth of 5-8 cm and whirled it up into the air. The soil fell upside down, covering the seed and incorporating crop residues.

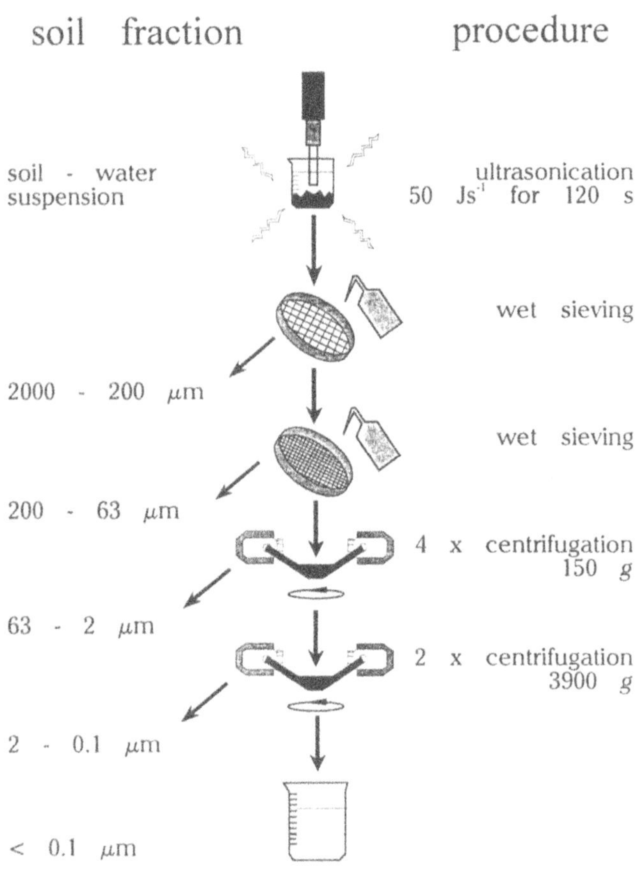

Figure 1: Flow chart of the dispersion and separation procedure to obtain different particle-size fractions (2000-200 µm, 200-63 µm, 63-2 µm, 2-0.1 µm and <0.1 µm) according to Stemmer *et al.* (1998).

Several days before minimum tillage, soil samples were collected at depths of 0-10 cm and 20-30 cm. Each type of tillage treatment had three replicate plots. On each plot, sixteen sub-samples were taken with a single gouge auger (cores 30 mm in dia). The field-moist samples were mixed and stored in plastic bags at -20°C. For the analysis, the samples were allowed to thaw at 4°C. The samples were sieved (<2 mm) and physically fractionated within two days.

Physical fractionation of the soils
The procedure involved the dispersion of soil samples by a low-energy sonication and separation of particle-size fractions by a combination of wet sieving and centrifugation as described by Stemmer *et al.* (1998) (Fig.1). Briefly, 35 g of field-moist soil were dispersed in 100 ml of cooled distilled water by a probe-type ultrasonic disaggregator (50 Js^{-1} for 120 s). Coarse and medium sand (2000-200 µm) and fine sand (200-63 µm) were separated by manual wet sieving with about 400 ml of cooled distilled water. Silt-sized particles (63-2 µm) were separated from the clay fraction (2-0 µm) by centrifugation at approximately 150 *g* for 2.0 min maximal speed at 15°C. The pellets were resuspended in water and centrifuged under the above-mentioned conditions three-times to purify the silt fraction. The combined supernatants were centrifuged at 3900 *g* for 30 min at 15°C to yield clay-sized particles (2-0.1 µm, according to an equispherical diameter and a particle density of 2.65 g cm^{-3}).
Soil microbial analysis
Biomass N: Ninhydrin-reactive N was measured according to a modified method of Amato and Ladd (1988). Briefly, 0.3 - 0.5 g of the moist fractions were fumigated with 0.1 ml of chloroform for 24 h at 25°C. Subsequently, the chloroform of the samples was removed. Samples and unfumigated controls were extracted with 5.0 ml of 2*M* KCl solution for 60 min on a rotatory shaker. After filtration, 2 ml of the filtrates were mixed with 0.5 ml of 0.4 *M* sodium citrate solution. Ninhydrin-reactive N was determined by a colorimetric procedure (Schinner *et al.* 1996).
Xylanase and invertase activity: 0.5-1.0 g of the moist fraction were incubated with 5.0 ml of a substrate solution (1.7%w/v xylan from oat spelts suspended in 2*M* acetate buffer, pH 5.5) and 5.0 ml 2*M* acetate buffer (pH 5.5) for 24 h at 50°C. Before incubation, only the clay fractions were mixed for at least 1 min with 0.7 g of quartz to improve the dispersion of the suspension. Reducing sugars released during the incubation period reduced potassium hexacyanoferrate (III) in an alkaline solution. Potassium hexacyanoferrate (II) was measured colorimetrically according to the Prussian blue reaction (Schinner *et al.* 1996). To measure invertase activity, 0.5-1.0 g of the moist fractions were incubated with 5.0 ml of 50 *mM* sucrose solution and 5.0 ml of 2 *M* acetate buffer (pH 5.5) for 3 h at 50°C. Reducing sugars were determined as described for xylanase activitiy (Schinner *et al.* 1996).
Urease activity: 0.5-1.0 g of the moist fractions were incubated with 1.5 ml of a 79.9 *mM* urea solution for 2 h at 37°C. Released ammonium was extracted with 13.5 ml of 2*M* potassium chloride solution, and determined colorimetrically by a modified Berthelot reaction (Kandeler and Gerber 1988).
Organic carbon and total nitrogen: Organic carbon and total nitrogen were measured by dry combustion in a Carlo Erba NA 1500 CN elemental analyser. Freeze-dried aliquots of the bulk soil and the fractions were pulverized in an agate mill. Carbonates in the fractions were removed within the silver capsules by the dropwise addition of 6 *M* hydrochloric acid until no further effervescence was observed and were oven-dried at 80°C for 2 h. This process (addition of hydrochloric acid and oven-drying at 80°C) was repeated twice to complete the reaction.

RESULTS AND DISCUSSION

Organic carbon, total nitrogen and enzyme activities in particle-size fractions
Particle size distributions of the four soils obtained by the centrifugation method after ultrasonication pretreatment are given in Table 3. Compared with the particle size method using sodium-pyrophosphate as the dispersion agent, we obtained similar amounts of the 2000-200 µm and 200-63 µm fraction, but a higher amount of 63-2 µm particles (Stemmer *et al.* 1998). Therefore, ultrasonication completely disrupted the macroaggregates (>250 µm) but

did not entirely separate the smaller fractions. About 54 % (Calcic Chernozem), but no significant amount (Haplic Chernozem) of the clay particles were found as preserved microaggregates in the silt-sized fraction (63-2 μm).
Recoveries of organic carbon and total nitrogen after fractionation exceeded 98% (Tab.4). Organic carbon and total nitrogen contents increased with diminishing particle size of the Calcic Chernozem, whereby C/N ratios were closer in the smaller than in the coarser particles (Tab.4). As demonstrated by microscopical and chemical investigation (Schulten *et al.* 1993), the high C/N ratio of the 2000-200 μm fraction can be attributed to the particulate organic matter, which consists mainly of plant residues. The decline of the C/N ratio towards the smaller fractions reflects the disappearance of plant debris and its replacement by microbial material of low C/N ratio (Ahmed and Oades 1984).
The distribution of enzyme activities in particle size fractions of the Calcic Chernozem varied with the enzyme assayed (Tab.4). Xylanase activity was mainly located in the coarse and medium sand fraction, invertase activity in the silt fraction, and urease activity in the silt and clay fraction (Tab.4). The recoveries of invertase and urease were in the range of 97-102 %, whereas the yield of xylanase activity exceeded the activity of the bulk soil to a great extent. This result clearly showed that the diffusion of the high-molecular weight substrate xylan in the bulk soil is hindered and that the fractionation procedure exposes new surfaces for enzyme-substrate interaction. In contrast, urea and sucrose as substrates for urease and invertase are soluble in the soil solution and are able to diffuse to the enzyme-organo-mineral-complexes. In addition, the high recovery of invertase and urease showed that the method used to disperse and separate particles preserves enzyme-organo-mineral-complexes (Tab.4). In contrast, Burns (1978) reported that invertase activity of dried soils was partly lost (average decrease, 23%) after dispersion of the soil by ultrasonic vibration in an organic solvent. An even higher loss of urease activity was determined by Tabatabai (1973) after a similar procedure involving ultrasonic vibration and by Mateos and Carcedo (1985) using a mechanical dispersion procedure.
The different predominance of the individual enzymes in particle-size fractions can be discussed on the basis of the location of soil microorganisms and their substrates. In aerobic environment, saprophytic fungi are the major decomposers of plant litter. They excrete relatively large amounts of xylanase and other exoenzymes directly into the microenvironment. Because xylanase is an adaptive enzyme, the xylanase content of soils appears to be primarily a function of the amount of substrate: more substrate gives rise to accelerated enzyme excretion (Burns 1978). The high xylanase activity and the high C/N ratio of the coarse fraction made it probable that this enzyme is mainly bound to and protected by the particulate organic matter.
In contrast, invertase activity was mainly located in the silt fraction (Tab.4). The importance of the 60-20 μm particles for invertase activity was demonstrated by Schinner and von Mersi (1990), who reported a close correlation between these parameters. In addition, invertase activity is correlated with the surface area of aggregates and with soil organic matter content (Burns 1978). Because invertase activity is bound to membranes and often to cell fragments of microorganisms (Schinner and von Mersi 1990), and because microbial biomass is most concentrated in the smaller fractions (for references see Tab.2), the invertase in the 63-2 μm particles of the Calcic Chernozem was probably mainly associated with the microbial biomass.
Urease activity was mainly located in the smaller fractions (63-2 μm and 2-0.1 μm). The similarity of the Michaelis-Menten-constants obtained for urease in soils and soil fractions (sand, silt, clay) (Tabatabai 1973) suggests that the urease of these fractions had the same origin or that the association of this enzyme with soil constituents influenced the K_m value. A similar thermal stability and pH optimum of urease activity in soil fractions (2000-100 μm, 100-50 μm, <50 μm) support this hypothesis (Mateos and Carcedo 1987).

Influence of tillage on enzyme activities in particle-size fractions

In the second experiment, we used soil samples (0-10 cm and 20-30 cm) from an experimental plot with different tillage treatment (conventional, minimum and reduced). The xylanase activity of the bulk soil was significantly higher in the top soil layer (0-10 cm) of the minimum and reduced tillage plots than in conventional plots (Fig.2). This was due to a

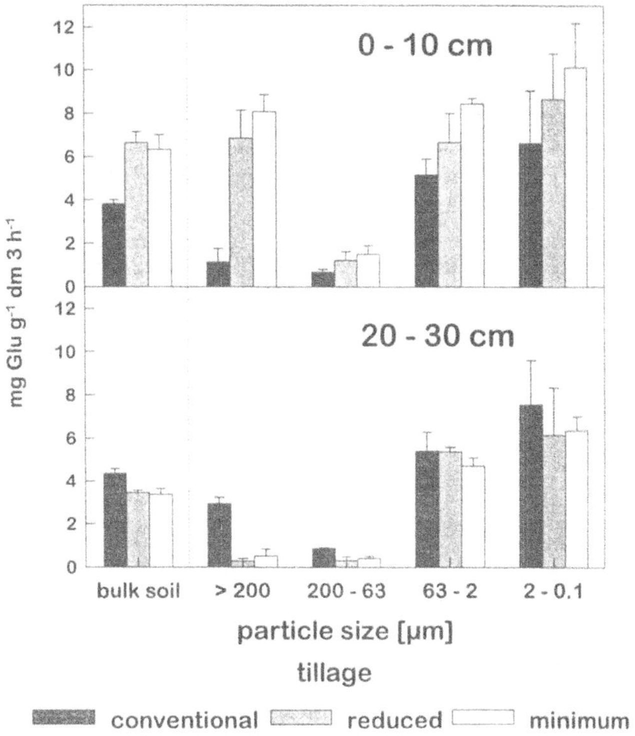

Figure 2: The influence of conventional, reduced and minimum tillage on the xylanase activity of the bulk soil and particle-size fractions in two soil layers (0-10 cm and 20-30 cm). Results are given as mean and SD of the three replicates.

shallow distribution of the remaining residues in the top layer (Lynch and Panting 1980, Saffigna *et al.* 1989, Kandeler and Böhm 1996). As shown for the Calcic Chernozem, xylanase activity was mainly located in the 2000-200 µm particles and decreased in the smaller fractions. The difference between xylanase activity in the tillage treatments was highest in the coarse sand fraction (Fig.2).

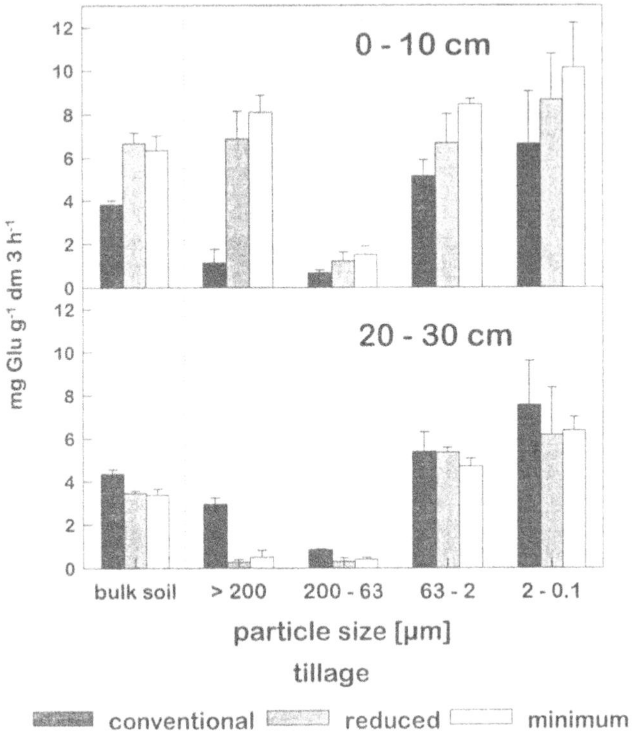

Figure 3: The influence of conventional, reduced and minimum tillage on the invertase activity of the bulk soil and particle-size fractions in two soil layers (0-10 cm and 20-30 cm). Results are given as mean and SD of the three replicates.

Table 4: Chemical and microbiological properties of the Calcic Chernozem (bulk soil) and its particle size fractions (Mean and SD of three replicate samples are given as absolute values and as percent of the bulk soil; standard deviations are given in parentheses)

soil fraction	C_{org} (mg g^{-1})	N_t (mg g^{-1})	C/N	xylanase (mg glucose g^{-1} 24h^{-1})	invertase (mg glucose g^{-1} 3h^{-1})	urease (µg N g^{-1} 2h^{-1})
bulk soil	15.0 (0.4)	1.61 (0.10)	9.3 (0.5)	0.74 (0.04)	2.54 (0.27)	20.5*
2000-200 µm	4.2 (0.3)	0.33 (0.02)	12.7 (1.0)	2.92 (0.59)	0.27 (0.03)	4.6
200-63 µm	7.4 (1.3)	0.68 (0.05)	10.9 (0.6)	1.61 (0.60)	0.77 (0.08)	2.9
63-2 µm	16.5 (0.6)	1.76 (0.07)	9.4 (0.2)	0.49 (0.11)	3.97 (0.07)	32.6
2-0.1 µm	37.4 (2.3)	4.35 (0.07)	8:6 (0.1)	1.14 (0.09)	1.62 (0.08)	131.4
	C_{org} (%)	N_t (%)	C/N	xylanase (%)	invertase (%)	urease (%)
bulk soil	100.0	100.0	-	100.0	100.0	100.0
2000-200 µm	4.1	3.0	-	57.6	1.6	4.3
200-63 µm	9.3	7.9	-	40.7	5.7	4.0
63-2 µm	62.8	62.4	-	37.8	89.2	44.0
2-0.1 µm	22.9	24.9	-	14.2	5.9	45.2
recovery	99.1	98.2	-	150.3	102.4	97.5

* mean of two replicate samples

Table 5: The influence of different tillage treatments on the microbial biomass N of a Haplic Chernozem (bulk soil) and its particle size fractions.

soil fraction	conventional tillage		reduced tillage		minimum tillage	
	0-10 cm	20-30 cm	0-10 cm	20-30 cm	0-10 cm	20-30 cm
bulk soil	12.4*	13.5	20.3	10.9	20.6	8.0
2000-200 μm	2.1	4.3	11.3	1.8	15.7	1.6
200-63 μm	2.0	0.6	5.3	1.2	4.6	0.8
63-2 μm	10.2	11.0	13.9	8.1	14.5	7.1
2-0.1 μm	28.4	41.2	74.2	30.4	65.2	31.8
2000-200 μm	2.0**	3.3	4.5	1.4	5.6	2.3
200-63 μm	5.4	1.2	6.6	3.6	6.0	1.0
63-2 μm	41.0	32.0	27.2	32.5	30.5	38.1
2-0.1 μm	51.6	63.5	61.7	62.5	57.9	58.1
sum of fractions	100.0	100.0	100.0	100.0	100.0	100.0
recovery (%)	82.3	96.2	99.9	90.3	96.0	103.5

* results are given as μg ninhydrin-reactive nitrogen g^{-1} soil (means of two replicates)

** values show microbial biomass in a given fraction, expressed as percent of the sum of microbial biomass in all four fractions

Minimum tillage considerably increased xylanase activity in these particles, providing strong evidence that the xylanase is mainly bound to the particulate organic matter. In the 20-30 cm layer, xylanase activity was equally distributed in the particle-size fraction of minimum and reduced tillage, but peaked in the coarse and medium sand fractions of the conventional tillage treatment (Fig.2). This result illustrated clearly the consequence of conventional plowing methods. Aside from mixing the soil, plowing also turned it over and buried the organic matter (Kandeler and Böhm 1996). Therefore, conventional tillage increased xylanase activity of the bulk soil and of all fractions in the 20-30 cm layer.

Invertase activity of the bulk soil was significantly higher in the top soil layer (0-10 cm) of the minimum and reduced tillage plots than of the conventional plots (Fig.3). The former tillage treatments caused a bimodal distribution of invertase activity in particle-size fractions of the 0-10 cm layer, whereas invertase activity of the conventional tillage treatment was concentrated in the smaller fractions (63-2 μm and 2-0.1 μm) and, therefore, followed the same pattern as in the Calcic Chernozem (Fig.3). Further studies should clarify whether the high invertase activity of the coarse sand particles of the minimum and reduced tillage treatment may be caused by differences in the amount of the light fractions consisting essentially of particulate organic matter. The mean invertase activity of a light fraction of various soils was 6.7 times that of the heavy fraction (Burns 1978). In the 20-30 cm layer, invertase activity was mainly found in the smaller fractions. An additional peak of invertase activity was detected in the coarse sand particles of the conventional tillage treatment (20-30 cm layer) (Fig.3).

Microbial biomass N measured as ninhydrin-reactive nitrogen was higher in the bulk soil (0-10 cm) of minimum and reduced tillage than of conventional tillage (Tab.5). The high recovery of biomass N after the fractionation reconfirms that we have chosen a very sensitive and careful procedure, as described also by the recent paper of van Gestel *et al.* (1996). Therefore, virtually no microbial cells were damaged during soil dispersion and separation. Highest concentrations of biomass N were found in the clay particles (2-0.1 μm), lowest in the fine sand particles (200-63 μm) (Tab.5). Van Gestel *et al.* (1996) explained almost completely (>99%) their corresponding results based on the clay and organic matter contents of the size fractions.

CONCLUSION

The physical fractionation procedure proposed by Stemmer *et al.* (1998) enabled us to estimate the chemical and microbiological characteristics of particle size fractions ranging from 2000-250, 250-63, 63-2 and 2-0.1 µm. The high mean recovery of the single microbiological properties supported previous studies in which the fractionation procedure yielded meaningful fractions. The importance of different fractions for the protection of microorganisms and enzyme activities bound to organo-mineral particles could be shown. In addition, xylanase activity of the coarse sand fraction was identified as a very sensitive indicator of different tillage treatments, apparently being even more sensitive than the xylanase activity of the bulk soil.

We conclude that physical fractionation of soils according to particle size, combined with chemical and microbiological methods, offers future potential to evaluate management-induced changes in SOM composition, microbial biomass and enzyme activities. Further studies should clarify whether the C/N ratio of the microbial biomass varies in the size fractions and whether the high biomass N in the clay particles was predominately of bacterial origin.

Acknowledgements: We thank Eva Kohlmann for technical assistance and determination of microbial biomass in soils and particle-size fractions. Thanks are due to Dr.Georg Dersch, who supported us with the technical data of the field experiment in Fuchsenbigl, Lower Austria. We thank Dr.Michael Stachowitz for linguistic help. Financial support was provided by an EU project (contract EV5V-CT94-0434) and by the Federal Ministry of Agriculture and Forestry, Vienna, Austria.

REFERENCES

Ahmed, M. and J.M. Oades, 1984. Distribution of organic matter and adenosine triphosphate after fractionation of soils by physical procedures. Soil Biol. Biochem., 16, 465-470.

Arshad, M.A., M. Schnitzer, D.A. Angers and J.A. Rippmeester, 1990. Effects of till, vs no-till in the quality of soil organic matter. Soil Biol. Biochem., 22, 595-599.

Boyd, S.A. and M.M. Mortland, 1990. Enzyme interactions with clays and clay-organic matter complexes. In: J.M. Bollag and G. Stotzky (Editors). Soil Biochemistry, Volume 6. Marcel Decker, New York, Basel, pp. 1-28.

Burns, R.G.(ed), 1978. Soil Enzymes. Academic Press, London. pp. 1-380.

Burns, R.G., 1982. Enzyme activity in soil: Location and possible role in microbial ecology. Soil Biol. Biochem., 14, 423-427.

Burns, R.G., 1986. Interaction of enzymes with soil mineral and organic colloids. In: P.M. Huang and M. Schnitzer (Editors). Interactions of Soil Minerals with Natural Organics and Microbes. Special Publication No.17. Soil Science Society of America, Madison, Wisconsin, pp. 429-451.

Christensen, B.T. and S. Bech-Andersen, 1989. Influence of straw disposal on distribution of amino acids in soil particle size fractions. Soil Biol. Biochem., 21, 35-40.

Degens, B. and G. Sparling, 1996. Changes in aggregation do not correspond with changes in labile organic C fractions in soil amended with ^{14}C-glucose. Soil Biol. Biochem., 28, 453-462.

Gupta, V.V.S.R. and J.J. Germida, 1988. Distribution of microbial biomass and its activity in different soil aggregate size classes as affected by cultivation. Soil Biol. Biochem. 20, 777-786.

Kanazawa, S., 1979. Studies on the plant debris in rice paddy soils. II. Microbial respiration and enzyme activities of fractionated plough layer of paddy soils. Soil Sci. Plant Nutr., 25, 71-80.

Kanazawa, S. and Z. Filip, 1986. Distribution of microorganisms, total biomass, and enzyme activities in different particles of a brown soil. Microb. Ecol., 12, 205-215.

Kandeler, E., 1990. Characterization of free and adsorbed phosphatases in soils. Biol. Fertil. Soils, 9, 199-202.

Kandeler, E. and H. Gerber, 1988. Short-term assay of soil urease activity using colorimetric determination of ammonium. Biol Fertil Soils, 6, 68-72.

Kandeler, E. and K.E. Böhm, 1996. Temporal dynamics of microbial biomass, xylanase activity, N-mineralization and potential nitrification in different tillage systems. Applied Soil Ecology 5, 221-230.

Kiss, S., M. Dragan-Bularda and D. Radulescu, 1975. Biological significance of enzymes accumulated in soil. Adv. Agron., 27, 25-87.

Ladd, J.N. and E.A. Paul, 1973. Changes in enzymic activity and distribution of acid-soluble, amino acid-nitrogen in soil during nitrogen immobilization and mineralization. Soil Biol. Biochem., 5, 825-840.

Ladd, J.N., R.C. Foster, P. Nannipieri and J.M. Oades, 1996. Soil structure and biological activity. In: G. Stotzky and J.M. Bollag (Editors). Soil Biochemistry, Volume 9. Marcel Dekker, New York, pp. 23-78.

Lensi, R., A. Clays-Josserand and L.J. Monrozier, 1995. Denitrifiers and denitrifying activity in size fractions of a mollisol under permanent pasture and continous cultivation. Soil Biol. Biochem., 27, 61-69.

Lynch, J.M. and L.M. Panting, 1980. Cultivation and the soil biomass. Soil. Biol. Biochem., 12, 29-33.

Marshall, T.J. and J.W. Holmes, 1988. Soil Physics, 2nd edition, Cambridge University Press, Cambridge.

Mateos, M.P. and S.G. Carcedo, 1985. Effect of fractionation on location of enzyme activites in soil structural units. Biol. Fertil. Soils, 1, 153-159.

Mateos, M.P. and S.G. Carcedo, 1987. Effect of fractionation on the enzymatic state and behaviour of enzyme activities in different structural units. Biol. Fertil. Soils, 4, 151-154.

Monrozier, L.J., J.N. Ladd, R.W. Fitzpatrick, R.C. Foster and M. Raupach, 1991. Components and microbial biomass content of size fractions in soil of contrasting aggregation. Geoderma, 49, 37-62.

Robert, M. and C. Chenu, 1992. Interactions between soil minerals and microorganisms. In: G. Stotzky and J.M. Bollag (Editors). Soil Biochemistry, Volume 7. Marcel Dekker, New York, pp. 307-404.

Rojo M.J., S.G. Carcedo and M.P. Mateos, 1990. Distribution and charcaterization of phosphatase and organic phosphorus in soil fractions. Soil Biol. Biochem., 22, 169-174.

Ruggiero, P., J. Dec and J.M. Bollag, 1996. Soil as a catalytic system. In: G. Stotzky and J.M. Bollag (Editors). Soil Biochemistry, Volume 9. Marcel Dekker, New York, Basel, pp. 79-122.

Saffigna, P.G., D.S. Powlson, P.C. Brookes, and G.A. Thomas, 1989. Influence of Sorghum residues and tillage on soil organic matter and soil microbial biomass in an Australian Vertisol. Soil Biol. Biochem., 21, 759-765.

Singh, S. and J.S. Singh, 1995. Microbial biomass associated with water-stable aggregates in forest, savanna and cropland soils of a sesonally dry tropical region, India. Soil Biol. Biochem., 27, 1027-1033.

Schinner, F. and W. von Mersi, 1990. Xylanase-, CM-cellulase- and invertase activitiy in soil: an improved method. Soil Biol. Biochem., 22, 511-515.

Schinner, F., R. Öhlinger, E. Kandeler and R Margesin (Editors), 1996. Methods in Soil Biology. Springer-Verlag, Berlin Heidelberg New York, pp. 1-425.

Schulten, H.R., P. Leinweber, and C. Sorge, 1993. Composition of organic matter in particle-size fractions of an agricultural soil. J.Soil Sci., 44, 677-691.

Sequi, P., G. Cercignani, M. de Nobili and M. Paglia, 1985. A positive trend among two soil enzyme activities and a range of soil porosity under zero and conventional tillage. Soil Biol. Biochem., 17, 255-256.

Skujins, J.J., 1978. History of abiontic soil enzymes research. In: R.G. Burns (Editor). Soil Enzymes. Academic Press, London. pp. 1-49.

Stemmer, M., M.H. Gerzabek and E. Kandeler, 1998. Organic matter and enzyme activity in particle size fractions of soils obtained after low-energy sonication. Soil Biol. Biochem. 30, 9-17.

Stotzky, G. and R.G. Burns, 1982. The soil environment: Clay-humus-microbe interactions. In: R.G. Burns and J.H. Slater (Editors). Experimental Microbial Ecology. Blackwell Scientific, Oxford. pp. 105-133.

Tabatabai, M.A., 1973. Michaelis constant of urease in soils and soil fractions. Soil Sci. Soc. Am. Proc., 37, 707-710.

Tabatabai, M.A. and M. Fu, 1992. Extraction of enzymes from soils. In: G. Stotzky and J.M. Bollag (Editors). Soil Biochemistry, Volume 7. Marcel Dekker, New York, pp. 197-227.

Van Gestel, M., R. Merckx and K. Vlassak, 1996. Spatial distribution of microbial biomass in microaggregates of a silty-loam soil and the relation with the resistance of microorganisms to soil drying. Soil Biol. Biochem., 28, 503-510.

Key words: Soil enzymes, microbial biomass, particle-size fractions, xylanase, invertase

ACTIVITY OF ß-GLUCOSIDASE IN THE PRESENCE OF COPPER OR ZINC AND MONTMORILLONITE OR AL-MONTMORILLONITE

Gabriella Geiger[1], Gerhard Furrer[1], Helmut Brandl[2] and Rainer Schulin[1]

1 Institute of Terrestrial Ecology, Department of Soil Protection, ETH Zurich, Grabenstrasse 11a, CH - 8952 Schlieren

2 Institute of Environmental Sciences, University of Zurich, Winterthurerstrasse 190, CH - 8057 Zurich

1. INTRODUCTION

Various studies have shown that the activities of many enzymes in soil are inhibited by trace elements (Tyler, 1974; Haanstra and Doelman, 1991; Doelman and Haanstra, 1986). Deng and Tabatabai (1995) found that cellulase activity in soil was inhibited by metals in the sequence: $Ag^+ > Cu^{2+}, Cd^{2+}, Hg^{2+}, Pb^{2+}, Al^{3+} > Zn^{2+} > Ni^{2+}$. Since cellulose is the major carbohydrate synthesized by plants ($26*10^9$ t C y^{-1}; as calculated from terrestrial primary production (Schimmel *et al.*, 1995) and cellulose content of terrestrial biomass (Brown, 1983)), its degradation represents an important part of the carbon cycle within the terrestrial biosphere. Nevertheless, only scarce information is available on the mechanisms involved in the inhibition of cellulases, e.g., the inhibition of ß-glucosidase by metals.

It is generally assumed that the hydrolysis reaction catalyzed by ß-glucosidase proceeds via an acid-base mechanism involving aspartic and glutamic acid. The first amino acid acts as a general acid catalyst, the second as a nucleophil (Figure 1).

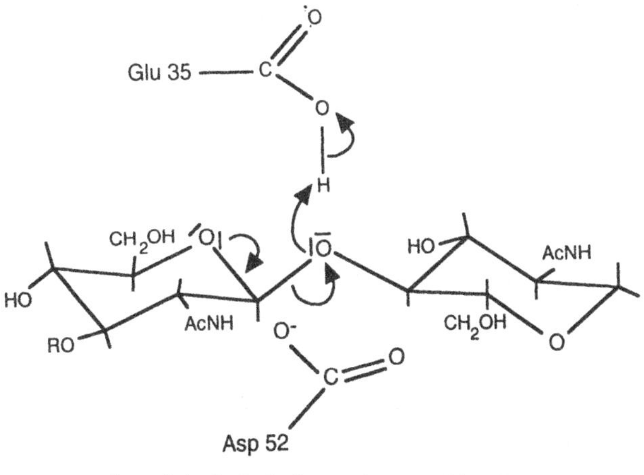

Figure 1. Schematic representation of the hydrolysis reaction, resulting in the cleavage of a polysaccharide chain. The arrows indicate the postulated movements of electron pairs (adapted from Béguin and Aubert, 1994).

The inactivation of enzymes by heavy metals is caused by complexation. This can lead to the denaturation of the enzymes or to the displacement of essential metal cations from active sites. Therefore the inhibition of enzyme activity by a metal is crucially dependent on the kind of amino acids included in the active part of the protein (Geiger *et al.*, 1998a).

Clay minerals, with their permanent negatively charged surfaces, can significantly reduce extracellular enzyme activity in soil by adsorption of positively charged enzymes. This may also cause a shift in the pH optimum of the catalytic activity towards alkaline values (Quiquampoix, 1987; Quiquampoix *et al.*, 1995).

Because clay surfaces do not only interact with enzymes, but also specifically bind metal cations, they may also reduce the toxicity of metals. In fact, the addition of montmorillonite or aluminum-coated montmorillonite (Al-montmorillonite) has been proposed for the remediation of agricultural soil polluted by heavy metals (Lothenbach *et al.*, 1997a). Such *in situ* treatments may successfully reduce the mobility and, therefore, the metal availability for plants. However, inhibitory effects on extracellular soil enzymes may still occur due to the formation of ternary enzyme-metal-mineral surface complexes.
The formation of surface complexes involving mineral surfaces, metal cations and enzymes may influence metal availability towards the enzymes, i.e., the enzyme activity can be reduced through the formation of dissolved or surface complexes involving the enzyme (Figure 2).

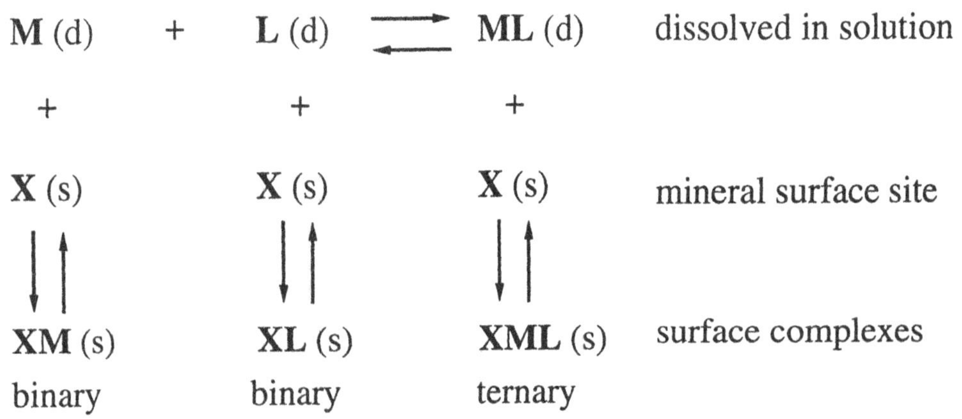

Figure 2 Schematic presentation of equilibrium complex formation reactions in an aqueous metal-ligand-mineral system. M, L and X symbolize metals, ligands and mineral surface sites, respectively; L also represents enzymes.

Ternary complexes may be favored by various factors including electrostatic interactions (Eliott and Huang, 1979). For example the adsorption of a metal-ligand complex at clay mineral surfaces is mainly caused by the opposite charge of the complex in the dissolved state and the surface of the mineral (Schindler, 1990). Bodenheimer and Heller (1967), have shown that the adsorption of neutral and basic amino acids on montmorillonite was enhanced at pH > 5 by the presence of copper. This observation was explained with the formation of ternary surface complexes involving the negatively charged mineral surface, the copper cation, and the amino acids. The same authors reported that acidic amino acids such as glutamic acid did not show any tendency for the formation of ternary complexes.

The aim of this work was to investigate the influence of copper and zinc on cellulose-degrading enzymes in a model soil system. In particular we studied the effect of (i) pH, (ii) copper or zinc, (iii) clay minerals, and (iv) clay minerals and a metal in combination on the activity of ß-glucosidase. Montmorillonite was chosen as representative of soil clay minerals. Al-montmorillonite is a modified form of montmorillonite with a smaller specific surface area and a lower permanent charge (Lothenbach *et al.*, 1997b). Copper and zinc were chosen since both are essential micro nutrients but in excess amounts are pollutants of environmental concern. Copper has a high chemical affinity to solid organic substances, whereas zinc has a higher solubility and tends to form weaker organic complexes than copper.

2. MATERIALS AND METHODS

2.1. Activity of ß-Glucosidase

The kinetics of the cellobiose cleaving reaction catalyzed by ß-glucosidase from almonds (EC 3.2.1.21, 5.8 U mg^{-1}, Mr ≈ 130 000, Fluka) were studied in 5 ml assays containing 25 mg of cellobiose, 4.3 U (approx. 10^{-6} M) of ß-glucosidase and 0.1 M Na acetate buffer. The reactions were carried out for 15 minutes at 25 °C. To terminate the reaction, the samples were placed in boiling water for 5 minutes.

2.2. Sample Preparation and HPLC Analysis

Prior to high pressure liquid chromatography (HPLC) analysis, samples were desalted by electrodialysis to remove ionic components of the enzyme assay because these interfered with the analysis of sugars. Thereafter, the samples were analyzed by a HPLC-system equipped with a refractive index detector as described in Geiger *et al.* (1998a).

2.3. Speciation Calculations

The concentrations of species in solution were calculated using the chemical speciation program MICROQL as described in Geiger *et al.* (1996). The input data (mass action laws, stability constants and concentrations of the chemical components) were taken from Geiger *et al.* (1998b).

2.4. Modeling of Chemical Speciation

The chemical speciation model is based on the assumption that the activity of an enzyme such as ß-glucosidase depends critically on the protonation state of the amino acids at the reactive center of the enzyme, in this case an aspartic acid (Asp) and a glutamic acid (Glu). Since the amino acids of the reactive site of ß-glucosidase are not in terminal positions (Béguin and Aubert, 1994), only the side chain carboxylic groups of Asp and Glu are able to play an important role in copper binding. The pK values of the two carboxylic acids were chosen in order to explain the pH optimum of enzyme activity (Geiger *et al.*, 1998b). The inactivation of ß-glucosidase by metals is thought to be mainly due to complexation of the amino acids at the reactive center. Since the two amino acids are sterically located near each other, they may not only form binary complexes but also a chelate complex with copper. There are no experimentally determined stability constants available for copper complexation at the reactive site of ß-glucosidase. Therefore, we used the stability constant of the simple monocarboxylic acid acetic acid for the binary complexes. For the ternary complex (Asp-Cu-Glu) we considered the structure of a chelate complex with a dicarboxylic acid. The copper activity effective in enzyme inhibition may be lowered by non-specific binding at any complexing group of the protein. It was found that this was not the case in the system chosen (Geiger *et al.*, 1998b) and we therefore only considered the reactive site for the calculations.

3. RESULTS AND DISCUSSION

3.1. Influence of Copper and Zinc on ß-Glucosidase Activity

Figure 3 shows the influence of 0.6 mM copper or zinc on almond ß-glucosidase activity in buffered assay mixtures. The influence of copper was strongest in the pH range 5-5.5, resulting in reduced activity by 90%. Close to pH 4, the effect of copper was weak, with a reduced activity of 30% at most. The presence of copper caused a shift in pH optimum towards a lower pH value. In contrast, 0.6 mM zinc did not shift the pH optimum. The maximum inhibition by zinc (40%) was much lower and less pH dependent in comparison to copper.

3.2. Modeling of the Interactions between ß-Glucosidase and Copper or Zinc

As mentioned above, the interactions of a metal cation with the amino acids of the reactive center (Rcenter) of an enzyme may cause strong inhibition of the enzyme activity. The critical amino acids of almond ß-glucosidase are glutamic and aspartic acid (Becher, 1977). To estimate the influence of copper and zinc on the activity of ß-glucosidase in dependence of pH, the chemical speciation of the involved amino acids and heavy metals has been calculated. Since the two amino acids are sterically located close to each other, the reaction center may not only form binary complexes ($CuRcenter^{+}$) but also a chelate complex ($CuRcenter_{chelate}$) with copper.

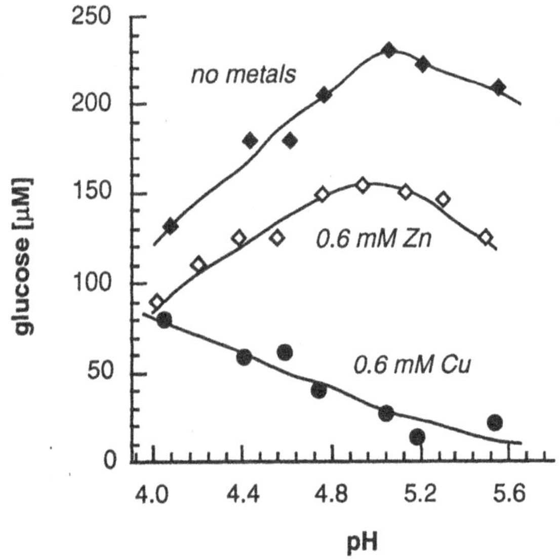

Figure 3. Influence of 0.6 mM copper and zinc on the activity of ß-glucosidase (4.3 U, 0.5 g l^{-1} cellobiose) at 25°C in 0.1 M Na-acetate buffer over a range of pH 4-5.5 expressed as rate of glucose production in the first 15 minutes of the experiments.

Figure 4 shows the calculated chemical speciation of 0.6 mM copper in presence of 10^{-6} M ß-glucosidase. From pH 2 to 6, the total amount of reactive centers complexed by copper increases continuously. This indicates that the enzymes in the protonated state do not interact with metals. At higher pH, the complexation of reactive centers by copper may cause an inhibition of ß-glucosidase activity.

Figure 5 shows the calculated chemical speciation of 0.6 mM zinc in presence of 10^{-6} M ß-glucosidase. The amount of reactive centers complexed by zinc is lower than in the case of copper due to the different affinities of the two metal cations for the reactive center. The calculated ratios of reactive centers complexed by copper or zinc correspond qualitatively very well with the experimentally found inhibition of the enzyme activity (compare Fig. 3).

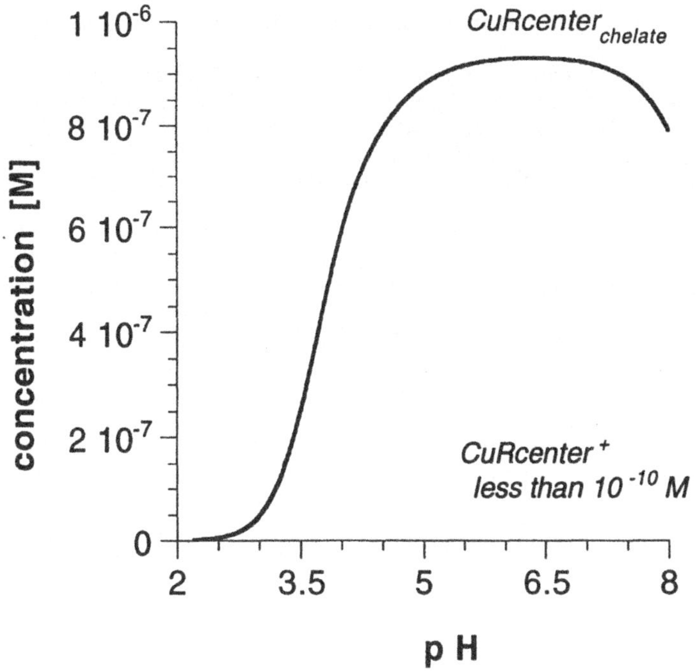

Figure 4. Concentration of reactive centers complexed by copper ($CuRcenter_{chelate}$ and $CuRcenter^{+}$) of ß-glucosidase (10^{-6} M) in presence of 0.6 mM copper in 0.1 M Na-acetate buffer over a range of pH 2-8.

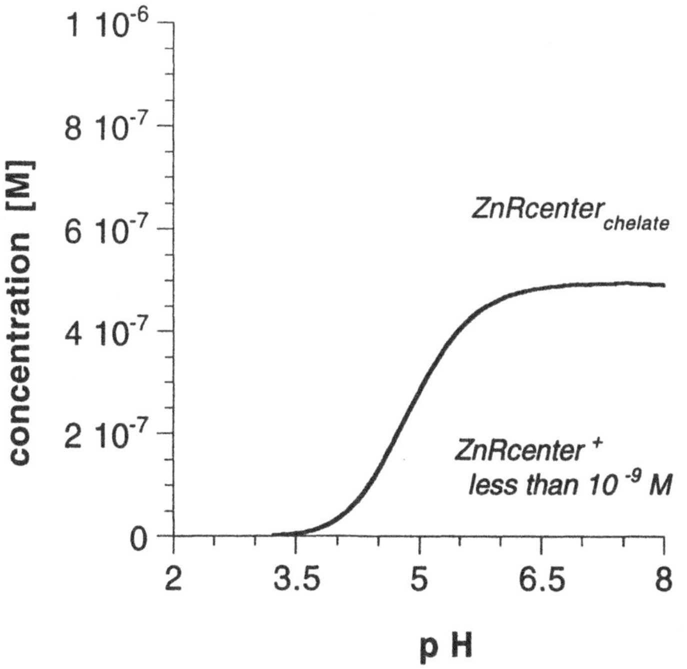

Figure 5. Concentration of reactive centers complexed by copper ($ZnRcenter_{chelate}$ and $ZnRcenter^{+}$) of ß-glucosidase (10^{-6} M) in presence of 0.6 mM zinc in 0.1 M Na-acetate buffer over a range of pH 2 - 8.

3.3. Effects of Clay Minerals in Combination with Copper or Zinc on ß-Glucosidase Activity

The activity of ß-glucosidase was significantly reduced in the presence of Al-montmorillonite and montmorillonite (Fig. 6). Montmorillonite inhibited the enzyme activity more strongly than Al-montmorillonite did. The different extent of inhibition on the activity of ß-glucosidase by the two montmorillonite compounds can be related to the difference in specific surface and net negative charge (Gianfreda *et al.*, 1991; Geiger *et al.*, 1998a).

In the presence of montmorillonite and Al-montmorillonite, the addition of copper further inhibited the ß-glucosidase activity strongly, whereas zinc only reduced the ß-glucosidase activity in Al-montmorillonite, but not in montmorillonite suspensions. Around pH 5.5, 0.6 mM copper reduced the ß-glucosidase activity by more than 90% in the presence of Al-montmorillonite, and between pH 4.5 and 5.5, the inhibitory effects of copper were similar to those in the clay-free suspensions (Fig. 3).
However, the combined inhibitory effects of each of the two clays together with copper in comparison with the mineral- and metal-free solution were the same. A similar observation was made for zinc. This means that the stronger inhibitory effect of montmorillonite alone as compared to Al-montmorillonite is compensated by a weaker metal effect in the presence of montmorillonite than in the presence of Al-montmorillonite. At pH below 5, the total inhibitory effect of copper was stronger in combination with both clays than without clay, whereas above pH 5 total enzyme inhibition was similar with and without clay (cmp. Figs. 3 and 6). For zinc-contaminated systems, enzyme inhibition in the presence of clay was stronger than without the two clays over the entire pH range studied, although the difference tended to disappear towards pH 5.6.

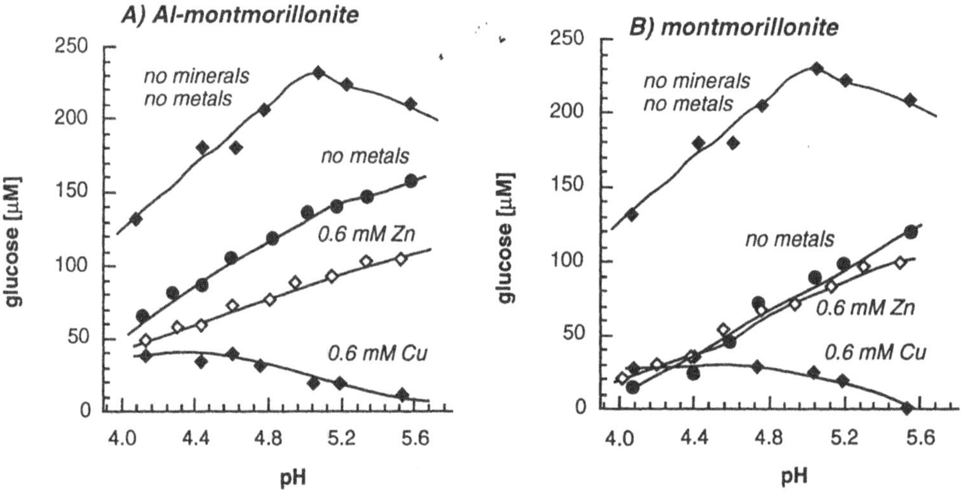

Figure 6. Influence of copper and zinc on the activity of ß-glucosidase (4.3 U, 0.5 g cellobiose l^{-1}) in the presence of (A) Al-montmorillonite (2 g l^{-1}) or (B) montmorillonite (2 g l^{-1}), expressed as rate of glucose production in the first 15 minutes of the experiments.

The different effects of the two metals in the two systems can be explained by the formation of ternary complexes between mineral surfaces, copper or zinc and organic ligands reducing metal activity in the solution. The enzyme complexed by copper is shown in Figure 4. The dominant species in the pH range 2-8 are not charged ($CuRcenter_{chelate}$). I.e., these species probably do not form stable surface complexes with both montmorillonite compounds. Less important are the positively charged copper-enzyme complexes ($CuRcenter^{+}$). Thus, in the presence of copper, ß-glucosidase probably forms more ternary complexes with montmorillonite than with Al-montmorillonite. Therefore, copper should reduce the activity

of the enzyme in solution less in a system with montmorillonite than in a system with Al-montmorillonite (Fig. 6).
In the pH range below 8, most of the zinc-enzyme complexes are uncharged ($ZnRcenter_{chelate}$) and only a small amount of the enzyme-zinc complexes are positively charged ($ZnHRcenter^{+}$). These species probably form stable complexes with the permanent negatively charged montmorillonite surfaces, but not with the positively charged aluminum sites. This suggests that ternary complexes are not likely to be formed at the Al-montmorillonite surface, whereas in the montmorillonite system, the formation of ternary surface complexes could prevent an inhibition of free enzyme by the addition of zinc.
In summary, for the estimation of the influence of a metal on enzyme activity in a system with clay minerals, the stability constants of the metal-enzyme complexes, the pH at the point-of-zero charge (pH_{PZC}) of the enzyme, the pH_{PZC} of the clay mineral and the charge of the metal-enzyme complex are of great importance. From our experimental findings and the modeling of chemical speciation we may conclude that the ß-glucosidase activity can be significantly affected by the formation of metal-enzyme complexes but also by the formation of binary and ternary enzyme-mineral surface complexes.

4. ACKNOWLEDGMENTS

This research was financially supported by the Swiss National Science Foundation (Project 2200-037701.93 and 200-043554.95).

5. REFERENCES

Becher, H.J., 1977. Isolierung und Untersuchung einer ß-Glucosidase aus dem Bittermandelemulsin. Dissertation, University of Köln.

Béguin, P. and J-P. Aubert, 1994. The biological degradation of cellulose. FEMS Microbiol. Rev., 13, 25-58.

Bodenheimer, W. and L. Heller, 1967. Sorption of -amino-acids by copper montmorillonite. Clay Minerals 7, 167-176.

Brown, D.E, 1983. In: Coughlan, M.P. Cellulose degradation by fungi. In: Fogarthy, V.M. and C.T. Kelly (Editors). 1990. Microbial enzymes and biotechnology. Elsevier, London and New York, 1-36.

Deng, S.P. and M.A. Tabatabai, 1995. Cellulase activity of soils: Effect of trace elements. Soil. Biol. Biochem., 27 (7), 977-979.

Doelman, P. and L. Haanstra, 1986. Short- and long-term effects of heavy metals on urease activity in soils. Biol. Fertil. Soils, 2, 213-218.

Eliott, H.A. and C.P. Huang, 1979. The adsorption characteristics of Cu(II) in the presence of chelating agents. J. Colloid Interface Sci., 70 (1), 29-45.

Geiger G., M. Gfeller, G. Furrer and R. Schulin, 1996. Soil bacteria sensitivity towards heavy metals- Experimental system optimisation using chemical speciation calculations. Fresenius' J. Anal. Chem., 354, 624-628.

Geiger, G., H. Brandl, G. Furrer and R. Schulin, 1998a. The effect of copper on the activity of cellulase and ß-glucosidase in presence of montmorillonite or Al-montmorillonite. Soil Biol. Biochem. (in press).

Geiger, G., G. Furrer, F. Funk, H. Brandl, and R. Schulin, 1998b. Heavy metal effects on ß-glucosidase activity influenced by pH and buffer systems. Soil Biol. Biochem. (submitted).

Gianfreda, L., M.A. Rao and A. Violante, 1991. Invertase (ß-fructosidase): Effect of montmorillonite, Al-hydroxide and $Al(OH)_x$-montmorillonite complex on activity and kinetic properties. Soil Biol. Biochem., 23 (6), 581-587.

Haanstra, L. and P. Doelman, 1991. An ecological dose-response model approach to short- and long-term effects of heavy metals on arylsulphatase activity in soil. Biol. Fertil. Soils, 11, 18-23.

Lothenbach, B., R. Krebs, G. Furrer, S. Gupta and R. Schulin, 1997a. Heavy metal immobilisation in soil by addition of montmorillonite, Al-montmorillonite, and gravel sludge: Batch and pot experiments. Eur. J. Soil Sci. (submitted).

Lothenbach, B., G. Furrer and R. Schulin, 1997b. Immobilization of heavy metals by polynuclear aluminium and montmorillonite compounds. Env. Sci. Technol., 31, 1452-1462.

Quiquampoix, H., J. Abadie, M.H. Baron, F. Leprince, P.T. Matumoto-Pintro, R.G. Ratcliffe, and S. Stainton, 1995. Mechanisms and consequences of protein adsorption on soil mineral surfaces. In: Horbett T.A., and J.L. Brash. (Edirors), Proteins at Interfaces II: Fundamentals and Applications, ACS Symp. Ser. 602, 321-333.

Quiquampoix, H, 1987. A stepwise approach to the understanding of extracellular enzyme activity in soil I. Effects of electrostatic interactions on the conformation of a ß-D-glucosidase adsorbed on different mineral surfaces. Biochimie, 69, 753 - 763.

Schimmel, D., I.G. Enting, M. Heinemann, T.M.L. Wigley, D. Raynaud, D.Alves and U. Siegenthaler, 1995. CO_2 and the carbon cycle. Climate change 1994. In: Houghton, J.T., L.G. Meira Filho, J. Bruce, Hoesung Lee, B.A. Callander, E. Haites, N. Harris and K. Maskell (Editors). Radiative Forcing of Climate Change and an Evaluation of the IPCC IS92 Emission Scenarios. Cambridge University Press, 35-71.

Schindler, P.W, 1990. Co-adsorption of metal ions and organic ligands: Formation of ternary surface complexes. Rev. Miner. 23, 281-305.

Tyler, G, 1974. Heavy metal pollution and soil enzymatic activity. Plant and Soil, 41, 303-311.

TRACE MINERAL AMENDMENTS IN AGRICULTURE FOR OPTIMIZING THE BIOCONTROL ACTIVITY OF PLANT-ASSOCIATED BACTERIA

Brion K. Duffy and Geneviève Défago

Phytopathology Group, Plant Sciences Institute, Swiss Federal Institute of Technology, Zürich CH-8092

1. INTRODUCTION

There is growing interest in the large-scale application of plant-associated, nonpathogenic bacteria for the biological control of soilborne plant diseases. This is largely due to a lack of other control options for soilborne diseases but also to agricultural trends towards sustainability and to public concern over the hazards of synthetic fungicides and fumigants. In the last twenty years, a diversity of plant-associated bacteria, particularly fluorescent pseudomonads, have been shown to effectively control a wide-spectrum of soilborne diseases in greenhouse trials and field plots of small to moderate scale. Commercialization of most biocontrol agents, however, has been hindered by variable performance and only a handful of bacterial products have been registered for agricultural application (Cook, 1993). The level and reliability of disease suppression must be optimized for biocontrol to become commercially feasible on a large-scale.

Numerous factors have been proposed as contributing to the variable performance of biocontrol agents, including inconsistent root colonization, absence of the target pathogen or presence of nontarget pathogens, inconsistent production or inactivation of antimicrobial compounds, loss of biocontrol activity through genetic mutation, and resistance of the pathogen to antimicrobial compounds (Mazzola et al., 1995; Weller and Thomashow, 1994; Weller, 1988). It has been suggested, however, that biocontrol agents perform much more consistently than generally thought as long as their performance is considered under similar site conditions rather than between widely divergent environments (*i.e.*, edaphic parameters), as has typically been the case (Cook, 1993). Recently, the biocontrol activity of *Pseudomonas fluorescens* (Ownley et al., 1991) and *Trichoderma koningii* (Duffy et al., 1997) has been correlated with specific soil chemical and physical components and variable levels of these key factors has been implicated as a primary reason for the "variable" performance of these agents at different field sites.

Soil factors, especially as related to mineral nutrition, exert a major influence on the deleterious activity of many plant pathogens. Management of macro- and microelements has been developed as a control strategy for several economically-important diseases (Elmer, 1995; Engelhard, 1989; Rengel et al., 1994; Schneider, 1985). In some cases, reduction or increase in disease with certain minerals (*e.g.*, Ca, Mn, NH_4- or NO_3-N, urea) can be attributed to indirect effects on the indigenous microbial community (Engelhard, 1989; Henis and Katan, 1975; Miranda et al., 1985; Smiley, 1978). For example, NaCl applications have been traditionally used to control Fusarium crown and root rot of asparagus. NaCl increases the proportion of Mn-reducing bacteria in the rhizosphere, thereby, increasing the supply of plant-available Mn^{2+} which

Effect of Mineral-Organic-Microorganism Interactions on Soil and Freshwater Environments
Edited by Berthelin, *et al*, Kluwer Academic / Plenum Publishers, New York, 1999

in turn optimizes host defenses and may exert a fungistatic effect on *Fusarium* spp. (Elmer, 1995). Along these same lines, it is plausible that mineral nutrition could be manipulated to improve the disease-suppressive activitiy of introduced biocontrol agents. Surprisingly though, little research has focused on the impact of minerals in biocontrol.

To determine the effect of minerals on biocontrol, we used the model strain *Pseudomonas fluorescens* CHA0. This strain was originally isolated from a soil naturally suppressive to black root rot of tobacco caused by *Chalara elegans* Nag Raj & Kendrick (synanamorph *Thielaviopsis basicola*) (Voisard et al., 1994). Seed or soil application of CHA0 effectively suppresses this and several other soilborne fungal diseases (Voisard et al., 1994). CHA0 produces several high-affinity metal-chelating siderophores (pyoverdine, pyochelin, and salicylic acid (Sal)) which contribute to nutrient competition with pathogens (Voisard et al., 1994). However, the primary mechanism of biocontrol for most diseases is the production of antimicrobial compounds such as pyoluteorin (Plt), 2,4-diacetylphloroglucinol (Phl), and hydrogen cyanide (HCN) (Voisard et al., 1994). Biosynthesis of these antifungal metabolites is controlled by a two-component regulatory system comprising the response regulator *gacA* (Laville et al., 1992) and a sensor kinase that is functionally similar to *apdA* from *P. fluorescens* Pf5 (D. Haas, C. Bull, and F. Carruthers, *personal communication*; Corbell and Loper, 1995). Mutation in either gene abolishes production of Phl and Plt antibiotics and HCN and greatly reduces the biocontrol activity of CHA0. Spontaneous mutation in these genes reduces the disease suppressive activity of inoculant and is an important problem in large-scale inoculum production of CHA0 and other biocontrol strains (B. Duffy, *unpublished data*; Voisard et al., 1994).

The objectives of this study were to determine the utility of mineral amendments to inoculant production media and to plant growth substrate for (i) maintaining the genetic stability, (ii) increasing the production of key secondary metabolites, and (iii) increasing the disease-suppressive activity of bacterial biocontrol agents. To detemine the effects on biocontrol, we used Fusarium crown and root rot of tomato. This disease is caused by *Fusarium oxysporum* Schlechtend.:Fr. f.sp. *radicis-lycopersici* Jarvis & Shoemaker (FORL), and is an increasingly important problem in Europe and North America (Hartman and Fletcher, 1991). It is particularly severe in soil-less greenhouse production. This pathosystem was selected for our studies because (i) mineral nutrition is easily manipulated in hydroponic culture and (ii) CHA0 provides only moderate control of this disease.

2. MATERIAL AND METHODS

2.1 Effect of Minerals on the Genetic Stability of Biocontrol Traits

Two experiments were used to determine the influence of minerals on mutation in CHA0. Mutants were easily distinguished from the wild-type based on altered colony phenotype (*i.e.*, enlarged colony size and darker orange color, lack of HCN and extracelluar protease in plate assays, increased fluorescent pigments on King's B agar). Phenotypic differences were routinely confirmed with complementation tests using cloned *apdA* and *gacA* genes. Filter-sterilized stock solutions of minerals were added individually at 1 mM (BH_3O_3, $CaCl_2$ x $2H_2O$, $CuSO_4$, $FeSO_4$ x $7H_2O$, LiCl, $MgSO_4$ x $7H_2O$, $MnCl_2$ x $4H_2O$, $Mo_7(NH_4)_6O_{24}$ x $4H_2O$, NaCl, $ZnSO_4$ x $7H_2O$) or 0.1 mM (CoCl) to standard inoculum production medium consisting of 0.8% nutrient broth plus 0.5 % yeast extract (Difco, Detroit) after media was autoclaved. In the first experiment, the effect of $CuSO_4$ on mutant accumulation was tested. Broth media (20 ml per 100 ml Erlenmeyer flask) with or without mineral amendment was inoculated with 10 µl overnight culture of CHA0 with no detectable mutants ($<10^{-5}$). After incubation for 12 days at 27 C with 140 rpm, dilution samples were plated on King's B agar and the percentage of mutants was quantified after 5 to 7 days of growth on agar. Total bacterial growth was determined as $\log_{10}$ colony-forming units (CFU). Treatments were arranged in a randomized-complete block (RCB) design with ten replicate broths and the experiment was repeated four times with similar results and data were pooled for final analysis.

A second experiment was designed to determine if minerals increased competitiveness of the wild-type. Fresh media with or without mineral amendments was inoculated with 10 μl from cultures taken at 12 days from the previous experiment with a low but detectable level of mutants. This mimics the scale-up of inoculum cultures used in commercial production where cultures of increasing size are used to seed larger volumes. After incubation for 24 h, cultures were sampled as described above and the proportion of wild-type per mutant was determined. Treatments were arranged in a RCB design with four replicate cultures. The experiment was repeated four times with similar results and data were pooled for final analysis. After ANOVA analysis, means were compared for significance using Fisher's protected (P=0.05) LSD (SAS Institute, Cary, NC).

2.2 Effect of Minerals on Biosynthesis of Antibiotics and Siderophores

Nutrient broth medium with or without mineral amendments was inoculated with wild-type CHA0 and incubated for 48 h at 24 C with 140 rpm. Cultures were acidified to pH 2 with 5 N HCl, extracted with an equal volume of ethyl-acetate, and metabolite production was quantified using HPLC methods as previously described (Maurhofer *et al.*, 1992). Maximum UV-absorbances were: 270 nm (Phl), 313 nm (Plt), and 300 nm (Sal). Metabolite yield was expressed relative to bacterial growth. Treatments were arranged in a RCB design with three replicate broths. The experiment was repeated once with similar results and data were pooled for final analysis. After ANOVA analysis, means were compared using Fisher's LSD. The influence of $ZnSO_4 \times 7H_2O$ on Phl and Plt production and growth was further evaluated at a range of concentrations (0 to 2 mM) with cultures incubated for 72 h.

2.3 Effect of Minerals on the Biocontrol Activity of CHA0

The effect of minerals on biocontrol of Fusarium crown and root rot was tested in a rockwool system. Nutrient solution (1/3 normal strength micronutrients) was amended with one biocontrol treatment (none, FORL alone, FORL plus CHA0) and one mineral treatment (none, Zn-EDTA, Cu-EDTA, $Mo_7(NH_4)_6O_{24} \times 4H_2O$, or a mixture of 1/3 each mineral). EDTA forms of zinc and copper were used because they are more stable in rockwool. The nutrient solution was designed for commercial hydroponics tomato production by the Office of Horticultural Production of Geneva, Switzerland and was composed of (mg/liter): $Ca(NO_3)_2 \times 4H_2O$, (955.5); $NH_4H_2PO_4$, (660); KNO_3, (450); $MgSO_4 \times 7H_2O$, (360); K_2SO_4, (174); KH_2PO_4, (78); Fe-EDDHA, (25); EDDHA, (4); $Na_2B_4O_7 \times 10H_2O$, (1.05); $MnSO_4 \times 7H_2O$, (1.02); $ZnSO_4 \times 7H_2O$, (0.43); $CuSO_4 \times 5H_2O$, (0.2); $Na_2MoO_4 \times 2H_2O$, (0.04). Inoculum of FORL was produced in malt broth and added at 10^6 conidia per ml nutrient solution; inoculum of CHA0 was produced in Luria broth and added at 10^7 CFU per ml nutrient solution. A one-time amendment of minerals at 33 ppm (based on Zn^{2+} or Cu^{2+}, or total $Mo_7(NH_4)_6O_{24} \times 4H_2O$) was made at the start of the experiment. Tomato cv. Bonnie Best seedlings pregerminated for 2 days were planted into rockwool blocks (Grodania A/S, Hedehusene, Denmark) saturated with nutrient solution (1 liter/18 seedlings). After two weeks in the growth chamber (22 C, 70 relative humidity, 16 h day) disease severity was assessed on a standard scale of 0 to 4 where 0=no symptoms and 4=plants were dead or nearly so. Treatments were arranged as a 3 x 5 factorial in a split-plot design with a main-plot of mineral treatment and sub-plot of biocontrol treatment. Treatments consisted of 18 plants and were replicated four times. The experiment was repeated twice with identical trends among the treatments, however, slight variation due to the use of different growth chambers prohibited pooling of the data. Results from a representative experiment are presented. Main effects and interactions were analyzed for significance with ANOVA and means were compared using Fisher's LSD when appropriate (SAS Institute).

3. RESULTS AND DISCUSSION

3.1 Trace Minerals Improve the Genetic Stability of CHA0

The quality of CHA0 inoculant is compromised by the accumulation of spontaneous mutants with defective *gacA* or *apdA* genes which regulate the production of antimicrobial compounds essential for optimal biocontrol activity. In standard medium, cultures inoculated with pure wild-type CHA0 (having no detectable mutants) became contaminated with as much as 1.1 % spontaneous mutants after 12 days incubation (Table 1). Medium amendment with 1 mM $CuSO_4$ reduced the accumulation of mutants to about 0.4 % without having a significant effect on total bacterial growth (Table 1).

Table 1. Copper-sulfate amendment of inoculum production medium to reduce accumulation of *apdA/gacA* regulatory mutants[1].

Culture medium	Total bacterial growth (log_{10} CFU per ml)	Percentage mutants
Nutrient broth medium	9.2	1.11
Nutrient broth plus 1 mM $CuSO_4$	9.1	0.46 *

[1]Cultures of 0.8% nutrient broth were inoculated with wild-type CHA0 and grown 12 days at 27 C and 140 rpm. Values represent the means of four trials. Each treatment was replicated 10 times. To determine percentage mutants, approximately 2 x 10^3 colonies were observed per replicate. Asterisk indicates means were significantly different according to a Studentized *t*-test (P=0.05).

Results from preliminary work where mutant accumulation was directly related to increasing incubation time up to 12 days (B. K. Duffy, *unpublished data*), suggested that mutant accumulation was due to favorable selective pressure in media rather than adaptive mutation or a mutational "hot-spot". Indeed when mixed cultures having a low but detectable level of mutants were used to seed fresh media, the mutants were more competitive and substantially outnumbered wild-type cells after 24 h incubation (Fig. 1).

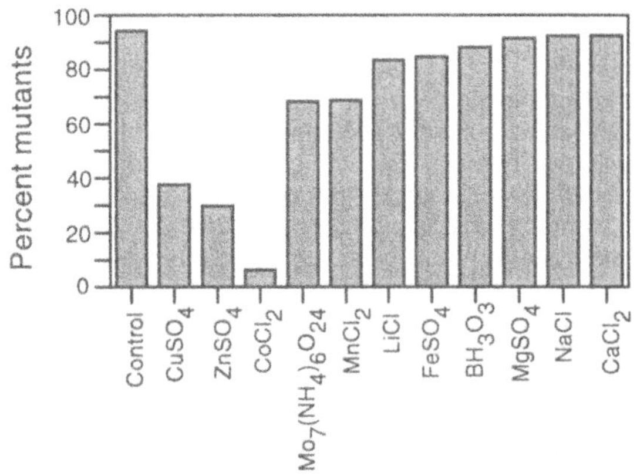

Figure 1. Trace-mineral amendments increase the competitiveness of wild-type CHA0 by reducing the accumulation of regulatory mutants during inoculum production. Asterisks indicate treatments with a significant reduction of mutants accumulation.

Amendment of media with Co^{2+}, Zn^{2+}, Cu^{2+}, Mo^{2+}, or Mn^{2+}, however, greatly improved the competitiveness of the wild-type reducing the overall percentage of mutants (Fig. 1). Clearly, the beneficial effect of these minerals is due to a positive influence on wild-type competition with mutants. Mutants were not more sensitive to inhibition by these minerals than the wild-type (B. K. Duffy, *unpublished data*). Possible explanations which remain to be investigated are that certain minerals alter membrane integrity or enzymatic functions of mutant cells, thereby limiting nutrient uptake or increasing sensitivity to wild-type metabolites.

3.2 Trace-minerals increase the production of antimicrobial metabolites by CHA0

Nutrient broth (0.8%) supported moderate Plt production in liquid medium, but little or no Phl and Sal (Table 2). Medium amendment with $ZnSO_4 \times 7H_2O$ yielded 36 times more Phl and 5 times more Plt, but had no effect on Sal production compared to the nonamended control. Amendment with $Mo_7(NH_4)_6O_{24} \times 4H_2O$ yielded 132 times more Phl and increased Sal production from 0 to 492 nM per 10^8 CFU, but had no effect on Plt production. CoCl increased Plt production 17 fold. Amendment with $MgSO_4 \times 7H_2O$ or $MnCl_2 \times 4H_2O$ increased Sal to 224.6 and 133.3 nM per 10^8 CFU, respectively. Metabolite production was not enhanced by the other six mineral amendments.

Table 2. Influence of trace mineral amendments on metabolite production[1].

Mineral	Metabolite yield (nM per 10^8 CFU) 2,4-diacetylphloroglucinol (Phl)	Pyoluteorin (Plt)	Salicylic acid (Sal)
None	0.9 c	249.3 c	0.0 c
$ZnSO_4$	34.3 b	1218.7 b	NS
$Mo_7(NH_4)_6O_{24}$	125.7 a	NS	492.0 a
$MgSO_4$	NS	NS	224.6 b
CoCl	NS	4479.1 a	NS
$MnCl_2$	NS	NS	133.3 bc

[1]Cultures of 0.8% nutrient broth with minerals were grown 48 h, extracted and analyzed with HPLC (see Material and Methods). Only those minerals with a significant influence on metabolite production are presented. Yield was expressed relative to bacterial growth. NS= not significantly different from the control according Fisher's LSD test.

Production of Phl was positively correlated with increasing concentration of $ZnSO_4 \times 7H_2O$ with 2 mM supporting the greatest yield (Fig. 2). However, total bacterial growth was substantially reduced at concentrations above 1.5 mM (*data not shown*). There was little growth inhibition at 1 mM or less compared with the nonamended control and these moderate concentrations would be more compatible with inoculum production where optimal biomass yield is a primary goal. Production of Plt was also stimulated by increasing Zn^{2+} concentration, although the effect was less than that observed with Phl (Fig. 2). Combining Zn^{2+} with certain carbon sources further enhanced antibiotic production by CHA0 and at the same time increased bacterial growth. For example, combination of Zn^{2+} plus 1% glycerol yielded 10 246.7 nM per 10^8 CFU Phl which was 72.2 fold more than glycerol alone and 298.9 fold more than Zn^{2+} alone; bacterial growth was increased 10 fold compared with the mineral alone (Duffy and Défago, 1996). In other systems, combination with carbon sources has been shown to enhance the beneficial effects of Zn^{2+} on production of phenazine antibiotics, bacterial cell survival, and protein protection (Crowe et al. 1987; Slininger *et al.*, 1996).

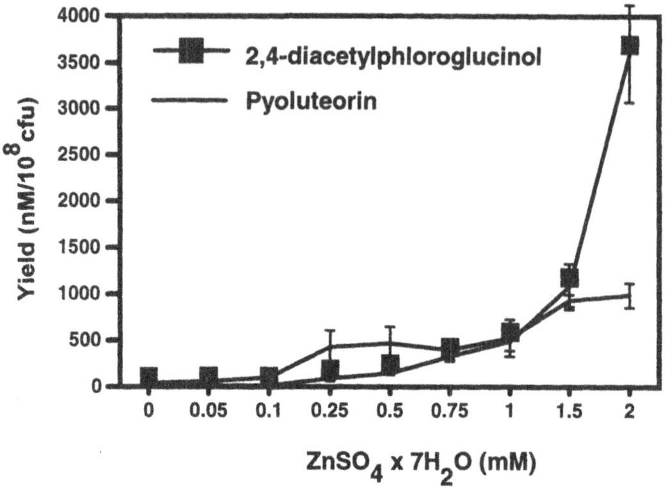

Figure 2. Positive relationship between antibiotic biosynthesis and increasing concentration of $ZnSO_4$ x $7H_2O$. Antibiotic yield was determined with HPLC and expressed relative to bacterial growth. Vertical bars represent standard error of six replicate broths from two experiments with similar results.

Our results with CHA0 are in accordance with studies using another model biocontrol strain, *P. fluorescens* 2-79 isolated from wheat grown in a take-all suppressive soil from Washington State, USA. Production of phenazine antibiotics is the primary biocontrol mechanism in this strain (Weller and Thomashow, 1994). As with many strains, the level of disease control varies considerably among soils even of similar texture and cropping history. This variability has been positively correlated with certain soil physical and chemical parameters. Biocontrol was positively correlated with zinc, ammonium-nitrogen, sulphate-sulphur, sodium, soil pH, and percentage sand; and negatively correlated with iron, manganese, total carbon, total nitrogen, cation exchange capacity, exchangeable acidity, and % organic matter (Ownley *et al.*, 1991). Adding Zn-EDTA (50 µg per g) to soil improved the performance of 2-79, probably because zinc increased production of phenazine antibiotic (Slininger and Jackson, 1992). Production of antimicrobial compounds, particularly phloroglucinol and phenazine antibiotics and HCN, has been determined to be the primary mechanism of biocontrol in most well characterized bacterial systems. Biosynthesis of antibiotic compounds and other secondary metabolites by biocontrol strains appears to be cell density dependent and involves coregulation by *gacA/apdA* and genes encoding autoinducer compounds (Chancey and Pierson, 1995; Pierson and Pierson, 1996). This means that plants may be largely unprotected until bacterial populations reach a sufficient density. Generally, a threshold population of 10^5 cfu per g root seems to be required for optimal autoinduction, production of important antimicrobial compounds, and effective biocontrol (Pierson and Pierson, 1996; Weller and Thomashow, 1996). This may have little impact on control of diseases like take-all of wheat which remains a threat throughout the growing season or for control of tobacco black root rot which has an epidemiologically critical secondary disease cycle. However, biocontrol of diseases with a rapid onset, such as Pythium damping-off, may greatly benefit from stimulating antibiotic production earlier or by providing a sufficient concentration of the active antimicrobial compound(s) with inoculum. In contrast, high antibiotic concentrations in inoculum may occasionally have a phytotoxic effect on bacterized planting material and under conditions that favor phytotoxicity it would be desirable to utilize media that limit accumulation of potentially deleterious compounds like phenazines, Phl, and Plt (Maurhofer *et al.*, 1992; Slininger *et al.*, 1996).

While the exact role of Zn^{2+} in biocontrol remains uncertain, it is an essential nutrient for bacteria, fungi, plants, and animals, it is known to be a component or catalyst of over 300 hundred enzymes and other proteins (Berg and Shi, 1996; Vallee and Auld, 1990), to regulate gene expression in the human pathogen *Pseudomonas aeruginosa* (Brumlik and Darzins, 1992; Olson and Ohman, 1992), to stabilize bacterial cell membranes and walls (Failla, 1977), and it is critical in many secondary metabolic processes (Weinberg, 1977). Amendment of growth media with Zn^{2+} increases production of siderophores, including pyoverdine and pyochelin, by *Azotobacter vinelandii* and *P. aeruginosa* presumably by inhibiting ferric reductase which decreases the ratio of Fe^{2+} to Fe^{3+} in the cell and leads to a sense of iron limitation (Höfte *et al.*, 1993; Huyer and Page, 1988; Huyer and Page, 1989). In addition to Sal, both Phl and Plt antibiotics of CHA0 have metal chelating activity and thus may be produced in response to nutrient limitation. Along similar lines, it may be that adverse environmental conditions (*i.e.*, presence of toxic metal ions) trigger production of certain compounds by CHA0 or physiological changes as a protective strategy against potential damage (Appanna and St. Pierre, 1996; Heipieper *et al.*, 1996; Kidambi *et al.*, 1995; Singh *et al.*, 1992). Knowing that $Zn^{2+}$and other minerals can also exert a negative influence on certain metabolic and physiological processes in pseudomonads (Makhzoum et al., 1995), the possibility cannot be discounted that positive effects of a mineral on the production of one metabolite could be related to the inhibition of another compound(s) or the disruption of critical enzyme systems.

3.3 Trace-minerals enhance the level of biocontrol activity of CHA0 against tomato root disease in hydroponic culture

Amendment with Zn-EDTA increased the level of disease suppression with CHA0 from approximately 22 to 51% (Table 3). Although the exact mechanism of action for Zn^{2+} in this system is not known, other studies in our laboratory suggest that Zn^{2+} works by reducing production of the phytotoxin fusaric acid by the pathogen (Duffy and Défago, 1997). While our finding that Zn-EDTA alone had no effect on Fusarium crown and root rot indicates that fusaric acid has only a limited role as a pathogenicity factor, we suggest fusaric acid may have ecological importance to the pathogen in interactions with antagonistic bacteria. We found that this phytotoxin repressed antibiotic production by CHA0 which reduces antagonism by this biocontrol agent (B. Duffy, *unpublished data*).

Table 3. Increased biocontrol activity of CHA0 against Fusarium crown and root rot of tomato in hydroponic culture[1].

	Disease severity (0 to 4)		
Mineral	FORL	FORL plus CHA0	Disease reduction
None	2.9 a	2.5 a	0.4 a
Zn-EDTA	3.2 a	2.2 b	1.0 b
$(NH_4)_6Mo_7O_{24}$	3.3 a	2.6 a	0.7 a
Cu-EDTA	2.5 b	1.3 d	1.2 d
Mineral mix	2.3 b	1.7 c	0.6 c

[1]Disease severity was assessed using a standard scale of 0 to 4 where 0 indicates no symptoms and 4 indicates plants were dead or nearly so. Biocontrol activity was evaluated as the reduction in disease severity rating from the control with only the pathogen (disease reduction = FORL - FORL plus CHA0).

Therefore, Zn^{2+} indirectly increased the level of antibiotic produced by CHA0 by repressing production of fusaric acid by the pathogen, and thereby increased its biocontrol activity. Recently, we confirmed this hypothesis by extracting metabolites from the rockwool assay. We found that while zinc had no direct effect on Phl production by CHA0, when it was added to a nutrient solution fusaric acid was repressed and Phl was produced (Duffy and Défago, 1997). In the plant assay, Zn^{2+} concentrations were approximately half that used in vitro, to avoid problems with phytotoxicity.

Interestingly, Mo^{2+} did not effect phytotoxin production (Duffy and Défago, 1997) nor did it improve biocontrol with CHA0. Amendment with Cu-EDTA or a mixture containing Cu-EDTA reduced disease when used alone and had an additive effect on biocontrol when used with CHA0 (Table 3). This was not surprising considering the well established fungitoxic activity of copper. Copper, though, also reduced production of fusaric acid by the pathogen albeit to a lesser extent than Zn-EDTA (Duffy and Défago, 1997), and this may have also contributed to its beneficial effect on biocontrol.

4. CONCLUSIONS AND PROSPECTIVE AGRICULTURAL APPLICATIONS

Two important agricultural applications that could potentially develop from our studies with minerals and bacterial biocontrol agents are (i) the use of mineral amendments to improve activity and (ii) selection of strains based on mineral content of target soils. Minerals could be incorporated at any stage during inoculant production to promote efficacy after application. During growth of the bacteria, minerals could limit accumulation of ineffective mutants and increase the concentration of antimicrobial compounds in the inoculant to protect plants until *in situ* production begins. Amendments to the inoculant formulation after bacterial growth or coapplication of minerals and inoculant to planting material may enhance bacterial growth, colonization, and beneficial activity *in situ*. However, consideration must be given to how minerals are applied since Zn^{2+} and other minerals can also have detrimental effects reducing the growth and activity of other beneficial microbes (Biro *et al.*, 1995; McIlveen *et al.*, 1975; Page *et al.*, 1996), increasing certain diseases, particularly those caused by *Fusarium* spp. (Jones *et al.*, 1989), and can be phytotoxic. Application methods that selectively provide minerals to the biocontrol agent will avoid these nontarget effects and more efficiently supply minerals to the bacteria.

Second, by identifying minerals that increase biocontrol activity *in vitro* and in the greenhouse we can better determine conditions that favor biocontrol. This should enable us to customize biocontrol treatments by selecting strains most likely to provide an economic level of control at specific locations. Understanding how introduced organisms react to soil amendments may indicate manipulations to cropping practices that will favor entire populations of indigenous beneficial organisms and lead to an even more sustainable level of disease suppression.

5. REFERENCES

Appanna, V. D., and M. St. Pierre, 1996. Aluminum elicits exocellular phosphatidylethanolamine production in *Pseudomonas fluorescens*. Appl. Environ. Microbiol., 62, 2778-2782.

Berg, J. M., and Y. Shi, 1996. The galvanization of biology: a growing appreciation for the roles of zinc. Science, 271, 1081-1085.

Biro, B., H. E. A. F. Bayoumi, S. Balazsy, and M. Kecskes, 1995. Metal sensitivity of some symbiotic N_2-fixing bacteria and *Pseudomonas* strains. Acta Biol. Hungar., 46, 9-16.

Brumlik, M. J., and A. Darzins, 1992. Zinc and iron regulate translation of the gene encoding *Pseudomonas aeruginosa* elastase. Mol. Microbiol., 6, 334-337.

Chancey, S. T., and L. S. Pierson, III, 1995. Phenazine gene expression in *Pseudomonas aureofaciens* 30-84 is regulated in part by a *gacA* homologue. Phytopathology, 85, 1186 (abstract).

Cook, R. J., 1993. Making greater use of introduced microorganisms for biological control of plant pathogens. Annu. Rev. Phytopathol., 31, 53-80.

Corbell, N., and J. E. Loper, 1995. A global regulator of secondary metabolite production in *Pseudomonas fluorescens* Pf-5. J. Bacteriol., 177, 6230-6236.

Crowe, J. H., L. M. Crowe, J. F. Carpenter, and C. A. Wistrom, 1987. Stabilization of dry phospholipid bilayers and proteins by sugars. Biochem. J., 242, 1-10.

Duffy, B. K., and G. Défago. 1997. Effect of trace-minerals on biocontrol of *Fusarium oxysporum* f.sp. *radicis-lycopersici* on tomato using *Pseudomonas fluorescens* CHA0. Phytopathology, 87:1250-1257.

Duffy, B. K., and G. Défago. 1996. Influence of minerals, C-source, and pH on antibiotic and salicylate production by *Pseudomonas fluorescens* biocontrol strain CHA0. Phytopathology, 86:S79.

Duffy, B. K., B. H. Ownley, and D. M. Weller, 1997. Soil chemical and physical properties associated with suppression of take-all of wheat by *Trichoderma koningii*. Phytopathology, 87, 1118-1124.

Elmer, W. H., 1995. Association between Mn-reducing root bacteria and NaCl applications in suppression of Fusarium crown and root rot of asparagus. Phytopathology, 85, 1461-1467.

Engelhard, A. W. (Editor), 1989. Soilborne Plant Pathogens: Management of Diseases with Macro- and Microelements. APS Press, St. Paul, MN.

Failla, M. L., 1977. Zinc: functions and transport in microorganisms. In: E. D. Weinberg (Editor). Microorganisms and Minerals. Marcel Dekker, New York, pp. 151-214.

Hartman, J. R., and J. T. Fletcher, 1991. Fusarium crown and root rot of tomatoes in the UK. Plant Pathol., 40, 85-92.

Heipieper, H. J., G. Meulenbeld, Q. van Oirschot, and J. A. M. de Bont, 1996. Effect of environmental factors on the *trans/cis* ratio of unsaturated fatty acids in *Pseudomonas putida* S12. Appl. Environ. Microbiol., 62, 2773-2777.

Henis, Y., and J. Katan, 1975. Effect of inorganic amendments and soil reaction on soil-borne plant diseases. In: G. W. Bruehl (Editor). Biology and Control of Soil-borne Plant Pathogens. APS Press, St. Paul, MN, pp. 100-106.

Höfte, M., S. Buysens, N. Koedam, and P. Cornelis, 1993. Zinc affects siderophore-mediated high affinity iron uptake systems in the rhizosphere *Pseudomonas aeruginosa* 7NSK2. Biometals, 6, 85-91.

Huyer, M., and W. J. Page, 1988. Zn^{2+} increases siderophore production in *Azotobacter vinelandii*. Appl. Environ. Microbiol., 54, 2625-2631.

Huyer, M., and W. J. Page, 1989. Ferric reductase activity in *Azotobacter vinelandii* and its inhibition by Zn^{2+}. J. Bacteriol., 171, 4031-4037.

Kidambi, S. P., G. W. Sundin, D. A. Palmer, A. M. Chakrabarty, and C. L. Bender, 1995. Copper as a signal for alginate synthesis in *Pseudomonas syringae* pv. *syringae*. Appl. Environ. Microbiol., 61, 2172-2179.

Jones, J. P., A. W. Engelhard, and S. S. Woltz, 1989. Management of Fusarium wilt of vegetables and ornamentals by macro- and microelement nutrition. In: A. W. Engelhard (Editor). Soilborne Plant Pathogens: Management of Diseases with Macro- and Microelements. APS Press, St. Paul, MN, pp. 18-32.

Laville, J., C. Voisard, C. Keel, M. Maurhofer, G. Défago, and D. Haas, 1992. Global control in *Pseudomonas fluorescens* mediating antibiotic synthesis and suppression of black root rot of tobacco. Proc. Natl. Acad. Sci. USA, 89, 1562-1566.

Makhzoum, A., J. S. Knapp, and R. K. Owusu, 1995. Factors affecting growth and extracellular lipase production by *Pseudomonas fluorescens* 2D. Food Microbiol. 12, 277-290.

Maurhofer, M., C. Keel, U. Schnider, C. Voisard, D. Haas, and G. Défago, 1992. Influence of enhanced antibiotic production in *Pseudomonas fluorescens* strain CHA0 on its disease suppressive capacity. Phytopathology, 82, 190-195.

Mazzola, M., D. K. Fujimoto, L. S. Thomashow, and R. J. Cook, 1995. Variation in sensitivity of *Gaeumannomyces graminis* to antibiotics produced by fluorescent *Pseudomonas* spp. and effect on biological control of take-all of wheat. Appl. Environ. Microbiol., 61, 2554-2559.

McIlveen, W. D., R. A. Spotts, and D. D. Davis, 1975. The influence of soil zinc on nodulation, mycorrhizae, and ozone-sensitivity of pinto bean. Phytopathology, 65, 645-647.

Miranda, C. H. B., N. F. Seiffert, and J. Döbereiner, 1985. Efeito de aplicacão de molibdênio no número de *Azospirillum* e na producão de *Brachiaria decumbens*. Pesq. Agropec. Bras. 20, 509-513.

Olson, J. C., and D. E. Ohman, 1992. Efficient production and processing of elastase and LasA by *Pseudomonas aeruginosa* require zinc and calcium ions. J. Bacteriol., 174, 4140-4147.

Ownley, B. H., D. M. Weller, and J. R. Alldredge, 1991. Relation of soil chemical and physical factors with suppression of take-all by *Pseudomonas fluorescens* 2-79. In: C. Keel, B. Koller, and G. Défago (Editors). Plant Growth-Promoting Rhizobacteria - Progress and Prospects. IOBC/WPRS Bull. Vol. 14, pp. 299-301.

Page, W. J., J. Manchak, and M. Yohemas, 1996. Inhibition of *Azotobacter salinestris* growth by zinc under iron-limited conditions. Can. J. Microbiol., 42, 655-661.

Pierson, L. S., III, and E. A. Pierson, 1996. Phenazine antibiotic production in *Pseudomonas aureofaciens*: role in rhizosphere ecology and pathogen suppression. FEMS Microbiol. Lett., 136, 101-108.

Rengel, Z., J. F. Pedler, and R. D. Graham, 1994. Control of Mn status in plants and rhizosphere: genetic aspects of host and pathogen effects in the wheat take-all interaction. In: J. A. Manthey, D. E. Crowley, and D. G. Luster (Editors). Biochemistry of Metal Micronutrients in the Rhizosphere. Lewis Publishers, Boca Raton, pp. 125-145.

Schneider, R. W., 1985. Suppression of Fusarium yellows of celery with potassium, chloride, and nitrate. Phytopathology, 75, 40-48.

Singh, S., B. Koehler, and W. Fett, 1992. Effect of osmolarity and dehydration on alginate production by fluorescent pseudomonads. Curr. Microbiol., 25, 335.

Slininger, P. J., and M. A. Jackson, 1992. Nutritional factors regulating growth and accumulation of phenazine 1-carboxylic acid by *Pseudomonas fluorescens* 2-79. Appl. Microbiol. Biotechnol., 47, 388-392.

Slininger, P. J., J. E. van Cauwenberge, R. J. Bothast, D. M. Weller, L. S. Thomashow, and R. J. Cook, 1996. Effect of growth culture physiological state, metabolites, and formulation on the viability, phytotoxicity, and efficacy of the take-all biocontrol agent *Pseudomonas fluorescens* 2-79 stored encapsulated on wheat seeds. Appl. Microbiol. Biotechnol., 45, 391-398.

Smiley, R. W., 1978. Antagonists of *Gaeumannomyces graminis* from the rhizoplane of wheat in soils fertilized with ammonium- or nitrate-nitrogen. Soil Biol. Biochem., 10, 169-174.

Vallee, B. L., and D. S. Auld, 1990. Biochemistry 29, 5647-5659.

Voisard, C., C. T. Bull, C. Keel, J. Laville, M. Maurhofer, U. Schnider, G. Défago, and D. Haas, 1994. Biocontrol of root diseases by *Pseudomonas fluorescens* CHA0: current concepts and experimental approaches. In: F. O'Gara, D. N. Dowling, and B. Boesten (Editors). Molecular Ecology of Rhizosphere Microorganisms: Biotechnology and the Release of GMO's. VCH, Weinheim, Germany, pp. 67-89.

Weinberg, E. D., 1977. Mineral element control of microbial secondary metabolism. In: E. D. Weinberg (Editor). Microorganisms and Minerals. Marcel Dekker, New York, pp. 289-316.

Weller, D. M., 1988. Biological control of soilborne plant pathogens in the rhizosphere with bacteria. Annu. Rev. Phytopathol. 26, 379-407.

Weller, D. M., and L. S. Thomashow, 1994. Current challenges in introducing beneficial microorganisms into the rhizosphere. In: F. O'Gara, D. N. Dowling, and B. Boesten (Editors). Molecular Ecology of Rhizosphere Microorganisms: Biotechnology and the Release of GMO's. VCH, Weinheim, Germany, pp. 1-18.

EFFECTS OF MECHANICAL STRESSES AND STRAINS ON SOIL RESPIRATION

C.W. Watts,[1] P.D. Hallett,[2] and A.R. Dexter.[1]

1. Silsoe Research Institute, Wrest Park, Silsoe, Bedford, MK45 4HS, UK
   Tel: (01525) 860000, Fax: (01525) 860156, e-mail: chris.watts@bbsrc.ac.uk
2. Scottish Crop Research Institute, Invergowrie, Dundee, DD2 5DA, UK

1. INTRODUCTION

Mechanical stresses and strains are imposed on soils by natural processes, such as wetting and drying cycles, and also by anthropogenic actions such as tillage and traction (Dexter, 1988). If the imposed stress exceeds a critical value, the pore structure of the soil changes (Mitchell, 1993). In a general context applicable to agricultural activities, compressive stresses lead to a bulk reduction in porosity, whereas shear and tensile stresses under certain conditions increase porosity. Since the pore space is also the habitat for soil micro-organisms, changes in pore properties induced by mechanical stresses and strains can have a marked effect on microbial activity. Numerous field studies have been conducted to evaluate the change in microbial activity resulting from the application of mechanical energy in the form of different tillage systems (Carter, 1992; Chan *et al.*, 1992; Beare *et al.*, 1994; Franzluebbers *et al.*, 1994). An increase in tillage intensity usually results in an increase in microbial activity measured as evolved CO_2. It has been suggested in these studies that tillage opens up pores thus exposing previously physically protected organic matter to attack by organisms. Other changes likely to influence microbial activity following tillage include changes to soil climate, water status and aeration (Jenkinson *et al.*, 1992). The depletion of the carbon pool in soil through tillage has significant implications for soil quality and greenhouse gas emission (Kern and Johnson, 1993).

Soil compaction, on the other hand, often decreases microbial activity in soils resulting in a poorer environment for plants (Smucker and Erickson, 1989). Compaction reduces the size of the largest pores in the soil (Eriksson *et al.*, 1974). As a consequence, the volume of air-filled pore space, which is controlled by the matric potential and pore size, is reduced, causing a shift to more anaerobic activity. A practical implication of compaction is reduced decomposition of plant residues by micro-organisms.

In order to gain a better understanding of the relationship between the change in pore structure through mechanical stress inputs and microbial activity, there is a need for controlled laboratory experiments to complement the findings obtained from field investigations. Such investigations would allow soil mechanics theory to be applied to soil biology which is an area that has been neglected in the past. To address this gap between disciplines, an initial investigation was conducted in which controlled compressive and shear stresses were imposed on undisturbed soil cores collected from sites with a long history of different management practices. The stress-strain behaviour of the mechanical loading was recorded in order to assess the effect of organic matter on soil mechanical properties. After the load was applied, CO_2 output was monitored using an electro-conductimetric respirometer linked to a data logging system.

2. MATERIAL AND METHODS

2.1 Soils

Soil samples were collected from one of the long-term, ley-arable experiments at Rothamsted in the UK. The site, known as Highfield, was prior to 1948 under permanent grass. Following that date, a range of contrasting management practices have been applied to the different parts of the site, in particular, different grass ley-arable rotations (Johnston, 1972). Samples were collected from the upper 0-100 mm horizon of three of the treatments:

(A) Permanent fallow: This area is kept as a bare-fallow and cultivated two or three times per year to prevent weed growth. The soil organic carbon (SOC) measured on this plot was 11 g kg^{-1}.

(B) Ley-arable rotation: This area is kept in a 3-year grass/clover, 3-year cereal rotation. Samples were collected during the second year of the cereal cycle. The SOC of this site was 21 g kg^{-1}.

(C) Permanent grass: This is an unbroken continuation of the original permanent pasture on this site. SOC was 32 g kg^{-1}.

Table 1. Physical and chemical characteristics of the soils on the Highfield site

	Treatment A	Treatment B	Treatment C
Plot number	P.F.	15/16	23/24
Cropping system	Fallow	Ley-arable rotn.	Permanent grass
*Particle size analysis, g kg^{-1}			
Sand (2 - 0.05 mm)	90	110	110
Silt (50 - 2 mm)	660	640	660
Clay (<2 mm)	250	250	230
* SOC, g kg^{-1}	11	21	32
Water content, g kg^{-1} at -50 hPa water potential	161	213	241
Dry bulk density, Mg m^{-3}	1.42	1.13	1.04

*Technique as described in British Standard 1377 (1975).

The soil on this site is classified as the Batcombe series (Clayden and Hollis, 1984) being a silt loam over clayey drift with siliceous stones. This soil series is approximately equivalent to Chromic Luvisols and Orthic Acrisols as classified by the F.A.O. soil classification system (Avery, 1980). Some important physical and chemical parameters are listed in Table 1.

2.2 Sample Collection and Preparation

Samples were collected in March 1996 using double, polypropylene rings of 52 mm inside diameter and 19 mm height. Pairs of these rings were taped together to form sampling cylinders 38 mm high. For samples requiring subsequent shearing, the tape was removed allowing the two halves to be displaced relative to each other. The moist condition of the site during sampling allowed the cylinders to be pushed directly into treatments A and B, but some trimming of the root mat with a sharp knife was necessary on treatment C. Approximately 40 samples were collected from each treatment.

The soil cores were transported to the laboratory and placed on a tension table at -50 hPa suction for 24 h. The samples were then slowly saturated before being returned to -50 hPa suction where they remained for 28 days. This procedure was designed to bring the samples to a similar water potential but allow sufficient time for any 'flush' of CO_2, due to change in water content, to pass (Anderson, 1982).

Equilibration to -50 hPa water potential, mechanical loading and respiration measurements were all made at a constant temperature of 25°C.

2.3 Application of Mechanical Stress to Soil Samples

The experimental procedure is outlined in Figure 1. Throughout the entire experiment the samples remained in their sampling cylinders, with the only disturbance being the stress applied deliberately during mechanical testing. Since the matric potential of the soil influences both the strength and microbial activity, each sample was tested at -50 hPa, immediately following its removal from the tension table.

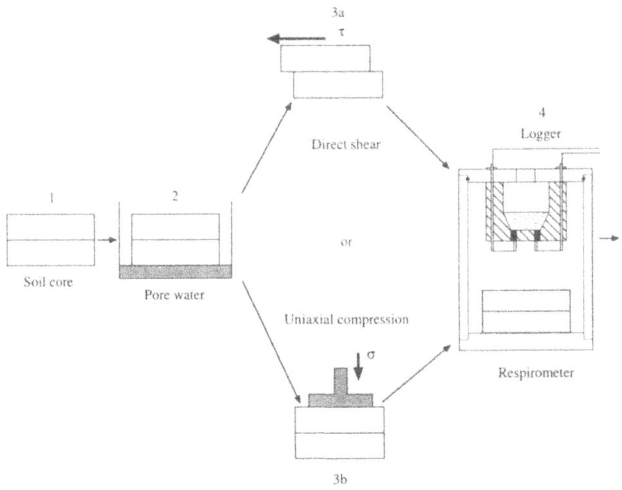

Figure 1. Illustration of the experimental procedure. Each soil sample remains in its original sampling tube throughout the experiment to minimise any unwanted additional disturbance.

In the direct shear test the two halves of the soil core are moved laterally with respect to each other (Koolan and Kuipers, 1983). The upper half is physically displaced, while the force required to maintain the lower half in its original position is measured using a 100 N load sensor. Shear stress, τ, is determined by dividing the shear force by the sample cross-sectional area. Shear displacement was applied using a conventional soil direct shear apparatus with a horizontal displacement rate of 0.02 mm s^{-1}. No normal stress was applied to the samples. The initial sample, from each soil management treatment, was loaded until fracture occurred, which is detectable as a drop in shear stress with displacement (Wood, 1990). From this data, nine different intensities of shear displacement were selected for subsequent tests on the remainder of the samples. Additional samples were not subjected to loading and were used to determine the basal respiration rate.

An identical set of samples was subjected to compressive loading. A vertical stress was applied to the upper face of each sample using a loading frame with a displacement rate of 0.07 mm s^{-1}. The resultant force was measured using a 2.5 kN load cell, and the applied stress, σ, was evaluated by dividing the applied load by the cross-sectional area of the core. Compressive strain, ε, was determined by:

$$\varepsilon = \frac{\Delta h}{h} \quad (1)$$

where h was the initial height of the sample, and Δh the change in sample height caused by loading. It is convenient to use ε since it normalises the data, thus removing any differences in the results caused by differences in initial porosity. One sample from each treatment was compressed until free water just started to appear around the piston. This was taken as the maximum compressive strain. Nine different strain values were then selected for each treatment for the subsequent tests.

2.4 Measurement of Soil Respiration

Following the application of either shear or compressive stresses, each sample was placed in an electro-conductimtric respirometer. These respirometers, based on a design by Chapman (1971) and subsequently updated by Watts *et al.* (1998), consisted of a perspex cylinder and base with a diameter of 75 mm and height of 155 mm. An airtight lid contained a conductivity cell into which an alkali solution (KOH) was injected, which reacts with CO_2 produced by the soil in the cylinder and converts the hydroxide into carbonate.

$$2KOH + CO_2 \leftrightarrow K_2CO_2 + H_2O \quad (2)$$

The ionic mobility of carbonate ions is 0.368 that of hydroxide ions. Therefore the reaction between CO_2 and KOH increased the cell resistance.

Cell conductivity was measured with platinum electrodes connected to a measuring circuit with an AC excitation voltage V_E (2.0 V, peak to peak, 125 Hz square wave) and a standard resistor R_S (10 kΩ) provided by a Delta-T DL2 logger. The cell resistance, R_X was determined as follows:

$$R_X = \frac{V_X . R_S}{V_E - V_X} \quad (3)$$

where V_X was the voltage measured across the respirometer. Twelve respirometers were coupled to the logger and readings of V_X were taken every 10 minutes for a periods of up to 48 hours.

Prior to the experiment, each respirometer was calibrated using successive known volumes of standard (6%) CO_2 gas. In this experiment, 5 ml of 0.05 M KOH solution was used which allowed up to 5000 μg of CO_2 to be adsorbed during the linear phase of the calibration curve.

3. RESULTS AND DISCUSSION

3.1 Initial Soil Conditions and Basal Respiration

Table 1 shows important differences between the three experimental soils, resulting from almost 50 years of contrasting soil management practices. Treatment C, permanent grassland had the highest organic carbon, the highest water content (at -50 hPa) and the lowest bulk density. By contrast, treatment A, permanent follow, had the lowest organic carbon, the lowest water content but the highest bulk density. There were also important differences in soil structure between treatments. Treatment C was a well aggregated granular structure bound together by a large number of grass roots, while treatment A had much smaller aggregates, tightly packed, with the occasional large crack but no roots.

For all the treatments reported, respiration reached a steady state, following minor initial perturbations during the first hour or so, with a linear increase in the quantity of CO_2 trapped by the electrolyte with time. Respiration rates were taken as the slope of these curves over time periods ranging from 4 to 48 hours. These values are listed for the different treatments in table 2.

The basal respiration rate, R_b, varied markedly between samples from the different treatments. Mean values for treatments A, B and C were 45.9(±1.3), 157.7 (±11.9) and 486.3 (±17.4) μg CO_2 h^{-1}, respectively. Numbers in brackets represent standard errors.

The differences in basal respiration between treatments probably reflect previous mechanical perturbations (such as tillage) which have been shown in previous studies to cause drastic reductions in active microbial biomass. In particular, levels of fungi reduce dramatically following cultivation since hyphae strands and roots are broken by the mechanical input (Beare *et al.*, 1993; Tisdall and Oades, 1980). The available organic carbon was also lower following increased levels of mechanical perturbations associated with the different soil management practices (Table 1) thus providing less substrate for microbial activity (Martyniuk and Wagner, 1978).

3.2 Direct Shear Loading

For one sample from each treatment, shear displacement was applied until the sample fractured (Figure 2(b)). Other samples were loaded to different fractions of these maximum

Table 2. Respiration rates, *R*, (mg CO_2 h^{-1}) associated with: (a) direct shear for a given displacement x mm, and (b) compressive loading for a given compressive strain, ε.

(a) Direct Shear

Treatment A			Treatment B			Treatment C		
x	R	(se)	x	R	(se)	x	R	(se)
0	48.6*	(±0.26)	0	150*	(±1.28)	0	522*	(±14)
0	46.6*	(±0.41)	0	181*	(±0.60)	0	457*	(±20)
0.10	63.2	(±0.54)	0	142*	(±0.20)	0.20	772	(±18)
0.20	68.3	(±0.20)	0.10	173	(±0.67)	0.39	955	(±39)
0.29	62.0	(±0.29)	0.20	212	(±1.07)	0.78	932	(±23)
0.39	65.0	(±0.79)	0.29	195	(±1.01)	1.17	906	(±21)
1.18	68.9	(±0.23)	0.39	206	(±1.01)	1.76	933	(±16)
1.57	67.1	(±0.32)	0.49	216	(±0.79)	2.51	1049	(±24)
2.75	78.5	(±0.26)	1.18	219	(±1.76)	2.75	1047	(±16)
3.53	79.6	(±0.91)	1.76	226	(±0.95)	4.00	1045	(±39)
			2.35	264	(±1.64)	7.76	1199	(±26)
			2.94	251	(±1.02)	9.50	1108	(±24)

(b) Compressive Loading

Treatment A			Treatment B			Treatment C		
ε	R	(se)	ε	R	(se)	ε	R	(se)
0	42.4*	(±0.15)	0	150*	(±0.28)	0	456*	(±13)
0	46.0*	(±0.10)	0	181*	(±0.60)	0	510*	(±25)
0.023	49.2	(±0.47)	0	142*	(±4.00)	0.009	510	(±24)
0.029	55.8	(±0.38)	0.037	302	(±1.30)	0.019	629	(±26)
0.046	65.7	(±0.44)	0.103	263	(±2.00)	0.037	577	(±11)
0.084	56.7	(±0.13)	0.115	208	(±1.20)	0.056	567	(±23)
0.116	58.7	(±0.19)	0.153	185	(±2.50)	0.084	844	(±17)
0.149	64.4	(±0.16)	0.184	242	(±1.90)	0.111	826	(±34)
0.171	58.8	(±0.19)	0.211	147	(±0.80)	0.149	618	(±19)
0.174	36.5	(±0.08)	0.264	143	(±1.60)	0.186	687	(±10)
0.219	36.1	(±0.55)	0.267	154	(±1.10)	0.279	862	(±10)
0.226	36.6	(±0.10)	0.326	163	(±5.50)	0.371	400	(±46)

* Samples receiving no loading represent base respiration R_b
(se) standard error of preceding coefficient

strains. Treatments A and B fractured at similar displacements (Table 2(a)). In treatment C, biological bridging agents such as roots and fungal hyphae are probably more numerous owing to no previous mechanical disturbance (Tisdall and Oades, 1980). These binding agents need to be broken before fracture could occur resulting in a large increase in shear displacement at fracture. In Figure 2(b), increasing shear displacement is shown to cause increasing shear stress. The shaded zone represents the approximate region of shear yielding, indicated by a lower rise in shear stress with shear displacement. In this region there is a transition from the primary elastic process (energy imparted to break inter-particle bonds) to a primary plastic process (energy expended to rearrange particles). The fallow soil, treatment A, appeared to yield first, and at the lowest displacement. This was because: (a) it was more plastic, (b) it had a higher density, resulting in a greater interparticle transmission of stress (Mitchell, 1993) and (c) lower organic matter levels which contribute to elastic energy dissipation. The grassland soil failed at the highest displacement for the opposite reasons. The stress level was lower, however, because of its aggregated structure, which caused the aggregates to roll over each other during shear, rather than interparticle rupture. Yielding is also dependant on mineral bonds which for these soils are very similar.

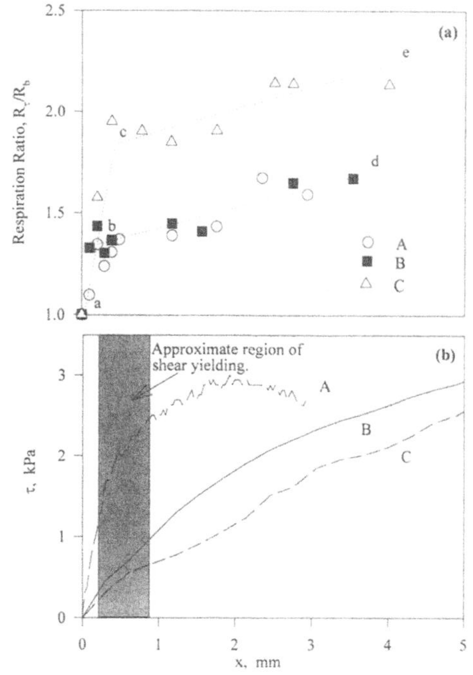

Figure 2. The effect of shear displacement, x on: (a) respiration ratio, R_τ/R_b : (b) shear stress, τ. the shaded area represents the approximate area of shear yielding. Treatment A = permanent fallow (SOC, 11g kg^{-1}), B = ley-arable rotation (SOC, 21g kg^{-1}) and C = permanent grassland (SOC, 32g kg^{-1}).

Respiration ratio was determined by dividing respiration values recorded following the application of stress (R_τ or R_σ), listed in table 2., by basal respiration (R_b) measured on undisturbed cores. Plotting these values against shear displacement, x, (Figure 2(a)), two distinct regions were evident for each of the three management practices. There was an initial steep rise in respiration, which coincides approximately with the region of elastic deformation, followed by a lower rate of increase in respiration ratio associated with plastic deformation. We were unable to identify any treatment difference in the rate of increase in respiration ratio in the elastic region, identified by a line of best fit, a,b,c. The two previously cultivated soils, treatments A and B, respiration ratios increased sharply, to around 1.4 during the elastic phase (a to b) and then continued more slowly to 1.6 during plastic deformation (b to d). However, for the uncultivated grassland samples, respiration ratio rose to 1.9 during elastic shearing (a to c) before increasing more slowly to around 2.1 during the plastic phase (c to e).

Under direct shear loading at low values of normal stress, new pore space is created in the shear zone as the sample dilates (Mitchell, 1993). Increases in pore space have been hypothesised to cause an increase in respiration, because of the exposure of previously protected organic materials to microorganisms (Rovira & Greacen, 1957) and the creation of more aerobic pores spaces. Mitchell (1993), reviewed research which showed that the rate of increase in new pore space with increasing shear displacement diminishes after yielding, which were reflected approximately, by the changes in the respiration ratio.

3.3 Compressive Loading

One sample from each treatment was compressed until free water just appeared around the piston. The stress-strain curves of these three samples are shown in Figure 3(b). During subsequent tests, the remaining samples were compressed to fractions of these maximum strain values (Table 2(b)) Estimates of overall and air-filled porosity, based on the sample initial and final volumes (at maximum compressive strain) are listed (Table 3), for each of the three treatments.

Table 3. The effect of maximum compression on total and air-filled porosity.

	Treatment A	Treatment B	Treatment C
Initial porosity	0.46	0.57	0.61
Initial air-filled porosity	0.23	0.33	0.34
Final porosity	0.36	0.38	0.38
Final air-filled porosity	0.13	0.05	0.04

The three stress-strain curves in Figure 3(b) appear very similar to each other. This at first glance is rather surprising. During crushing, three factors influence the soils behaviour: (a) large pores close first at a lower stress than smaller pores (Davis *et al.*, 1973), (b) the greater the initial porosity the lower the expected stress for a given strain and (c) as pores are closed or reduced in size, water is redistributed throughout the sample.
For treatment C, we would have expected factors (a) and (b) to result in a less steep rise in stress, for a given strain, because of its greater porosity compared with the other treatments, (Table 3). However, mechanical response to compression at low initial water potentials are

dominated by the pore water pressure as water is redistributed within the sample (Mitchell, 1993).

Figure 3(a) represents the effect of increasing compressive strain on the respiration ratio. Most respiration ratio values appeared greater than unity for strain values between 0 - 0.2. At strain values greater than 0.2, respiration values fell below one. To help illustrate the changing trend in respiration ratio with increased strain, data from each treatment was fitted using a second order regression, giving:

for treatment A, $$\frac{R_\sigma}{R_b} = 7.0\,\varepsilon - 3.62\,\varepsilon^2 + 1 \qquad r^2 = 0.651\,; \qquad (4)$$

for treatment B, $$\frac{R_\sigma}{R_b} = 5.4\,\varepsilon - 19.0\,\varepsilon^2 + 1 \qquad r^2 = 0267\,; \qquad (5)$$

and for treatment C, $$\frac{R_\sigma}{R_b} = 6.9\,\varepsilon - 19.1\,\varepsilon^2 + 1 \qquad r^2 = 588\,; \qquad (6)$$

(r^2 represents the coefficient of determination)

These equations give maximum respiration ratios for treatments A, B and C at strain values of 0.10, 0.14 and 0.18 respectively.

Soil compresses at the expense of larger pores first (Davis *et al.*, 1973; Eriksson *et al.*, 1974), which causes a change in pore size distributions. With increasing strain the soil becomes less aerobic as fewer pores remain air filled and gas diffusion is lowered. This interpretation of soil compression explains changes in respiration caused by high compressive strain values but not at the onset of loading (Figure 3(a)).

The increase in the respiration at the onset of loading can be explained by the effect of loading and unloading of the soil under compression. Soil has a resilience to compression which is dependant on packing and bond energy between particles and also on its previous stress history (Mitchell, 1993). A soil taken from under a tractor wheel way, for example, would require a higher stress to compress it than a soil taken from under pasture. The previous stress history defines the pre-consolidation stress which is characterised by the deflection point at which compressive strain increases at a greater rate with increasing stress.

If the compressive strain is removed before the pre-consolidation stress is exceeded, the soil will recover most of its volume, or rebound. Once the pre-consolidation stress is exceeded, the soil deforms plastically at the expense of the largest pores which are not recovered upon unloading. Since pores cause a highly anisotropic distribution of applied stress, some isolated zones of shear or tensile failure will result, even at very low applied stresses. Some macro-cracks may be formed under tension from pores orientated perpendicular to the applied compressive stresses. The compressive stress also causes the pore water to be redistributed and forced out of zones of interparticle contact. Upon unloading, this water becomes reabsorbed (Day and Holmgren, 1952).

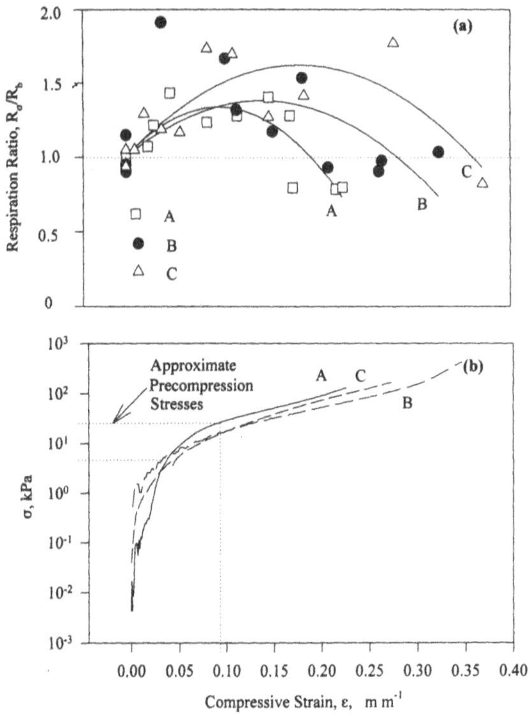

Figure 3. The effects of compressive strain, ε on: (a) respiration ratio, R_σ/R_b; (b) compressive stresses, τ. Treatment A = permanent fallow (SOC, 11g kg^{-1}), B = ley- arable rotation (SOC, 21g kg^{-1}) and C = permanent grassland (SOC, 32g kg^{-1}).

This movement of pore water within the sample will redistribute both soil organisms and easily decomposable substrates like amino acids which may have been trapped as integrangular pore water.

Gas diffusion is reduced most dramatically following the destruction of continuous pore space by compression, which Davis *et al.*, (1973) found occurred at an air filled porosity of around 0.15. For all three treatments, this is similar to values at peak respiration, reported above, beyond which their respiration ratios declined .

In summary, compression has a three-fold effect on respiration: (a) some new pore space is created by shear and tensile stresses which may expose previously - protected organic material, (b) the pore water becomes redistributed which may provide a flush of organic material, and (c) after the pre-consolidation stress is exceeded, the air filled porosity and hence gas diffusion is diminished.

4. CONCLUSIONS

Using soil of the same origin but with a relatively large, management-induced difference in organic carbon, the following points were noted:

(1) The basal respiration of soils from under contrasting soil management practices, varied by a factor greater than 10, at a water potential an temperature of -50 hPa and 25°C respectively

(2) The application of direct shear stress to soils at the same water potential and temperature resulted in an initial sharp rise in respiration associated with soil elastic deformation. Further displacement of the samples resulted in a plastic deformation and an associated reduction in the rate of increase in respiration ratio. Shear stresses were responsible for an up to 50% increase in respiration of previously cultivated soils (ley-arable rotation and permanent fallow) and an almost doubling of respiration rate from grassland soil.

(3) Compressive stress results in an initial rise in respiration rates, with a peak value being reached at a point approximately equal to the pre-consolidation stress (associated with the soils previous stress history). Further strain results in a reduction of respiration rate as air filled porosity and gas diffusion rates decline.

5. ACKNOWLEDGEMENTS

The authors would like to thank IACR - Rothamsted for permission to sample Highfield plots. The work was funded by the Biotechnology and Biological Sciences Research Council.

6. REFERENCES

Anderson, J.P.E., 1982. Soil respiration. In: A.L. Page, R.H. Miller and D.R. Keeney (Editors) Methods of Soil Analysis, Part 2 - Chemical and Micro biologic Properties, 2nd Ed. Am. Soc. Agronomy, Madison, Wisconsin, pp. 831-877.

Avery, B.W., 1980. Soil classification for England and Wales (Higher categories). Soil Survey Technical Monograph, No. 14, Harpenden, UK.

Beare, M.H., B.R. Pohlad, D.H. Wright and D.C. Coleman, 1993. Residue placement and fungicide effects on fungal communities in conventional and no-tillage soils. Soil Sci. Soc. Am. J., 57, 392-399.

Beare, M.H., M.L. Cabrera, P.F. Hendrix, and D.C. Colman, 1994. Aggregate protected and unprotected organic matter pools in conventional and no-tillage soils. Soil Sci. Soc. Am. J. 58, 787-795.

British Standard 1377, 1975. Methods of testing soil for civil engineering purposes. British Standard Institutes, London. 134 pp.

Carter, M.R., 1992. Influence of reduced tillage systems on organic matter, microbial biomass, macro-aggregate distribution and structural stability of surface soil in a humid climate. Soil Tillage Res., 23, 361-372

Chan, K.Y., W.P. Roberts and D.P. Heenan, 1992. Organic carbon and associated soil properties of a red earth after 10 years of rotation under different stubble and tillage practices. Aust. J. Soil. Res., 30, 348-353.

Chapman, S.D., 1971, A simple conductimetric soil respirometer for field use. Oikos, Copenhagen, 22, 348-353.

Clayden, B. and J.M. Hollis., 1984. Criteria for differentiating soil series. Soil Survey Technical Monograph No.17, Harpenden, UK.

Davis, P.F., A.R. Dexter and D.W. Tanner, 1973. Isotropic compression of hypothetical and synthetic tilths. J. Terramech., 10, 21-34.

Day, P.R. and G.C. Holmgren, 1952. Microscopic changes in soil structure during compression. Soil Sci. Soc. Am. Proc., 16, 73-77.

Dexter, A.R., 1988. Advances in the characterisation of soil structure. Soil Tillage Res., 11, 199-238.

Eriksson, J., I. Håkansson, and B. Danfors, 1974. The effect of soil compaction on soil structure and crop yields. Swed. Inst. Agric. Engng., Uppsala, Bull. 354, 101pp. (English translation by J.K. Aase).

Franzluebbers, A.J., F.M. Horis and D.A. Zuberer, 1994. Long-term changes in soil carbon and nitrogen pools in wheat management systems. Soil Sci. Soc. Am. J., 58, 1639-1645.

Jenkinson, D.S., D.D. Harkness, E.D. Vance, D.E. Adams and A.F. Harrison, 1992. Calculating the net primary production and annual input of organic matter to soil from the amount and radiocarbon content of soil organic matter. Soil Biol. Biochem., 24, 295-308.

Johnston, A.E., 1972. The effects of ley- arable rotations on the amount of soil organic matter in the Rothamsted and Woburn ley-arable experiments. Rothamsted Annual Report for 1972, Part 2, 131-159.

Kern, J.S. and M.G. Johnson, 1993. Conservation tillage impacts on national soil and atmospheric carbon levels. Soil Sci. Soc. Am. J., 57, 200-210.

Koolen, A.J. and H. Kuipers, 1983. Agricultural soil mechanics. Advanced series in agricultural sciences No. 13. Springer-Verlag, Berlin.

Martyniuk, S. and G.H. Wagner, 1978. Quantitative and qualitive examination of soil micro flora associated with different management systems. Soil Sci., 125, 343-350.

Mitchell, J.K., 1993. Fundamentals of soil behaviour. John Wiley & Sons Inc., New York

.Rovira, A.D. and E.L. Greacen, 1957. The effect of aggregate disruption on the activity of micro organisms in the soil. Aust. J. Agric. Res., 8, 659-673

Smucker, A.J.M. and A.E. Erickson, 1989. Tillage and compaction modifications of gaseous flow and soil aeration. In: W.E. Larson *et al.* (editors) Mechanics and related processes in structural agricultural soils, Kluver Academic Publishers, 205-225.

Tisdall, J.M. and J.M. Oades, 1980. The effect of crop rotation on aggregation in a red-brown earth. Aust. J. Soil Res., 18, 423-433.

Watts, C.W., S. Eich, and A.R. Dexter, 1998. Effects of mechanical energy inputs on soil respiration at aggregate and field scales. Soil Tillage Res. (in press)

Wood, D.M., 1990. Soil behaviour and critical state soil mechanics. Cambridge University Press, Cambridge, UK

INHIBITORY ACTIVITY OF STRAINS OF THE GENUS *ARTHRINIUM* ON *ASPERGILLUS* SPECIES IN VINEYARD SOILS OF REQUENA (SPAIN)

H. AISSAOUI, M. AGUT, A. AISSAOUI and M.A. CALVO

Microbiology. Faculty of Veterinary Science.
Universitat Autònoma de Barcelona. 08193 Bellaterra (Spain).

INTRODUCTION

The *Arthrinium* genus include 28 species. They have been described as constituents of a wide variety of habitats although they are considered fundamentally phytoparasites and specifically, a species of Gramineae.

In recent years, investigators have detected some of the secondary metabolites which are produced and accumulated by various strains of *Arthrinium*. The capacity of strains of the *Arthrinium* genus to inhibit yeast development *in vitro* was indicated by Traxler *et al.* (1977). Antibiotic activity on bacteria and on moulds was also demonstrated *in vitro* by Calvo *et al.* (1982), Larrondo and Calvo (1989, 1990) and Larrondo *et al.* (1995).

The main purpose of this study was to investigate the antifungal capacity of some strains of the *Arthrinium* genus on *Aspergillus* species when they are both inoculated in soils.

MATERIALS AND METHODS

All the species used in this investigation are deposited in the culture collection of the Faculty of Veterinary Science (FVB) of the Universitat Autònoma de Barcelona (Spain). The study was undertaken with 3 strains of the *Arthrinium* genus: *Arthrinium serenensis* (A2) (FVB 548), *Arthrinium phaeospermum* (Ae) (FVB 570) and *Arthrinium aureum* (Aa) (FVB 574). The moulds on which the antimicrobial activity was tested were *Aspergillus niger* (An) (FVB 1004), *Aspergillus fumigatus* (Af) (FVB 1005) and *Aspergillus terricola* (At) (FVB 1006).

The fungal strains used to perform this study were isolated from vineyard soil from Requena, Spain. This soil was characterized previously (Aissaoui *et al* 1996).
Suspensions of each mould under study were prepared from a heavily sporulating culture on 2% malt extract agar following the methodology described by Booth (1971), then the concentration of the suspension was determined. Tubes containing 5 g of sterilized soil were enriched with 1 ml of 2% malt extract broth. One ml of a suspension of each mould under study was added to different tubes; which were also inoculated with one species of *Arthrinium* and one species of *Aspergillus*. The tubes were incubated at 28°C for 1 to 7 days (t_1 to t_7). Every day, aliquots of each tube were analyzed to determine the number of colony forming units/g (UFC/g) of the different moulds inoculated.

We also studied the influence of different concentrations of conidia inoculated. The study was carried out with three proportions of fungal conidia of one species of *Aspergillus*:

Effect of Mineral-Organic-Microorganism Interactions on Soil and Freshwater Environments
Edited by Berthelin, *et al*, Kluwer Academic / Plenum Publishers, New York, 1999

one species of *Arthrinium* (1:1, 1:2 and 2:1). Each assay was carried out in triplicate. The different concentrations of fungal suspensions used in these trials are shown in Table 2. The results obtained were analyzed with the Student's t test. The methodology used is shown in Figure 1.

Table 1. *Conidia* concentration of the suspensions (CFU/ml).

STRAIN	CFU/ml
Arthrinium aureum (Aa)	80×10^4
Arthrinium phaeospermum (Ae)	96×10^4
Arthrinium serenensis (A2)	97×10^4
Aspergillus niger (An)	90×10^6
Aspergillus fumigatus (Af)	84×10^6
Aspergillus terricola (At)	64×10^6

Table 2. Concentrations of fungal suspensions.

Proportions of conidia suspensions	CFU/g 10^2					
	x			y		
	An	Af	At	Aa	Ae	A2
1 x: 1 y	90	85	64	80	96	97
1 x: 2 y	90	85	64	160	140	180
2 x: 1 y	170	160	140	80	95	97

x: Strains of *Aspergillus* under study.
y: Strains of *Arthrinium* investigated.

RESULTS

Table 1 gives the results corresponding to the conidia concentration of the suspensions of the microorganisms under study.

Figure 2 shows the data for the survival of the different strains of *Arthrinium* inoculated in vineyard soil of Requena. Figures 3 to 5 summarize the results obtained after the inoculation of pure cultures of the *Aspergillus* strains in soil and after inoculation with one strain of *Arthrinium*. We indicate the median values. Figure 6 shows the percentage reduction of strains of *Aspergillus* inoculated in soil at the same concentration as strains of *Arthrinium*. No significant difference existed among triplicate.

DISCUSSION

After their preparation, the *Aspergillus* conidia suspensions had about 9×10^7 CFU/g, whereas the *Arthrinium* ones reached values on the order of 9×10^4 CFU/g. The difference between these concentrations is a consequence of the higher capacity of the *Aspergillus* strains to produce conidia compared to *Arthrinium* under the same conditions.

Like other moulds grown in a discontinuous culture, without renovation of nutriments (Righelato, 1975), *Arthrinium serenensis* and *Arthrinium aureum* follow a sigmoidal pattern of development in soils. After 6 days of incubation, both moulds started their phase of decelerated growth. We did not detect this point in the development of *Arthrinium phaeospermum* during the period of this investigation. On the other hand, the *Aspergillus* strains under study remained viable throughout the study.

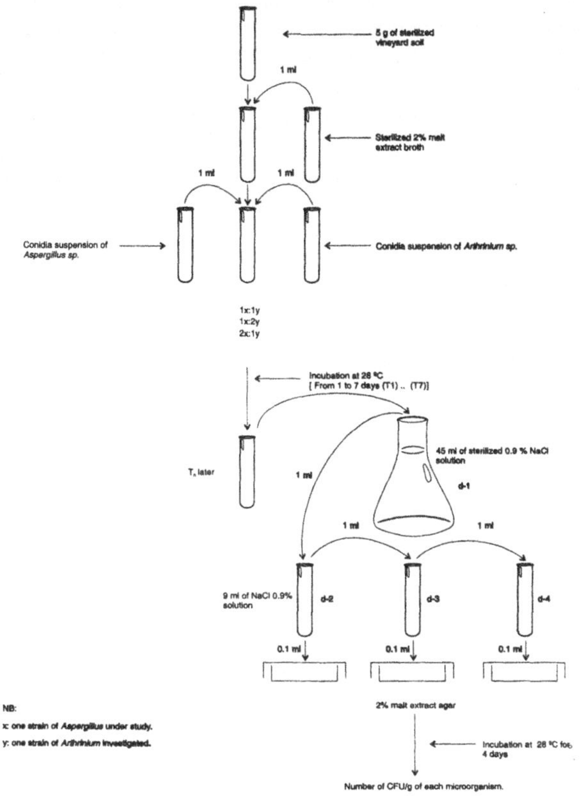

Figure 1. Methodology used to study the influence of different concentrations of *Conidia* inoculated.

When we inoculated one strain of *Aspergillus* together with one strain of *Arthrinium*, the results show that the inhibitory capacity of the *Arthrinium* genus against *Aspergillus* is highest when the rate of the inoculum is 1:1. The inhibitory capacity is detected after 168 hours of incubation. Inhibition can be greater than 80%.

When *Aspergillus fumigatus* or *Aspergillus terricola* were inoculated in the soil together with *Arthrinium aureum*, the number of CFU/g of the *A. fumigatus* decreased more than 99% after four days of incubation. However for *A. terricola* this decrease did not happen; only after seven days of incubation the number of CFU/g decreased slightly. The number of CFU/g of *A. niger* showed a reduction by 70% after 4 days of incubation and reached 96% on the sixth day, coinciding with the greatest development of this strain when it is inoculated in pure culture in the soil. Considering these results we can say that *Aspergillus niger* is more resistant to *Arthrinium aureum* than the other *Aspergillus* strains studied. This behaviour can be explained by the higher concentration of melanine in the conidia of *A. niger*.

CONCLUSIONS

The *Arthrinium* strains under study have the capacity to produce metabolites that can inhibit the development in vineyard soils of the *Aspergillus* strains investigated. These results are in accordance with those obtained *in vitro* in previous studies (Larrondo *et al.* 1995).

We can conclude that *Arthrinium aureum* of the strains studied has the greatest inhibitory effect on the development of the *Aspergillus* species investigated. *Aspergillus niger* was most able to resist the inhibitory action of the *Arthrinium* strains investigated.

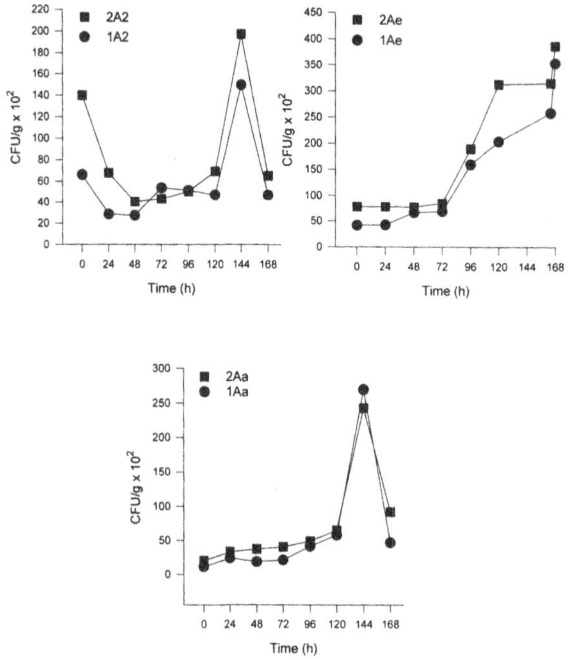

Figure 2. Evolution of *Arthrinium serenensis* (A2), *Arthrinium phaeospermum* (Ae) and *Arthrinium aureum* (Aa) (CFU/g) inoculated in vineyard soil (two concentrations of pure cultures are indicated.

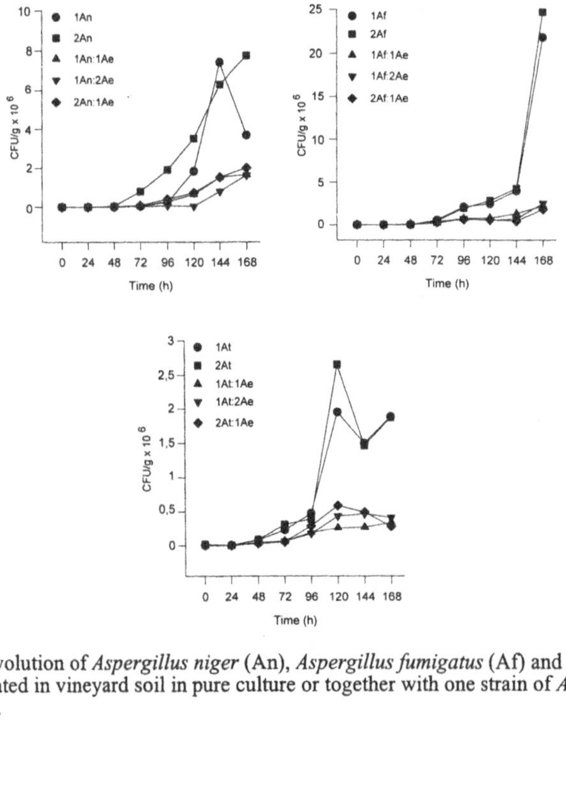

Figure 3. Evolution of *Aspergillus niger* (An), *Aspergillus fumigatus* (Af) and *Aspergillus terricola* (At) (CFU/g) inoculated in vineyard soil in pure culture or together with one strain of *Arthrinium phaeospermum* (Ae).

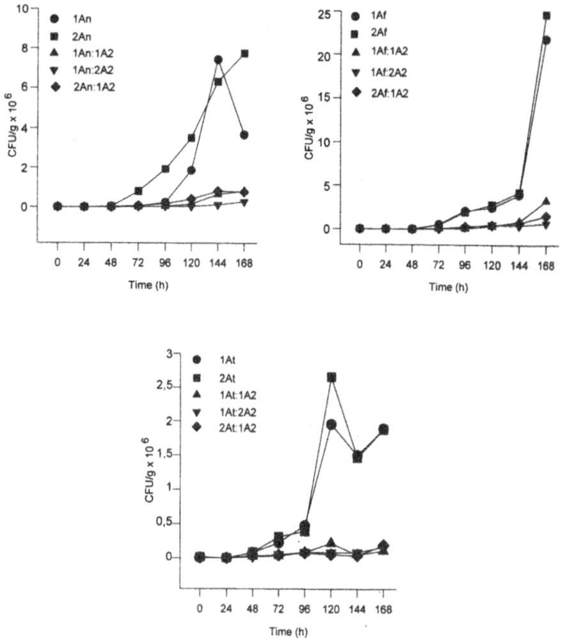

Figure 4. Evolution of *Aspergillus niger* (An), *Aspergillus fumigatus* (Af) and *Aspergillus terricola* (At) (CFU/g) inoculated in vineyard soil in pure culture or together with one strain of *Arthrinium serenensis* (A2).

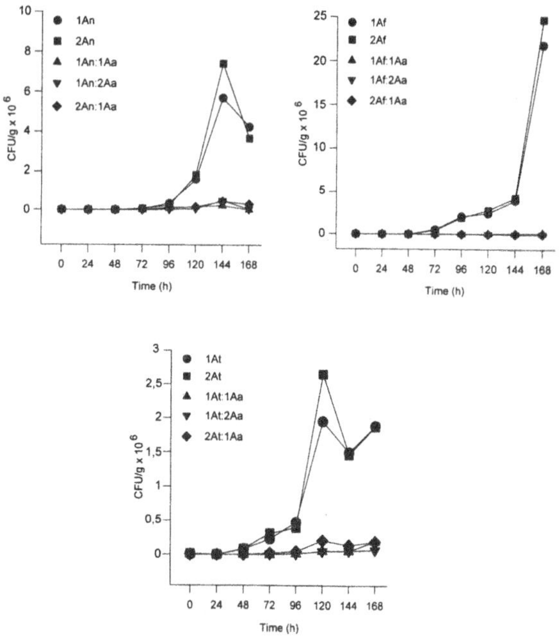

Figure 5. Evolution of *Aspergillus niger* (An),(An), *Aspergillus fumigatus* (Af) and *Aspergillus terricola* (At) (CFU/g) inoculated in vineyard soil in pure culture or together with one strain of *Arthrinium aureum* (Aa).

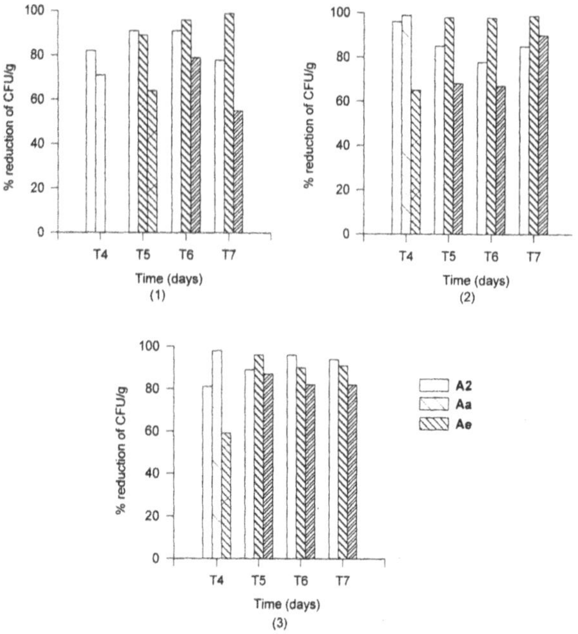

Figure 6. Percentage reduction of *Aspergillus niger (1),* Aspergillus fumigatus (2) and *Aspergillus terricola* (3) (CFU/g) inoculated in vineyard soil at the same concentration (1:1) as *Arthrinium* sp.

REFERENCES

Aissaoui, H., M.A. Novella, M. Agut and M.A. Calvo, 1996. Densidad y distribución de hongos filamentosos en suelos de viñedos de Requena (Valencia). IV Congreso de la SECS, pp. 131-136.

Booth, C. (Editor), 1971. Methods in Microbiology, Volume 4. Commonwealth Mycological Institute, Kew, Surrey, pp. 33-34.

Calvo, M.A., M.A. Vía, A. Busqué and R.M. Calvo, 1982. Pouvoir antibiotique d'*Arthrinium aureum* et *Arthrinium phaeospermum*. Cryptog. Mycol., **3**, 145-150.

Larrondo, J. and M.A. Calvo, 1989. Influence of vitamins on the inhibiting activity of some strains of the *Arthrinium* genus. Microbios Letters, **40**, 17-18.

Larrondo, J. and M.A. Calvo, 1990. Influence of the addition of mineral salts on the inhibitory activity of strains of the *Arthrinium* genus. Microbios, **63**, 17-20.

Larrondo, J., R.M. Calvo, M. Agut and M.A. Calvo, 1995. Inhibitory activity of strains of the genus *Arthrinium* on *Aspergillus* and *Penicillium* species. Microbios, **82**, 115-126.

Righelato, R.C., 1975. Growth kinetics of mycelial fungi. In: The Filamentous fungi, Volume I, Industrial Mycology. Ed. Edward Arnold, London, pp. 79-103.

Traxler, P., J. Gruner and J.A.L. Auden, 1977. Papulacandins, a new family of antibiotics with antifungal activity. I. Fermentation, isolation, chemical and biological characterization of papulacandins A, B, C, D and E. J. Antibiot., **30**, 289-296.

STUDY OF MICROFUNGAL SPECIES IN A CALCAREOUS SOIL TREATED WITH SEWAGE SLUDGE

C. Anaya, J. Forgas, M. Agut and M. A. Calvo

Microbiology. Faculty of Veterinary Science.
Universitat Autònoma de Barcelona. 08193
Bellaterra, Spain.
(Tel. (34)(3)581-17-48 – Fax (34)(3)435-42-17 – email : ST00003@campus.uab.es)

1. INTRODUCTION

Sewage sludge is a by-product of the secondary treatment of municipal sewage. The practice of disposing of sewage sludge by application to land has several potential benefits. Sewage sludge has a favorable soil conditioning property due to its large concentration of humified organic matter, and it also contains substantial concentrations of the macronutrients nitrogen and phosphorus (Harrison, 1992; Freedman, 1995).

In the present work, calcareous soils of a quarry enriched with sewage sludge obtained from a wastewater treatment plant were under study to determine the relationship between the concentration of sludge added to the soils and the fungal diversity of them. The quarry is located in Alcover (Tarragona, Spain). The sludge come from a plant located in Lérida (Spain).

2. MATERIAL AND METHODS

Eight different concentrations (0, 10, 20, 30, 40, 60, 80 and 100% v/v) of sludge were added to the A and B horizons of the studied soils (Typic Xerochrept). We prepared a parcel (12 m x 4 m) for each concentration. Each parcel was divided in to three lots (3m x 3m) to form 24 subparcels. Samples were collected from the subparcels every month for a year (from June 1995 to May 1996). The samples were taken to the laboratory and seeded on 2% malt extract agar using the dilution-plate technique. Plates were incubated at 28°C for five days. Once the incubation period was completed, the colony-forming units per gram (CFU/g) of each studied sample were counted. Analysis of variance (ANOVA) (SAS, 1995) was used to detect differences between the CFU/g in the control and the different concentrations. We also looked at the fungal diversity in the four seasons of the year. Differences were considered significant at $p < 0.05$. Colonies were isolated by sight and with the help of the stereoscopic microscope. The culture medium for isolation was also 2% malt extract agar (Calvo *et al.*., 1979).

The genus and species identification was performed using pure cultures and microcultures, according to the classical criteria employed in fungal taxonomy (Raper and Thom, 1949; Raper and Fennell, 1965; Boot, 1971; Ellis, 1971, 1976; Carmichael *et al.*, 1980; Domsch *et al.*, 1980; Arx, 1981; Stolk and Samson, 1983).

Effect of Mineral-Organic-Microorganism Interactions on Soil and Freshwater Environments
Edited by Berthelin, *et al*, Kluwer Academic / Plenum Publishers, New York, 1999

3. RESULTS

The number of CFU/g determined in the studied soil samples is represented in Figure 1. We have detected significant differences among CFU/g values depending on the concentrations of sludge added.

Figure 1 shows the results corresponding to each of the four seasons of the year. The greatest number of CFU/g found in the samples corresponds to an 80% concentration of sludge added to the soil in autumn (45 x 10^{-4} CFU/g).

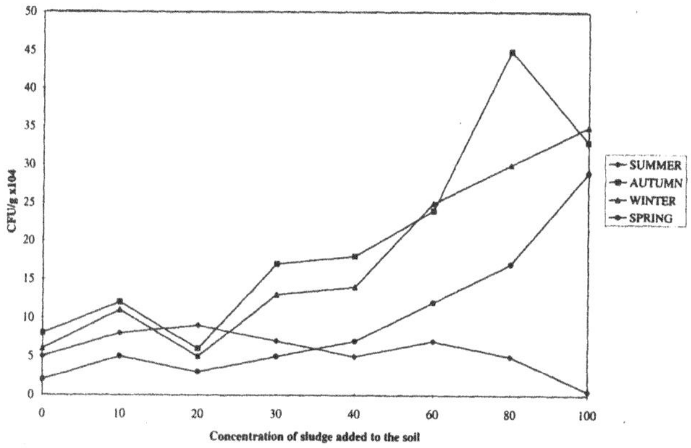

Figure 1. CFU/g found in the soil samples.

Figure 2 summarizes the average number of different microfungal species isolated and identified in the twelve sample collections. The highest result was obtained in the samples collected in December 1995 and the smallest fungal diversity in those sample collected in August 1995

In Figure 3, the number of micofungal species identified in the different concentrations of sludge added to the soil is represented for each of the seasons of the year.

Table 1 shows the species isolated and identified in the samples collected in each season of the year. During the whole study we identified 24 different species. Species that could be isolated in the samples in each of the four seasons of the year include: *Acremonium strictum, Alternaria alternata, Cladosporium cladosporioides, Cladosporium herbarum, Epicoccum purpurascens, Fusarium moniliforme, Fusarium oxysporum, Geotrichum candidum, Mucor mucedo, Penicillium chrysogenum, Penicillium meleagrinum, Penicillium viridicatum, Rhizopus arrhizus* and *Trichoderma viride*

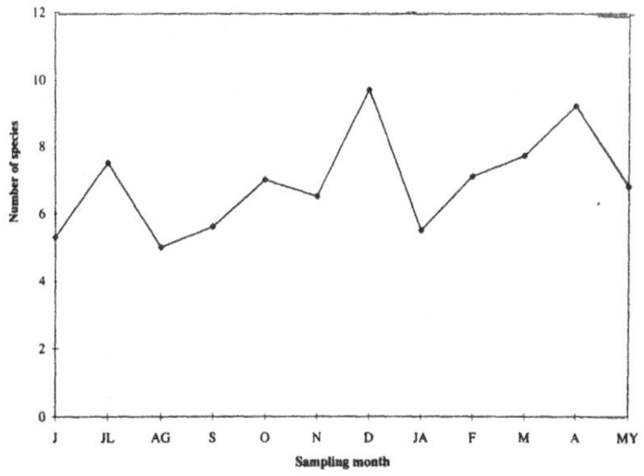

Figure 2. Number of different microfungal species identified.

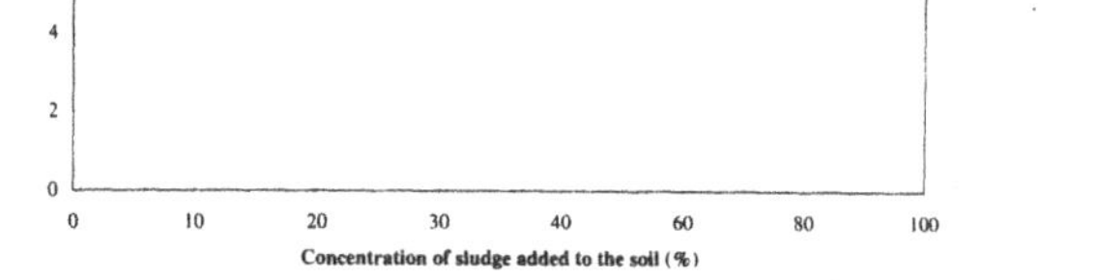

Figure 3. Number of different microfungal species identified in the concentrations of sludge added to the soil.

Table 1. Fungal species isolated in the samples collected in summer (su), autumn (a), winter (w) and spring (s).

	0	10	20	30	40	60	80	100
Acremonium roseum	w	w-s	w	su-w-s	w-s	w	w-s	w
Acremonium strictum	su-a-w-s	a-w-s	a-w-s	a-w-s	a-w-s	a-w-s	a-w-s	a-w-s
Alternaria alternata	su-a-w	su-a-w	su-a-w	su-a-w	su-a-s	su-a	su-a-w	w
Aspergillus chevalieri	su-a-w	su-a	su-a		su	a	su	su-w
Aspergillus niger	s	w					su	w
Aspergillus ochraceus	su				a	a	a	w
Cladosporium cladosporioides	su-a-w	su-a-w-s	su-a-w	a-w-s	a-w-s	w	w-s	a-w-s
Cladosporium herbarum	su-s	su-s	su-s	su-s	su-a-s	su-s	su-w-s	w-s
Epicoccum purpurascens	w-s	su-a-w-s	a-w-s	a-w-s	su-a-w-s	su-a-w-s	w-s	w-s
Fusarium moniliforme	a-w-s	a-w-s	a-w-s	su-w-s	su-a-w-s	a-w-s	su-a-w	su-a-s
Fusarium oxysporum	w	w	w-s	su-a-w-s	su-w-s	a-w	su-a-w-s	su-w-s
Geotrichum candidum	su-a-w-s	su-a-w-s	su-a-w-s	su-a-w-s	su-a-w-s	su-a-w-s	su-a-w-s	su-a-w-s
Mucor mucedo	su-a-w-s	su-a-w-s	su-a-w-s	su-a-s	su-a-w-s	su-a-w-s	su-w-s	su-a-w-s
Paecilomyces variotii		a-w	a			s	a	
Penicillium chrysogenum	su-s	su	su	su-a	su	su-w	su-a-w-s	
Penicillium decumbens		a-s	a			a-s	s	a-s
Penicillium meleagrinum	su-a-w-s	su-a-w-s	su-a-w-s	su-a-w-s	su-a-w-s	su-a-w-s	su-a-w-s	su-a-w-s
Penicillium oxalicum	su-s	su-s	su-s	su	su-s	su-s	su-a-s	s
Penicillium viridicatum	su-a-w-s	su-w-s	su-a-w-s	su-w	su-a-w-s	su-a-w-s	a-w	
Rhizopus arrhizus	su-a-w-s	su	su	su-s	su	su	su-s	
Rhodotorula glutinis						s		s
Scopulariopsis nivea		s	su-a			s	s	s
Trichoderma viride	s	su	su-w		s	s	w	
Ulocladium chartarum			w		a	a	a	w

4. DISCUSSION

All the species isolated in this study have also been recorded in other soil studies (Domsch *et al.*, 1970; Larrondo *et al.*, 1989; Agut *et al.*., 1995).

The number of different fungi isolated varied depending on the concentration of sludge added to the soil. Except for the samples collected in summer, the fungal counts were higher when we added a higher percentage of sludge to the soil. However, this caused the fungal diversity to decrease.

Geotrichum candidum was the only species isolated in all the samples. *Ulocladium chartarum* was only found in samples collected during winter. *Scopulariopsis nivea* was isolated in all seasons except winter. *Acremonium strictum* was not found in the autumn samples, and *Paecilomyces variotii* was not isolated in the summer ones.

Fungi of the *Penicillium* genus were isolated in all the samples. The species of the *Penicillium* genus detected were: *Penicillium chrysogenum, Penicillium decumbens, Penicillium meleagrinum, Penicillium oxalicum* and *Penicillium viridicatum.*

Among those species, *Penicillium decumbens* and *Penicillium oxalicum* did not appear in the winter samples. *Penicillium decumbens* does not also appear in the summer samples.

Some authors (Kramer *et al.*., 1960; Calvo *et al.*, 1980) indicate that the species of P*enicillium* are more frequently isolated during the months between July and November; rain is often cited as a decisive factor for their presence.

Only three species of *Aspergillus* were identified in this study: *Aspergillus chevalieri, Aspergillus niger* and *Aspergillus ochraceus. Aspergillus chevalieri* and *Aspergillus ochraceus* were not found in the samples collected during the spring. *Aspergillus niger* was not isolated in the autumn ones. These data are in accordance with others showed by other authors (Kramer *et al.*., 1960; Calvo *et al.*., 1979, 1980).

In view of the results obtained in this study, *Geotrichum candidum, Mucor mucedo* and *Penicillium meleagrinum* are the species that could be considered constituents of the soil and sludge microbiota throughout the year. *Acremonium strictum, Penicillium viridicatum* and *Rhizopus. arrhizus* are only present in the soil.

The species which could be isolated in each one of the concentrations of sludge added to the soil are the following: *Acremonium roseum, Acremonium strictum, Alternaria alternata, Cladosporium cladosporioides, Cladosporium herbarum, Epicoccum purpurrescens, Fusarium moniliforme, Fusarium oxysporum, Geotrichum candidum, Mucor mucedo, Penicillium meleagrinum* and *Penicillium oxalicum.* The rest of them present a wide distribution depending on the concentration of sludge added.

5. CONCLUSIONS

Only three species were isolated in all the concentrations of sludge added to the soil and also in the soil and in 100% sludge: *Geotrichum candidum, Mucor mucedo* and *Penicillium meleagrinum.*

In general, the fungal counts were higher when a higher percentage of sludge was added to the soil, however this caused the fungal diversity to decrease.

6. REFERENCES

Agut, M., M. Bayó, J. Larrondo and M. A. Calvo, 1995. Keratinophilic fungi from soil of Brittany, France. Mycopathol., 00,1-2.

Arx, J. A. von, 1981. The genera of fungi sporulating in pure culture 3"de. J. Cramer.

Boot, C. 1971. The genus *Fusarium*. CAB. Kew England.

Calvo, M. A., J. Guarro, G. Suarez and C. Ramirez, 1979. Airborne fungi in the air of Barcelona (Spain). II. The genus *Alternaria*. Mycopathol., 69, 137-142.

Calvo, M. A., J. Guarro, G. Suarez and C. Ramirez, 1980. Airborne fungi in the air of Barcelona (Spain). I. Two years of study (1976-1978). Mycopathol., 71(1), 89-93.

Carmichael, J. W., W. B. Kendick, I. L. Conners and L. Sigler , 1980. Genera of Hyphomycetes. Edmonton. University of Alberta Press. Domsch, K. H., 1980. Compendium of soil fungi. 2. Academic Press, London.

Domsch, K. H. and W. Gams, 1970. Pilze aus Agrarböden. Ed. Gustav Fischer Verlag. Stuttgart. Germany.

Ellis, M. B., 1971. Dematiaceous Hyphomycetes. CMI. Kew England.

Ellis, M. B., 1976. More Dematiaceous Hyphomycetes. CMI. Kew England.

Freedman, E., 1995. Environmental Ecology. The ecological effects of pollution, disturbance, and other stresses. Second edition. Academic Press, San Diego, California.

Harrison, R. M., 1992. Understanding our environment: an introduction to environmental chemistry and pollution. Second edition. Royal Society of Chemistry, Cambridge.

Kramer, C. L., J. M. Pady and C. T. Rogerson, 1960. Kansas aeromycology. V. *Penicillium* and *Aspergillus*. Mycologia, 52(4), 545-551.

Larrondo, J. V. and M. A. Calvo, 1989. Fungal density in the sands of the Mediterranean coast beaches. Micopathol., 108, 185-193.

Raper, K. B. and D. I. Fenneell, 1965. The genus *Aspergillus*. Williams and Wilkins Co., MD.

Raper, K. B. and C. Thom. 1949. The manual of *Penicillia*. Williams and Wilkins Co., MD.

Stolk, A. C. and R. A. Samson, 1983. The Ascomycetes. Genus *Eupenicillium* and related *Penicillium* anamorphs. Studies in mycology 23, CBS.

Key words: Fungi, Diversity, Sewage Sludge-Amended Soils

TWO-STEP BIOREMEDIATION OF SOILS CONTAMINATED WITH CHLOROAROMATICS

P. Rosenbrock[1], R. Martens[1], F. Buscot[1], J.C. Munch[2]

[1] FAL - Federal Agriculture Research Center, Institute of Soil Biology, Bundesallee 50, D-38116 Braunschweig, Germany

[2] GSF-National Research Center for Environment and Health GmbH, Institute of Soil Ecology, Ingolstädter Landstraße 1, D-85764 Oberschleißheim, Germany

1. INTRODUCTION

Chloroaromatics are widely used as solvents, pesticides, flame retardants and dielectric fluids. Polychlorinated dibenzo-p-dioxins and dibenzofurans are formed as undesirable byproducts of chemical manufacturing. Thus, high amounts of chlorinated aromatics have been released into the environment. These compounds are toxic and resistent to biological degradation. One reason for the persistence of chloroaromatics in soils is their low bioavailability, which is caused by hydrophobicity and strong sorption to soil organic matter.

Under aerobic conditions, biodegradation of highly chlorinated aromatics by an electrophilic attack of bacterial oxygenases occurs only to a minor extent or not at all because of the negative inductive effect of the chlorine substituents (Knackmuss, 1975).

On the contrary, under anaerobic conditions reductive dechlorination of chloroaromatics can occur and has been described in a variety of anoxic habitats e. g., anaerobic biofilm reactors (Fathepure and Vogel, 1991), sediments (Linkfield *et al.*, 1989; Nies and Vogel, 1990; Abramowicz *et al.*, 1993; Alder *et al.*, 1993; Adriaens and Gribic′-Galic, 1994) or sediment slurries (Bosma *et al.*, 1988).

The idea of these investigations was to take advantage of the biological process of reductive dechlorination to clean up soils contaminated with chloroaromatics, resulting in a bioremediation strategy with two steps.

In the present work we investigated the two separate parts of such a two-step bioremediation process.

In the first step of the strategy, the chloroaromatics should be dechlorinated under anaerobic conditions by the reductive dechlorination reaction. A chloride will be exchanged for a hydrogen in a biological process. In experiments with a pure culture of *Desulfomonile tiedjei,* it could be shown that 3-chlorobenzoate functions as an electron acceptor and organic substrate serves as an electron donor in the process of reductive dechlorination (Dolfing, 1990; Mohn and Tiedje, 1990).

In the second step of the strategy, the dechlorinated molecules should be mineralized aerobically. Unchlorinated aromatics can be degraded by bacteria (Fortnagel *et al.*, 1989; Robinson *et al.*, 1990; Trzesicka-Mlynarz and Ward, 1995) and by lignolytic white rot fungi. The irregular structure of lignin, containing variable aromatic substructures, is similar to that of many pollutants. This knowledge led to the presumption that the nonspecific lignolytic system produced by white rot fungi might also degrade aromatic

Effect of Mineral-Organic-Microorganism Interactions on Soil and Freshwater Environments
Edited by Berthelin, *et al*, Kluwer Academic / Plenum Publishers, New York, 1999

xenobiotics. Many investigations were done in this context. The white rot fungus *Phanerochaete chrysosporium* degrades unchlorinated aromatics (Bumpus *et al.*, 1985; Hammel *et al.*, 1986; Bumpus, 1989) and even some chlorinated aromatics (Bumpus *et al.*, 1985; Mileski *et al.*, 1988; Yadav *et al.*, 1995) in a fluid culture system.

In our investigations soil samples from two different agricultural sites were used. In investigations on the anaerobic process of dechlorination, the soil samples were spiked with ^{36}Cl-labeled hexachlorobenzene (^{36}Cl-HCB) as a model polychlorinated compound. Afterwards the soil samples were amended with organic electron donors and saturated with water. The added substrates were oxidized by the soil microflora, resulting in depletion of the remaining oxygen from the water-saturated soil and establishment of an anaerobic environment in which the anaerobic indigenous microflora should dechlorinate the HCB.

In investigations on the aerobic phase, [U-^{14}C]-dibenzo-p-dioxin ([U-^{14}C]-DD) served as the nonchlorinated model compound. The soils were spiked with [U-^{14}C]-DD and amended with an organic electron donor to stimulate the indigenous degrading microflora or inoculated with straw-substrate colonized by lignolytic white rot fungi, i.e., the *Phanerochaete chrysosporium* DSM 9620, *Pleurotus* sp. Florida DSM 11191, *Dichomitus squalens* DSM 9615 and an unidentified lignolytic soil fungus PRT87 DSM 11497 isolated from a dioxin-contaminated soil.

2. MATERIALS AND METHODS

2.1 Soil Characteristics

Two agricultural soils from Baden-Württemberg, Germany, were used. The sites had been contaminated by different industrial dioxin emittents (Table 1). The samples were designated according to their origin sites, i.e., Eppingen and Maulach. After sampling the A-horizon, the soils were stored at 9°C with no further treatment. Before the experiments were started the soils were sieved (2mm pore diameter) and preincubated for 14 days at room temperature (20°C) with a water content of about 30% of their water holding capacity (WHC).

Table 1: Characteristics of investigated soils from Baden-Württemberg, Germany.

	Eppingen	Maulach
WHC* [%]	60	110
clay [%]	12.3	7.6
silt [%]	61.0	36.8
sand [%]	26.7	55.6
pH [$CaCl_2$]	7.1	6.5
C_{org} [%]	1.5	6.7
N [%]	0.12	0.62
dioxin contamination [ng kg^{-1}] TE **	248	5580

* water holding capacity; ** toxicity equivalents

2.2 Materials

The uniform ^{36}Cl-labeled hexachlorobenzene (^{36}Cl-HCB) was synthesized by K. Haider (Federal Agricultural Research Centre, Institute of Plant Nutrition and Soil Science, Braunschweig, Germany). The radioactive purity of the ^{36}Cl-HCB dissolved in toluene was tested by scanning a thin layer chromatogram of the labeled compound (TLC aluminum sheets, silica gel, F_{254} S, Merck, Darmstadt, Germany; developed in cyclohexane:chloroform (80:20)) with a thin layer scanner (LB 2723, Berthold, Germany) and comparison with unlabeled HCB (99%, Aldrich, Steinheim, Germany) as a reference. Only one radioactive peak was registered on the TLC plate, indicating a high purity of the ^{36}Cl-labeled HCB. The radioactive area had the same R_f value (0.51) as the unlabeled HCB detected on the TLC plate by UV light. The radioactivity of the^{36}Cl-HCB-toluene solution amounted to 3900 cpm μl^{-1} with an ^{36}Cl-HCB concentration of 10 μg μl^{-1} toluene. All radioactivity measurements of ^{36}Cl-HCB were based on cpm (counts per minute) instead of the usual dpm (disintegrations per minute), since no ^{36}Cl-standard could be purchased to correct cpm values for quenching effects. It was found that the measuring efficiency for all counted sample types was identical. This allowed a direct relation between measured activities in the extracts and water phases and the original applied radioactivity.

[U-^{14}C]-dibenzo-p-dioxin ([U-^{14}C]-DD) was purchased from Amersham International plc, Braunschweig (Germany). It had a specific radioactivity of 40 mCi mmol^{-1} and a radiochemical purity of 98.6%. Unlabeled dibenzo-p-dioxin (DD) for blending the [U-^{14}C]-DD was a gift of R.M. Wittich (Gesellschaft für Biotechnologische Forschung (GBF), Braunschweig, Germany).

Radioactivity of liquid samples was measured after mixing of aliquots with 14 ml of a scintillation cocktail (Rotiszint ecoplus, Roth, Karlsruhe, Germany) in a liquid scintillation counter (LSC) (Beckman LS 1801, Fullerton, USA). Unless otherwise stated, this scintillation cocktail was used.

All other chemicals were of analytical grade.

2.3 Spiking of Soil

For the investigations of HCB dechlorination 15.75 μl ^{36}Cl-labeled HCB-toluene solution (61425 cpm, containing 157.5 μg ^{36}Cl-labeled HCB) were applied to a 250 mg portion of dry and pulverized soil. For the investigations of DD mineralization 20μl of DD-toluene solution (2.0 10^6 dpm, containing 4.178 μg [U-^{14}C]-labeled and 25.822 μg unlabeled DD) were applied to 400 mg dryed pulverized soil.

After evaporation of toluene overnight, the ^{36}Cl-HCB-spiked soil was carefully mixed with 5.0 g soil (dry weight) to reach a ^{36}Cl-HCB concentration of 30 μg g^{-1} soil. The [U-^{14}C]-DD-spiked soil was mixed with 30 g soil (dry weight) to reach a DD concentration of 1μg g^{-1} soil.

Spiking the soil in this way seemed to be the best way to simulate a "natural" sorption to soil and ensure an extensive distribution of the xenobiotic. The possible loss of radioactivity during evaporation of toluene overnight was measured by absorption of ^{36}Cl-labeled HCB and [U-^{14}C]-DD in paraffin-coated glass wool and cellulose stoppers, respectively, placed in the neck of the vial used for spiking. Afterwards the glass wool and the cellulose stoppers were extracted with toluene and radioactivity was measured in the LSC.

2.4 Design of Experiments

The experiments for the reductive dechlorination of ^{36}Cl-HCB were carried out in 20 ml LSC glass vials. The ^{36}Cl-HCB-spiked soil samples (5 g, dry weight) were mixed with the electron donors, i.e., glucose (0.5%, w/w), lactate (0.5%, w/w), wheat straw (1%, w/w, particle size <1mm) and cut grass (2%, w/w). All substrates were added under sterile conditions, soluble substrates as sterile distilled water solutions, and straw as sterilized and cut grass as unsterilized solid substrate. Distilled water was added to saturate the soil with an excess of 5 ml to obtain a supernatant. The LSC vials were sealed with silicon septa lids. Three replicates per treatment and per investigated soil were set up. All assays were performed in darkness at 30°C without shaking. For the sterile control the soils were sterilized by autoclaving (121°C) three times for 30 minutes with an incubation of 24 h at 20°C between each autoclaving.

Dechlorination was monitored by measuring the release of $^{36}Cl^-$ into the aqueous phase. After homogenizing the waterlogged soil samples by shaking followed by a 24h-sedimentation phase, a 2 ml aliquot of the supernatant was taken with a syringe (needle 0.42 mm diameter) through the silicon septum without opening the vessel. After sampling the removed supernatant volume was replaced by 2 ml distilled water. One ml of the aliquot was centrifuged (3 min, 13500 x g) to precipitate soil particles and the supernatant was analyzed directly in the LSC to quantify the radioactivity. If radioactivity was detected in the sample, the second part of the aliquot was used to investigate whether the radioactivity accounted for ^{36}Cl-HCB residues and/or lipophilic or hydrophilic ^{36}Cl-HCB metabolites or $^{36}Cl^-$ dissolved in the aqueous phase. To determine the content of ^{36}Cl-HCB or lipophilic metabolites the supernatant of the centrifuged aliquot (1 ml) was extracted three times with 0.5 ml toluene. The radioactivity of the united toluene fractions was measured in the LSC and was attributed to ^{36}Cl-HCB or lipophilic metabolites. The proportion of possible hydrophilic ^{36}Cl-HCB metabolites in the toluene-extracted aqueous aliquot was determined by precipitating the $^{36}Cl^-$ with $AgNO_3$ as $Ag^{36}Cl$. First the aliquot was acidified with 300 µl 1M HNO_3 and 50 µl of a 0.1M NaCl solution were added to reach the solubility product of AgCl. Then the non-labeled and the $^{36}Cl^-$ were precipitated with 100 µl 0.1M $AgNO_3$. After centrifugation for 10 min at 13500 x g the radioactivity in the supernatant was measured. Any detected radioactivity was attributed to hydrophilic metabolites of ^{36}Cl-HCB.

Radioactivity in the toluene extract and in the supernatant after $^{36}Cl^-$ precipitation was substracted from the radioactivity determined directly in the untreated aqueous phase. The calculated difference in radioactivity was attributed to the release of $^{36}Cl^-$.

At first, samples were taken 24 hours after the start of the experiments, proving that no radioactivity was in the supernatant at this time. Afterwards, samples of the supernatant were taken every four weeks for analysis.

The experiments for mineralization of DD were carried out in 250 ml glass vessels. For the degradation experiments spiked 30 g (dry weight) soil samples were used. Soil moisture content was adjusted at 45% of the WHC in soil Eppingen and 60% of the WHC in soil Maulach to reach a similar crumbly texture. In amended assays solid substrates were carefully mixed into the spiked soil samples. Sterile wheat straw (2.5 g, 121°C, 20 min) was added. 7.5 g of the compost (dry weight, cut grass / biowaste 1:1, ripe compost rotted ten weeks, sieved <1cm, moisture content of 50%) was originally added. In the inoculated assays spiked soil samples were added into the straw-fungus substrate without mixing. Three replicates per treatment and per investigated soil were done. The soils of the sterile controls were prepared by autoclaving (121°C) three times for 30 minutes with an incubation time (20°C) of 24 hours between autoclaving.

The assays were incubated for 10 weeks at 25°C in the dark. Aeration occurred with CO_2-free, moistened air flowing through an NaOH/CaO-filled tube and bottles with 2N NaOH and distilled water. Air flux through the vessels was in the range of 5 ml min^{-1}. After passage of the flask, air was conducted through a column filled with paraffin-coated glass wool to trap volatilized [U-^{14}C]-DD or its lipophilic metabolites. $^{14}CO_2$ formed by the mineralization of [U-^{14}C]-DD was trapped in 25 ml 2N NaOH. The NaOH tube was replaced weekly and radioactivity was determined by measuring an aliquot of the NaOH mixed with the scintillation cocktail in the LSC. Aliquots (0.5 ml) of the NaOH samples of the first, third, sixth and tenth samplings were investigated for their possible content of [U-^{14}C]-DD or ^{14}C-labeled metabolites. The aliquot was mixed with 0.3 ml 20% H_2SO_4 to remove $^{14}CO_2$. After 24 hours the radioactivity of the aliquot was determined again. Three ml of a H_2O-methanol solution (2:1) was added first to prevent a precipitation of $NaSO_4$. Afterwards the sample was mixed with 14 ml of a scintillation cocktail containing methanol (22%) and Qickszint 212 (78%) (Roth, Karlsruhe, Germany) and radioactivity was determined in the LSC.

For preparation of the soil fungus substrate, the *Phanerochaete chrysosporium* DSM 9620, *Pleurotus* sp. Florida DSM 11191 and *Dichomitus squalens* DSM 9615 were obtained from the collection of F. Zadrazil (Federal Agricultural Research Centre, Institute of Soil Biology, Braunschweig, Germany). The unidentified strain called PRT87 DSM 11497 was isolated from the dioxin-contaminated soil of Eppingen using a wheat straw agar (wheat straw with a particle size <1mm, 2 g l^{-1}, agar 15 g l^{-1}, pH 6). This procedure allows the isolation of fungi able to degrade the degradation-resistant contents of straw, mainly cellulose and lignin. Afterwards the isolate was screened with respect to its potential for production of lignolytic enzymes, i.e., lignin peroxidase, manganese peroxidase and laccase. Strain PRT87 was superior in production of manganese peroxidase (data not shown)

Phanerochaete chrysosporium, *Pleurotus* sp. Florida and *Dichomitus squalens* were maintained on 1.5% maltextract agar, PRT87 on modified Basal III medium. This contained 0.2% wheat straw as the sole carbon source, added to the basal medium of Tien and Kirk (1988), which was further modified by addition of 15 g l^{-1} agar (pH 6).

Before the degradation assays the fungi were precultivated on 1.5% maltextract agar plates (pH 6) for 14 days at 20°C to get a sufficient amount of inoculum. Two round pieces (7mm diameter) of these cultures were cut out from the margin of the actively growing mycelia and were put onto the autoclaved (121°C, 30 min) straw (5 g, 15 ml distilled water) (33) in a 250ml bottle. The straw-fungus culture was incubated for 14 days at 25°C in the dark. After this incubation phase the fungi had completely colonized the straw.

3. RESULTS

Dechlorination of ^{36}Cl-HCB in naturally oxic soils under anaerobic conditions monitored for 140 days by release of $^{36}Cl^-$ is shown in Figures 1 and 2. The extent of dechlorination of HCB varied considerably depending on both the investigated soil and the soil treatment.

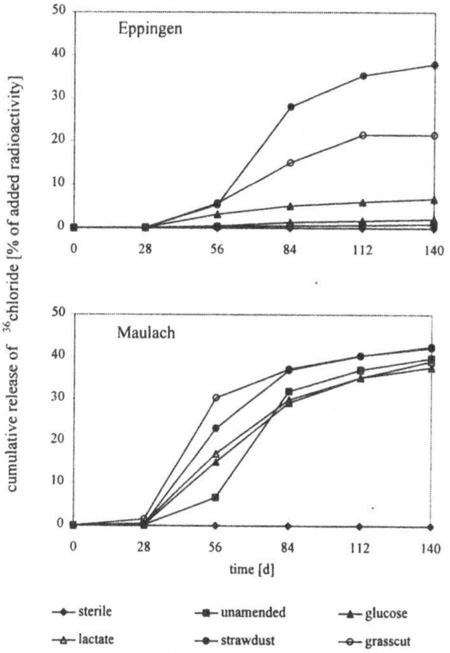

Figure 1: Course of reductive dechlorination of ^{36}Cl-labeled hexachlorobenzene measured as release of ^{36}chloride in percent of added radioactivity in soil under anaerobic conditions with addition of different carbon sources.

Extraction of the supernatant samples showed that neither ^{36}Cl-HCB nor ^{36}Cl-HCB metabolites were detectable in the aqueous phase, indicating that measured radioactivity in the supernatant corresponded only to dechlorination to $^{36}Cl^-$. In the sterilized controls no dechlorination occurred during the 140-day incubation. At the beginning of the experiment all biotic treatments of both soils showed a lag-phase of at least 28 days during which no detectable dechlorination occurred.

The soil Maulach showed a high dechlorination of about 40% within 140 days in the unamended as well as in the amended assays, independent of the electron donor (Fig. 1 and 2), but amendment with electron donors caused a faster dechlorination compared with unamended soil (Fig. 1). The investigations with the soil Eppingen showed greater differences between the treatments. No dechlorination occurred without amendment with an organic substrate. Addition with a easily degradable carbon source such as glucose or lactate caused only a minor dechlorination. The greatest release of $^{36}Cl^-$ was observed after an amendment with straw, a heavily degradable substrate that ensured a long-term source of electrons.

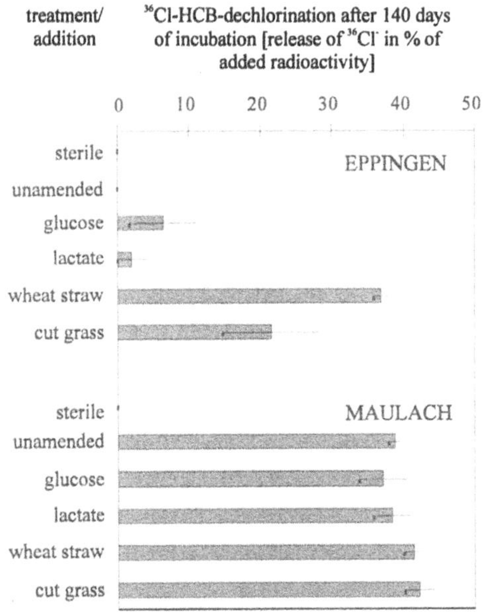

Figure 2: Reductive dechlorination of ^{36}Cl-labeled hexachlorobenzene measured as release of ^{36}chloride in percent of added radioactivity in soil under anaerobic conditions with addition of different carbon sources after an incubation of 140 days.

Mineralization of [U-^{14}C]-DD could be observed in each of the investigated soils (Fig. 3) with no long-term lag-phase. Radioactivity trapped in the NaOH corresponded only to $^{14}CO_2$. After removing $^{14}CO_2$ no radioactivity could be detected that could be attributed to [U-^{14}C]-DD or ^{14}C-labeled metabolites. In the sterile controls, no $^{14}CO_2$ formed by mineralization of [U-^{14}C]-DD was detected. In the unamended assays, the endogenous DD mineralization potential was shown (Fig. 3). Soil Maulach was characterized by an almost double endogenous DD degradation potential compared to soil Eppingen.

In contrast to soil Maulach, in samples of soil Eppingen with a minor content of organic carbon (Table 1), addition of organic substrate led to an increase of the total mineralization (Fig. 3).

Inoculation with lignolytic fungi as straw-fungus substrate enhanced DD biodegradation in both soils compared with the addition of straw alone. But the extent of increase and the influence of different fungi species on biodegradation varied depending on the type of soil.

In soil Eppingen inoculation with unidentified strain PRT87 led to the greatest DD mineralization (Fig. 3); this was caused by an extremely high mineralization of more than 20% within the first seven days of incubation.

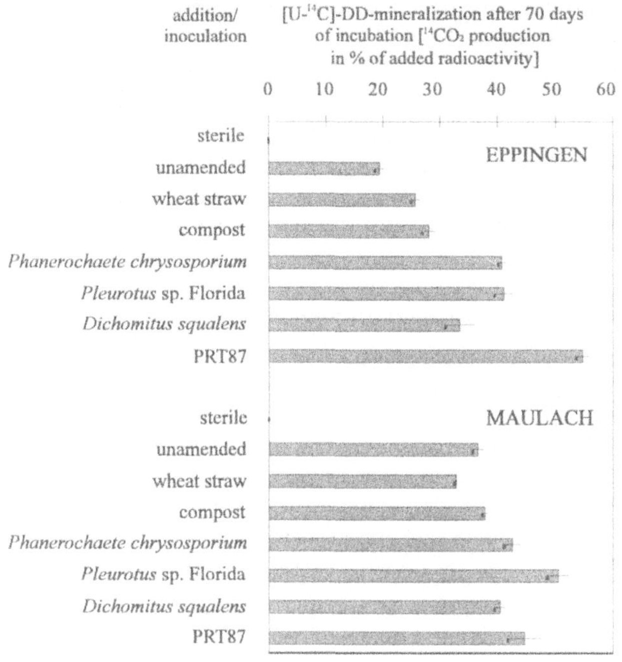

Figure 3. Mineralization of ^{14}C-labeled dibenzo-p-dioxin measured as $^{14}CO_2$ production in percent of added radioactivity in soil under aerobic conditions with addition of different carbon sources or inoculation with lignolytic fungi after an incubation of 70 days.

4. DISCUSSION

4.1 Dechlorination of ^{36}Cl-Hexachlorobenzene

Dechlorination of HCB was observed previously in naturally anaerobic habitats or in enrichment cultures that originate from such habitats (Beurskens *et al.*, 1993; Holliger *et al.*, 1992). The present experiments demonstrated the possibility to adjust anaerobic conditions in naturally oxic soils, suitable for a reductive dechlorination of HCB. The investigation of two soils and different amendments shows that the extent of dechlorination depends both on the type of soil and on the type of added electron donor.

Lag periods prior to onset of dechlorination of chloroaromatics are typically on the order of weeks or months (Linkfield *et al.*, 1989; Williams, 1994). Such acclimation periods prior to dechlorination can be attributed to processes such as exhaustion of preferential substrates, induction of new protein synthesis or genetic changes. In our experiments a lag-phase of at least 28 days generally occurred before dechlorination started. Both soils have an indigenous dechlorinating microflora and we suggest that dechlorination starts soon after establishing an anaerobiosis but in a minor and nondetectable extent because of the low density of the degrading microflora.

Adding organic electron donors often results in an increase of dechlorination. The addition of glucose enhanced dechlorination of PCBs in sediments (Nies and Vogel, 1990) and dechlorination of PCP in anaerobic sludge blanket reactors (Hendriksen *et al.*, 1992). Addition of complex carbon sources also increased dechlorination of trichlorophenol (Madsen and Aamand, 1992) and PCBs (Abramowicz *et al.*, 1993). Contrary to our experiments with soil Maulach, addition of electron donors did not enhance the absolute extent of dechlorination, but the dechlorination process was accelerated when an additional substrate was applied. This result could be explained by the high content of organic matter in this soil (6,7% C_{org}, Table 1), since soil organic matter represents a source of long-term, slowly flowing carbon. In soil Maulach organic matter served sufficiently as a carbon source for a growing population, but mobilization of this carbon source seemed to be a lengthy process. In the investigations with soil Eppingen the addition of the more degradation resistant carbon source straw caused higher dechlorination compared with addition of easily degradable glucose or lactate. No amendment of substrate led to no HCB dechlorination in this soil, which is characterized by a low organic matter content (1,5% C_{org}, Table 1) which is apparently not sufficient to serve as an electron donor for establishment of anaerobiosis and during dechlorination. This interpretation is consistent with the observations that dechlorination of PCBs in anaerobic sediments with a high organic content could not be enhanced by amendment with electron donors, whereas adding electron donors to sediments with a low organic content increased the dechlorination rate (Alder *et al.*, 1993).

4.2 Mineralization of [U-^{14}C]-Dibenzo-p-dioxin

The amount of soil organic matter is related to the density of soil microflora (Anderson and Domsch, 1989), and thus propably related to the density of the degrading microflora. Corresponding to these assumptions, in the unamended assays, the endogenous DD mineralization could depend on the content of organic matter of soils. The organic-rich soil Maulach mineralized about twice as much DD as soil Eppingen, which had a lower organic matter content. However, other investigations of mineralization of lipophilic xenobiotics in soils have showed a correlation between content of soil organic matter and adsorption of the xenobiotic (Weißenfels *et al.*, 1993). Strong adsorption results in decreased bioavailability and minor biodegradation of the xenobiotic.

In contrast to soil Maulach, in soil Eppingen application of straw increased DD mineralization. This could also be explained by the amount of organic carbon. Addition of straw initiated growth of degrading microflora in the organic-poor soil Eppingen, whereby in soil Maulach the soil organic matter was sufficient to ensure growth of indigenous microflora. Addition of compost increased DD mineralization in both investigated soils compared with the unamended assays (Fig. 3). This could be explained by inoculation with the indigenous compost microflora that is generally characterized by a high degradation efficiency. Rotted ripe compost possesses considerable abilities to mineralize recalcitrant molecules such as PAH (Martens, 1982).

Application of white rot fungi to soil enhanced biodegradation of recalcitrant xenobiotics such as pentachlorophenol (Mileski *et al.*, 1988), PAH (Hüttermann *et al.*, 1988; Morgan *et al.*, 1993) and 3,4-dichloroaniline (Morgan *et al.*, 1993). These indications following, we tested the influence of lignolytic fungi on mineralization of DD. Preincubation of fungi on substrate such as straw has three advantages: i. production of a sufficient amount of biomass for inoculation, ii. lignin-containing substrate serves as natural support from which fungi could grow into soil and iii. straw substrate provide a wide C:N ratio that stimulates the production of lignolytic enzymes by white rot fungi (Reid, 1979). Investigations into the growth of the white rot fungus *Phanerochaete chrysosporium* into soil have shown that wheat straw influenced fungal growth into soil best, compared with other lignin-containing substrates such as hay or wood chips (Morgan *et al.*, 1993).

All tested fungi, including known lignolytic white rot fungi and the unidentified lignolytic soil born isolate PRT87, increased DD mineralization compared with uninoculated soils, but the influence of the fungi on DD degradation in the investigated soils differed. The lignolytic soil fungus PRT87 caused the best mineralization result in soil Eppingen, from which it was isolated. In soil Maulach the influence of this fungus and of all other tested white rot fungi on DD degradation was minor. However it could be expected that the high endogenous mineralization potential of soil Maulach combined with the fungal mineralization potential would cause a higher mineralization result. This could be deduced to an inhibition of the inoculated fungus by the indigenous microflora, which is well developed in the organic-rich soil Maulach. Ali and Wainwright (1994) found a growth inhibition of *Phanerochaete chrysosporium* in soil amended with pepton and starch. They suggested, that soil amending stimulates a bacterium antagonist to *Phanerochaete chrysosporium* that produced in vitro a fungal growth inhibiting "antifungal agent". Another explanation for this phenomenon could be probably the formation of "bound residues". Bollag *et al.* (1980) showed that the lignolytic enzyme laccase (phenoloxidase) of the soil fungus *Rhizoctonia praticola* mediates oxidative cross-coupling between the phenolic constituents of humus and 2,4-dichlorophenol. Thus, a higher amount of organic matter should intensify the formation of bound residues.

5. CONCLUSION

In these investigations it was shown that a partial dechlorination of HCB in soils can be achieved by adjusting anaerobic conditions in a naturally oxic soil. But it also was demonstrated that a treatment to reach a satisfying dechlorination strongly depends on the particular soil. The two investigated soils differed in their requirements for reaching the necessary conditions for dechlorination. In the organic matter-rich soil Maulach water saturation was sufficient to get conditions promoting dechlorination, whereas the indigenous microflora of the soil Eppingen needed additional electron donors for dechlorination, probably because of the minor organic matter content of this soil. Heavily degradable substrates seemed to be the best way to ensure a long-term source of electrons. Reductive dechlorination is the velocity-determining part of a two-step bioremediation process. More research will be necessary to accelerate this anaerobic treatment.

The results of the study of the aerobic mineralization demonstrate that both investigated soils are provided with an indigenous DD mineralization potential. In the organic-poor soil, addition of organic substrate enhanced mineralization. The extent of DD mineralization could be enhanced by inoculation with white rot fungi or a lignolytic soil fungus, but the influence of lignolytic fungi differed between the investigated soils.

6. ACKNOWLEDGMENTS

This work was supported by Projekt Wasser-Abfall-Boden (project number PW 94 161) from the Forschungszentrum Karlsruhe GmbH with the help of Baden-Württemberg (Germany) and the Commission of the European Union. We are grateful to Prof. K. Haider and to Dr. R.M. Wittich for synthesizing the ^{36}Cl-labeled hexachlorobenzene and dibenzo-p-dioxin, respectively. We thank Dr. F. Zadrazil, who placed the strains of the white rot fungi at our disposal, and S. Pretzer for excellent technical assistance.

7. REFERENCES

Abramowicz, D.A., M.J. Brennan, H.M. Van Dort and E.L. Gallagher, 1993. Factors influencing the rate of polychlorinated biphenyl dechlorination in Hudson River sediments. Environ. Sci. Technol., 27, 1125-1131.

Adriaens, P. and D. Grbic'-Galic, 1994. Reductive dechlorination of PCDD/F by anaerobic cultures and sediments. Chemosphere, 29, 2253-2259.

Alder, A.C., M.M. Häggblom, S.R. Oppenheimer and L.Y. Young, 1993. Reductive dechlorination of polychlorinated biphenyls in anaerobic sediments. Environ. Sci. Technol., 27, 530-538.

Ali, T.A. and M. Wainwright, 1994. Growth of *Phanerochaete chryosporium* in soil and its ability to degrade the fungicide benomyl. Bioresource Technology, 49, 197-201.

Anderson, T.H. and K.H. Domsch, 1989. Ratios of microbial biomass carbon to total organic carbon in arable soils. Soil. Biol. Biochem., 21, 471-479.

Beurskens, J.E.M., C.G.C. Dekker, J. Jonkhoff and L. Pompstra, 1993. Microbial dechlorination of hexachlorobenzene in a sedimentation area of the Rhine river. Biogeochemistry, 19, 61-81.

Bollag, J.-M., S.-Y. Liu and R.D. Minard, 1980. Cross-coupling of phenolic humus constituents and 2,4-dichlorophenol. Soil Sci. Soc. Am. J., 44, 52-56.

Bosma, T.N.P., J.R. van der Meer, G. Schraa, M.E. Tros and A.J.B. Zehnder, 1988. Reductive dechlorination of all trichloro- and dichlorobenzene isomers. FEMS Microbiol. Ecol., 53, 223-229.

Bumpus, S.A., M. Tien, D. Wright and S.D. Aust, 1985. Oxidation of persistent environmental pollutants by a white rot fungus. Science, 228, 1434-1436.

Bumpus, J.A., 1989. Biodegradation of polycyclic aromatic hydrocarbons by *Phanerochaete chrysosporium*. Appl. Environ. Microbiol., 55, 154-158.

Dolfing, J., 1990. Reductive dechlorination of 3-chlorobenzoate is coupled to ATP production and growth in an anaerobic bacterium, strain DCB-1. Arch. Microbiol., 153, 264-266.

Fathepure, B. Z. and T.M. Vogel, 1991. Complete degradation of polychlorinated hydrocarbons by a two-stage biofilm reactor. Appl. Environ. Microbiol., 57, 3418-3422.

Fortnagel, P., H. Harms, R.M. Wittich, W. Franke, S. Krohn and H. Meyer, 1989. Cleavage of dibenzofuran and dibenzodioxin ring systems by a *Pseudomonas* bacterium. Naturwissenschaften, 76, 222-223.

Hammel, K.E., B. Kalyanaraman, B. and T.K. Kirk, 1986. Oxidation of polycyclic aromatic hydrocarbons and dibenzo[p]dioxins by *Phanerochaete chrysosporium* ligninase. J. Biol. Chem., 261, 16948-16952.

Hendriksen, H.V., S. Larsen and B.K. Ahring, 1992. Influence of a supplemental carbon source on anaerobic dechlorination of pentachlorophenol in granular sludge. Appl. Environ. Microbiol., 58, 365-370.

Holliger, C., G. Schraa, A.J.M. Stams and A.J.B. Zehnder, 1992. Enrichment and properties of an anaerobic mixed culture reductively dechlorinating 1,2,3-trichlorobenzene to 1,3-dichlorobenzene. Appl. Environ. Microbiol., 58, 1636-1644.

Hüttermann, A., J. Trojanowsky and D. Loske, 1988. Verfahren zum Abbau schwer abbaubarer Aromaten in kontaminierten Böden bzw. Deponiestoffen mit Mikroorganismen. German patent, 3731816.

Knackmuss, H.-J., 1975. Über den Mechanismus der biologischen Persistenz von halogenierten aromatischen Kohlenwasserstoffen. Chemiker Zeitung, 99 (5), 213-219.

Linkfield, T.G., J.M. Suflita and J.M. Tiedje, 1989. Characterization of the acclimation period before anaerobic dehalogenation of halobenzoates. Appl. Environ. Microbiol., 55, 2773-2778.

Madsen, T. and J. Aamand, 1992. Anaerobic transformation and toxicity of trichlorophenols in a stable enrichment culture. Appl. Environ. Microbiol., 58, 557-561.

Martens, R., 1982. Concentrations and microbial mineralization of four- to six-ring polycyclic aromatic hydrocarbons in composted municipal waste. Chemosphere, 11, 761-770.

Mileski, G.J., J.A. Bumpus, M.A. Jurek and S.D. Aust, 1988. Biodegradation of pentachlorophenol by the white rot fungus *Phanerochaete chrysosporium*. Appl. Environ. Microbiol., 54, 2885-2889.

Mohn, W.W. and J.M. Tiedje, 1990. Strain DCB-1 conserves energy for growth from reductive dechlorination coupled to formate oxidation. Arch. Microbiol., 153, 267-271.

Morgan, P., S.A. Lee, S.T. Lewis, A.N. Sheppard and R.J. Watkinson, 1993. Growth and biodegradation by white-rot fungi inoculated into soil. Soil Biol. Biochem., 25, 279-287.

Nies, L. and T.M. Vogel, 1990. Effects of organic substrates on dechlorination of Arochlor 1242 in anaerobic sediments. Appl. Environ. Microbiol., 56, 2612-2617.

Reid, I.D., 1979. The influence of nutrient balance on lignin degradation by the white-rot fungus *Phanerochaete chrysosporium*. Can. J. Bot., 57, 2050-2058.

Robinson, K.G., W.S. Farmer and J.T. Novak, 1990. Availability of sorbed toluene in soils for biodegradation by acclimated bacteria. Wat. Res., 24, 345-350.

Tien, M. and T.K. Kirk, 1988. Lignin peroxidase of *Phanerochaete chrysosporium*. In: M. Tien and T.K. Kirk (Editors). Methods in Enzymology, Vol. 161B, 238-249.

Trzesicka-Mlynarz, D. and O.P. Ward, 1995. Degradation of polycyclic aromatic hydrocarbons (PAHs) by a mixed culture and its component pure cultures, obtained from PAH-contaminated soil. Can. J. Microbiol., 41, 470-476.

Weißenfels, W. D., H.-J. Klewer and F. Berger, 1993. Mikrobielle Abbaubarkeit und Biotoxizität von polyzyklischen aromatischen Kohlenwasserstoffen (PAK) in Böden. BioEngineering, 4, 29-34.

Williams, W.A., 1994. Microbial reductive dechlorination of trichlorobiphenyls in anaerobic sediment slurries. Environ. Sci. Technol., 28, 630-635.

Yadav, J.S., J.F. Quensen III, J.M. Tiedje and C.A. Reddy, 1995. Degradation of polychlorinated biphenyl mixtures (Aroclor 1242, 1254, and 1260) by the white rot fungus *Phanerochaete chrysosporium* as evidenced by congener-specific analysis. Appl. Environ. Microbiol., 61, 2560-2565.

Zadrazil, F. and H. Brunnert, (1981). Investigation of physical parameters important for the solid state fermentation of straw by white rot fungi. Eur. J. Appl. Microbiol., 11, 183-188.

TEST METHOD FOR DETERMINING THE ACID PRODUCTION POTENTIAL OF SULFUR TREATED SOILS

P. Cantin[1], A. Karam[2] and R. Guay[1]

[1]Département de Biologie Médicale, Faculté de Médecine, and [2]ERSAM, Faculté des Sciences de l'Agriculture et de l'Alimentation, Université Laval, Ste-Foy (Québec), CANADA, G1K 7P4

INTRODUCTION

The disposal of sulfur waste products from industry in soil can have significant environmental impact. With increased concerns regarding environmental pollution, a greater effort has been made to investigate the effect of sulfur waste products on some chemical properties of soils, especially pH and levels of toxic metals such as aluminium, cadmium, manganese, nickel, copper and zinc. Sulfur waste products are considered as an environmental hazard because of their potential to generate mineral acids.

It is well known that heterotrophic and autotrophic bacteria, e.g. organisms belonging to the genus *Thiobacillus*, can oxidize many forms of sulfur to sulfuric acid, especially elemental sulfur (S^0), metal sulfides and sulfur compounds such as thiosulfate and polythionates (Kelly and Harrison, 1984). This acidity solubilizes heavy metals which reache groundwater, and have further detrimental effects on vegetation.

Several static and kinetic procedures have been developed and are used in Canada and the United States by research organizations and commercial laboratories to predict the acid production, as well as the acid consumption characteristics of mine tailings and waste rocks (Coastech Research Inc., 1991).

The static test is an acid-base procedure which aims to provide a quantitative measure of the net neutralization potential of a mine tailing sample. This procedure compares the maximum acid generation potential based on the stoichiometry of complete sulfur oxidation with the capacity of the sample to consume acid (Sobek et *al.*, 1978).

The kinetic test is designed to determine if sulfide oxidizing bacteria can produce more acid from oxidation of pyrite (mine tailing sample) than the amount of acid that can be consumed by an

Effect of Mineral-Organic-Microorganism Interactions on Soil and Freshwater Environments
Edited by Berthelin, *et al*, Kluwer Academic / Plenum Publishers, New York, 1999

equal quantity of the same mine tailing sample (Duncan and Bruynesteyn, 1979). This is an active test which involves inoculating the sample with a bacterial culture.

However, in the case of the soil materials, there is no specific method to predict wether acid drainage will be generated in sulfur contaminated soils. Prediction of acid production from biological oxidation of sulfur compounds in soil environments is likely to differ than that predicted in mining environments.

The objective of this study was to develop a predictive method based on static and kinetic procedures to evaluate the acidification potential of five soils supplemented with S^0.

MATERIAL AND METHODS

Soil Sampling and Analysis

Four agricultural soils (S15, S23, S28 and S30) and one soil from an industrial site (C7) have been collected for the purpose of this study. The soils were air-dried, ground and passed through a 2-mm sieve prior to analysis. Soil pH was determined in distilled water at a 1:3 soil:water ratio (pH_w) and organic matter was determined using the modified Walkey-Black method (McKeague, 1978). Total sulfur (S_{tot}), including both organic and inorganic S, was determined after incineration in a LEKO S analyzer. Particle size analysis was performed by the hydrometer method on separate samples of the <2-mm soil fraction (McKeague, 1978). Calcium carbonate equivalent was determined with the acid-neutralization method (Allison and Moodie, 1965). The general properties of the soil samples are listed in Table 1.

Table 1. Selected properties of the studied soil samples.

Soil no	Texture	pH_w	Organic matter (%)	S_{tot} (%)	$CaCO_3$ equivalent (%)
C7	Loamy sand	7.6	7.5	0.12	3.3
S23	Loam silty clay	7.8	2.9	0.03	5.3
S28	Clay	7.0	5.2	0.04	4.9
S15	Loam silty clay	7.1	3.6	0.02	2.5
S30	Loamy sand	6.4	1.6	<0.01	2.9

Microorganisms and Culture Media

The bacteria *T. thiooxidans* ATCC 55128 and *T. thioparus* ATCC 23645 were cultivated in basal alts liquid culture media adjusted to pH 4 (BS4) and pH 7 (BS7) respectively (Barton and Shively, 1968). Both BS4 and BS7 media were supplemented with 0.5% (W/V) of elemental sulfur (S^0) as energy source, and following inoculation, were incubated with gyratory shaking at 30^0C. One mL of 7-day cultures of both bacteria were used to inoculate soil suspensions for the kinetic step.

For both static and kinetic procedures, the BS7 culture medium was modified to lower its buffer capacity. This modified medium (BS7-m) contained per liter: 0.3 g KH_2PO_4, 0.2 g Na_2HPO_4, 0.4 g $(NH_4)_2SO_4$, 0.5 g $MgSO_4 \cdot 7H_2O$, 0.20 g $CaCl_2$ and 0.01 g $FeSO_4 \cdot 7H_2O$ and the pH was adjusted to 7.0.

Prediction Procedure

Static Step

The acid titration measures the resistance to pH change as hydrogen ion (H^+) is added to soil in presence of BS7-m without bacteria. The aim of the static step is to determine the pH_{stat} which represents the pH at which the whole sulfur content of the soil is oxidized to sulfuric acid. The total soil S is converted stoichiometrically to the sulfuric acid equivalent. This value, referred to as S_{stat}, is applied to the acid titration curve of the soil. The pH which corresponds to S_{stat} is called pH_{stat}. S_{stat} values for 0.5 and 2% S^0 treated soils were respectively 30 and 120 cmoles H^+/kg soil.
Acid titration curves were obtained for each soil sample as follows: 10 g of soil sample were shaken for 1 min with 100 mL of BS7-m culture medium (without addition of bacteria). Incremental aliquots (0.25 mL) of 1N H_2SO_4 were successively added to the soil suspension. After the equilibrium period (24 h) under aerobic conditions, the pH value in the supernatant was measured. This step was repeated until the total amount of added acid achieved 50 cmoles H^+/kg soil.

Kinetic step

The objective of the kinetic step is to assess the biological oxidation of soil sulfur species and to determine the pH_{kin}. It follows a rather simple approach: a soil sample is suspended in presence of active bacterial cultures. The bacteria oxidize the S^0 into sulfuric acid and this causes the pH to rapidly decrease in the soil-culture medium. When the pH value reaches a constant minimum, the pH is recorded and noted as kinetic pH (pH_{kin}).
Biological sulfur oxidation curves were obtained for each soil as follows: 10 g of soil sample was shaken for 1 min with 100 mL of BS7-m culture medium. S^0 was added at rates of 0, 0.5 and 2.0% (W/W) to simulate sulfur-contaminated soils from some areas of Quebec Province, Canada. Soil suspensions were inoculated with cultures of *T. thioparus* and *T. thiooxidans* and incubated with gyratory shaking at 30°C. The pH of the suspensions was recorded every 2 or 3 days up to 6 weeks of incubation period.
The advantage of the kinetic procedure is to ascertain the oxidizable sulfur content of a soil specimen. The bacterial species used in the kinetic test, *T. thioparus* and *T. thiooxidans*, are the most performing sulfur oxidizing bacteria (in terms of yield and kinetics of oxidation of sulfur compounds) in a culture medium. These organisms are preferred for testing even if heterotrophic bacteria are considered by some researchers to be most effective in agricultural soils because of their relatively large numbers compared to autotrophic bacteria (Janzen and Bettany, 1987; Lawrence and Germida, 1988). The use of the two species has multiple advantages. First, these organisms are the most potent sulfur oxidizers in natural environments: *T. thiooxidans*, for example, can withstand a pH under 1, and still maintain its viability (Kelly and Harrison, 1984). Second, the two species can join forces and oxidize sulfur over a wide range of pH from 7.0 to 0.5 (Cantin, 1995). Thirdly, these aerobic and chemolithotrophic bacteria have very simple growth requirements: sulfur as an energy source, atmospheric CO_2 as a carbon source, a basal salts solution as culture medium and O_2 as electron acceptor. In the presence of soils that would be suspected to contain pyrite (FeS_2), from natural or anthropogenic origin, one could also add *T. ferrooxidans* to the bacterial inoculum. This species shares the same simple growth requirement as the two others, and can oxidize both Fe^{2+} and S^{2-}, leading to a more accurate evaluation of the acid production potential of these soils.

RESULTS AND DISCUSSION

The acid titration curves (Fig. 1) for soils with initial $pH_w \leq 7.1$ (S15, S28 and S30) followed the typical pattern of buffer curves for mineral soils (Neilsen *et al.*, 1995); these show a weak buffer capacity between pH 7 and 4, but exhibit well defined buffer plateau below pH 3. However, the titration curves for the soils with initial $pH_w > 7.1$ (S23 and C7) showed a high buffering capacity up to 20 cmoles H^+/kg soil. The pH of the soil S23 (highest carbonate content) declined rapidly prior to reach a buffered plateau (pH 2.9) due to the rapid dissolution of carbonate ions. The soil C7 (highest organic matter content) exhibited the highest buffering due to the high initial soil pH. These data indicate that the acidification potential of the five soils was controlled by the solubility of the carbonate alkaline ions, organic matter content and the concentration of the hydrogen ion added to the soil suspension over the pH range 7.0-2.0.

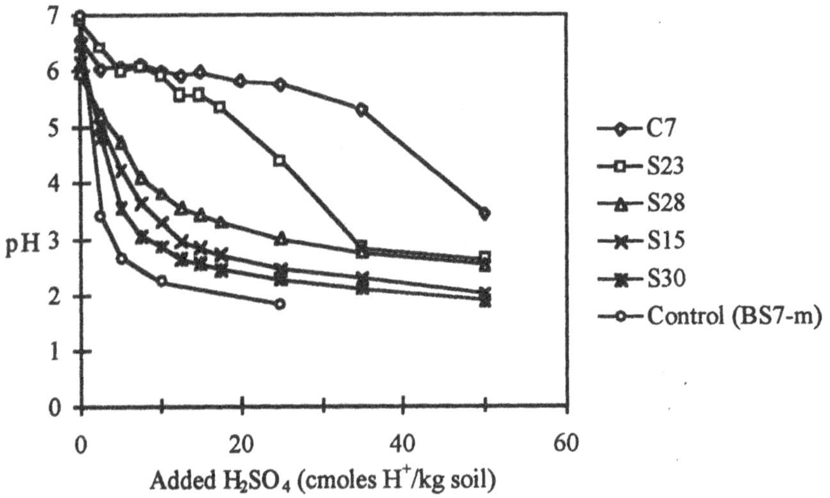

Figure 1. Acid titration curves of soils

Changes in pH in soil samples treated with three levels of S^0 application and inoculated with bacteria are illustrated in Fig. 2. The overall acidification of the soil samples was dependent of the initial soil pH, S^0 application rate and incubation time. Generally, pH decreased with increasing levels of S^0. In addition to the effect of S^0 level on pH, there was an obvious decrease in pH with time of exposure in soil (Fig. 2).

Biological S^0 oxidation curves differed among soils throughout the entire range of incubation time. As shown in Fig. 2, acidification was more delayed in the C7 and S23 soils (highest initial soil pH) than in other soils. These observations are in agreement with the observations of Arkesteyn (1979) concerning the effect of pH on pyrite oxidation. The soils without added sulfur showed a slight pH increase indicating that no sulfuric acid had been produced. At the 2% level of S^0 application, the pH of the soil samples shifted to about 2.0-1.5. At the 0.5% level of S^0 application, the magnitude of the pH drop varied among soil samples. In the case of C7 (pH_w 7.6), the pH decreased slowly but was maintained above 6. However, with S30, the less buffered soil (loamy sand) with the lowest initial pH_w, the pH value of the soil suspension decreased to about 2.0. In all cases, the pH of the S^0 treated soil suspensions had a tendency to stabilize between 10 and 40 days and was termed pH_{kin} (Table 2).

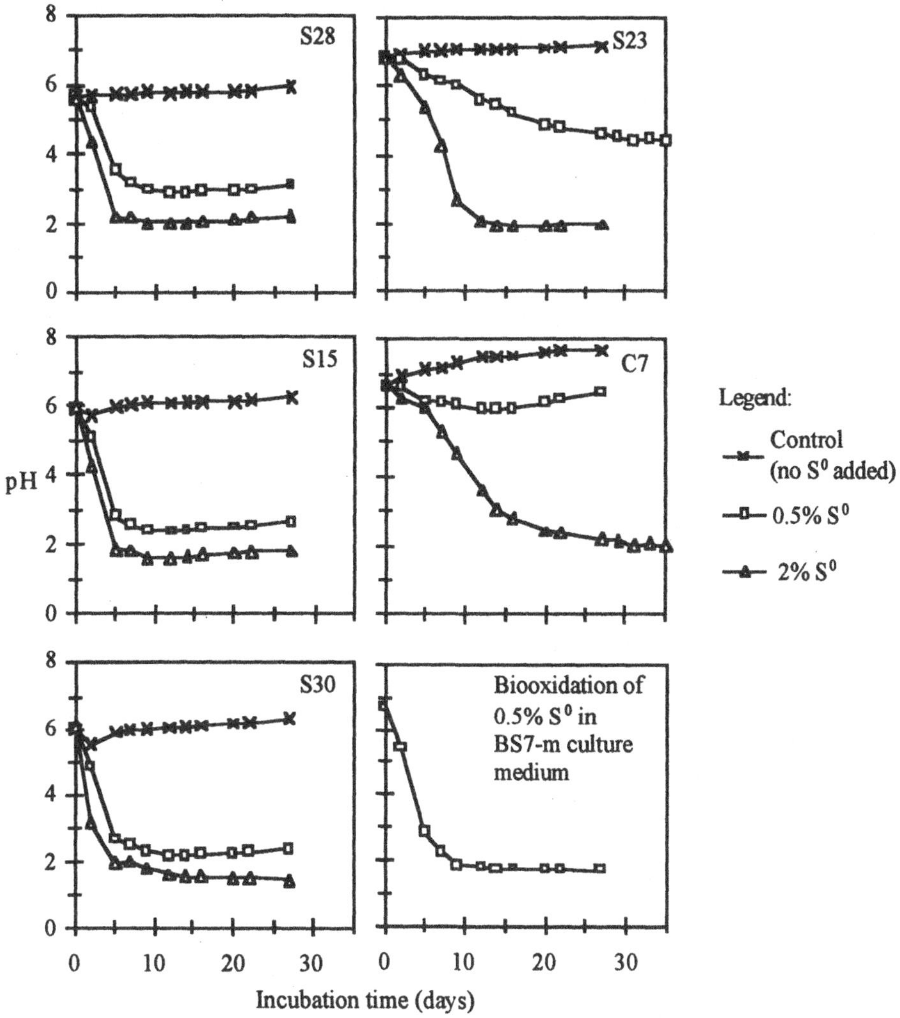

Figure 2. Biological S^0 oxidation curves

Table 2. Predicted pH_{stat} values derived from both total sulfur content of five soil samples and titration curves, and pH_{kin} values obtained from growth curves of bacteria in soil suspensions supplemented with elemental sulfur

Soil no.	Rate of added S^0 (%)	pH_{stat}	pH_{kin}
C7	0	6.6	[b]6.7
	0.5	5.6	5.9
	2	[a]<1.9	2.0
S23	0	6.9	6.9
	0.5	3.6	4.4
	2	<2.7	1.9
S15	0	6.2	5.9
	0.5	2.4	2.4
	2	<2.0	1.6
S28	0	6.0	5.7
	0.5	2.9	2.9
	2	<2.5	2.0
S30	0	6.5	6.0
	0.5	2.2	2.2
	2	<1.9	1.4

[a] The indicated pH_{stat} value is obtained from the end point of the titration curve.
[b] The pH_{kin} values for the control S^0 treatment (0%) were taken at the time 0 of the incubation period.

Predicting Resistance to Acidification from Static Procedure, Equivalent Sulfuric Acid and Kinetic Procedure

In order to properly interpret pH_{stat} and pH_{kin}, a third parameter was defined, namely the minimal pH (pH_{min}). The pH_{min} would be selected on the basis of the futur use of the S-contaminated land. For general agricultural purposes, the pH_{min} was choosen as 5.5, in order to reduce the activity or solubility of aluminium and manganese. It is well-known that aluminium toxicity is probably the most important growth-limiting factor in many mineral agricultural soils, particularly those having pH values below 5.0 to 5.5 (Tisdale *et al.*, 1985). On the other hand, a pH_{min} value of 5.5-5.6 could be satisfactory from the standpoint of minimum toxicity and adequate availability of micronutrients in many mineral soils (Tisdale *et al.*, 1985).

Assessment of sensitivity of soils to acidification is illustrated in Fig. 3. If the pH_{stat} value is higher or equal to pH_{min}, the sample is not considered as a potential acid producer, and no kinetic procedure is then required. This means that the amount of sulfuric acid that can be potentially produced from the oxidation of total soil S is not sufficient to overcome the acid neutralizing capacity of the soil, and consequently bring the pH under the minimal pH value. If the pH_{stat}

value is lower than pH_{min}, a microbiological kinetic procedure is performed out to assess the biologically oxidizable forms of sulfur. The same type of comparisons are then made. If the pH_{kin} is higher than pH_{min}, the soil sample is not considered as a potential acid producer. It further means that the biologically oxidizable fraction of total soil S is not sufficient to overcome the acid neutralizing capacity of the soil, and bring the pH under the minimal pH value. On the other hand, if the pH_{kin} is lower than pH_{min}, the sample will be considered an acid producer and particular attention will be directed to its management.

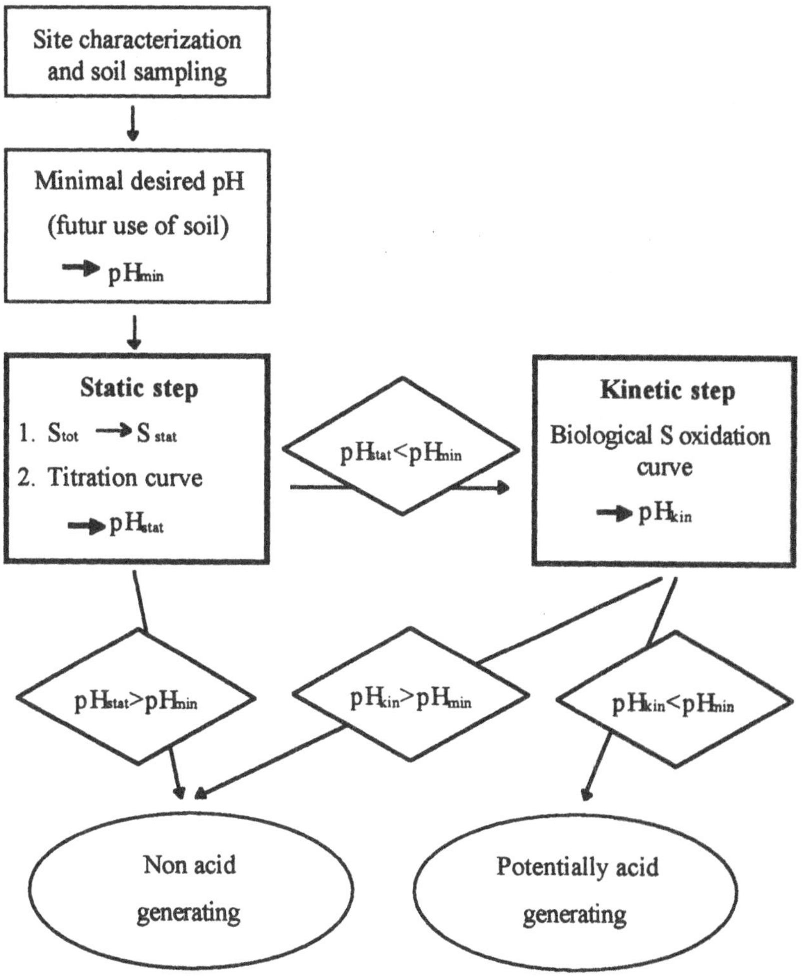

Figure 3. Steps in the determination of the sensitivity of soils to acidification

In the present study, given the selection of a pH_{min} value of 5.5, the data of either pH_{stat} or and pH_{kin} show that all soil samples tested are, *per se*, non-acid producers (without added S^0). However, for the S^0 treated soils (0.5% S^0), most soils (except C7) exhibit pH_{stat} and pH_{kin} lower than pH_{min}, and should then be considered acid producers. Finally, since none of the five soils treated with 2% S^0 exhibited pH_{stat} and pH_{kin} higher than pH_{min}, they should be considered as potentially acid producers. Thus, only the static step would be sufficient to predict the acid production potential of these five S^0 treated soils.

CONCLUSION

The proposed method for the determination of acid production potential of soils treated with S^0 takes into account both data of acid titration curves and sulfuric acid equivalent in the static procedure, and of biological S oxidation curves in the kinetic procedure. New parameters have been defined from both procedures and interpreted using appropiate rules. The results suggested that acid production capacity of S^0 treated soils would be influenced by initial soil pH and S^0 application rate. In general, S^0 treatments decreased considerably the initial soil pH values when incubated with sulfur oxidizing bacteria. The proposed testing method could be used for assessing the acid production potential of soils containing high sulfur concentration.

ACKNOWLEDGEMENTS

This research was supported by the Fonds de Recherche et de Développement Technologique en Environnement, Ministère de l'Environnement et de la Faune du Québec.

REFERENCES

Allison, L.E. and C.D. Moodie, 1985. Carbonate. In: C.A. Black (Editor). Methods of soil analysis, part 2, Chemical and microbiological properties. American Society of Agronomy, pp. 1379-1396.

Arkesteyn, G.J.M.W., 1979. Pyrite oxidation by *Thiobacillus ferrooxidans* with special reference to the sulphur moiety of the mineral. Antonie van Leeuwenhoek J. Microbiol. Serol., 45, 423-435.

Barton, L.L. and J.M. Shively, 1968. Thiosulfate utilization by *Thiobacillus thiooxidans* ATCC 8085. J. Bacteriol., 95, 720.

Cantin, P., 1995. Lixiviation du phosphore de l'apatite par l'acide sulfurique produit par l'oxydation microbiologique du soufre élémentaire. Mémoire de maîtrise, Université Laval, Canada.

Coastech Research Inc., 1991. Acid Rock Drainage Prediction Manual (1991) CANMET - MSL Division, MEND Project 1.16.1. Département de l'énergie, mines et ressources, Canada.

Duncan, D.W. and A. Bruynesteyn, 1979. Determination of acid production potential of waste materials. Mtg. Soc. AIME, Paper A-79-29.

Janzen, H.H. and J.R. Bettany, 1987. Oxidation of elemental sulfur under field conditions in central Saskatchewan. Can. J. Soil Sci., 67, 609-618.

Kelly, D.P. and P. Harrison, 1984. Genus *Thiobacillus*. In: Bergey's manual of systematic bacteriology (vol. 3). William and Wilkins Co., Baltimore, USA.

Lawrence J.R. and J.J. Germida, 1988. Relationship between microbial biomass and elemental sulfur oxidation in agricultural soils. Soil Sci. Soc. Am. J., 52, 672-677.

McKeague, J.A., 1978. Manuel de méthodes d'échantillonnage et d'analyse des sols (2e édition). Comité Canadien de Pédologie, Ottawa, Ontario.

Neilsen, D., P.B. Hyot, P. Paromchuk, G.H. Neilsen and E.J. Hogue, 1995. Measurement of the sensitivity of orchard soils to acidification. Can. J. Soil Sci., 75, 391-395.

Sobek, A.A., W.A. Schuller, J.R. Freeman and R.M. Smith, 1978. Field and laboratory methods applicable to overburden and mine soils. EPA 600/2-78-054.

Tisdale, S.L., W.L. Nelson and J.D. Beaton, 1985. Soil fertility and fertilizers (Fourth edition). Macmillan Publishing Company, New-York.

Wainwright, M., 1984. Sulfur oxidation in soils. Adv. Agron., 37, 349-396.

USE OF PYROPHOSPHATE TO EXTRACT EXTRA- AND INTRACELLULAR ENZYMES FROM A COMPOST OF MUNICIPAL SOLID WASTES

J. C. Rad(*), M. Navarro-González and S. González-Carcedo

Laboratory of Pedology & Agricultural Sciences. Faculty of Food Technology & Chemistry. University of Burgos. 09080 Burgos. CASTILE (SPAIN). E-mail: Salva@cid.cid.ubu.es
(*) Author for correspondence.

1. INTRODUCTION

The agronomic utilization of organic wastes such as sewage sludge or municipal solid waste composts may be an important way to recycle nutrient elements and to increase the fertility of soils. The application of organic wastes improves soil structure (Pagliai and De Nobili, 1993) and increases soil microbial biomass (Perucci, 1990, 1992) and the levels of enzymatic activities (Giusquiani *et al.*, 1995; Serra-Wittling *et al.*, 1996). In general, higher activities were found in organic amendments than in soils, reflecting the higher microbial activity of the compost compared to soil (Martens *et al.*, 1992).
The organic part of urban solid wastes contains non-humic substances such as carbohydrates, lipids and proteins from plant, animal or microbial origin (He *et al.*, 1992). During composting, organic components of MSW are decomposed and transformed through complex chemical and microbiological reactions into humic-like substances (Inbar *et al.*, 1990). The contents of C, H, and O of humic acids (HA) from MSW-composts fall in the range for corresponding HA's from soils, but their N contents are significantly higher from the incorporation of proteins into the HA-fraction during composting (González-Vila and Martin, 1985).
Enzymatic activities play an important role in the process of organic matter decomposition in soils (Burns, 1983; Sinsabaugh *et al.*, 1991) and in the fermentative steps of a composting process, making possible their use as indicators of compost maturity (Foster *et al.*, 1993; Herrmann and Shann, 1993). Changes in enzymatic activities showed that the age of compost of organic wastes could be determined by microbial and enzymatic activity levels, older composts having less activity than younger ones (García *et al.*, 1992). Perucci (1990) found that the addition of municipal solid waste-composts to soil increased the global microbial activity and some soil enzyme activities permanently during a 1-year incubation experiment depending on the MSW-compost application rate. Serra-Wittling *et al.* (1995) showed that soil enzymatic activities were increased in a non-additive way with the ratio of compost added to soil, and only intracellular enzymes like dehydrogenases were significantly correlated with C mineralization.
A number of extracellular enzymes catalyze the hydrolytic degradation of biopolymers into their constitutive monomers, sugars, amino acids or nucleotides, that are thereafter mineralized. These extracellular soil or compost enzymes were released from microbial cells and could be physically adsorbed onto inorganic colloids or bond to organic matter by ionic or covalent bounds, becoming persistent in soil or compost even under adverse conditions (Sarkar *et al.*, 1989; Boyd and Mortland, 1990). Compost organic matter is a humic-like material (Inbar *et al.*,

1990) that makes possible the existence of humus-enzyme complexes with similar properties to those found in soils.

Extracellular soil enzymes can be extracted in high yields only when humic matter is also put ino solution (Tabatabai and Fu, 1992). The extraction of soil organic matter using alkali reagents or ultrasonic vibration can cause pronounced lysis of soil organisms. However, Nannipieri *et al.* (1974), using 0.1 M pyrophosphate, extracted about 30 to 40% of the total soil urease activity. Under these conditions, the number of ureolytic microorganisms was unaffected, so it is assumed that the extracted enzymes were extracellular. Other enzymes such as phosphatases, urease, casein- and BAA-hydrolases and soluble organic matter were also extracted with 0.1 M (pH 7) sodium pyrophosphate (Nannipieri *et al.*, 1980). The aims of this work are to determine the state of hydrolytic enzymes in a MSW-compost, study the influence of pH and pyrophosphate concentration on their extraction, and quantify by ultrafiltration their extra- and intracellular forms.

2. MATERIAL AND METHODS

2.1. Physico-Chemical Properties of Municipal Solid Waste-Compost

MSW was sampled after six months of composting from the dumping place of Burgos town (Castile, Spain), 2 mm sieved, milled with agate balls in a Pulverissete 6 system and stored frozen at -18°C. The physical, chemical and enzymatic properties of the MSW-compost are summarized in Table 1.

Table 1. Physical and chemical properties of the municipal solid waste compost used in this study. All results are given on a dry weight basis (oven dry).

Ash		77.7%	
Density		2.53 Kg l^{-1}	
Dry matter		96.5%	
pH		7.78	
Organic Matter		23.46%	
N_t		0.576 g N Kg^{-1}	
P_t		0.376 g P Kg^{-1}	
K_t		0.480 g K_2O Kg^{-1}	
Cd	5.5 mg Kg^{-1}	Cu	251.8 mg Kg^{-1}
Pb	626.6 mg Kg^{-1}	Ni	87.8 mg Kg^{-1}
Cr	130.6 mg Kg^{-1}	Fe	20725.9 mg Kg^{-1}
Mn	353.1 mg Kg^{-1}	Al	14404.3 mg Kg^{-1}
Acid Phosphatase (pH 5.6)		0.871 µmol pNP g^{-1} h^{-1}	
Alkaline Phosphatase (pH 10.5)		10.921 µmol pNP g^{-1} h^{-1}	
β-D-Glucosidase (pH 5.6)		0.152 µmol pNP g^{-1} h^{-1}	
CM-Cellulase (pH 6)		2.932 µmol glu g^{-1} 24 h^{-1}	
Casein-Protease (pH 7.5)		37.125 µmol tyr g^{-1} 24 h^{-1}	

2.2. Enzyme Extraction

Five g of MSW-compost were suspendedand in 100 ml $Na_2H_2P_2O_7$ at different concentrations (0.2, 0.1 and 0.05 M) and pH (2-11), mixed five hours with vibrational agitation (200 obsc min^{-1}) and centrifuged (21000 *g* for 30 min at 4°C). Suspended particles were discarded by filtration.

2.3. Fractionation by Ultrafiltration

MSW-extract (50 ml) was fractionated by ultrafiltration with tangential flow (Minitan System, Millipore) through a 0.45 µm membrane filter plate (HVLP type, Durapore®) and washed with a continuous flux of distilled water (200 ml) until clear filtrate was obtained. The retentate was the UF_3 fraction and contained mainly cell-bound enzymes. The eluate was ultrafiltrated through a membrane filter plate (PTGC type, Polysulfone) with a 10 KD exclusion limit. The filtrate was concentrated to 50 ml and thereafter washed with a continuous flux of 300 ml distilled water -sufficient volume to assure the complete disappearance of salts and low molecular weight compounds. The new retentate (UF_2 fraction) contained cell-free molecules. The last eluate, containing the lowest MW compounds and inorganic salts, was concentrated to 50 ml in a rotary evaporator at 40°C (UF_1 fraction). A schematic diagram of the fractionation is presented in Fig. 1.

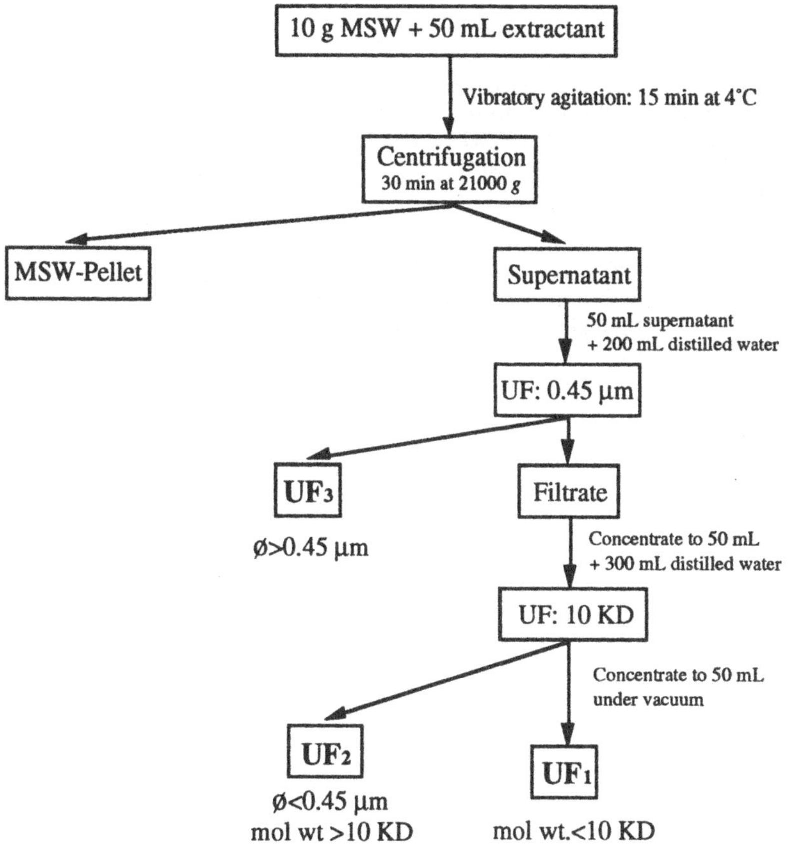

*Extractant: $Na_2H_2P_2O_7$ at different concentrations and pH.

Figure 1. Extraction and ultrafiltration of enzymes extracted from MSW-compost.

2.4. Enzymatic Assays

Phosphatase activities in extracts were assayed using 110 mM *p*-nitrophenyl phosphate (*p*NPP) as artificial substrate, in 0.2 M citrate buffer at pH 5 (Moss, 1984), and 0.1 M 2-amino-2-methyl-1-propanol with 1 mM $ZnSO_4$ and 2 mM $MgSO_4$ at pH 10.5 (Bretaudiere and Spillman, 1984) for acid and alkaline phosphatases, respectively. After 1 h of incubation at 37°C, the *p*-nitrophenol (*p*NP) liberated as a consequence of the enzymatic hydrolysis was measured colorimetrically at 410 nm in alkaline medium.

CM-cellulase was assayed with 2% (w/v) carboxymethylcellulose (CMC) in 0.2 M acetate buffer (pH 4.8) (IUPAC, 1984). The increase in reducing sugars was measured by the Somogy-Nelson method after 24 h of incubation at 37°C (Nelson, 1944). To avoid microbial proliferation 0.1% Na_3N was added to the buffer.

β-D-glucosidase was measured with 25 mM *p*-nitrophenyl-β-D-glucopyranoside (*p*NPG) as artificial substrate in 0.2 M acetate buffer (pH 5.5). The *p*-nitrophenol (*p*NP) released after 1 h of incubation at 37°C was measured at 410 nm in 0.1 M tris-hydroxy-methyl-amino-methante (THAM) buffer, pH 12 (Hayano and Katami, 1977).

Casein-proteases (endo- and exopeptidases) were tested using casein as substrate in 0.1 M tris-phosphate buffer (pH 7.5) with 0.1% Na_3N. Amino acids liberated after 24 h of incubation at 37°C were measured by the Folin method after TCA precipitation of non-hydrolyzed casein (Walter, 1984).

Soluble proteins were measured by the Folin method (Lowry *et al.*, 1951). Triplicate assays were made for all determinations. For enzymatic analysis, two controls, in which the substrate was added after the incubation, were run for each sample.

3. RESULTS

The use of 0.2 M sodium pyrophosphate produced different patterns for the extraction of extra- and intracellular enzymes from MSW-compost. Extractable extracellular proteins (UF_2) were mainly solubilized at near-neutral or alkaline pH values with two extraction maxima at pH 6 and 9 (Fig. 2). Cell-bound proteins (UF_3) showed uniform extraction levels independent of the pH value of pyrophosphate solution.

Extracellular alkaline phosphatase had its highest extraction at pH 6; the intracellular form of this enzyme displayed a wider extraction range, between pH's 6 to 10.7 (Fig. 3). Acid phosphatase (AcPA) in the UF_2 fraction had the highest yield of extraction at pH values below 5 and a second peak of extraction at pH 8 (Fig. 3). The ratio between acid and alkaline phophatase activities was less than unity in the pyrophosphate extract, but this ratio was higher than unity in the MSW-compost.

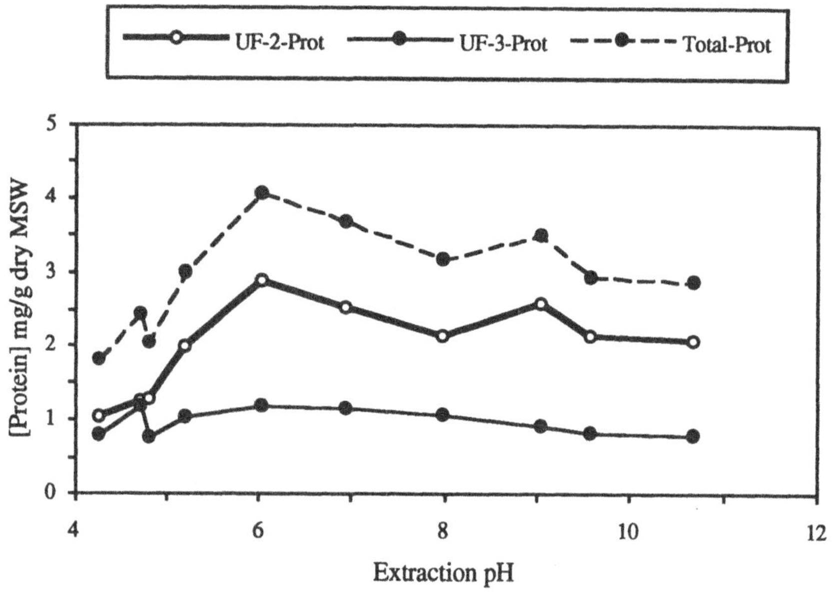

Figure 2. Extraction of intra- and extracellular proteins from MSW-compost using 0.2 M sodium pyrophosphate.

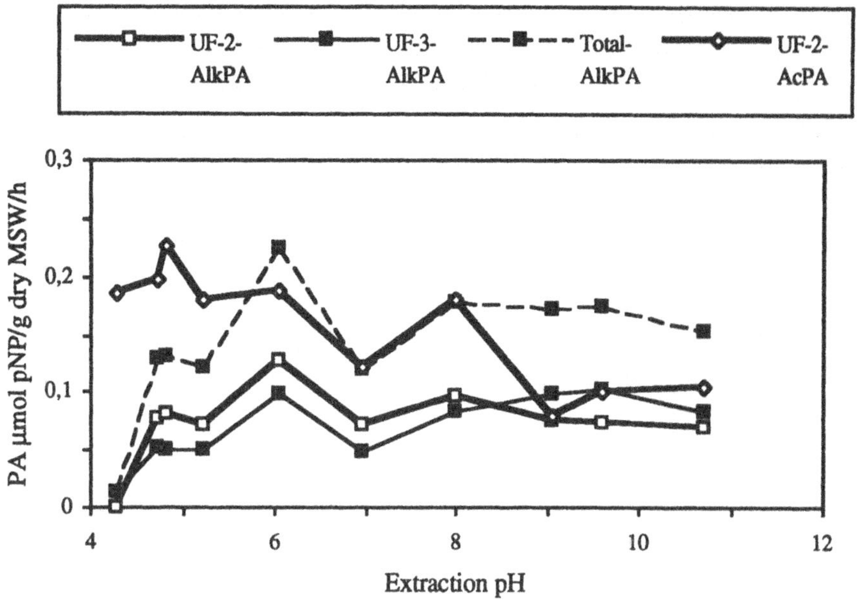

Figure 3. Acid (AcPA) and alkaline phosphatase (AlkPA) extraction from MSW-compost using 0.2 M sodium pyrophosphate.

The total amount of activity extracted with 0.2 M pyrophosphate was 26% and 2% of the acid and alkaline PA contained in the compost, respectively. For cellulolytic and proteolytic enzymes the activity displayed by the extracted enzymes was higher than those in the compost sample. The enzymes related to cellulose hydrolysis -β-D-glucosidase (β-GA) and CM-cellulase (CelA)- showed different extraction profiles for their extracellular forms and very similar profiles for their intracellular forms (Fig. 4 and 5). Extracellular β-GA and CelA in the UF_2 fraction had activities two- or three-folds higher than enzymes of the UF_3 fraction, as reflected by the ratio between the activities of both fractions ($r_{2/3}$). Both enzymes had two extraction maxima in the UF_2 fraction: cellulase at pH 7 and 9 and β-glucosidase at pH 6 and 10.7. Enzymes in the intracellular fraction showed a wide and similar range of extraction pH.

The pH extraction profile for extracellular casein-proteases showed a maximum at pH 9, while intracellular proteases had their highest extraction at alkaline pH values (Fig. 6). The values of the $r_{2/3}$ ratio for the casein-proteases were near unity, except for pH lower than 5 where the extraction of intracellular enzyme was predominant.

The extraction of proteins at 0.1 and 0.05 M pyrophosphate concentrations was low in the UF_3 and did not affect the extraction levels of UF_2 enzymes. Two maxima located at pH 6 and 9 appeared for the UF_2 fraction. The extraction of acid phosphatase was less affected by the variation in ionic strength, but extracellular alkaline phosphatase was increased five-fold by the use of more dilute extracting solutions. Sodium pyrophosphate (0.05 M) at pH 9 was the best extractant for the total of extra- and intracellular AlkPA.

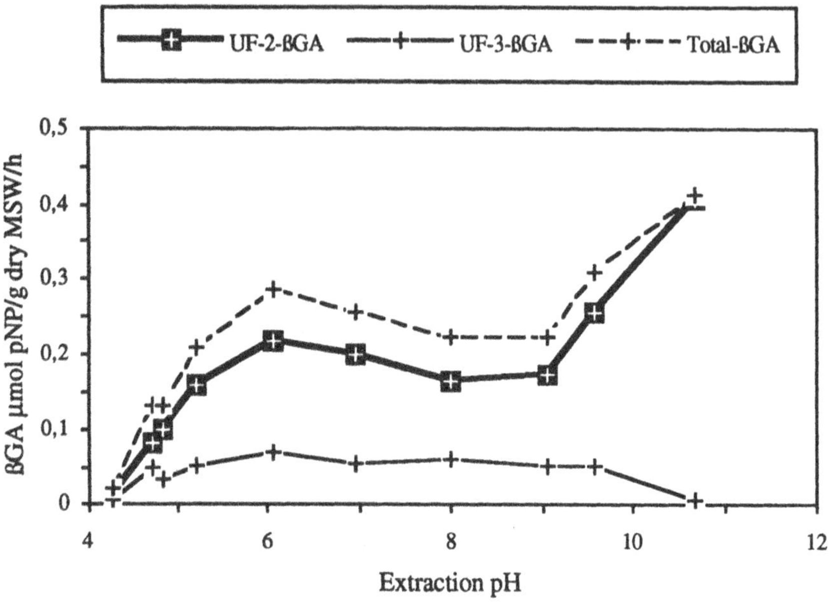

Figure 4. Extraction of β-D-glucosidase from a MSW-compost using 0.2 M sodium pyrophosphate.

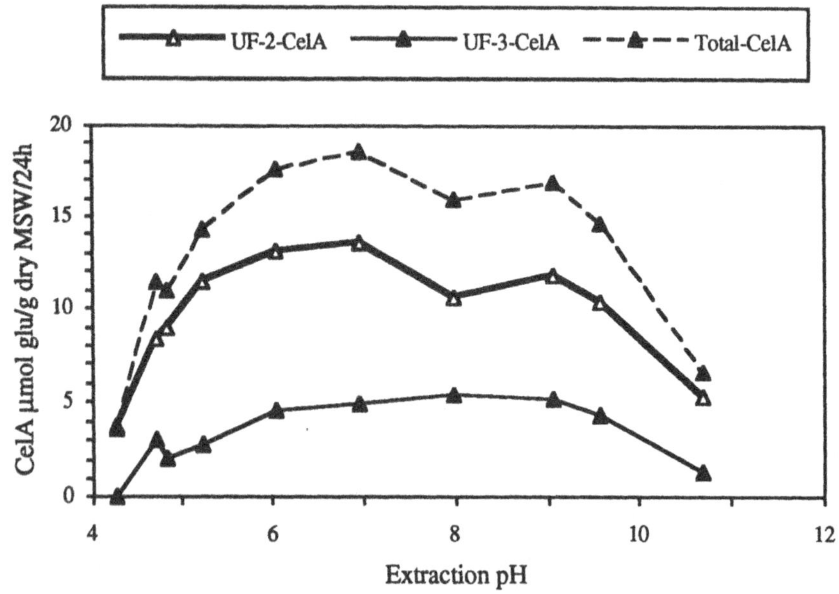

Figure 5. pH profile for the extraction of CM-cellulase (CelA) from MSW-compost using 0.2 M Na-pyrophosphate.

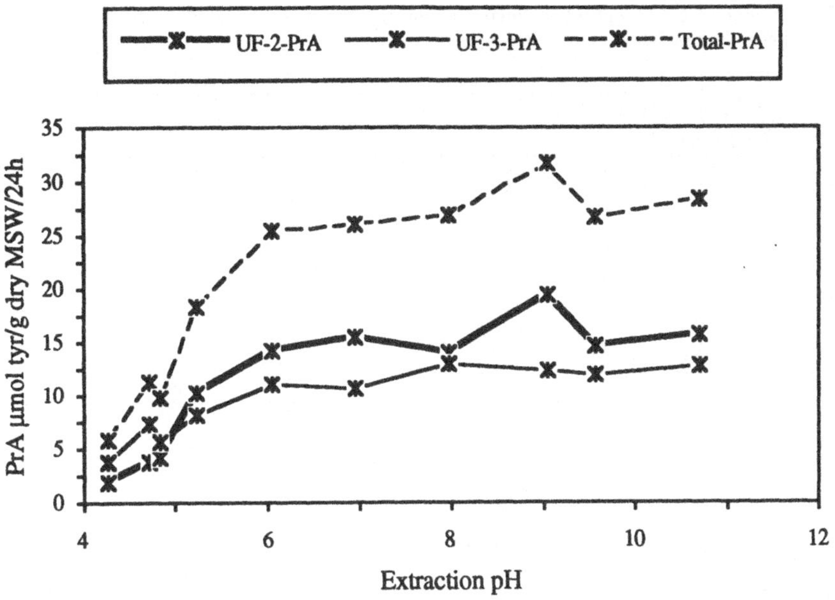

Figure 6. Extraction of casein-proteases (PrA) from MSW-compost using 0.2 M sodium pyrophosphate.

For β-GA, the highest extraction was obtained with 0.1 M pyrophosphate (pH 7) for the extracellular form, but the use of 0.05 M pyrophosphate was best for the sum of β-GA in both fractions. The extraction pattern for extracellular cellulases and proteases confirmed the existence of two extraction maxima at pH 7 and 9 for CelA and the maximum at pH 9 for PrA. In this case, by reducing the ionic strength of the extractant, high amounts of enzyme extraction were obtained for extra- and intracellular cellulases and especially, for proteases.
Positive correlations appeared between enzymatic activities in the UF_3: 0.910 ($p \leq 0.0001$) for the pair AlkPA β-GA, 0.875 ($p \leq 0.0001$) for AlkPA PrA and 0.854 ($p \leq 0.0001$) for β-GA PrA. In the UF_2 fraction, positive correlations existed between protein values and all enzymatic activities tested, mainly with PrA (0.815, $p \leq 0.0001$). Between enzymatic activities, correlations were lower than those found in UF_3 fraction: 0.811 ($p \leq 0.0001$) for β-GA PrA and 0.786 ($p \leq 0.0001$) for both AlkPA β-GA and AlkPA PrA. Comparing protein values in both fractions, there was no significant correlation between extra- and intracellular proteins or between the different enzymes. The intracellular activity of CelA was better correlated with extracellular values of proteins and enzymes in the UF_2 fraction (0.902, $p \leq 0.0001$ with AlkPA; 0.859, $p \leq 0.0001$ with PrA; 0.678, $p \leq 0.001$ with βGA and 0.624, $p \leq 0.003$ with proteins) than with the corresponding intracellular values in the UF_3.

4. DISCUSSION

Hydrolytic enzymes and proteins were successfully extracted from compost using 0.2 M pyrophosphate at different pHs, but with different yields for each one. Only small amounts of alkaline phosphatases were extracted using the optimal conditions tested (11%), but extracted acid phosphatase accounted for 39% of compost enzyme activity. The other enzymes tested showed higher values of enzymatic activity in the ultrafiltered extracts than those displayed by the compost sample. In the extraction of soil humus-enzyme complexes, increases in enzymatic activities as consequence of their solubilization may appear due to the disappearance of sterical hindrance, the increase in the accessibility of active centers, and the absence of diffusional barriers for substrate or products. The highest increases appeared for cellulases and proteases which act upon substrates of low solubility and high MW's (Nannipieri *et al.*, 1982, 1985; McClaugherty and Linkins, 1988).

Table 2. Extraction of extra- and intracellular proteins and enzymes from a MSW-compost using increasing concentrations of sodium pyrophosphate.

Protein (mg g^{-1} dry MSW)							
0.1 M $Na_2H_2P_2O_7$				0.05 M $Na_2H_2P_2O_7$			
pH_e	UF_2	UF_3	$r_{2/3}$	pH_e	UF_2	UF_3	$r_{2/3}$
5.95	2.105	0.760	2.77	6.05	2.157	1.068	2.02
6.89	2.865	0.709	4.04	6.83	2.496	1.263	1.98
7.75	2.352	0.760	3.09	7.57	2.465	1.140	2.16
8.97	2.465	0.678	3.64	8.78	2.660	1.006	2.64
9.56	2.609	0.401	6.51	9.37	1.962	1.273	1.54
Acid phosphatase (µmol pNP g^{-1} dry MSW h^{-1})							
0.1 M $Na_2H_2P_2O_7$				0.05 M $Na_2H_2P_2O_7$			
pH_e	UF_2	UF_3	$r_{2/3}$	pH_e	UF_2	UF_3	$r_{2/3}$
5.95	0.219	0.147	1.49	6.05	0.206	0.135	1.53
6.89	0.174	0.158	1.10	6.83	0.112	0.113	0.99
7.75	0.150	0.140	1.07	7.57	0.139	0.195	0.71
8.97	0.142	0.156	0.91	8.78	0.101	0.150	0.67
9.56	0.125	0.191	0.65	9.37	0.125	0.174	0.71
Alkaline phosphatase (µmol pNP g^{-1} dry MSW h^{-1})							
0.1 M $Na_2H_2P_2O_7$				0.05 M $Na_2H_2P_2O_7$			
pH_e	UF_2	UF_3	$r_{2/3}$	pH_e	UF_2	UF_3	$r_{2/3}$
5.95	0.657	0.078	8.42	6.05	0.425	0.367	1.16
6.89	0.677	0.064	10.58	6.83	0.498	0.470	1.06
7.75	0.534	0.067	7.97	7.57	0.546	0.441	1.24
8.97	0.469	0.076	6.17	8.78	0.568	0.454	1.25
9.56	0.499	0.089	5.61	9.37	0.254	0.378	0.63
β-D-glucosidase (µmol pNP g^{-1} dry MSW h^{-1})							
0.1 M $Na_2H_2P_2O_7$				0.05 M $Na_2H_2P_2O_7$			
pH_e	UF_2	UF_3	$r_{2/3}$	pH_e	UF_2	UF_3	$r_{2/3}$
5.95	0.474	0.038	12.43	6.05	0.377	0.126	2.99
6.89	0.495	0.010	49.50	6.83	0.379	0.141	2.69
7.75	0.426	0.038	11.21	7.57	0.334	0.134	2.49
8.97	0.308	0.036	8.56	8.78	0.280	0.104	2.69
9.56	0.289	0.032	9.03	9.37	0.297	0.105	2.83
CM-cellulase (µmol glu g^{-1} dry MSW 24 h^{-1})							
0.1 M $Na_2H_2P_2O_7$				0.05 M $Na_2H_2P_2O_7$			
pH_e	UF_2	UF_3	$r_{2/3}$	pH_e	UF_2	UF_3	$r_{2/3}$
5.95	10.168	11.328	0.90	6.05	13.313	5.246	2.53
6.89	8.757	9.993	0.88	6.83	18.559	10.492	1.77
7.75	9.468	8.832	0.96	7.57	12.766	10.638	1.20
8.97	11.150	9.186	1.21	8.78	16.866	11.955	1.41
9.56	9.029	9.876	0.91	9.37	13.804	9.259	1.49
Casein-protease (µmol tyr g^{-1} dry MSW 24 h^{-1})							
0.1 M $Na_2H_2P_2O_7$				0.05 M $Na_2H_2P_2O_7$			
pH_e	UF_2	UF_3	$r_{2/3}$	pH_e	UF_2	UF_3	$r_{2/3}$
5.95	22.068	8.275	2.67	6.05	18.165	16.336	1.11
6.89	24.112	6.487	3.72	6.83	19.690	16.817	1.17
7.75	23.764	8.630	2.75	7.57	20.382	17.593	1.16
8.97	24.030	8.173	2.94	8.78	22.916	17.368	1.32
9.56	24.009	7.580	3.16	9.37	20.924	17.500	1.20

Molecular fractions: UF_2 (Ø<0.45 µm, MW>10 KD) and UF_3 (Ø>0.45 µm), $r_{2/3}$ ratio of protein and enzyme concentrations between both fractions.

The use of more diluted extracting solutions showed little effect on proteins and acid phosphatase extraction, but caused the mobilization of large amounts of extracellular proteases and alkaline phosphatases with a positive effect on the activity displayed by the UF_3 fraction. The use of diluted extracting solutions would diminish the denaturing effect on extracellular enzyme stability and avoid cell lysis. Pérez-Mateos *et al.* (1988) obtained the highest extraction yield of extracellular catalase from a brown acidic soil diminishing the concentration of pyrophosphate to 0.01 M at neutral pH. In our research, the extraction of proteins was accomplished with the extraction of humic-like materials that interferred with the measurement of proteins by the Folin method. Lerch *et al.* (1993) showed the accuracy of this method for the determination of proteins in sludges after extraction in water, Triton X-100 solution or 1 M NaOH; García *et al.* (1997) applied this method for the study of extractable soil proteins. The global measurement must be properly called Folin-positive compounds. The correlation in the UF_2 fraction between these values and enzymatic extracted activities revealed the existence of associative mechanisms between compost soluble humic materials and extracellular enzymes. Rad and González-Carcedo (1995) found that the ionic strength of the extractant changed the molecular distribution of humus-phosphatase complexes extracted from MSW-compost, possibly by the ionic nature of this association.

Extracted enzymes and proteins were also fractionated by exhaustive ultrafiltration dividing them between extracellular and cell-bound enzymes. The elimination of soluble salts attained an important increase in total and specific activity of extracted enzymes. Positive correlations between proteins and enzymatic activities in the extracellular fraction UF_2 would indicate the existence of associations between organic matter and enzymatic proteins. These enzymatic complexes had the highest extraction yields at different pH, reflecting differences in their solubilities. There were two extraction maxima located at acidic (5-6) and slightly alkaline (7-9) pH values. Acid and alkaline phosphatases were mainly extracted at acid pH (5 and 6, respectively), proteases at pH 9, CM-cellulases at both pH 7 and 9 and β-glucosidase at the most alkaline pH tested 10.5. In soils, extractable enzymes are closely associated with soluble proteins and humic compounds, making possible the existence of a multi-enzyme humic complex (Wirth, 1992). Our results indicated that in MSW-compost it is possible to find two kinds of extracellular humus-enzyme complexes of different solubility and affinity for binding extracellular enzymes.

The intracellular fraction (UF_3) had a broad extraction pH pattern and high correlation coefficients between enzyme activities were found. The association of enzymes to cell membranes or to exocellular polymers associated with co-extracted microorganisms may contribute to a simultaneous extraction. The correlation found between extracellular cellulases and cell-bound proteins and enzymes suggests the existence of an association between extracellular cellulases and cellulolytic microorganisms that may be partly broken in the ultrafiltration process. Coughlan and Ljungdahl (1988) reported in anaerobic cellulolytic bacteria the existence of cellulosomes, a cluster of cellulases and proteins linked to microbial membranes.

5. CONCLUSION

Extracellular hydrolytic enzymes were produced during composting in high yields but they were thereafter degraded in the maturation steps. The remaining extracellular enzyme activity was extracted with 0.05 M sodium pyrophosphate at two different pH values: acidic (5-6) or neutral-alkaline (7-9). Under these mild alkaline conditions, high levels of cell-bound enzymes were extracted altogether.

Compost extracellular enzymes could be stabilized by their association to humic-like materials produced during organic mater decay. Their presence in mature compost may be important after compost addition to soil because these enzymes promote mineralization of compost organic matter and nutrient bioavailability.

Future perspectives in the enzymology of compost may be the inoculation with microorganisms with high production of extracellular enzymes or the addition of highly adsorbent supports, like clays, ashes or metal hydroxides, that contribute to enzyme immobilization.

6. REFERENCES

Boyd, S. A. and M. M. Mortland, 1990. Enzyme Interactions with Clays and Clay-Organic Matter Complexes. In: J.-M. Bollag and G. Stotzky (Eds). Soil Biochemistry, Vol. VI. Marcel Dekker, New York, pp. 1-28.

Bretaudiere, J. P. and T. Spillman, 1984. Alkaline Phosphatases. In: H. U. Bergmeyer, J. Bergmeyer and M. Graßl (Eds). Methods of Enzymatic Analysis, Vol. IV. Verlag Chemie, Weinheim (Germany), pp. 75-92.

Burns, R. G., 1983. Extracellular Enzyme-Substrate Interactions in Soil. In: J. H. Slater, R. Wittembury and J. W. T. Wimpenny (Eds). Microbes in Their Natural Environment. Cambridge University Press, London, pp. 249-298.

Coughlan, M. P. and L. G. Ljungdahl, 1988. Comparative Biochemistry of Fungal and Bacterial Cellulolytic Enzyme Systems. In: J. P. Aubert, P. Begin and J. Millet (Eds). Biochemistry and Genetics of Cellulose Degradation. Academic Press, New York. pp. 11-30.

Foster, J. C., W. Zech and E. Würdiger, 1993. Comparison of chemical and microbiological methods for the characterization of the maturity of composts from contrasting sources. Biol. Fertil. Soils, 16, 93-99.

García, C., T. Hernandez, F. Costa, B. Ceccanti and C. Ciardi, 1992. Changes in ATP content, enzyme activity and inorganic nitrogen species during composting of organic wastes. Can. J. Soil Sci., 72, 243-253.

García, C., A. Roldan and T. Hernandez, 1997. Changes in microbial activity after abandonment of cultivation in a Semiarid Mediterranean environment. J. Environ. Qual., 26, 285-291.

Giusquiani, P. L., M. Pagliai, G. Gigliotti, D. Businelli and A. Benetti, 1995. Urban waste compost: Effect on physical, chemical and biochemical soil properties. J. Environ. Qual., 24, 175-182.

González-Vila, F. J. and F. Martin, 1985. Chemical structural characteristics of humic acids extracted from composted municipal refuse. Agric. Ecosyst. Environ., 14, 267-278.

Hayano, K. and A. Katami, 1977. Extraction of β-glucosidase activity from pea field soil. Soil Biol. Biochem., 9, 349-351.

He, X. T., S. J. Traina and T. J. Logan, 1992. Chemical properties of municipal solid waste composts. J. Environ. Qual., 21, 318-329.

Herrmann, F. R. and J. R. Shann, 1993. Enzyme activities as indicators of municipal solid waste compost maturity. Compost Sci. Utilization, 1, 54-63.

Inbar, Y., Y. Chen, and Y. Hadar, 1990. Humic substances formed during the composting of organic matter. Soil Sci. Soc. Am. J., 54, 1316-1323.

IUPAC, Commission on Biotechnology, 1984. Measurement of Cellulase Activities. In: T. K. Goshe (Ed). Methods of Enzymology. Indian Institute of Technology, New Delhi, India.

Lerch, R. N., K. A. Barbarick, P. Azari, L. E. Sommers and D. G. Westfall, 1993. Sewage sludge proteins: I. Extraction methodology. J. Environ. Qual., 22, 620-624.

Lowry, O. H., N. J. Rosebrough, A. L. Farr and R. J. Randall, 1951. Protein measurement with the Folin phenol reagent. J. Biol. Biochem., 193, 265-275.

Martens, D. A., J. B. Johanson, and W. T. J. Frankenberger, 1992. Production and persistence of soil enzymes with repeated addition of organic residues. Soil Sci., 153, 53-61.

Moss, D. W., 1984. Acid Phosphatases. In: H. U. Bergmeyer, J. Bergmeyer and M. Graßl (Eds). Methods of Enzymatic Analysis, Vol. IV. Verlag Chemie, Weinheim (Germany), pp. 92-106.

McClaugherty, C. A. and A. E. Linkins, 1988. Extractability of cellulases in forest litter and soil. Biol. Fertil. Soils, 6, 322-327.

Nannipieri, P., B. Ceccanti, D. Bianchi and M. Bonmati, 1985. Fractionation of hydrolase-humus complexes by gel chromatography. Biol. Fert. Soils, 1, 25-29.

Nannipieri, P., B. Ceccanti, S. Cervelli and E. Matarese, 1980. Extraction of phosphatase, urease, proteases, organic carbon and nitrogen from soil. Soil Sci. Soc. Am. J., 44, 1011-1016.

Nannipieri, P., B. Ceccanti, S. Cervelli and P. Sequi, 1974. Use of 0.1 M pyrophosphate to extract urease from a podzol. Soil Biol. Biochem., 6, 359-362.

Nannipieri, P., B. Ceccanti, C. Conti and D. Bianchi, 1982. Hydrolases extracted from soil: Their properties and activities. Soil Biol. Biochem., 14, 257-263.

Nelson, N., 1944. A photometric adaptation of the Somogyi method for determination of glucose. J. Biol. Biochem., 153, 375-380.

Pagliai, M. and M. De Nobili, 1993. Relationships between soil porosity, root development and soil enzyme activity in cultivated soils. Geoderma, 56, 243-256.

Pérez-Mateos, M., S. González-Carcedo and M. D. Busto-Núñez, 1988. Extraction of catalase from soil. Soil Sci. Soc. Am. J., 52, 408-411.

Perucci, P., 1990. Effect of the addition of municipal solid-waste compost on microbial biomass and enzyme activities in soil. Biol. Fertil. Soils, 10, 221-226.

Perucci, P., 1992. Enzyme activity and microbial biomass in a field soil amended with municipal refuse. Biol. Fertil. Soils, 14, 54-60.

Rad, J. C. and S. González-Carcedo, 1995. Different location of acid and alkaline phosphatases extracted from a compost of urban refuse. In: M. De Bertoldi, P. Sequi, B. Lemmes and T. Papi (Eds). The Science of Composting, Vol.1. Blackie A&P, London, pp. 286-293.

Sarkar, J. M., A. Leonowicz and J.-M. Bollag, 1989. Immobilization of enzymes on clay and soils. Soil Biol. Biochem., 21, 223-230.

Serra-Wittling, C., S. Houot and E. Barriuso, 1995. Soil enzymatic response to addition of municipal solid-waste compost. Biol. Fertil. Soils, 20, 226-236.

Serra-Wittling, C., S. Houot and E. Barriuso, 1996. Modification of soil water retention and biological properties by municipal solid waste compost. Compost Sci. Utilization, 4, 44-52.

Sinsabaugh, R. L., R. K. Antibus and A. E. Linkins, 1991. An enzymic approach to the analysis of microbial activity during plant litter decomposition. Agric. Ecosyst. Environ., 34, 43-54.

Tabatabai, M. A. and M. Fu, 1992. Extraction of Enzymes from Soils. In: G. Stotzky and J.- M. Bollag (Eds). Soil Biochemistry, Vol. VI. Marcel Dekker, New York, pp. 197-227.

Walter, H. E., 1984. Method with Haemoglobin, Casein and Azocoll as Substrate. In: H. U. Bergmeyer, J. Bergmeyer and M. Graßl (Eds). Methods of Enzymatic Analysis., Vol. V. Verlag Chemie, Weinheim (FRG), pp. 270-277.

Wirth, S. J., 1992. Detection of soil polysaccharide endo-hydrolase activity profiles after gel permeation chromatography. Soil Biol. Biochem., 24, 1185-1188.

ADSORPTION OF METHYLENE BLUE BY RED MUD, AN OXIDE-RICH BYPRODUCT OF BAUXITE REFINING

M. Arias, E. López, A. Nuñez, D. Rubinos, B. Soto, M.T. Barral and F. Díaz-Fierros

Departamento de Edafoloxía e Química Agrícola, Facultade de Farmacia. Universidade de Santiago de Compostela. España.

1. INTRODUCTION

The use of organic basic dyes for the determination of the surface adsorption in different materials is one of the oldest ways of measuring the adsorption capacity or the amount of exchange sites of small-size particles and colloidal material (Mäkitie and Erviö, 1966). Methylene blue (MB) is an organic cationic dye with formula $C_{16}H_{18}N_3SCl.3H_2O$. The molecule can be regarded approximately as a rectangular volume of dimensions 17.0 x 7.6 x 3.3 Å. The projected area of the molecule has been given by several authors (quoted in Hang and Brindley, 1970; Taylor, 1985). Estimated values ranged between 130-135 $Å^2$.

Methylene blue replaces natural cations irreversibly (Taylor, 1985). This is unlike the reversible exchange of inorganic cations. Plesh and Robertson (1948) considered that MB is adsorbed in two ways: (a) irreversible exchange to an amount equivalent to the CEC and (b) reversible adsorption of complete molecules, in agreement with the Freundlich isotherm governing physical adsorption.

Adsorption of methylene blue (MB) by clay materials has been used for estimation of the clay content and determination of cation exchange capacity (CEC) and surface area of soils and clays (Mäkitie & Erviö, 1966; Hang & Brindley, 1970; Wang & Wang, 1987; Savant, 1994). MB adsorption has been adopted as a standard test in France for determining the presence of clay in aggregates (Taylor, 1985). Most works on MB adsorption refer to clay minerals such as kaolinite, montmorillonite or illite. However data MB adsorption on oxide-rich materials are scarce (Casanova, 1986).

In this work, methylene blue adsorption on red mud - a by-product of the bauxite refining - is studied for the purpose of investigating its adsorbent capacity for coloured organic compounds. This property could be useful for removing dyestuffs from industrial wastewater.

Effect of Mineral-Organic-Microorganism Interactions on Soil and Freshwater Environments
Edited by Berthelin, *et al*, Kluwer Academic / Plenum Publishers, New York, 1999

2. MATERIAL AND METHODS

The red mud used, obtained from Alumina-Aluminio factory in San Ciprián (Lugo, Spain), is a silty-clay material with 37.2% Fe_2O_3, 20.1% TiO_2, 12.4% Al_2O_3, 6.3% CaO, 4.6% Na_2O, 3.8% SiO_2, and 0.1% MgO. Main minerals are Rutile, Hematite, Goethite, Ilmenite, Magnetite, Bayerite and zeolite-type minerals.

The dye used was Methylene Blue C.I. 52016 from Panreac, with molecular weight of 319.85. Average water content of this reactive product was 14.97%, which means that there were 3 molecules of water hydrating each molecule of MB.

Batch and column experiments were carried out by the following procedures:

2.1. *Batch experiments*: 0.5 g of red mud were suspended in 25 cm^3 of 0.01 M NaCl. An aliquot of a solution containing 1.6 mg l^{-1} MB was added to this suspension and the volume completed to 50 cm^3 with distilled water. Final MB concentrations ranged between 2.5 x 10^{-3} M and 3.12 x 10^{-4} M and represent 80.1 to 0.6 mg of MB per g of red mud (0.19 to 25 cmol of MB per kg of red mud). The suspensions were adjusted at pH 7 and left stirring for 6, 48 or 192 h, before centrifugation. Concentration of MB remaining in solution was determined colorimetrically (Hitachi U-3200 Spectrophotometer). An optimum wavelength for the MB standard used was determined, so that wavelength variations attributable to the presence of monomer and/or dimer molecules in solution could be largely eliminated.

2.2. *Column experiment*: a column consisting in a 10 cm^3 polypropylene syringe was used; it was packed with aggregated (0.50-0.25 cm diameter) red mud (with 8% $CaSO_4$) and a pore volume (PV) of 5.9 cm^3; 121 PV of a 1.02 x 10^{-4} M MB solution was percolated through it. Mean residence time was 3 h. Previously, column red mud was washed by pumping distilled water through the column until values of 45 μS cm^{-1} for electrical conductivity and 8.1 for pH were reached in the percolates. A peristaltic pump (Minipuls 3, Gilson Model M312), was used to impulse distilled water and after the MB solution through the column; a refrigerated fraction collector (Gilson FC 203B) was also used to sample the percolates. TOC and MB concentrations in the percolates (sampled each 0.6-0.7 PV) were determined using a Shimadzu TOC-5000 analyzer and an Hitachi U-3200 Spectrophotometer, respectively.

Furthermore, CEC of red mud was measured using both the ammonium acetate method (as referred in Gillman et al., 1983) and the MB method (referred in Taylor, 1985).

3. RESULTS

The red mud CEC measured by the MB method was 0.66 $cmol_c kg^{-1}$, clearly lower than that resulting from the ammonium acetate method (10.8 $cmol_c kg^{-1}$).

Overall results show some differences in MB adsorption for the three residence times in batch experiments. MB adsorption capacity of red mud did not change significantly -between a range of 0 to 35 mg of MB per g of red mud- when contact time was increased from 6 to 48 or 192 h (Fig. 1). However, differences became clear when the MB-added amount was 80 mg per g of red mud; in this case the highest rise took place when contact time increased from 6 to 48 h. The MB-added/MB-adsorbed ratio increases with increasing MB-added concentrations, with a plateau between 2 and 10 cmol kg-1 (6.4 and 32.0 mg g^{-1}), probably related to MB-CEC (Fig. 1). Maximum adsorption, which occurred for the 25 cmol kg^{-1} MB solution after 192 h, was 2.0 cmol kg^{-1}, which is well under the ammonium-acetate-CEC of red mud (10.8 $cmol_c kg^{-1}$), but is higher than the MB-CEC (0.66 $cmol_c kg^{-1}$).

Spectrophotometric determinations of MB in the percolates from the column experiments showed an almost total retention of MB for the first 65 PV displacements (Fig. 2). The initial amount of MB in the percolating solution was not reached in the percolates even after 121 PV. However TOC values were 3102 times higher than C equivalents of MB in the percolate when 14 PV of the MB solution had passed through the column; the ratio TOC/MB-Carbon decreased progressively until a steady value of about 3.5 is achieved.

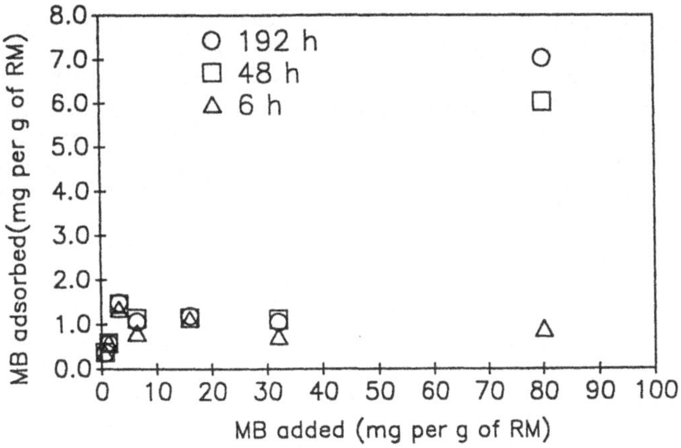

Figure 1. Methylene blue (MB) adsorption on red mud (RM) for different contact times and different amounts of MB added.

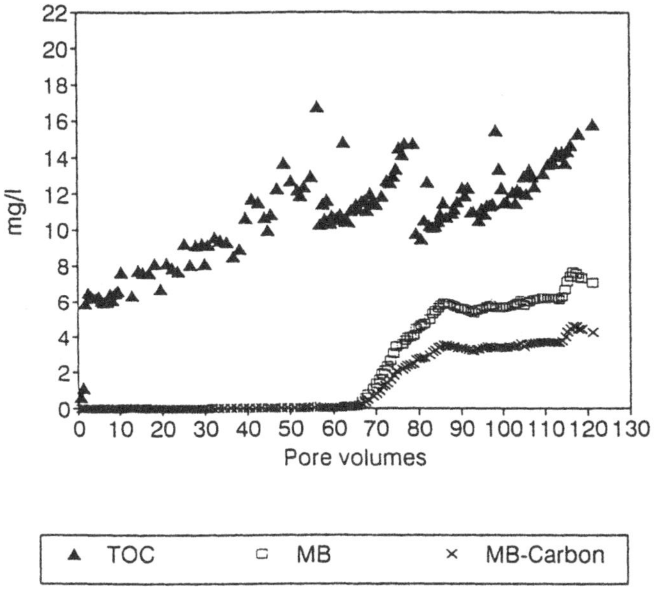

Figure 2. Total organic carbon (TOC), methylene blue (MB) determined by spectrophotometry and calculated MB-Carbon concentrations (mg l^{-1}) in the percolates from the column experiment.

4. DISCUSSION

Figure 1 shows that, if the contact time is not higher than 6 h, although the amount of MB added is increased MB adsorption on red mud do not rise (it is not higher than 1.5 mg MB per g of red mud, which means 0.41 cmol kg^{-1}). However, the rise is evident when the contact time is 48 or 192 h and the MB-added amount is 80 mg g^{-1}. These data suggest that some adsorption reactions take place following rapid kinetics, but also that there are another kind of slow kinetic reactions -at the same time dependent on the MB concentration- possibly associated to multilayer physical adsorption phenomena (Bergman and O'Konski, 1963; Mäkitie and Erviö, 1966).

In column experiments TOC values were higher than C equivalents of MB determined by spectrophotometry in the percolate. The possibility that some MB could be oxidated to non coloured substances on the surface of oxidic components of the red mud would aid to explain the differences between TOC and MB behaviour and deserves further research.

In the MB-CEC method, contact time is 1 h, so that the value of 0.66 $cmol_c kg^{-1}$ is not in opposition to the 0.41 cmol kg^{-1} value corresponding to maximum adsorption for the 6 h contact time. Moreover, MB-CEC value is clearly lower than the ammonium acetate one; this fact may be contrasted with that pointed out by Wang and Wang (1987), constating that CEC values in bentonite and shales, measured by means of MB titration methods, were consistently lower than these generated through conventional Ca/Mg and K/NH4 exchange. Savant (1994) and Kitsopoulos (1997) also found MB-CEC values lower than CEC values determined by exchange with ammonium acetate. Large molecular size of MB can cause a steric hindrance effect responsible for the lower CEC values.

Red mud usually contains variable amounts of zeolite-type minerals, such as sodalite, nosean or cancrinite (Leiteizen et al., 1978). The red mud used in this study contains approximately 4% of zeolitic components which represent 80% of the total CEC. Kitsopoulos (1997) compared the MB absorption and the ammonium acetate saturation methods for determination of CEC values of zeolite-rich tuffs and observed that the values of CEC given MB method were 1-28% of the ammonium acetate values for zeolitic materials. The reason why MB methods gives lower values than ammonium acetate relates to the zeolite structure, characterized by the presence of channels of specific sizes. The size of the MB molecule is higher than the size of the channels of common zeolitic minerals. Kitsopoulos (1997) suggests that this immobility of the MB can be exploited and be used as a "molecular probe" technique for quantitative analysis of zeolite-clay mixtures.

5. CONCLUSIONS

The MB adsorption on red mud shows little variation between 6 and 192 h of contact time when the MB-added amount is lower than 35 mg per g of red mud. When the MB-added amount is 80 mg per g of red mud, differences between MB-adsorbed amounts for 6 h or for 48 and 192 h of contact time became clear. Maximum adsorption (2.3 cmol kg^{-1}) corresponds to 25 cmol kg^{-1} MB concentration and contact time of 192 h.

MB retention in the red mud column was almost total during the first 65 PV displacements. The initial amount of MB was not reached in percolates even after 121 PV have passed through the column.

Red mud ammonium-acetate-CEC (10.8 $cmol_c kg^{-1}$) was clearly higher than the MB-CEC (0.66 $cmol_c kg^{-1}$).

6. ACKNOWLEDGEMENTS

This work was supported by a grant (XUGA 27101B94) from the Government of Galicia (Xunta de Galicia).

7. REFERENCES

Bergmann, K. and O'Konski, C.T., 1963. A spectroscopic study of methylene blue monomer, dimer, and complexes with montmorillonite. J. Phys. Chem., 67, 2169-2177.

Casanova, F.J., 1986. O ensaio do azul de metileno na caracterizaçao de solos lateriticos. 21 Reuniao Annual de Pavimentaçao. Salvador-Ba, 277-285.

Gillman, G.P., Bruce, R.C., Davey, B.G., Kimble, J.M., Searle P.L. and Skjemstad, J.O., 1983. A comparison of methods used for determination of cation exchange capacity. Commun. Soil Sci. Plant Anal., 14, 1005-1014.

Hang, P.T. and Brindley, G.W., 1970. Methylene blue absorption by clay minerals. Determination of surface areas and cation exchange capacities (Clay-organic studies XVIII). Clays and Clay Minerals, 18, 203-212.

Kitsopoulos, K.P., 1997. Comparison of the methylene blue absorption and the ammonium acetate saturation methods for the determination of CEC values of zeolite-rich tuffs. Clay Minerals, 32, 319-322.

Leiteizen, M.G., Pashkevich, L.A., Firfarova, I.B. and Tsekhovolskaya, D.I., 1978. Characteristics of sodium hydroalumino-silicates forming during bauxite digestion. In: Research Inst. Non-ferrous Metal (Editor). Proc. 2nd Conf. VAMI and FKI. Leningrad, pp. 167-179.

Mäkitie, O. and Erviö, R., 1966. Comparative studies on the cation exchange properties of mineral soils by the methylene-blue adsorption method and by the ammonium acetate method. Annales Agriculturae Fenniae, 5, 260-266.

Plesh, P.H. and Robertson, R.H.S., 1948. Adsorption on to ionogenic surfaces. Nature, 161, 1020-1021.

Savant, N.K., 1994. Simplified methylene blue method for rapid determination of cation exchange capacity of mineral soils. Commun. Soil Sci. Plant Anal., 25, 3357-3364.

Taylor, R.K., 1985. Cation exchange in clays and mudrocks by methylene blue. J. Chem. Tech. Biotechnol., 35A, 195-207.

Wang, M.K. and Wang, S.H., 1987. Evaluation of methylene blue tests for determining CEC of bentonite and shales. Journal of the Chinese Agricultural Chemical Society, 25, 387-397.

THE TRANSFORMATION OF WATER QUALITY: FROM SOIL CONTRIBUTION TO WATER TREATMENT

Jean-Luc BERSILLON[1,2], Bruno LARTIGES[2], Fabien THOMAS[3] and Laurent MICHOT[3]

[1] LYONNAISE DES EAUX - CIRSEE. 38 rue du Président Wilson, 78230 LE PECQ, France – E-Mail : bersillo@ensg.u-nancy.fr

[2] INPL - ENSG, L.E.M.-GRESD, Boite Postale 40, 54501 VANDŒUVRE CEDEX, France – E-Mail : bsl@ensg.u-nancy.fr

[3] CNRS, UMR 7569, L.E.M.-GRESD, Boite Postale 40, F-54501 VANDŒUVRE CEDEX, France – E-Mail: fthomas@ensg.u-nancy.fr; michot@ensg.u-nancy.fr

INTRODUCTION: THE IMPORTANCE OF ORGANIC MATTER IN WATER TREATMENT

Organic Matter (OM) is present in all natural waters. It is responsible for many problems when a natural water is treated for drinking purposes. Inconveniences such as tastes and odors may occur in water due to the presence of minute amounts of OM. Specific compounds such as pesticides may also occur in natural waters. Finally, OM by-products may be undesirable (halomethanes) or the mere presence of OM may impair the treatment process performances. For all these reasons, OM has been a major concern in the water treatment community for several decades. The following aspects of OM occurrence in waters have been addressed by the water treatment community:

- Chemical characterization
- Fate of OM along a treatment line
- Influence of OM occurrence on treatment performances
- Influence of OM occurrence on water quality maintenance

Interestingly, it was only a few years ago that the water treatment community " discovered " soil sciences and that interactions between these 2 fields started to be fruitful.

ASSESSING ORGANIC CONTENT IN NATURAL WATERS

Organic material in natural waters never exceeds a few $mg.l^{-1}$ as C. The quantity of organic material is therefore quite small and for convenience purposes, OM was often measured as lumped — and unsatisfactory — parameters such as oxydability as measured by the $KMnO_4$ consumption in either acidic or alkaline medium, as Total Organic Carbon (TOC) which is quantitative or as specific UV Absorbance (e.g. $\lambda = 254nm$) which is an easy, non destructive measurement that can be correlated to TOC. Among these measurements, only TOC refers directly to a mass and can be used for material balance purposes. The major drawback in the use of this measurement is that it is a lumped parameter and therefore does not contain any information about the nature of the phenomena that are quantified.

Effect of Mineral-Organic-Microorganism Interactions on Soil and Freshwater Environments
Edited by Berthelin, *et al*, Kluwer Academic / Plenum Publishers, New York, 1999

Figure 1 illustrates the knowledge that the water treatment community had about organics up to the late 1980s. Standards have been required for the analytical identification of selected molecules while process management lead to the measurement of other organics that were not part of the standards. Natural Organic Compounds (NOMs) were recognized as influential material (detrimental) in the performance of water treatment processes, as shown in table 1.

Table 1: Detrimental effect of different categories of NOMs on selected water treatment processes (after, MALLEVIALLE, 1994)

	Sequestration (metal or pesticides)	Bacteria regrowth (distribution systems)	Competitive Adsorption	Competitive Oxidation
Proteins	+	+++	+	+
PAH	+++	+	++	++
Polysaccharides	0	+	+?	0
Amino Sugars	0	?	?	?

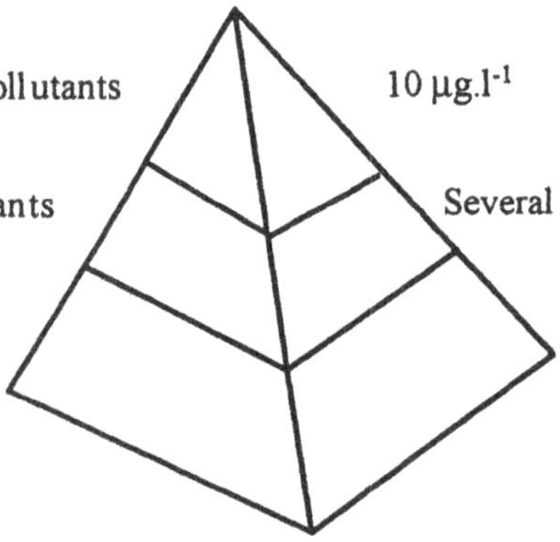

Figure 1 : Representation of organic matter present in natural waters

It was as late as 1987 (GADEL *et al.*, 1987 mentioned in BRUCHET *et al.*, 1990) that the Pyrolysis - Gas Chromatography - Mass Spectrometry (Py-GC-MS) technique was usable to analyze *semi - quantitatively* NOMs and to allow some sort of description of aqueous NOM chemical properties other than the amount of related carbon and titrimetric data on their ion exchange capacity. This made it possible to examine the relative removal of the different classes of organics through a treatment line and to either suspect or deduct their influence on water treatment performances (BERSILLON, 1988; BRUCHET *et al.*, 1994).

When examining standard material and actual NOMs in natural waters, it becomes difficult to consider aqueous NOMs as a simple mixture of a limited number of compounds and the classical reference to SCHNITZER's fulvic acid (SCHNITZER *et al.*, 1972) to describe NOMs is no longer applicable. Also, the use of "standard" or "purified" fulvic material to design new mechanistic process models to describe and perhaps predict water resource evolution or water treatment performance becomes problematic. The actual NOM complexity and the questionable reference to purified fulvic material in experimental work on either water resource or water treatment process design is illustrated by the Table 2. It is interesting to compare these data to OADES' (1989) where the " model " molecule shows a carboxy-polyphenolic unit grafted with a sugar and an amino acid that polymerizes to give an entity having a bundle structure averaging 50 nm in diameter.

Table 2: Py-GC-MS Analysis of 2 commercial sources of fulvic material and 2 water resources.

	INRA FA	Contech FA	Douchy Spring	Houlle River
Polysacharides	61	20	45	31
Proteins	25	8	11	13
Amino sugars	1	0	9	1
PAH	2	62	11	26
Undefined & losses	11	10	24	29

SOIL CONTRIBUTION TO WATER QUALITY

Soils in general constitute a compulsory step between rainwater and water resources (either groundwater or surface waters). As a first approximation, rain water may be considered as relatively pure, even though it contains minute amounts of suspended solids and dissolved compounds. It will encounter the different soils and will react with this complex environment. Besides natural compounds, either inorganic or organic, material coming from human activity will be encountered as well. These inputs come from:

- agricultural practices (fertilizers, pesticides),
- landfill sites (by leaching),
- urban and traffic deposits, mobilized in either suspended or dissolved form by runoff waters.

Since water trickles through soils, it is bound to be in contact with solids with it is not in equilibrium. Interactions between dissolved species and with solids determine a quite complex process that can be categorized with respect to the size of the species. This is illustrated in figure 2.

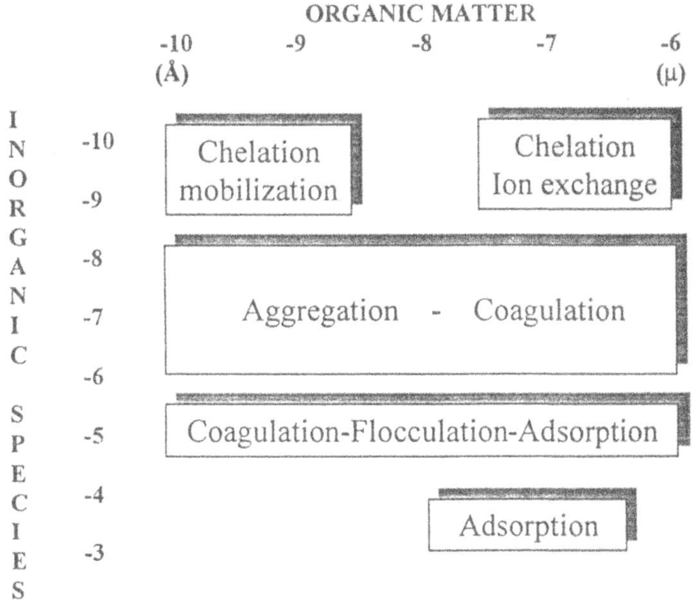

Figure 2: Mechanisms involved in the interactions between inorganic and organic compounds and water in soils.

The elementary processes involved in the interaction between water, organic and inorganic compounds are:

- *complexation*, specially the formation of organo-mineral complexes (chelates) that may be responsible for either metal mobilization or immobilization. This concerns a large part of the organic compounds, regardless of their size and the dissolved inorganics. Large organic compounds may be responsible for exchange or immobilization of smaller anthropic organics such as pesticides (CALVET *et al.*, 1980). Ion exchange occurs at oxyhydroxide and clay surfaces;

- *coagulation and flocculation*, that concern larger inorganic species such as oxyhydroxide polycations and colloids that result from hydrolysis. These species react with a broad spectrum of organic compounds to form larger aggregates that slow down the dispersion of both classes of species;
- *adsorption*, that concerns large immobile species immobilizing smaller organic species;
- *filtration and capture*, which is a mechanical straining of the water and that contributes to the quality evolution of both water and soil.

All these phenomena occur at once and are related to each other. Added to the rather unpredictable character of rainfall, this results in a quasi impossibility to accurately predict water resource quality over a long period of time. Such a prediction may be crucial when a decision is needed to invest in treatment equipment. A quite new modeling technique is now under investigation to take into account the uncertainties related to stochastic processes such as rainfall patterns and incomplete knowledge of soil characteristics. The results of such models are expressed as a set of scenarios to which the statistics of occurrence are associated. An example of this technique is given by FOUSSEREAU *et al.* (1993).
If decision aids may progress through a wider application of stochastic modeling, any improvement in resource management needs a better understanding of the complex phenomena occurring in the soils.

WATER DEPOLLUTION PROCESSES

Water supplied to human consumption is standardized. The standards correspond to an extensive set of parameters corresponding to sanitary, toxicological and aesthetic requirements. Natural waters seldom comply with these requirements and treatment is very often necessary. The minimum treatment is chlorination, in order to insure continuous disinfection of water and ducts during distribution. The use of this treatment may require a pretreatment so that chlorine consumption is kept as low as possible or that chlorination by-product formation is hindered.
The most widely used water treatment processes are meant to remove suspended matter, microorganisms (including viruses) and dissolved organic compounds. Among these processes, coagulation, sand filtration, carbon adsorption, ion exchange and carbonaceous and nitrogenous pollutant biodegradation are of special interest within the frame of this report.

Coagulation

The coagulation process is used to enhance particle removal. The practice consists in the addition of a hydrolysing salt — aluminum or ferric — to a turbid water. This addition results in the formation of millimeter range aggregates that are easily separated from the water by settling or sand filtration. Coagulation aids may be used such as the addition of polymerized silica or organic polymers.
Referring to the diagram presented in figure 2, this process can be considered as a disturbance of the ionic equilibrium of a diluted suspension and as the addition or the formation of smaller aggregates. Therefore, complexation, ion exchange, coagulation and adsorption are likely to occur between the added species or their hydrolysis by-products and a broad spectrum of organic compounds. Depending on the nature and structure of the added chemicals, immobilization or dissolution may occur, as shown by MOLIS *et al.* (1996). In this work, it is shown that polymerized aluminum species known as very efficient to remove natural organic matter, are completely dissolved when exposed to selected organics such as salicylic acid.

Activated carbon adsorption

This operation is used to improve the water quality by the removal of dissolved organic compounds, either of natural or anthropic origin. Activated carbon is used either in powdered or granulated form. If used in the powdered form, activated carbon is added prior to a separation process. The granulated form is used as filtering beds. Due to its nature, activated carbon tends to adsorb non polar hydrophobic material. Pesticides tend to adsorb quite well on this material but natural organic material adsorbs as well. As already discussed earlier, NOMs are present in waters at concentrations in the order of mg/L whereas pesticides occur at the $ng.l^{-1}$ to the $\mu g.l^{-1}$ order of magnitude. Also, competition or co-adsorption may occur between NOMs and pesticides.

Ion exchange

Resins are used in order to soften or to partially demineralize waters. This process is not very common except for small water resources or as point-of-use or point-of-entry treatments. The nature and texture of the material has also an effect on the organic content of the water as minute amounts of the compounds constituting the resin may leak into the water.

The comparison of the order of magnitude of the retention potential of a soil with the engineered processes used in water treatment is given in the table 3. The values quoted for the soil correspond to an average soil of the western European region. The figures quoted for the coagulation process correspond to a direct filtration of a moderately polluted water treated with 25 $mg.l^{-1}$ of alum. It is interesting to note that the organic content of the solid phase (carbon / suspended solids) once the treatment is completed roughly matches the value found in soils. As far as the ion capacity of a resin is concerned, there is again a similarity in the values of ion exchange capacity with these expected from a soil.

Table 3: Comparison of soil content and water treatment reactor capacity for selected unit processes.

	Soil content or retention capacity (per m^3 of soil)	Plant capacity (per m^3 of reactor)
Coagulation and sand filtration		
Al Hydroxide (g)	-	120
Carbon content (g)	30000	50
Suspended Solids (g)	650000	500
Adsorption		
Carbon content (g)	30000	37000
Ion exchange		
Ion exchange capacity (eq)	2000	1000 to 4000

Therefore, it can be concluded that the unit processes that are responsible for the transformation of rainwater into either surface water or groundwater are the same as those used in the treatment of natural continental waters for human consumption purposes. This is true up to the order of magnitude of some of the characteristics of these processes such as sorption capacities or ion exchange capacities. The difference is due to the fact that water treatment involves the use of newly formed solids on a water containing organic and inorganic materials whereas the soil - water exchange processes involve a rather pure water seeping through a leaching solid.

Biological treatment of carbonaceous and nitrogenous pollutants

Soils contain living organisms that are responsible for the transformation of the material used for their metabolism. This makes the soil a biological reactor that processes carbonaceous and nitrogenous compounds. These are most important in the carbon and the nitrogen cycles. In European countries, a large part of the water supply resources are constituted by groundwater (60% in France, 50% in UK). In these countries, nitrate pollution still represents a crucial problem: in addition to the present nitrate value, the trend that these values follow can be alarming. Until recent years, an approximate rate of increase ranged from 0.3 $mg.l^{-1}.yr^{-1}$ in Beauce (Eure) to 4 $mg.l^{-1}.yr^{-1}$ in Brittany (Finistère). The highest annual increase often occurs where the nitrate concentrations are already high (>30 $mg.l^{-1}$). It is estimated that the nitrate concentration is likely to increase during the next 30 years, although not everywhere. This situation made it compulsory to equip already existing water resources (mainly wells) with a nitrate removal process. Ion exchangers can be installed, with a major drawback: nitrate is

transferred to the ion exchange resin which is regenerated. The regeneration process produces nitrate containing brines that must be treated through a wastewater treatment plant equipped with a biological denitrification step. In this process, bacteria are maintained in such conditions that they metabolize nitrate to produce molecular nitrogen. This process has been redesigned to remove nitrate from natural sources, with concentrations ranging from 50 to several hundred $mg.l^{-1}$. The species involved in this engineered process e.g. *Thiobacillus Denitrificans* and *Pseudomonas sp.* are also involved in the nitrogen cycle in soils.
Ammonium is also a nuisance since it is responsible for chlorine consumption and possibly the formation of chloramines. Even though this effect is considered as desirable by some practitioners, it is necessary to lower its concentration. This is performed by a biological process involving *Nitrosomonas* and *Nitrobacter*. The process is autotrophic and forms nitrate as a by-product. It is practiced in wastewater, often in combination with a biological denitrification process or as such on selected water resources presenting a good overall quality except for ammonium content.

THE ULTIMATE WATER TREATMENT RESIDUES: SLUDGES

Sludges from wastewater treatment plants may be disposed on fields as fertilizers. Such a disposal makes the soil the main contributor to the recycling or the treatment of one of the ultimate wastes of human activity. Extensive programs of this nature have been implemented in Vexin (northwestern end of the Parisian region) where part of the sludges from the Achères wastewater plant are disposed of (ROBERT, 1996). Heavy metal concentration and flux standards exist in France (NF U44-041) and in Europe as a recommendation. Despite these standards, this practice encounters resistance from farmers as there is no guarantee provided by wastewater operators on nitrogen, phosphorus and potassium content Sanitary quality of these sludges must also be controlled because microbial colntamination and hazard are possible. This lack of information on the quality of this potential fertilizer does not allow a precise use of this product. Furthermore, the agricultural community may not accept to manage this waste without any return from the community. Last, some cumulative effects on the environment cannot be ruled out.

CONCLUSIONS

Water treatment and soil sciences have been ignoring each other for a long time. However, both disciplines study or use similar unit processes, under very different but complementary conditions. A first object of common interest is constituted by the soluble fraction of the soil organic matter that is responsible for the hindering of treatment processes efficiency.
Besides organic matter characterization which is expected to yield information about how to improve water treatment performances, a better knowledge of the interactions between biological, organic and inorganic material should help in a better management of water resources, specially through artificial recharge.

REFERENCES

Bersillon, J.L.,1988. Fouling Analysis and Control. p.234 - 247 - In Cecille,L & Toussaint, J.C. (Ed) *Future Industrial Prospects in Membrane Processes.* Elsevier Interscience, Oxford, UK. 307p.

Bruchet, A., Rousseau, C., Mallevialle, J.,1990. Pyrolysis - GC - MS Investigating High Molecular Weight THM Precursors and Other Refractory Organics. JAWWA,**82**(9), p66-74.

Bruchet, A. ,1994. Transformation of NOM During Water Treatment. p 127 - 138 In *Natural Organic Matter - Origin, Characterization and Removal.* AWWARF Ed. 243p.

Calvet,R., Terce, M., Arvieux, J.C.,1980. Adsorption des pesticides par les sols. Ann. Agron., **34**, 4, 1-127.

Foussereau, X., Hornsby, A.G., Brown, R.B.,1993. Accounting for Variability within Map Unit when Linking a Pesticide Fate Model to Soil Survey. Geoderma, **60**, p.257-276.

Gadel,F., Bruchet, A.,1987. Application of Pyrolysis - Gas Chromatography - Mass Spectrometry to the Characterization of Humic Substances Resulting from Decay of Aquatic Plants in Sediments and Waters. Water Research, **21**,10, p1195.

Mallevialle, J.,1994. Why is Natural Organics Problematic? p 1 - 18 In *Natural Organic Matter - Origin, Characterization and Removal.* AWWARF Ed. 243p.

Molis, E., Thomas, F., Bottero, J.Y., Barres, O.,1996. Chemical and Structural Transformation of Aggregated Al_{13} Polycations Promoted by Salicylate Ligands. Langmuir, **12**, 3195 - 3200.

Oades, J.M.,1989. An introduction to organic matter in soils, p.89-160. In Dixon, J.B. and Weed, S.B. (eds) *Minerals in Soil Environments* . Soil Science Soc. Am. Madison, 2nd edition.

Robert, M.,1996. Le sol: Interface dans l'environnement, ressource pour le développement. Masson Ed., Paris, 244p.

Schnitzer, M., Khan, S.U.,1972. *Humic Substances in the Environment* . Marcel Dekkern New York

INDEX

ABOUT THE EDITORS

Jacques Berthelin is Research Director at the Centre National de la Recherche Scientifique (CNRS, Director of the Centre de Pédologie Biologique associated to the University Henri-Poincaré (Nancy I), and Vice-Chairman of the IUSS Working Group M.O. on Mineral-Organic-Microorganism Interactions of Soil Minerals with Organic Components and Microorganisms.

Address: Centre de Pédologie Biologique—C.N.R.S.
17, rue Notre-Dame-des-Pauvres
B.P. 5, 54501 Vandœuvre-lès-Nancy Cedex,
France
Tel. (33) 3 83 51 08 60
Fax (33) 3 83 57 65 23
E-mail: bertelin@cpb.cnrs-nancy.fr

P. M. Huang is Professor of Soil Science at the University of Saskatchewan, Saskatoon, Canada, and Chairman of IUSS Working Group M.O. on Mineral-Organic-Microorganism Interactions of Soil Minerals with Organic Components and Microorganisms.

Address: Department of Soil Science.
University of Saskatchewan
51 Campus Drive,
Saskatoon SK S7N 5A8 Canada
Tel. (1) 306 966-6823
Fax (1) 306 966-6838

Jean-Marc Bollag is Professor of Soil Biochemistry and Co-director of the Center for Bioremediation and Detoxification, Environmental Resources Research Institute, at the Pennsylvania State University, and Vice-Chairman of the IUSS Working Group M.O. on Mineral-Organic-Microorganism Interactions of Soil Minerals with Organic Components and Microoganisms.

Address: Laboratory of Soil Biochemistry
Center for Bioremediation and Detoxification
129 Land and Water Building
The Pennsylvania State University
University Park, PA 16802 USA
Tel. (1) 814 863 0843
Fax (1) 814 865-7836
E-mail: JMBOLLAG@PSU.EDU

Francis Andreux is Professor of Soil Science and head of the Group Geosol at the University of Bourgogne, Dijon.
Address: University de Bourgogne
Centre des Sciences de la Terre
Equipe de G ochimie des Interfaces Sol-Eau
6, Boulevard Gabriel
21000 Dijon, France
Tel. (33) 3 80 39 63 64
Fax (33) 3 80 39 63 87
E-mail: fandreux@satie.u-bourgogne.fr

SPRINGER NATURE

GPSR Compliance

The European Union's (EU) General Product Safety Regulation (GPSR) is a set of rules that requires consumer products to be safe and our obligations to ensure this.

If you have any concerns about our products, you can contact us on ProductSafety@springernature.com

In case Publisher is established outside the EU, the EU authorized representative is:

Springer Nature Customer Service Center GmbH
Europaplatz 3
69115 Heidelberg, Germany

The manufacturer's authorised representative in the EU is Springer Nature Customer Service Centre GmbH, Europaplatz 3, 69115 Heidelberg, Germany. If you have any concerns regarding our products, please contact ProductSafety@springernature.com

Printed and bound by CPI Group (UK) Ltd, Croydon, CR0 4YY
10/10/2026
02231229-0011